627.38
Tob   Tobiasson, Bruce O.

           Marinas and small
           craft harbors

DUE DATE

ONONDAGA COUNTY
PUBLIC LIBRARY
The Galleries of Syracuse
447 S. Salina St.   OCT 07 '98
Syracuse, NY 13202-2494

WITHDRAWN

MARINAS and Small Craft Harbors

Bruce O. Tobiasson, P.E.
Waterfront Design Associates

Ronald C. Kollmeyer, Ph.D.
Oceanographic Studies, Inc.

VNR VAN NOSTRAND REINHOLD
_____ New York

Disclaimer

The intent of this book is to provide information which the authors have generated or obtained from other sources that are considered to be reliable. No presumption is made to guarantee the accuracy or the completeness of the information or its appropriateness to solve any given engineering or scientific problem. The supplying of this information does not constitute a rendering of engineering or other professional services and neither the authors or Van Nostrand Reinhold shall be held liable for any omissions, errors, or damages resulting from the application of the material and information contained in this book.

Copyright © 1991 by Van Nostrand Reinhold

Library of Congress Catalog Card Number 91-14287

ISBN 0-442-00233-5

All rights reserved. No part of this work covered by the copyright hereon may be reproduced or used in any form or by any means—graphic, electronic, or mechanical, including photocopying, recording, taping, or information storage and retrieval systems—without written permission of the publisher.

Manufactured in the United States of America

Published by Van Nostrand Reinhold
115 Fifth Avenue
New York, New York 10003

Chapman and Hall
2-6 Boundary Row
London, SE1 8HN, England

Thomas Nelson Australia
102 Dodds Street
South Melbourne 3205
Victoria, Australia

Nelson Canada
1120 Birchmount Road
Scarborough, Ontario M1K 5G4, Canada

16 15 14 13 12 11 10 9 8 7 6 5 4 3 2

Library of Congress Cataloging-in-Publication Data
Tobiasson, Bruce O.
    Marinas and small craft harbors / Bruce O. Tobiasson : Ronald C. Kollmeyer.
       p.   cm.
    Includes bibliographical references and index.
    ISBN 0-442-00233-5
    1. Marinas—Design and construction. 2. Harbors—Design and construction. I. Kollmeyer, Ronald C. II. Title.
TC328.T63   1991
627'.38—dc20
                                            91-14287
                                              CIP

Contents

Foreword by *C. Allen Wortley*	xiii
Foreword by *Neil W. Ross*	xv
Preface	xvii
Symbols and Abbreviations	xix
Reference Maps	xxiv

Part 1 — Development Overview   1

1. **INTRODUCTION**   3
   - 1.1 Scope and Purpose   6
   - 1.2 Historical Aspects   9
   - 1.3 What is a Marina?   11
   - 1.4 What is a Small Craft Harbor?   12
   - 1.5 Marina Ownership   13
   - 1.6 Profile of the Marina User   14
   - 1.7 Urban Waterfronts   16
   - 1.8 The Full Service Marina   18
   - 1.9 Marina Trends   19
   - 1.10 Marinas Around the World   24

2. **FINANCIAL CONSIDERATIONS**   28
   - 2.1 Market Studies   28
   - 2.2 Business Plan   31
   - 2.3 Capital Costs   35
   - 2.4 Financial Ratios   37
   - 2.5 Financing Sources   38
   - 2.6 Loan Approval Considerations   38
   - 2.7 Seasonal Slip Rentals Versus Dockominium/Long-term Lease   39
   - 2.8 Profit Centers   40
   - 2.9 Typical Income and Cash Flow Projection   41
   - 2.10 Developing a Marina Annual Budget   42

3. REGULATORY CONSIDERATIONS 45

    3.1   Regulatory Agencies   46
    3.2   Public Trust Doctrine   53
    3.3   Littoral and Riparian Rights Issues   55
    3.4   Length of Approval Term   58
    3.5   Preapplication Meetings   60
    3.6   Scientific Studies   61

    Part 1. Information Sources   63

Part 2 — Site Evaluation and Assessment 65

4. SITE SELECTION 67

    4.1   Exposure   69
    4.2   Marina Development Goals   71
    4.3   Water Dependency   74
    4.4   Public Access   74
    4.5   Environmental Considerations   75
    4.6   Site Surveys   76
    4.7   Commercial Maritime Conflicts   79
    4.8   Automobile Traffic Assessment   85

5. WINDS AND STORMS 87

    5.1   Wind Systems   88
    5.2   Sources of Wind Information   89
    5.3   Wind Velocities of Interest   91
    5.4   Data Collection and Usage   96
    5.5   Uses for Wind Data   101

6. WAVE CLIMATE 103

    6.1   The Anatomy of Water Waves   103
    6.2   Estimating Wave Activity at the Site   110
    6.3   Determination of Detailed Wave Activity for Design   117
    6.4   A Discussion of Wave Prediction Difficulties   124
    6.5   Design Wave Determinations   130
    6.6   Federal Emergency Management Agency Maximum Waves   132

7.	WATER RELATED SITE CONDITIONS		135
	7.1	Ship and Boat Wakes	135
	7.2	Coastal Tides	144
	7.3	Water Currents	158
	7.4	Lake Surface Elevation Changes	164
	7.5	River Floods	166
	7.6	Ice	167
8.	PERIMETER PROTECTION		172
	8.1	Wave Protection Concepts	174
	8.2	How Systems Work	175
	8.3	Wave Protection Devices	180
	8.4	Protection Device Survivability	196
9.	PLANNING THE MARINA BASIN		199
	9.1	Basin Planform	200
	9.2	Wave Activity Within the Marina Basin	211
	9.3	Water Quality and Marina Basin Water Exchange	215
	9.4	Marina Alteration of the Sediment Transport Regime	219
	Part 2. Information Sources		225

Part 3 — Engineering Design   **229**

10.	VESSEL CONSIDERATIONS		231
	10.1	Boat Design	231
	10.2	Boat Length	232
	10.3	Boat Beam	234
	10.4	Boat Profile Height and Windage	236
	10.5	Boat Weight	238
	10.6	Boat Freeboard	239
	10.7	Sailboat Mast Height	240
	10.8	Multihulls	242
11.	SELECTION OF DOCK TYPES		243
	11.1	Water Level Change Effects	243
	11.2	Fixed Dock Systems	245
	11.3	Floating Docks	251

11.4	Special Considerations for Large Yacht Docks	261
11.5	Special Types of Finger Floats and Mooring Devices	262
11.6	Cost	265
11.7	Tax and Insurance Issues	267

12. FACILITY LAYOUT 268

12.1	Channel Entrance Design	268
12.2	Perimeter Conditions	270
12.3	Fairways	271
12.4	Berth Sizing and Boat Density	275
12.5	Structure Sizing	277
12.6	Water Depth	279
12.7	Upland Boat Storage	280
12.8	Fuel Service Facilities	282
12.9	Sewage Pumpout Facilities	285
12.10	Service Docks	289
12.11	Special Dock Layout Configurations	290

13. MATERIALS OF CONSTRUCTION 293

13.1	Wood	293
13.2	Concrete	301
13.3	Steel	305
13.4	Stainless Steels	310
13.5	Aluminum	310
13.6	Fiberglass	312
13.7	Stone	314
13.8	Synthetic Materials	316

14. CORROSION AND MATERIAL DEGRADATION 319

14.1	Sea Water Environment	319
14.2	Fresh Water and Terrestrial Environment	325
14.3	Wood Preservatives	326
14.4	Coatings	328

15. DESIGN LOAD CRITERIA 331

15.1	Dead Load	331
15.2	Live Loads	332
15.3	Wind	337
15.4	Current	342
15.5	Boat Wake	343

	15.6	Boat Impact	344
	15.7	Hydrostatic Loads	346
	15.8	Load Transfer	349
	15.9	Safety Factors	352
	15.10	Special Case—Rowing Dock	353
16.	**DESIGN AND CONSTRUCTION**		355
	16.1	Suggested Planning and Design Guidelines	355
	16.2	Simple Dock Structure Analysis	364
	16.3	Dock Hardware	376
	16.4	Dock Utilities	377
	16.5	Dredging	385
	16.6	Handicapped Access	391
	16.7	Geotechnical Considerations	405
	16.8	Failure Analysis	406
17.	**MOORING SYSTEMS**		415
	17.1	Pile Supported Mooring Systems	416
	17.2	Pile Guide Systems	421
	17.3	Fixed Cantilevered Systems	422
	17.4	Bottom Anchored	423
	17.5	Swing Boat Moorings	434
	Part 3. Information Sources		441

Part 4—Operations and Management — 443

18.	**HAUL-OUT AND BOAT HANDLING SYSTEMS**	445
	18.1	Straddle Hoists	446
	18.2	Marine Railways	455
	18.3	Vertical Elevators	458
	18.4	Cranes	460
	18.5	Hydraulic Trailers	462
	18.6	Launching Ramps for Hydraulic Trailers	463
	18.7	Automobile Trailers	469
	18.8	Launching Ramps for Automobile Trailers	470
	18.9	Yard Dollies	475
	18.10	Boat Stands	475
	18.11	Forklifts	476

	18.12 Stacker Cranes	479
	18.13 Dry Stack (Rack) Storage	480
	18.14 Boat Mast Handling Equipment	487
19.	**UPLAND FACILITIES AND AMENITIES**	**489**
	19.1 Parking	489
	19.2 Restrooms and Showers	492
	19.3 Laundry Rooms	494
	19.4 Marina Office	495
	19.5 Ship's Store or Chandlery	497
	19.6 Boat Brokerage and New Boat Sales	498
	19.7 Food Services	499
	19.8 Swimming Pools and Other Amenities	500
	19.9 Waste Oil Tanks	501
	19.10 Fuel Storage	502
20.	**OPERATION**	**508**
	20.1 Administrative Plan	508
	20.2 Staffing and Personnel	509
	20.3 Marina Management Systems	510
	20.4 Insurance	513
	20.5 Service Orientation	514
	20.6 Security and Surveillance	515
	20.7 Concierge Services	517
	20.8 Operations Manual	518
	20.9 Storm Management Plan	519
	20.10 Emergency Procedures	522
	20.11 Suggested Rules and Regulations for Marina Users	524
	20.12 Signage	528
21.	**MAINTENANCE**	**532**
	21.1 Maintenance Plan	532
	21.2 Staffing	534
	21.3 Routine Maintenance Checklist Items	535
	21.4 Reserve Fund	539
	21.5 Concluding Remarks	540

Part 4. Information Sources540

Appendix 1	Conversion Factors	543
Appendix 2	Useful Information	545
Appendix 3	Associations and Organizations	573
Index		578

TO

Sue, Kristina, Marilyn, Susan, and Rick,
our families, who have shared our love of the sea
and fostered the opportunities for our professional growth.

Foreword

MARINAS and Small Craft Harbors defines the state-of-the-art in marina and small craft harbor planning, design, and construction. Its authors are experienced teachers and practicing professionals. Their contribution through this book will be recognized for many years to come—both here in North America and in the many countries throughout the world now developing recreational facilities.

Our engineering outreach program at the University of Wisconsin-Madison has for nearly two decades offered an annual state-of-the-art conference on marina planning, design, and construction. The conference has been built on meeting educational needs of professionals practicing in this field. The authors, Bruce Tobiasson and Ron Kollmeyer, have for many years participated as guest faculty at these conferences. By having done so, they have sharpened their instructional skills, broadened their knowledge base about state-of-the-art practice, and developed a compassion to help others in the field.

MARINAS and Small Craft Harbors is a very comprehensive and readable work and belongs in the hands of all those people responsible for planning and building waterfront facilities in the 1990s and beyond. I commend the authors for their excellent work.

Madison, Wisconsin   *C. Allen Wortley*
                             Professor of Engineering

Foreword

Over the past three decades, marina facilities have evolved rapidly from primarily Mom and Pop companies, using homemade docks, into much more sophisticated businesses providing a broad range of services to a wide variety of boats, in an increasingly complex regulatory environment. The International Marina Institute estimates that, in the United States alone, around 10,000 marina facilities have been developed, including pure marinas, boatyards, dockominiums, yacht clubs, public docks, state and military run marinas.

During the 1960-1980s era of marina growth and change, a comprehensive reference book for marina developers, owners, consultants, and regulators was not available. That void has now been filled with the publication of *MARINAS and Small Craft Harbors* by Van Nostrand Reinhold. The team of Bruce Tobiasson and Ronald Kollmeyer has created an outstanding book, which I believe now becomes THE MARINA DESIGN GUIDEBOOK.

There are many good features in this book, so many that readers will be both surprised and pleased with this resource. I've singled out a few for special comment.

Readability. There is almost nothing about marinas which is technologically simple. Yet Tobiasson and Kollmeyer have written in a simple style which makes complex concepts and factors easy to understand even for those just beginning to consider marina issues. Dr. Kollmeyer, author of Chapters 5 through 9 and sections of Chapter 17, did an outstanding job of converting complicated oceanographic theories into plain language. Professional engineer Tobiasson deserves highest credit for writing the rest of the book and for standardizing the writing style so all can understand it.

Balance. Many reports about marinas tend to skew one way or another, become very pro or con, proclaim the best and worst, or promote buying this but not that product. This book has judicially avoided taking sides, by presenting facts and alternatives for readers to judge what is best. As the authors often say. "Every marina is site specific" and must be planned, developed, constructed, and operated for each market and environment. Yet, in this book can be found all the common factors which exist, or must be designed into almost every marina worldwide.

Useful Information. There is so much good, practical information crammed into this book, that people seeking answers about marinas will find guidance and facts for almost every professional question. This is the only book I've seen which covers the waterfront from marina concept, through financing, permitting, site assessment, design planning, construction, to operation, management, and maintenance.

Charts, Photos and Illustrations. A picture is worth a thousand words. There are so many excellent pictures, charts, and illustrations throughout the text, that concepts and design alternatives can be quickly seen and understood. Fortunately they are here, for if they were absent, this book would be ten times this size with words of description.

References. At the end of each part of the book, the authors have listed key references and information sources for use by those needing more details on particular questions. I give the authors credit for realizing that this book would not and could not give every detail. As professionals they chose to give a comprehensive overview of all the important issues and alternatives, then to refer readers to more technical sources.

Rules of Thumb. With courage and determination to give us a handy reference, Tobiasson included a very useful marina design "Rules of Thumb" in the appendix. His absolutely unique collection of seventy-six rules are the common sense guides which tend to become lost whenever any art becomes a quantified and standardized technology. For all the good science available and finite engineering calculations via computers today, there are many questions about marinas which are still best answered with a somewhat loose, but darn practical formula. Today, in our hard driving desire to satisfy environmental regulations, predict storm surge probability, quantify boat operator behavior, and have all variables measured, all marina development would run aground, but for the use of these rules of thumb. Over time, as new information emerges, some of these marina thumb rules will be revised, while others will be just as valid in the twenty-first century as they were in the seventeenth century.

Readers should know that the two authors each devoted the better part of one year out of their professional wage earnings to research, write, edit, and present this book for our use. They both deserve our highest regard and appreciation for completing this monumental project. This book is good. Bruce Tobiasson and Ronald Kollmeyer have, by this task, moved the marina industry a quantum leap forward.

As a marina reference book, I recommend that this be on the desk of every marina owner, manager, consultant, design engineer, and government regulator. Anyone professionally involved with marinas needs this book.

Wickford, Rhode Island, USA   *Neil W. Ross*
                                      President
                                      International Marina Institute

Preface

The significant increase in private recreational boat sales in the past twenty years has resulted in a great demand for recreational boat dockage space, often unavailable at any price. This unfulfilled demand provides an attractive opportunity for placement of investment capital for marina and small craft harbor development. Marina and small craft harbor development is, however, capital intensive and the risk of failure significant. Successful development requires careful market analysis, appropriate engineering and scientific investigation, thoughtful material selection and responsive operation.

Today's boating men and women require safe, convenient and comfortable marina facilities. Although this generation reportedly has considerable discretionary income to afford a recreational boating pursuit, they often have limited time to participate, and therefore require facilities that maximize onboard time and boat utilization.

The design and construction of marinas and small craft harbors has, in the past, been essentially a do-it-yourself industry. Many marina owners derive from a water oriented background and translate this experience to marina development. Today, however, most "good" marina and small craft harbor locations are occupied and new development often requires venturing into exposed or difficult to build sites. Site difficulty combined with myriad rules and regulations for waterfront development and rapid advances in state-of-the-art facility components, demands professional input for successful marina and small craft harbor development.

This book attempts to put marina and small craft harbor design, development, and operation into a logical and rational sequence. Site evaluation is a critical step in new or rehabilitated marina development. Market factors influence profitability and the extent of possible development. Design criteria must be understood to make rational layout and material choices. Selection of docks and components can significantly affect long term suitability as well as first cost. Configuration and character of the upland facilities and other marina amenities significantly influence the success of marina development. Marina operation and services can make or break the best designed marina.

The successful marina and small craft harbor developer will need to become well versed in all aspects of development and operation. This book provides a basic review of parameters necessary for knowledgeable marina development. It is not a substitute for prudent thinking or experience but can be useful as a primer for intelligent design, development and operation.

Many people have contributed directly or indirectly to the successful completion of this work. Bruce Tobiasson is indebted to Paul S. Crandall and Kenneth M. Childs, Jr. who provided the opportunity for early professional growth in the field of waterfront engineering and whose continued counsel is highly valued, and to Arthur R. (Pete) Stagg, Jr. who helped provide the necessary physical and mental strength to pursue life's difficult goals. Ron Kollmeyer gives special thanks to the U.S. Coast Guard which took him off the streets of New York City and put him on the water. As co-authors, we are both grateful to Professor C. Allen Wortley, University of Wisconsin, who has furthered our education and that of many others by his knowledgeable and timely "Docks and Marinas" conferences and his personal interest in fostering the initiation of this book. H. L. (Harry) Burn who continues to challenge the state-of-the-art in marina design and William H. Koelbel, Neil W. Ross, Paul E. Dodson, and John W. Fenton, colleagues who deserve special mention as pioneers and leaders in the improvement and quality of marina design, education, and operation. Special thanks are extended to William H. Koelbel, Professor Eugene Spinazola, and Rebecca Ramsey Ruopp who reviewed the text and provided a wealth of constructive comment.

We also offer acknowledgements to the many clients who over the years have trusted their projects to our ability and have therefore invested heavily and hopefully wisely in the experience presented in this book. To them and the many others who have unselfishly provided guidance and assistance along the way we offer our many thanks.

Symbols and Abbreviations

A	area, cross sectional area acceleration of gravity
A_p	projected area
B	boat beam, width
Bm	boat beam at waterline
C	wave speed
CATV	cable television system
C_D	drag coefficient
CF	cubic feet
C_f	frictional drag coefficient
c	ice strength coefficient
cm	centimeters
coef.	coefficient
c.g.	center of gravity
cu	cubic
D	pile diameter boat draft
D_A	draft aft, vessel
D_F	draft forward, vessel direction factor, wind pressure
d	water depth
deg	degrees
dia	diameter
E	energy end area, boat profile
e	base of Napierian logarithmic system

SYMBOLS AND ABBREVIATIONS

F	degrees Fahrenheit
	fetch length
	force
FDD	freezing degree days
F_b	bending stress, allowable
Fd	drag force
Fc	current force
f	friction factor
	frequency
ft.-lbs.	foot-pounds of force
f_b	bending stress, actual
g	grams
	acceleration of gravity at earth's surface
gal	gallons
gpm	gallons per minute
gps	gallons per second
H	wave height
Hb	height of breaking wave
Hc	controlling wave height
H_s	significant wave height
H_{10}	highest 10 percent of waves in height
Hcomb	combination of wave heights
I.D.	inside diameter, pipe or pile
in.-lbs.	inch-pounds of force
KE	kinetic energy
k	kip, 1,000 pounds
kts	knots
L	length
	wave length
lbs.	pounds
LCG	location of center of gravity

LOA	length over all, for vessels
LWL	length along waterline, vessel
LT	long ton, 2,240 pounds
M	mass
	bending moment
MHW	mean high water
MHHW	mean higher high water
MLW	mean low water
MLLW	mean lower low water
mg/l	milligrams per liter
mph	miles per hour
NGVD	National Geodetic Vertical Datum
PMA	primary market area
PVC	polyvinylchloride
pcf	pounds per cubic foot
psi	pounds per square inch
psf	pounds per square foot
S	section modulus
	boat shielding factor
SWL	still water level
T	wave period
TSF	tons per square foot
T_s	significant wave period
t	thickness, plate
V	velocity
VHF	very high frequency marine radio telephone
v_f	fastest mile wind speed
v_m	fastest observed one minute wind speed
$	dollars, U.S.
Σ	summation of following numerical expression
%	percent

Acronyms

AASHTO:	American Association of State Highway and Transportation Officials
ABYC:	American Boat and Yacht Council
ACEC:	Areas of Critical Environmental Concern
ACI:	American Concrete Institute
ACOE:	Army Corps of Engineers (United States)
AFFF:	Aqueous Film Forming Foam
AISC:	American Institute of Steel Construction
AISI:	American Iron and Steel Institute
API:	American Petroleum Institute
APR:	Area of Preservation or Restoration
ASCE:	American Society of Civil Engineers
ASTM:	American Society for Testing and Materials
AWPA:	American Wood Preservers' Association
CERC:	Coastal Engineering Research Center (U.S. Army Corps of Engineers)
CRREL:	Cold Regions Research and Engineering Laboratory (U.S. Army)
CZM:	Coastal Zone Management Agency
DMA:	Defense Mapping Agency
EPA:	Environmental Protection Agency
FEMA:	Federal Emergency Management Agency
FWHA:	Federal Highway Administration
ICOMIA:	International Council of Marine Industry Associations
IMI:	International Marina Institute
ITE:	Institute of Traffic Engineers
JFK:	John F. Kennedy International Airport, New York
LOS:	Level of Service
NACE:	National Association of Corrosion Engineers
NAS:	National Academy of Sciences
NAVFAC:	U.S. Navy Facilities Engineering Command
NAVSEA:	U.S. Navy Sea Systems Command
NFPA:	National Fire Protection Association
	National Forest Products Association
NMMA:	National Marine Manufacturers Association
NOAA:	National Oceanic and Atmospheric Administration
NOS:	National Ocean Survey
OCZM:	Office of Coastal Zone Management
PIANC:	Permanent International Association of Navigation Congresses

RMA:	Robert Morris Associates (financial ratios)
SPM:	Shore Protection Manual (U.S. Corps of Engineers)
SNAME:	The Society of Naval Architects and Marine Engineers
USCG:	U.S. Coast Guard
USCGS:	U.S. Coast and Geodetic Survey

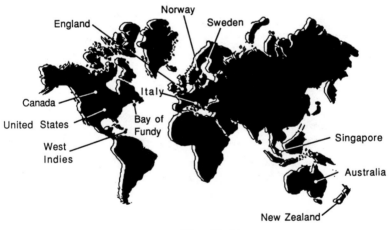

Reference maps showing principal locations referred to in the text.

PART 1
DEVELOPMENT OVERVIEW

1
Introduction

Over twenty-five years of marina design experience, by the authors, has lead to an intimate observation of the marina industry and its growth during this time. The most frequent comment by people entering the industry is "where do I find information about....?" The lack of available information on marina planning, design, construction and operation is due in part, to the infancy of the industry and the maturing of marinas as full partners in the business of recreational boating. Early marina development projects involved only a minimal design effort, usually just enough engineering to prepare regulatory applications with an occasional site visit by the designer to respond to problems arising in the field during construction. Most of the engineering effort was directed toward providing structural solutions and simple line drawing presentations.

Today, however, the marina is a complex facility requiring major capital investment and sophisticated scientific and engineering expertise. The awakening of environmental awareness and the subsequent desire to protect the fragile coastal environment has required development of new areas of study and understanding for marina planning. Coincident with the desire to protect and enhance the environment has been the rapid change in recreational boating itself. The development of low cost, mass produced boats and accessories has opened the once rather exclusive field of recreational boating to the full economic spectrum of the population. Boat ownership and use is no longer a sport of the idle rich, it is "everyman's sport" and it is depicted to be a healthful leisure pursuit. Recreational boating is very much a national pastime in the United States with as many as 72,000,000 Americans (1988), nearly a quarter of the population, using the nation's waterways more than once per year. The escalation in boat ownership by large numbers of the population has created an overwhelming demand for marina facilities (see Fig. 1-1).

The demand for marina facilities has resulted in a boom time for marina development in many areas. Marinas which were low or no profit operations and are now profitable, along with the many financially successful marinas,

4 PART 1/DEVELOPMENT OVERVIEW

Figure 1-1. Kongen Marina, Oslo, Norway. This large, modern marina is representative of the type of small boat facility for which demand is increasing throughout the world. *(Photo courtesy: SF Marina System AB, Kungalv, Sweden)*

may consider upgrading and expansion. The perceived opportunity to realize a desirable rate of return on investment has brought new players into the marina industry. Many of the new players are not recreational boaters and have little or no prior experience in waterfront development. The result has often been creation of marina complexes that appear, on paper, as desirable facilities but in reality miss the mark in terms of providing safe and practical berthing and related boating services.

The services that marinas provide and how they relate to the overall picture of getting people to the water may be described as suggested by Neil Ross, President of the International Marina Institute, as an hourglass (see Fig.1-2). The upper bowl of the hourglass represents the large population who desire to pursue active recreation on the water. The lower bowl represents the water bodies to which access is desired. The constricted center tube of the hourglass, represents the limited shoreline and associated water access facilities through which persons desiring to use the water must pass to participate in water related activities. It is in this narrow center conduit that the marina tenuously exists. Marinas are water dependent facilities and

INTRODUCTION 5

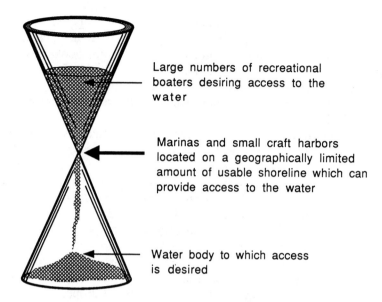

Figure 1-2. Hourglass analogy of access provided by marinas to water bodies.

generally they must be located on or near a water body to fulfill the need of access to the water. There is a finite amount of waterfront available and it now has many competing uses. Marina development must vie with often greater financial and political uses such as residences and offices, to secure its place on the shore. Recent, renewed, interest in waterfront property has resulted in as much as 80% of the tidal flowed and inland coastline, of the United States, to be held in private ownership, with access to the waterfront often denied the public. In many cases, the 20% of shoreline held in public ownership is providing public access to the water, but not always in sufficient quantity to meet the demands of recreational boating. The primary access on to the water is through privately owned marinas and other privately sponsored boating access facilities. Statistics on the international distribution of public and private access to waterways is not readily available, but personal observation by the authors indicates that most recreational boating facilities are privately sponsored.

Waterfront development encompasses many separate entities which are often interrelated in design and function. The term marina conveys an association with recreational boating, however, the facility may include commercial docking space and services, moorings, boat repair services, new or used boat sales, water transportation components and other related waterfront activities. The planning of appropriate marina facilities must address not only the many components of the single proposed marina but also the uses and constraints of surrounding activities. To adequately cover the full spectrum of marina and allied use activity, this book seeks to include not only the marina itself but many of the activities associated with small craft harbor development. A knowledgeable understanding of marina and small craft harbor development will lead to well planned and engineered facilities.

Apart from providing direct access to the water, marinas play an important role in the local, regional, national, and international economy. Recreational boating industry estimates for 1988 indicate that in excess of $17,000,000,000 was spent, in the United States alone, on boating related products and services during the year. It is further projected that nearly one half of this amount is spent at marinas. Recreational boating is certainly not limited to the United States and North America as demonstrated by significant boating activity in Scandinavia, Europe, the Middle East, Australia and New Zealand. Japan, Singapore, Turkey, Brazil, Mexico and many other countries are rapidly expanding recreational boating facilities to meet the increased demand for access to the water. Recent events in the eastern European community, with the potential for unprecedented political and financial stability in the region, may result in explosive growth in recreational water related activities. To support the demand for access to the water, many new marina and small craft harbor facilities must be developed by both public and private interests.

1.1 SCOPE AND PURPOSE

The successful development of a modern marina requires knowledge in a wide range of professional and practical disciplines. The scope of these disciplines includes: financial issues, regulatory constraints, site selection and civil works, climactic effects, environmental effects, protection considerations, vessel characteristics, layout parameters, materials of construction, design loads, haul-outs, upland facilities, maintenance, and operation. A knowledge of all of these topics is necessary to develop the complete marina.

The intent of this book is to define the state-of-the-art in marina and small craft harbor planning, design, and construction as it has developed since the late 1940s. The book addresses a wide range of marina related topics intended to provide a general knowledge of the many facets of the industry. It provides a lot of general information, some specific information, "rules of thumb" on general design concepts, and includes information sources for those desiring further pursuit of specific areas of interest. The industry is constantly changing and therefore this book is not to be considered as a final judgement on all that is or will be, but it is to be considered rather as a bridge between the unknown and that first step on the unexplored island of marina development.

No one reference text can fully explore all the details, considerations, and design techniques of such a diverse industry. The goal of this book is to establish a background of marina history, development, and current design trends to provide a basic technical understanding of the marina facility along with the information sources to allow the reader to further investigate those areas of particular interest and study.

An important use of this book is foreseen as a reference text to provide information on specific subjects of interest, well beyond an initial reading. To facilitate access to specific information, the text is developed in four Parts. **Part 1, Development Overview,** presents an introduction to the marina development industry and business issues relating to marina development. Chapter 2 addresses marina market considerations, business planning, and marina financing strategies. Chapter 3 describes regulatory agencies, public trust doctrine, and required approvals associated with marina development. **Part 2, Site Evaluation and Assessment,** focuses on information necessary to properly consider a potential marina site and evaluate the physical parameters that affect site development. Chapter 4 discusses development goals, public access, water dependency regulations, basic environmental factors, required site surveys, maritime conflicts, and traffic assessment associated with marina development. Chapter 5 investigates the mechanisms of winds and storms and discusses how this information relates to marina design. Chapter 6 focuses on wave climate, what waves are, how they are generated, and the analysis and prediction of waves and how they impact marinas. Chapter 7 reviews other water related effects such as boat wakes, tides, currents, river floods and ice. Chapter 8 considers how to use the knowledge of water related environmental effects to develop a perimeter protection plan. Chapter 9 addresses the marina basin and related wave activity, water quality, water exchange, and sediment transport within and around the marina. **Part 3, Engineering Design,** focuses on taking the information developed in Part 2 and creating the marina design. Chapter 10 provides an insight into vessel design as it affects marina development. Chapter 11 explores the types of dock structures available and their applications. Chapter 12 provides the parameters for intelligent marina layout and the provision of necessary support facilities. Chapter 13 discusses construction materials and how their selection relates to marina design. Chapter 14 considers the corrosion and degradation of materials used in fresh water, sea water, and terrestrial environments. Chapter 15 addresses specific structural design loads that are relevant to marina construction. Chapter 16 presents basic planning and design guidelines and utilizes the design loads of Chapter 15 to demonstrate basic structural analysis techniques. Design considerations associated with dock hardware, utilities, site dredging, provision of handicapped access, geotechnical issues, and comments on marina failures are discussed. Chapter 17 addresses considerations for mooring systems for marina structures including piled structures, bottom anchored structures, and individual swing boat moorings. **Part 4, Operations and Management,** discusses operation, management and maintenance of marinas. Chapter 18 provides information on numerous haul-out and boat handling facilities and design guidance in developing and operating these facilities. Chapter 19 moves on to the upland with consideration of parking requirements, marina patron amenities, marina office and ship's store facilities, other desirable site facilities, and on-site fuel storage issues. Chapter 20 focuses on marina management in terms of administration, staffing, service orientation, security, and procedures for handling storms and other emergencies. Suggested rules and regulations for marina users are also presented. Chapter 21 concludes the book with a discussion on maintenance, including preparation of a maintenance plan, staffing for maintenance operations, maintenance checklist items, and funding for maintenance operations.

Each Part contains several chapters on specific topics relating to the overall subject matter in the Part. Chapters are further divided into sections, which may also be subdivided. The intent is to provide a framework that allows rapid location of specific information in the text. A comprehensive index is also provided to facilitate location of specific topics of interest. The appendix is presented in three sections. Appendix 1 provides conversion factors to assist in converting units used in the text to metric system equivalents. Appendix 2 contains significant additional useful information not directly given in the text. A particularly helpful section of this appendix is the compilation of "Rules of Thumb and Other Useful Information" into a single source area. Each entry is generally quoted as presented in the body of the text with reference to the chapter and section where further information on the particular subject may be found. Also included is a "Marina Design Checklist," which is a convenient reference when beginning a project to catalog known project information and highlight information that must be obtained. A checklist entitled "Typical Preliminary Parameters to be Determined for Dredge Material Analysis" provides a handy reference for basic parameters of dredge material that are generally required to make preliminary assessments on dredge material quality. Appendix 3 provides a listing of associations and organizations which may provide assistance or information on a wide range of topics associated with marina development.

Mathematical units used in the text are generally those associated with traditional use in the United States or relevant to a specific topic if more universal units are generally accepted. In many cases preference is given to traditional nautical units, such as knots, nautical miles, fathoms, etc. To assist practitioners familiar with units other than as presented, Appendix 1, as discussed above, provides conversion factors for common units.

The book is based on the personal knowledge of the authors, each having over 25 years of professional waterfront design and investigation experience.

The authors acknowledge the contributions of many others who have prepared papers on marina subjects, presented lectures on marina issues, and made available pertinent data on marina development, all of which forms the core of experience the authors hope to convey. Both authors, through different career paths, have provided professional expertise to a variety of waterfront development projects and scientific investigations throughout the world. In addition to their international experience in marina development projects, they have presented state-of-the-art lectures on various aspects of marina development to marina industry conferences and technical institutions on five continents. Observations of marina development techniques obtained from world-wide travel has shown that marina development has a prominent thread of commonality that transcends regional differences. For the most part, therefore, the principles and concepts presented in this book are transferable world-wide. Regional considerations must be appreciated however, and traditional local techniques, and custom observed.

1.2 HISTORICAL ASPECTS

Marina and small craft harbor facilities have existed since man's earliest ventures on the sea. Vessels of all types have sought refuge in protected embayments and rivers to allow the safe mooring of the vessel and transfer of passengers and cargo. As far back as the 1600s and 1700s docks and piers were constructed to provide direct connection of a vessel to the land rather than require the use of small boats to ferry between vessels anchored offshore and the land. Early docking facilities were generally used by fishermen and maritime commerce. As recreational boating came into being, generally recognized to have begun in the early 1900s, the protected harbors used by fishermen and maritime commerce, were sought out for recreational boat berthing. With the increased popularity of recreational boating, in the mid 1900s, specialized facilities were developed to handle boats and to provide related services. Eventually communities of recreational boaters were established. In most cases these communities of recreational boats were not organized and were randomly disbursed.

It is only recently that marinas as specific entities have been defined. The term "marina" was first used to define a recreational boat facility in 1928 by the National Association of Engine and Boat Manufacturers in the United States. The word derives from the Italian word "marina," which means "small harbor." During the 1930s and early 1940s, recreational boating (yachting) was still a sport primarily pursued by the wealthy in large and often ornate sail or power yachts, or by people of moderate income in small, 8 to 30 foot, day cruising type boats. Technology that evolved from the war effort of the 1940s together with the spirit, goals and broadened perspective of returning military personnel is considered to be the major impetus to the rapid escalation of recreational boating as we know it today. The technological advances in use of materials such as wood, concrete, steel, aluminum and synthetics provided the basis for consideration of mass production of recreational boats at costs affordable by the average working person. Welding of steel components became widely available and allowed lighter and stronger fabrications, of the boats themselves and of developing manufacturing processes that allowed boats to be constructed from other materials. An example of the latter would be the ability to fabricate steel molds for casting fiberglass and plastic hull shells. The outboard motor also became affordable in the post war boom and in combination with low cost fiberglass, aluminum, and steel hulls made boating even more accessible to a broader population.

With the influx of returning service personnel into the peacetime work force, it was difficult to generate enough work to satisfy the supply of workers. Recreational boat construction and marina development became an attractive marketplace for enterprising entrepreneurs. Many of the existing older marinas were conceived by returning service personnel and many today are still run by these people or their families. Following the 1940s, recreational boating continued to grow at a relatively constant pace throughout the 1950s, 1960s and early 1970s. Since the mid 1970s, a major change has taken place that has altered the definition of the traditional marina facility. By the mid 1970s the generation of "war babies" had achieved early middle age and survived the tumult of the 1960s and the police action ("war") in Southeast Asia. They had developed families and worked long enough to earn moderate and sometimes meaningful incomes and had enough leisure time to enjoy a relatively high standard of living. With some discretionary income and available leisure time, recreational boating became a desirable and healthful way to fulfill a desire for active recreation in an interesting and challenging environment.

The "mom and pop" marinas, popular from the 1940s to the 1970s (see Fig. 1-3), were no match for the explosion of interest in recreational boating. Virtually all marinas needed to be expanded to fill the growing demand. Many small marina operators continued to want to maintain the compact organization afforded by sole proprietorships or small to medium size operations. This desire to remain small, however, often became overshadowed by economic considerations associated with waterfront land values. The desire by people to be on or near the water caused a spiraling effect in the escalation of waterfront property values, often depriving small marina operators of their ability to remain traditional small businesses. Today the competing interests for valuable waterfront property are creating major hurdles for new, intelligent marina development. Marinas today, mean many different things to the marina operator, the recreational boater, and the waterfront developer.

INTRODUCTION   11

Figure 1-3. The forerunner of today's modern marina. Typical wooden, fixed docks, of the type constructed in the mid 1940s and still in use today. *(Photo courtesy: Bayreuther Boat Yard, Inc., Niantic, CT)*

1.3   WHAT IS A MARINA?

The original concept of a marina as a small harbor providing facilities for recreational boats has changed. Marinas, today, may not even be in a harbor or may, in fact, have to create a harbor to exist (see Fig. 1-4). There are also "dry land marinas" with no direct access to the water. Marinas may be stand alone facilities, may be part of a boatyard facility or may provide facilities as an amenity to a larger upland project, either business or residential. Marinas are often not just places to park boats, but they may provide restaurants, new and used boat sales, boat repair services, shopping malls, theaters, night clubs and other excitement and amenities desired by boating people.

In general, however, marinas are construed to be facilities associated with the waterfront that provide services and storage capacity for 10 or more boats in slips, at moorings or in racks (dry storage). The variety of definitions of marinas is as varied as the types of marinas that exist. Marina operation and design is fraught with the imposition of variables, from the size and type of vessels accommodated, the individual personalities of the tenants or owners, the vagaries of weather and environment, and the ever changing requirements of the cognizant regulatory agencies.

A definition of marina design that acknowledges its complexity was given by William H. Koelbel, marina consultant, at a marina planners conference

12 PART 1/DEVELOPMENT OVERVIEW

Figure 1-4. Racine, Wisconsin. This small craft harbor on Lake Michigan illustrates how the original concept of marinas has expanded since their first use on the lake in the 1920s. This man-made harbor was rehabilitated from a commercial use into a major recreational facility including 920 boat slips, moorings, yacht club, and community recreational amenities. *(Photo courtesy: EWI Engineering Associates, Inc., Middleton, WI)*

in Hauppauge, New York in February 1989. Mr. Koelbel said, "Marina design is to deal with an infinite number of variables, in a constantly moving medium, dealing with a significant number of opinions to design an absolutely perfect facility for a constantly changing complement of vessels."

1.4 WHAT IS A SMALL CRAFT HARBOR?

The terms "marina" and "small craft harbor" are often used interchangeably to describe facilities for berthing and servicing recreational boats. The use of small craft harbor in this book is to define a broader spectrum of usage than just the berthing and servicing of recreational boats. Small craft harbors (see Fig. 1-5), may include arrays of single point or cluster moorings, boat launching ramps exclusive of any other facilities, water transportation docks and facilities, yacht clubs, swimming associations, fishing fleet accommodation, commercial boat berthing and mooring, and any other traditional use of water frontage. Many of the principles associated with marina design discussed in this book relate directly to small craft harbors as well as to marinas.

INTRODUCTION 13

Figure 1-5. Scituate, Massachusetts. Small craft harbor, as used in this book, may include multiple marinas, single point moorings, haul-out facilities, boat launching ramps, commercial boat berthing, and private residential docks as depicted in this photograph. *(Photo credit: William F. Johnston, Quincy, MA)*

The intent is to present a broad spectrum of investigation and discussion to provide guidance in the development of marina and other small craft harbor facilities.

Some examples of information that is common to development of marinas or small craft harbor facilities include the chapters on regulatory considerations, site selection, winds and storms, wave climate, location and environmental site conditions, perimeter protection, and marina basin effects. Materials of construction, corrosion and material degradation, and mooring systems may also provide guidance in development of certain small craft harbor facilities.

1.5 MARINA OWNERSHIP

Marina ownership has generally been held by the owner of the associated upland. In some cases land was leased to others to operate and maintain the facility. Another form of ownership was and still is by municipality or other governmental agency. Interestingly, the armed services of the United States operate numerous marinas around the world and until recently and perhaps still may be, the largest single marina owner in the world.

Today marinas are owned and operated by a diverse group of entities. Individual ownership is still the largest single group but other forms of ownership such as corporations, cooperatives, condominium (dockominium) associations, or a combination of these forms of ownership, are fast catching up. A number of organizations, both privately and publicly held, have been buying marina properties to operate as a chain with common management and ownership. The economies of scale in hiring, central management, purchasing and marketing have apparently provided adequate return on investment for these entrepreneurs. Many developers are looking to develop the marina and then sell it off, generally to slip holders in the form of condominium (dockominium) ownership. The association formed by the dockominium owners then becomes the common ownership entity. Cooperative ownership is similar except that the individual purchases shares in a common association rather than individual ownership of the specific berth rights.

In good economic times, it is often possible for governmental agencies to foster ownership of marina facilities as a means of providing public services. It appears, however, that in poor economic times the marina facility will often suffer from the redistribution of funds to higher priority areas of concern. The vagaries of public funding priorities and the generally low slip fee, and therefore revenue return, associated with public facilities often results in maintenance inattention and difficulty in facility upgrading over time.

1.6 PROFILE OF THE MARINA USER

The types of users traditionally associated with marinas have included: "old salts," whose love of the sea required access to the water in some type of vessel, albeit humble or otherwise; sport fishermen who used a marina to park their boat between fishing trips; and individuals who liked to make short or long cruises and explore new horizons at a leisurely pace. Today's marina users may include all of the above plus a new generation of hyper, high power individuals whose aspirations of success are often expressed in "physical possessions," namely boats and related accessories. This group has the apparent economic wherewithal to purchase sizable vessels and marina services, but often is pressured for time to enjoy this leisure pursuit. Another expanded group of marina users is the great middle class, who have entered the recreational boating market in large numbers due, in part, to the creation of package boat deals, offered at reasonable cost by manufacturers that have consolidated boatbuilding, engines, trailers and accessories under a single corporate structure. The economies of this production method coupled with mass manufacturing processes and reasonable quality have created affordable boats that can be purchased by the average worker.

Marina users are generally boat owners, so a look at boat ownership data provides some insight into the generalized profile of marina users by category. For the sake of discussion potential marina users are categorized into four groups. The first group is traditional, hard working, fun loving, outdoor type blue collar workers and might be labelled as the Bluejean Boaters. A second group is the upwardly mobile, middle income, young, suburbanites called the Go-For-It Group. A third group is the upper middle income group who have left the urban pace and favor the good life at a more leisurely pace while enjoying the fruits of their labor. This group is designated as the High Rollers. A fourth group is the Golden Oldies who have retired or reduced their work effort and enjoy cruising and noncompetitive sailing.

The Bluejean Boaters are generally in the range of 25 to 45 years of age, have median incomes in the $30,000 to $50,000 range and are the hard working middle class. They own boats in the range of 12 feet to 30 feet. They are predominantly blue collar workers with high school educations and are very conscience of their position in life and seek to join a higher social status. They equate boat ownership to status but have limited discretionary income to fulfill their greatest desires in terms of boat size. This group comprises a high percentage of people with fishing interests. They prefer, because of limited income, to work on their own boats rather than hire outside labor. They are the beer and cooler crowd but will be good supporters of marina facilities and generally good customers, if somewhat slow on payments.

The Go-For-It group are young (20 to 35 years old), middle income, suburban, sports enthusiasts who are the largest single group of boat owners and a group with significant market clout. They are competitive, often the mainstay of sail regattas and power boat contests. Their income ranges between $30,000 and $60,000. They own boats in the 25 feet to 50 feet range and represent the largest group of powerboaters in the over 30 foot size. They enjoy associating with others of common interest and are the club and organization joiners. Their interests go beyond boating and they spend considerable time participating in or attending other sporting events. They are status sensitive and select a marina to enhance their social image. They are loyal and make good marina tenants.

The High Rollers are hard working, sophisticated, social climbers 35 to 60 years old. They have incomes from $50,000 into the six figure range. They have often moved out from their urban roots but still desire the glitz and high life associated with urban living. They frequent urban and near urban marinas but use their boats in all areas. They maintain connections to the urban lifestyle and desire its camaraderie. They prefer sporty powerboats or classy sailboats and often pay the cost for professional care. They are good marina customers, valuing quality service and goods. They spend some time doing their own work to help justify boat ownership but leave the difficult work to others. They want to be associated with a first class marina and desire amenities in keeping with their lifestyle such as pools, lounges, tennis courts, etc. A large percentage are college graduates and have worked hard to enjoy the good life. This group has no particular focus in regional or national placement, but may be found scattered throughout the country. Their boats range from 16 to greater than 100 feet. Their time spent boating is limited so they look for accessible and easy to use facilities that represent the nature of their station in life.

The Golden Oldies are a group of retired or semi-retired individuals, often holdovers from past association with one of the other groups. They are money conscious, often living on fixed incomes, but are fully involved in boating as a lifestyle. Boats will be 30 feet or greater, usually with accommodations for sustained cruising. They will be good marina tenants but will not be big spenders and will attempt to do much work themselves. Experience has taught them, however, to use specialized experts when their knowledge is lacking. They may tend to move around to sample new sights and sounds and get restless staying too long in one area. Many will take extended cruises but return to homeports for connection with their roots. Income levels vary greatly.

Marinas, therefore, must be developed to provide a variety of facilities and services to accommodate a potentially diverse clientele. Once just a parking space for a boat, marinas must now provide safe and desirable boat parking spaces (berths), clean and functional heads and showers, and special services for time restricted guests, such as boat housekeeping services, telephone and television capability, weather information, provisioning services, safe efficient fueling stations, sewage pumpout services and other concierge services. As marinas become the equivalent of second homes, the marina must generate an atmosphere conducive to the summer cottage tenant rather than remain simply a place to park, service and store a boat.

1.7 URBAN WATERFRONTS

Urban waterfronts are singled out for special mention for several reasons. Only recently has there been a major thrust toward a renaissance of our major urban waterfronts. As recently as the mid 1970s, urban waterfronts were, for the most part, decaying relics of a once busy maritime commerce trade. The changing patterns of waterborne commerce and its decline left many urban areas with unattended piers and derelict infrastructure. The cost to maintain or improve these facilities became staggering to the owners, with needed repairs often occurring simultaneously with declining revenues. In an effort to offset declining revenues, many parcels of prime waterfront land were sold to the highest bidder by troubled traditional waterfront

INTRODUCTION   17

industries or financially strapped public agencies. While traditional waterfront commerce was declining in some areas, business office space and residential units were at a premium. Urban waterfronts became the new "place to be." This transition has been occurring worldwide.

An outgrowth of the resurgence of waterfront offices and residences is the desire by the users of these facilities to have recreational boating facilities close to their working and living environment. Marinas have become desirable assets to modern urban waterfronts. Revitalized cities that once considered their waterfront as a derelict liability have begun to view the waterfront as a vital part of the urban experience and a place to be used and enjoyed by all the public (see Fig. 1-6). Urban marinas have also become a focus point for providing public access to the waterfront in the form of pedestrian boardwalks, water transportation docking facilities, seasonal and transient

Figure 1-6. Shipyard Quarters Marina, Charlestown (Boston), Massachusetts. This major urban marina, developed at the former Charlestown Navy Yard, represents a trend in providing boating facilities with the revitalization of urban waterfronts. In this case, the navy yard was turned over to the city of Boston as an urban renewal project including: residential, business, commercial, medical research facilities and the proposed relocation site of the New England Aquarium. From the waterside, the marina forms the focal point of the renewal project. *(Photo credit: William F. Johnston, Quincy, MA)*

18 PART 1/DEVELOPMENT OVERVIEW

Figure 1-7. Rowes Wharf, Boston, Massachusetts. This new marina and commercial boat berthing facility is part of a major urban waterfront revitalization project in Boston, Massachusetts. The marina was required to provide seasonal and transient boat slips, dinghy landing floats, complete access to the public around the residential and office buildings, and provide for the berthing and operation of one of the city's major water transportation terminals. *(Photo courtesy: Atlantic Marina Services Corporation, Long Beach, CA)*

boat berthing, and provision of dinghy landing facilities for boaters that may anchor in designated anchorage areas in busy ports (see Fig. 1-7).

Marinas are often an important component of waterfront revitalization but there is a price to be paid for development in an urban environment. Some issues that are addressed later include: dealing with an old infrastructure that may greatly impact development; security in an urban environment; effects of wakes from passing commercial and recreational maritime traffic; mooring difficulty in deep water berths; unalterable, massive fixed piers; toxic bottom sediments; and high land and development costs.

1.8 THE FULL SERVICE MARINA

The following list offers an overview of services and amenities that might be considered in developing a full service marina facility depending on local conditions and customs. The order of the list is not intended to imply any ranking of importance.

- Adequate water depth for draft of boats
- Secure boat tie up system
- High capacity electrical system
- Fresh water
- Fuel, oil, propane, alcohol, kerosene
- Sewage pumpout
- Fire protection and fire fighting equipment
- Telephone on docks
- Cable TV access on docks
- Ample and close auto parking
- Security
- Dock locker boxes
- Dock carts
- Dinghy rack or other storage area
- Clean and ample heads and showers
- Laundry or laundry service
- Ice—block and cubes
- Fish cleaning station (away from docks)
- Bicycle racks
- Freezer lockers
- Information on outside services
- Vending machines, soda, candy
- Trash receptacles
- Waste oil disposal containers
- Ship store/parts
- Marine VHF monitoring
- Dockhands to assist in docking
- Pay telephones
- Mail and package acceptance
- Message board
- Weather condition board
- Marina landscaping
- Picnic area
- Swimming pool
- Deicing system (when appropriate)
- Daily newspaper availability
- Library or book exchange
- VCR tape library
- Recreation or lounge area
- Tennis courts
- FAX and office services
- Posted marina rules
- Concierge services
- Boat haul-out facilities
- Engine/mechanical shop
- Hull/carpentry shop
- Paint shop
- Fiberglass repair
- Rigging shop
- Electronics sales and service

1.9 MARINA TRENDS

The rapid growth of the marina industry has fostered new thinking on the composition of marinas and the services they provide. Once stand alone facilities, marinas may be combined with many related and unrelated activities. Providing a recreational component to residential development is widely observed. This concept has been carried further in the development of marina villages. These villages are often self sufficient communities having their own shopping malls, religious facilities, entertainment complexes and even post offices. The villages cater to the boating life style and provide the serious boater with direct connection between their berthed boat, residence, and community life (see Fig. 1-8). Marina villages, of course, require large amounts of undeveloped land upon which a village may be sited. The waterfront shoreline will often be reconfigured to provide as great an amount

20 PART 1/DEVELOPMENT OVERVIEW

Figure 1-8. Davidson Landing, Lake Norman, North Carolina. This marina facility represents the concept of a marina village, where boat berthing is provided in conjunction with upland residences and community activities. *(Photo courtesy: MEECO Marinas, Inc., McAlester, OK)*

of usable waterfrontage as possible. Canals were employed in early village development but today a more irregular, natural landscape is desired. Wetlands and island structures may be included to provide visual relief from the otherwise geometric configurations of residences and marina structures. In most marina villages, the marina becomes an extension of the residential component, with berthing directly in front of the residences rather than in a distant formally configured marina facility.

Many existing marinas have been forced to upgrade facilities to provide more berths, bigger berths, and better upland amenities (see Fig. 1-9). In addition to increasing the number of berths in the marina, the average berth length was increased from about 28 feet to 40 feet, with substantial provision for megayachts and transient small boat dockage. A new 50 ton straddle hoist and a hydraulic trailer for moving the boats around the large storage yard were added to facilitate boat service and repair operations. A completely rebuilt, state-of-the-art fuel dispensing facility and fire protection system were incorporated into the expansion together with all new and upgraded utility systems. A public amenity was provided in the form of a public boardwalk along the waterfront. The marina enjoys a state-of-the-art dock system and first class facilities.

Figure 1-9. Bay Pointe Marina, Quincy, Massachusetts. Older, established marinas such as shown in the upper photograph can often be upgraded to include expansion of berthing facilities and upland amenities. The lower photograph shows the marina after planned expansion, with a new layout that accommodates larger boats, provides modern fuel dispensing and haul-out services, and improves the upland by demolition of unsightly buildings and construction of a new shoreline boardwalk. *(Photo credit: William F. Johnston, Quincy, MA)*

22 PART 1/DEVELOPMENT OVERVIEW

Another marina development trend is in the sale or long-term lease of marina slips. In this scenario, usually an association is formed to operate and manage the facility. In some cases, these associations become like a small community with group activities and community spirit. These marinas are frequently the best maintained and enjoy a high degree of popularity. Others, however, retain a more formal "business only" attitude with slip ownership being the only common thread and no or limited marina community contact. In areas where boat slips are in high demand, some slips are being purchased or leased purely for investment purposes, which may alter the traditional camaraderie associated with recreational boating. The subject of dockominiums and long term leasing is discussed further in Chapter 2.

Dry stack storage is another recent development that has become very popular (see Fig. 1-10). The finite capacity of the available waterfront for boating facilities has promoted the use of dry stack (rack) storage facilities. Dry stack storage facilities may be a component of a large marina offering in-water berthing as well as dry stack storage or they may be stand alone facilities that occupy a parcel of upland area that has limited waterfront access. These facilities are routinely handling small cruising boats to 30 plus feet in length and high performance boats to 40 feet in length. The racks may

Figure 1-10. Queensland, Australia. Modern, dry stack storage facilities provide high density boat storage with only minimal waterfront land utilization. Boat sizes accommodated in dry stack storage range from 18 feet to 40 feet (high performance type boats).

be free standing, exposed to the weather or may be enclosed within or part of an enclosed building structure. The benefits of out of the water, covered storage appeals to many boaters. This may be especially true in areas of severe heat or cold. The protection offered the covered boat may considerably lengthen the life of the boat while reducing yearly maintenance costs and effort. In many areas, small boat rack storage is the only way to go since the in-water marinas have been developed for larger boats to the exclusion of the smaller craft. Discussion of rack storage facilities is continued in Chapter 18.

Consideration for the provision of wave and boat wake protection has escalated in the past few years. The state-of-the-art in wave and wake attenuation is rapidly developing and the affordability of providing such protection is becoming more realistic. Marinas have evolved from simple boat parking facilities to vacation retreats demanding minimum discomfort to boaters from sea state activity while at the marina as well as providing safe and secure boat berthing. There is likely to be significant improvement in wave and wake protection during the next decade. Wave and wake attenuation is of such importance that several chapters are devoted to an understanding of the effects of wind, wave, and boat wake, and how the marina designer may undertake to reduce the effects of these forces.

The growth of the marina industry has also prompted a closer scrutiny by health, building, and fire officials on such issues as boat sewage handling, potable water plumbing, electrical safety and fire protection. Environmental regulations regarding the storage and handling of fuels has forced many marinas out of the fuel supply business. New fuel storage and dispensing operations will be required to meet stringent regulations to protect the environment, often at great capital cost. Access to adequate fuel supplies may become a major boating issue in the 1990s. Marinas able to accept the high cost of providing fuel may be extremely well positioned to capture a full complement of tenants.

Marina haul-out facilities have greatly improved in the last twenty years. Slow and awkward marine railways have given way to fast, safe, and mobile haul-out by straddle hoist lifts. Hydraulic trailers have allowed haul-out access to areas previously unavailable by use of simple sloping ramps or beaches. Trailers have also allowed boat owners to move their boats safely over roads to areas away from the waterfront for storage or repair. Hydraulic trailers are now available for boats up to 60 feet in length. There will probably be innovations in this area to handle even larger boats.

Boat repair facilities have become more modern and have the ability to handle a variety of work on a rapid turnaround basis. Rather than being a spin-off of automotive design and repair, boat repair has developed its own industry with specialized equipment, tools and parts. Long delays during peak boating seasons due to unavailable parts or service are infrequent in

24 PART 1/DEVELOPMENT OVERVIEW

today's recreational boating industry. The increased volume of work has allowed individuals to specialize in full time marine repair, resulting in a core of knowledgeable mechanics and craftsmen committed to the recreational boating market. In many areas, however, there is still a shortage of qualified personnel. Development of profitable marinas should encourage more people to train in boat repair and help relieve the present shortage.

Of all the marina trends observed, the largest growing trend is the involvement of regulatory agencies in the marina business. Although necessary to protect the common well being and often with well intentioned purpose, regulatory involvement has in many areas stifled the ability of the marina entrepreneur to provide those services and facilities demanded by the public which the regulators serve. Some of the problem lies in the marina industry which has done little to defend its positions, actions, and goals in marina development. Education is the common thread of both public and private interests to promote a mutual goal of providing adequate, safe, low cost, and accessible boating facilities. It is hoped that this book will help educate all the players in the marina development industry and provide some of the knowledge that will assist in making sound development decisions that can benefit both the public and private interests and keep boating a successful leisure industry.

1.10 MARINAS AROUND THE WORLD

It is interesting when travelling around the world to observe the local marinas and look for new and innovative techniques. Although, overall, marinas do not differ greatly, construction materials will vary depending on local availability, but with wood, concrete or metal the usual materials of choice. Only in the United States is there substantial use of fiberglass and plastic materials. Concrete seems to be a favorite choice of construction material around the world, probably due to its ability to be created from local raw materials, its durability, and the relative ease with which it can be cast into varying shapes by unskilled labor. Wood also is prevalent where timber is readily available. Some countries that have access to tropical hardwoods make good use of the material. Taking advantage of the materials resistance to rot and marine borer activity.

Probably the greatest difference found in marinas around the world is in boat berthing arrangements. Most of the world uses a Mediterranean moor type of tie-up, which is the method by which one end of the boat is secured to a dock and the opposite end secured by an offshore mooring or by the boat's own anchor (see Fig. 1-11). No finger docks are involved. A variation on this theme provides outboard tie piles which control the outshore end of the boat through mooring lines. In this arrangement boat boarding is executed over

Figure 1-11. Guest harbor mooring in Berg, Sweden, on the Gota Canal. A universal form of boat mooring facility known as the Mediterranean mooring. The system uses a fixed or floating dock to which one end of the boat is tied, while the other end of the boat is secured to a set mooring or the boat's own anchor. No finger floats are used between the boats.

the boat end secured to the dock. Often short gangways are carried onboard the boat and rigged to the dock to allow safe access to the dock deck. In the true Mediterranean moor, boats are often side by side, requiring fenders to keep the boats apart. Some boat shifting may be necessary to undock a boat in the tight confines encountered. In cases where the boat must use its own anchor to support the outshore end, there is often fouling with adjacent boat's anchor lines causing great confusion when a boat prepares to depart the dock.

In North America, and elsewhere in the world, many marinas have finger docks to provide boat separation, secure tie-up, and easy boarding from the side of the boat. Americans have even carried the finger dock concept further by frequently providing a finger dock on each side of the boat (see Fig. 1-12). This technique has been especially helpful in areas where marinas are subject to wave or boat wake action. The fingers on either side allow the boat to be tightly secured to each finger, preventing the boat from breasting against any dock surface. As the boat gyrates to accommodate incoming wave action, it is free to move separate from the dock system.

26 PART 1/DEVELOPMENT OVERVIEW

Figure 1-12. Shipyard Quarters Marina, Charlestown, Massachusetts. A North American marina using a floating dock system with finger floats on each side of the boat. The ability to tie the boat securely to the dock system but not in contact with the docks, provides protection to the boats and docks in areas with wind and wave action. This system also eliminates the need for intermediate tie piles between boat berths.

Marina mooring systems also vary around the world. Piling still appears to be the preferred mooring system, however, bottom anchored cable or chain systems are also quite prevalent. The type of mooring system depends upon the availability of materials and the composition and capacity of the bottom subsoil. Chain or cable bottom anchored systems are used extensively in Scandinavia, where high rock bottom topography precludes the easy use of piles. If piles were to be used they must have sockets drilled into the rock to set the pile. This is usually quite expensive.

Americans have the highest demand for utility services. Only the newest and most modern marinas in other parts of the world will have full electric service, potable water, telephone and cable television available at each marina berth. In the United States these requirements are typical at most marinas. The penalty, of course, is that marinas in the United States cost more to construct, with the cost being passed on to the boating public in the form of higher slip rental fees.

With the exception of New Zealand and the United States there appears to be very few attempts to provide comprehensive fire protection systems at marinas. It is only recently that the United States has demanded the inclusion of comprehensive fire protection systems. This area is currently under considerable scrutiny and we may expect to see new guidelines issued for marina fire protection in the near future.

In warm climate countries, there is a relatively high percentage of liveaboard marina tenants as compared to marinas in colder climates. A tolerable climate, relatively inexpensive cost of living and the independence associated with being able to sail off when the mood strikes makes living aboard in a marina a desirable lifestyle for many people.

Marina development, as it exists today, has progressed a long way from its early formal beginnings in the late 1920s. Observation of marinas around the world indicates that there are more similarities than differences in marina development, marina design techniques, and marina materials application, irrespective of the marina location. It is now time to consolidate the lessons of the past and advance the state-of-the-art in marina development. The recent growth experienced in recreational boating and the current focus on environmental concern provides the impetus to create more and better marinas that achieve the goals of access to the water while protecting and enhancing the environment.

2
Financial Considerations

Financing and marketing marina projects are critical to the success of the project. Although we strongly recommend professional assistance in preparing the financial package, the following material will provide a sound basis for understanding the work to be accomplished and the information required to be obtained and analyzed to assure a firm grasp of the financial ramifications of a marina project.

2.1 MARKET STUDIES

Market or area studies play two important roles in the early stages of marina planning and design. The market study will reveal the need and financial potential for the project. The study will also assist the designer in the conceptual phases of marina layout by providing guidance on the desirable boat mix for the area. Without a solid grasp of the market, any conceptual layout and related cost estimating may be seriously in error and corrupt the potential for a financially responsible project.

The focus of the market study is to develop all pertinent facts associated with the business, to allow intelligent forecasting of future business requirements and goals, and to understand the competition.

Numerous firms specialize in the preparation and execution of market studies for a wide variety of needs. Marinas are special, however, and it is wise to select a firm that has past experience in marina studies or demonstrates an understanding of the market in the proposed project location. Reporting and survey techniques may vary widely with different firms, but the bulk of the data collected and reported should be relatively consistent. A basic marketing study should address the following topics: site description, seasonality of market, marina market inventory, current slip rental fees or dockominium sale prices, and absorption rates and trends.

Site Description

The market study should include a narrative and graphics describing and locating the project. Important considerations include: proximity to a major metropolitan area; primary transportation access e.g., highways, train routes, bus routes, airports and water transportation facilities; proximity to a major water body and the use of that water body. The site should be physically described, presenting information on topography, vegetation, upland acreage, water frontage, water depths, abutting land use, site zoning and general characteristics of the neighborhood. Any special or unusual characteristics should be mentioned, such as the existence of a railroad track on site, or the site being adjacent to a hospital, etc. The site description should provide the reader of the marketing study with a visual picture of the site and its environs to help assess the validity of the conclusions drawn in the study.

Seasonality of Market

Recreational boating is generally a seasonal activity. Even in climates that offer year around weather conducive to boating, boating activity is greater in some periods of the year than in others. Summer weather may be too hot or too wet, winter weather may be mild but too windy, etc. The effect of seasonal boating attitudes may greatly influence marina development. A marina in a hot climate which does not have covered slips may suffer a hot weather slump compared to a facility that can handle the hot weather environment. In colder climates, a secondary revenue source for the marina may be in boat hauling and storage, with this revenue being used to offset the in-water capital cost of the facility. Conversely, if no winter on-land storage is available then the summer rental may have to provide an adequate revenue stream for capital cost recovery.

Marina Market Inventory

A survey should be conducted of all marina and small boat facilities in the area. The extent of the Primary Market Area (PMA) is difficult to define but generally it lies within a radius of 25 miles of the site. Within the PMA, the number of existing and proposed boat slips should be inventoried. Proposed, but unbuilt, developments may require some clarification as to potential for completion and date of expected completion. The sponsor's proposed project itself is generally excluded from the slip count at this level of study. Within the boat slip inventory, a further breakdown of slips by length or boat size should be made and the type of boat described such as: recreational, commercial, sport fishing charter, other charter, etc. Facilities providing boat services, fueling, haul-out, repair, storage, etc. should also be characterized. Sources for locating information or the information itself can be obtained directly from marina managers, local and state permitting agencies, the Corps of Engineers, local newspapers and trade publications. Some states may release boat registration data by region, which may also be a source of area boat ownership statistics. Boat cruising guide books can also provide information on marina statistics and services provided in the area.

The purpose of the boat slip inventory is to develop a basic understanding of the existing and proposed competition for providing boat berthing in the Primary Market Area. The more comprehensive the study, the better the analysis of marina performance at the selected site will be.

Current Slip Rental Fees or Dockominium Sale Prices

Slip rental prices or dockominium sale prices may vary widely with the location, access and condition of the various marinas in the Primary Market Area. A survey should be conducted to determine the current slip pricing for all types of marinas within the area. The pricing data should be brought to a common base such as dollars per foot of rentable or saleable slip length. The survey should also include the history of pricing if available, including information on whether slip fees have escalated recently or the market been stable and for how long. Current slip rental fees may tell an important story for the potential dockominium developer. If slip fees are low, the attractiveness of slip purchase by dockominium or other form of ownership may not be desirable to potential users. If slip fees are high and future increases evident, an atmosphere is created where purchase of slips may be more financially attractive than seasonal rental of the slip. As is often expressed in real estate development, "location, location, location" is the single most important ingredient in successful development. The same is true in marina development and the proposed or existing site should be evaluated accordingly when assessing the impact of slip lease value or outright sale.

Absorption Rates and Trends

The projected absorption rate for a new marina i.e., the amount of slips that may be constructed and occupied in a given year, is a most important consideration in financial planning. If a reasonable absorption rate cannot be sustained during the projected build-out of the facility, then successful recovery of the capital development cost may be unachievable. The ever changing social and economic fabric of a community may make absorption rate estimating very difficult. A primary source of data for determining absorption rates is to look at how existing local marinas have performed in the recent past. Here again, the numbers determined by survey must be tempered by assessment of the quality, pricing and location of the facilities surveyed, with the same criteria being applied to the proposed facility. New boat sales statistics and local boat dealer assessment of current and future boating sale trends can be very useful in projecting absorption rates. It is also important to look beyond the boating community when developing marketing statistics to assess regional development and how that development, or lack thereof, will impact future marina boat berthing needs. Certainly an area having a sound economic base with a high percentage of boating age residents with discretionary income will provide a sounder base for marina development than an area that has a poor economy.

The goal of developing an absorption rate projection is to evaluate the extent to which a marina facility can be initially constructed and expanded within the constraints of capital cost recovery. Many marinas will be developed in phases so that the initial phase can become revenue producing, before the subsequent phases are constructed. With a phased plan each phase should be planned so it can stand alone and not have to rely on complete build-out to make the project financially feasible. The first phase must cover the initial development costs in case future development does not occur. The absorption rate information along with the marketing boat mix data will provide the marina designer with the tools to properly conceptualize a viable facility.

The completed market study will form the basis for preparation of a business plan that demonstrates the capacity of the project team to execute the project that the market study has revealed to be viable.

2.2 BUSINESS PLAN

A business plan is a road map created to allow a knowledgeable manager to navigate his way toward predetermined business goals. The plan may simply be an operating guide used by management to track progress of its ongoing business in relation to envisioned goals, or it may be used as a substantiating document in a loan application package. To secure a loan from a lending institution a business plan should contain substantial financial data. The lender will be interested in the nature of the business, the management team, the financial requirements, and the soundness of the investment. The more comprehensive the submission and thorough the analysis, the better the chances for a financing approval. The lender will become, in essence, a partner to the venture and will demand a full and honest disclosure of the project. The creation of a business plan is a mental exercise requiring a business owner to view the future of the business while operating in the present. Most business plans are conceived for a future time period of from one to five years. The well prepared plan will establish goals and objectives to use as milestones for evaluating business performance along the way. An important feature of a good business plan is its ability to flag areas where timely action and business modification can prevent unexpected business losses. Conversely, appropriate tracking of the plan may indicate better than expected business success and the need to infuse more capital or other resources into the business to continue its positive movement.

Business Plan Outline

A business plan may take many different forms, but all should contain several major types of information. The following is a general outline for a representative business plan.

Cover: Business name
Principal's of the business
Address and telephone number

Statement of Purpose:

The statement of purpose is a form of executive summary that should address the following issues to allow the reader to understand why the plan was prepared.

Purpose of the plan:
Business guidance document or financial proposal for loan application?
Nature of the business:
Type of ownership (corporation, sole proprietorship, partnership, Subchapter S corporation, or other business form).
Products or services offered
Market(s) served
History of the business
Year established
Continuity of ownership
Major events (mergers, buyouts, takeovers)
Plans for growth and expansion:
New products or services
New markets, distribution and outlets
Personnel expansion
Financial information: (for loan documentation)
Loan requested by whom
Size of loan request
Purpose of loan
Benefits of loan to the business
Justification for loan
Repayment potential

Details of the Business:

Description of the business:
Type of business (service, manufacturing, merchandising. etc.)
Form of ownership (corporation, sole proprietorship, partnership, etc.)
Operating status (new business, take over of existing business, or expansion of existing business)
History of the business or management experience if a new venture

FINANCIAL CONSIDERATIONS 33

    Hours of operation and seasonality of business
    Profitability of business: current and/or projected
    Potential for business success
    Trade agreements and credit status
    If a business take over, how was purchase price determined
Location of the business:
    Business address and community description including zoning
    Physical features of the land and or structures
    Property ownership or lease agreements
    Required property renovations or expansion plans
    Relation of location to business success and operating costs
    Property appraisal conforming to local financial institution standards
Market:
    Description of the Primary Market Area (PMA)
    Description of the market characteristics
    Estimated market share for the business and the growth potential
    Product/service pricing and relation to market pricing
Competition:
    Description of the local competition
    Similarities and differences between competition and the business
    Status of the competition's business health
    Lessons learned from observation of competition's operations
    How will the business be better than the competition
Management:
    Resumes of key individuals
    Business experience of the management team
    Experience in marina type business (managerial and operational)
    Management and operational chain of command
    Management payroll
    Other available resources (legal, accounting, secretarial)
Personnel:
    Staffing plan, current and projected
    Basic job descriptions and skill levels
    Manpower needs (full time, part time, seasonal)
    Salaries, wages, fringe benefits and overtime
    Continuing education, in-house or contracted

Financial Information:
    Operating statement: (primary accounting tool)
        Current year and three to five year projection
        Sales by division or other convenient grouping
        Total sales
        Cost of sales

Total cost of sales
Gross profit
Operating expenses
  Supplies
  Freight and postage
  Salaries
  Utilities
  Expand as needed for your business
Total operating expenses
Other expenses (debt repayment, etc.)
Total (all expenses)
Net income or loss (before taxes)
Taxes
Net income or loss (after taxes)
Balance sheet:
  Current assets
    Cash
    Accounts receivable
    Inventory
    Supplies
    Prepaid expenses
  Fixed assets
    Buildings
    Land
    Fixtures
    Equipment
    Vehicles
    Leasehold improvements
    Total fixed assets
  Current liabilities
    Accounts payable
    Long-term debt, current liability
    Total current liabilities
  Long-term liabilities
    Bank loans payable
    Notes payable
    Other loans payable
    Total long-term liabilities
  Total liabilities
  Net worth (owner's equity)
  Total liabilities and net worth

Breakeven analysis:
  Determine the value of the following categories
    Sales in dollars (or other money units)
    Cost of sales in dollars
    Gross profit in dollars
    Fixed expenses in dollars
  Divide gross profit by cost of sales to determine gross profit as a percentage of sales
  Divide fixed expenses by gross profit as a percentage of sales expressed as a decimal to obtain breakeven
  Loan or investment request documentation: (if purpose of business plan)
    Statement of need for loan or investment funding
    Distribution of funding (working capital, equipment, inventory, etc.)
    Specific description of use of funds and costs
    Statement on why securing the funding will make the business more profitable

A business plan should at least include the above items but may be more comprehensive to suit the particular business requirements. A complete business plan will provide a set of financial forecasts and budget guidelines appropriate for the business. Its preparation will force analysis of business practices, highlight potential pitfalls along the way, assist in obtaining capital infusion when necessary, and provide the tools to be a better manager. It is advisable to review the business plan with a professional or knowledgeable colleague to critique its validity and appropriateness.

2.3 CAPITAL COSTS

Assessing the capital cost of a proposed project is often an early step in the financial decision process to determine the value of proceeding with the project. Most development teams or individual entrepreneurs go into a project with some idea of how much investment they can afford to put into the project. Experience has shown that ballpark cost estimates for marine construction can be way off base and seriously compromise the completion of a project down the road. It is important to define realistic cost estimates very early in the conceptual design process. Several examples of marina construction elements that are difficult to assign cost to are presented along with a list of major capital cost categories to assist in preparation of a meaningful preliminary cost estimate.

In marina projects, the single most difficult cost number to define is the in-water work, specifically docks and dredging. A final assessment of attributable cost can only be generated at the completion of a final design, which is often many months away from when go, no-go financial decisions must be made. The best insurance to avoid unanticipated major capital cost is to thoughtfully prepare preliminary design concepts to identify major "red flags" and develop a responsive design criteria. A thorough and detailed preliminary design will greatly assist in preparing a meaningful cost estimate.

An example of a major capital cost that can get out of hand is dredging. To adequately estimate dredging costs, preliminary testing and analysis should be conducted to identify variations in subsoil characteristics, location of bedrock or nondredgable bottom, and to identify hazardous materials in the subsoils which could result in expensive material handling and disposal.

Another area in marina design where capital costs should be carefully considered is in selection and purchase of dock systems. There are hundreds of dock manufacturers around the world that are happy to sell product. The question is, what system best suits the needs and goals of the project while being cost effective? Often during the early stages of a project a ballpark estimate is input into financial planning for a dock system without due regard to its capacity to withstand the local environment, its ability to accept imposed utility loads and its mooring requirements. The result is a major cost increase during final design or construction bidding to accommodate the site requirements that should have been analyzed and accounted for in the conceptual and preliminary design process. Intelligent planning is the safeguard to financial success.

Each project will incur specific capital costs and therefore each project must be analyzed for its particular requirements. As a guide the following generalized list of typical capital costs that may be anticipated in marina projects may be useful in initial planning.

- Land acquisition costs
- Engineering and scientific studies
- Legal costs
- Dredging or filling
- Bulkheads or slope stabilization
- Docks and piers
- Dock mooring systems (piles, anchors)
- Sewerage or septic systems
- Marina office and dockmaster buildings
- Perimeter protection (wave attenuation)
- Utility installation
- Haul-out facilities and equipment
- Boat repair shops
- Boat storage buildings
- Rack storage facilities and equipment
- Auto parking areas
- Workboats and equipment

2.4 FINANCIAL RATIOS

Financial ratios are analytical tools used to assess the financial viability of a business in relation to others in a similar industry. Unfortunately marinas, in the past, have not fallen into suitable categories used in standard business analysis. It has only been since the late 1970s that sufficient statistical data has been available on marinas to allow rational development of financial ratio assessment. The reasons for past lack of data is that marinas are small in number and often not easily categorized. Some marinas are full service, some offer boatyard capabilities, some only provide boat parking, etc., so meaningful financial analysis was nonexistent.

As a result of the lack of standard financial yardstick data about marinas, many lending institutions elected to refrain from providing commercial loans to marinas, while other lenders treated marinas as akin to nominal retail outlets such as hardware stores or car dealerships. Comparing marinas to hardware stores (marine chandleries) or car dealerships (boat sales) did not provide effective or meaningful financial data to assure the lenders of a profitable and secure investment.

The financial ratio is a method that allows comparison of the financial posture of similar types of businesses by developing a standard ratio of one financial account value divided by another. The attempt is to compare the proverbial apples to apples. The lack of definitive data on marina performance led the University of Rhode Island, Sea Grant Marine Advisory Service to initially compile financial ratios for marinas in the late 1970s. The data has since been updated and formally published as a working guide in 1988 by the International Marina Institute. The author of the text is Robert A. Comerford, Ph.D., Associate Dean, The College of Business Administration, University of Rhode Island. Subsequent to Dr. Comerford's work, Robert Morris Associates (RMA) announced in early 1990, that they added marinas to their categories of financial analysis in their annual studies. RMA also includes boat yards and boat dealers in their financial coverage. Other institutions such as Dunn & Bradstreet and the Troy Almanac may follow the RMA lead. The inclusion of marinas in the RMA book will provide marina developers and lenders with recognized, up-to-date information on marina performance on an annual basis. The recognition of marinas as a specific category in the RMA statistics heralds a coming of age of marinas as an essential component of the marine recreation industry.

Even with the recent addition of marina financial statistics to the RMA profiles, it is valuable to look at Dr. Comerford's work as establishing the basis of understanding of what financial performance means and how various data is developed. In order to initiate meaningful comparison of marina performance a financial data bank was created to catalog financial responses from nearly 130 operating marina facilities. The data is currently biased to the east coast of the United States where the largest response to the request for financial information was received. The northwest region of the United States was unrepresented. The current analysis is not international in scope but some conclusions of comparability may be drawn by knowledgeable data users. The accumulated data was analyzed to reflect a series of financial ratios that are generally accepted for other business evaluations. Specifically, the selected ratios include: current, quick, earnings-to-interest (coverage ratio), return-on-net-worth, return-on-total-assets, debt-to-worth, inventory and receivables turnover, sales-to-working capital, sales-to-net-fixed-assets, sales-to-total-assets, and cash-flow-to-current-maturity-of-long-term-debt.

A detailed analysis of each ratio is beyond the scope of this book and the interested reader is directed to Part 1- Information Sources for reference to the publication cited. The current availability of meaningful financial statistics and on going update of the data is a very important asset for developing the financial performance of an existing or proposed marina facility.

2.5 FINANCING SOURCES

The key to obtaining financing is to be creative. Obviously it is wise to investigate the traditional sources for financing but you should also consider other sources. Following is a partial list of potential funding sources.

- Commercial banks
- Savings and loan associations
- Mutual savings banks
- Venture capital funds, public and private
- Industrial or recreational development bonds
- Money market funds (short-term debt)
- Capital markets (long-term debt)
- Pension funds
- Life insurance companies
- Credit unions
- Real estate investment trusts
- Investment banking institutions
- Limited partnerships
- Family or friends

2.6 LOAN APPROVAL CONSIDERATIONS

Many factors influence the successful application and ultimate acceptance for loan approval. Intangible considerations such as applicants appearance, demeanor, and life style may influence a loan officer's feeling toward a specific loan request. In addition to preparing a complete, factual and comprehensive loan application package, it is often an advantage to learn about the lending organization, who its loan officers are, the types of loans made, and their business outlook. With this information a quality proposal can be made addressing known requirements and interests of the lender. The following list presents some of the key points that lenders will look for and evaluate in processing a loan application.

FINANCIAL CONSIDERATIONS

- Quality of principals, capital available, collateral pledged, credit worthiness
- Documented proposal and feasibility study
- Business Plan
- Appraisal
- Life insurance policies on principals
- Business receivables collection policy
- Committed insurance coverage
- Ability to cover "points" or other front end fees
- Valid business contracts from performance bonded contractors
- Economic posture and future outlook of marina industry as lending risk

2.7 SEASONAL SLIP RENTALS VERSUS DOCKOMINIUM/LONG-TERM LEASE

Traditionally marinas have operated on a seasonal or perhaps year around payment basis for berthing arrangement. In areas with widely differing climate conditions, there is often a seasonal rate, higher for the good weather boating season and lower for winter berthing or storage. Lease arrangements generally follow the seasonal period selected with only a few marinas offering multiyear leases. Facilities for transient boaters regularly charge fees on a daily, weekly, monthly or seasonal basis. Fees are established on a per slip basis, on linear footage of the slip, on square footage of the slip or some other convenient method. The most general method is to apply a cost per linear foot of slip with a minimum fee for a particular slip. A 30 foot slip would get a minimum of the per foot fee times 30 feet but may get more if the berthed boat exceeds the 30 foot minimum. A 26 foot boat in a 30 foot slip would pay the minimum of 30 feet times the going rate.

Slip fees have often been based on the local going rate rather than on a sound business cost recovery basis. For this reason many marinas were and are borderline successful. Many facilities have been unable to pay their bills at the going rental rate and have dropped from the business. The upswing in recreational boating during the 1980s has resulted in a more businesslike approach to determination of slip fees. Certainly local pricing must still be considered, but marina entrepreneurs are now adjusting their business practices to capture the desired boating clientele and charging fees commensurate with the service provided. In many cases this approach has allowed marinas not only to survive and flourish but also expand and improve services. Marinas are service oriented businesses which in today's environment can assess and collect a premium for dedicated services. The wise marina owner will capitalize on this fact and create a facility in such a manner as to attract the type of client he needs to meet his revenue stream.

A new form of marina ownership was introduced to the industry during the 1980s. The concept of long-term leasing, cooperative or outright ownership became popular in many areas of the world as limited waterfront space and a rapidly growing participation in recreational boating created slip shortages. The renaissance of urban waterfronts often displaced low profile marinas to the benefit of high rise commercial and residential structures. Escalating land values required significant investment to create new marina facilities, the cost of which could often not be reasonably recovered with traditional seasonal slip fees. The individual ownership concept in a market with not enough slips, allowed for early and significant capital cost recovery by the new generation of marina constructors. A marina slip was determined to be a mortagable interest which allowed the average (maybe above average) boater to acquire appropriate long-term financing with the guarantee of berthing availability for the foreseeable future. In many cases, the mortgage and maintenance costs are close to seasonal rental fees with the boater having the benefit of equity interest in the slip.

The success of dockominium or long-term lease ownership of slips brought about a renewed interest by Federal, state and local licensing and permitting authorities. In many states the coastal waters below high or low water are held in the public interest by the government body. Many regulators believe that long-term lease or sale of marina slips violates public trust doctrine by leasing or selling water space reserved for the general public. A result of this belief has been numerous enactments or clarification of permitting statues that either forbid occupation of public trust waters, authorize the use of public trust waters at a prescribed fee and for a fixed time frame or provide other limitations or compensatory mitigation to preserve the public interest. From the financial perspective, a potential marina developer or an owner considering marina conversion must become familiar with the current policy and input this data into any financial plan. To assume dockominium type development, with its early investment recovery potential, only to find later in the process that this type of development is not permitable can compromise the entire project.

2.8 PROFIT CENTERS

In developing a financial projection for marina revenue sources (pro forma), an area often overlooked is profit centers that may be incorporated into the marina in addition to traditional berthing and storage plans. Boatyard operations may not be compatible with all marina scenarios but some of the traditional boatyard services may be incorporated to provide revenue enhancement. Engine repairs and minor hull work fall into this category and can be quite profitable. Sale and installation of accessories, especially electronics, is another area where good return may be anticipated. Concierge services, providing specialized amenities not directly related to the operation of the marina, may contribute to revenue enhancement by charg-

FINANCIAL CONSIDERATIONS   41

ing a service fee for the effort expended. Some concierge services include: arranging for floral delivery, tickets for entertainment events, grocery shopping, laundry services, babysitting services, limousine service, airline flights, hotel/motel accommodation arrangements, etc.

Traditional profit centers include: slip rental, boat storage, automobile parking, laundry receipts, ice and other vending services, transient docking or moorage fees, electric service charges (where power reselling is allowed), boat repair services, boat housekeeping services, telephone use fees, and subleased real estate.

2.9 TYPICAL INCOME AND CASH FLOW PROJECTION

To determine the financial performance of a proposed project it is necessary to develop an income and cash flow projection. The projection may be presented in many ways and formats. Generally several income and cash flow projections will be created to reflect differing anticipated potential

Table 2-1. Typical income and cash flow projection

10 Year Pay Back
100% Occupancy

INCOME	1986	1987	1988	1989	1990
Slip rental	180,000	180,000	202,500	270,000	300,000
Transient Dockage	10,000	12,000	12,000	12,000	12,000
Utilities	4,500	5,000	5,500	7,800	8,300
Parking	4,500	4,500	4,500	4,500	4,500
Equipment & Supplies	2,500	3,000	3,500	5,200	5,700
Vending & Ice	6,000	6,500	7,000	9,000	9,000
Miscellaneous	2,500	3,000	3,500	4,000	4,000
TOTAL INCOME	210,000	214,000	238,000	312,500	343,500
EXPENSES					
Rent - land	12,000	12,000	12,000	13,500	15,000
Utilities	9,000	10,000	11,000	15,600	17,200
Payroll	39,000	39,000	42,500	55,000	61,500
Employee taxes & benefits	9,400	9,400	10,200	13,200	14,760
Insurance	2,300	2,300	2,800	3,800	3,800
Maintenance & repair	2,500	3,000	5,000	7,000	9,000
Accounting & legal	1,000	1,000	1,000	1,500	1,500
Telephone	1,100	1,100	1,200	1,500	1,500
Advertising & promotion	2,500	2,000	2,000	2,000	2,000
Supplies	3,000	3,300	3,600	5,100	5,100
Miscellaneous	6,000	6,200	6,500	7,000	7,500
Interest	84,300	75,000	65,000	56,000	47,000
Depreciation (Under ACRS)	100,000	148,000	141,000	138,000	136,500
TOTAL EXPENSES	272,100	312,300	303,800	319,200	322,360
Federal Net Income (Loss)					
Distributed to Partners	-62,100	-98,300	-65,800	-6,700	21,140
Adjustment for State depreciation	38,000	21,000	13,500	10,500	9,000
State Net Income (Loss)					
Distributed to Partners	-24,100	-77,300	-52,300	3,800	30,140
Net Profit (Loss)	-62,100	-98,300	-65,800	-6,700	21,140
Less Principal Payments	-50,000	-50,000	-50,000	-50,000	-50,000
Add: Depreciation	100,000	148,000	141,000	138,000	136,500
Increase in Cash	-12,100	-300	25,200	81,300	107,640

scenarios. One variable may be the term of payback on the investment. Frequent terms are five years or ten years. The actual desired term of payback will be established as a requirement of the financial investors. Another variable might be the assumption of slip occupation during different years in the program. It is unrealistic, in most cases, to assume that a marina will be fully occupied during its first year of operation. The market studies and absorption rates for the area will help determine a reasonable projection to be used for yearly slip occupation during the first few years of operation. It is best to be conservative in estimating the occupancy rate to avoid overly high expectations for the project. The income and cash flow projection is generally performed for a specific term of years, such as the first five years from opening day. In this case, occupancy rates may be increased each year to full occupancy in a specified number of years, perhaps year three or four. An example of an income and cash flow projection is shown in Table 2-1 for a typical small marina. The dollar amounts are fictitious and presented only to show an anticipated progression of income and expenses.

2.10 DEVELOPING A MARINA ANNUAL BUDGET

As the project matures, it will be necessary to develop a marina operating budget. Each marina will have its own list of line items appropriate for its operation. A large and full service marina may have a budget similar to Table 2-2. The line item list should be as comprehensive as possible to assure that all items are covered to minimize future expense surprises not covered in the request for operating capital. It is also prudent to maintain a full listing each year and indicate by a zero mark those items not needed that particular year. In this fashion items not required in a particular year will not be overlooked in subsequent years.

Table 2-2. Typical, hypothetical, marina budget

General Ledger Account Name	General Ledger Account No.	Budget Amount
INCOME:		
Rental	3010	$ 0
Parking Summer	3020	31,600
Parking Winter	3021	4,900
Laundry	3050	1,700
Ice	3051	6,500
Vending Machines	3052	3,500
Slip Summer	3100	775,600
Slip Winter	3110	96,900
Slip Transient	3120	84,000
Moorings	3130	70,000

1990

FINANCIAL CONSIDERATIONS 43

Table 2-2. *Continued*

General Ledger Account Name	General Ledger Account No.	Budget Amount
Boat Repair	3140	550,000
Electric Summer	3150	61,600
Electric Winter	3151	3,000
Telephone	3160	1,500
Real Estate	3170	38,000
Miscellaneous	3300	15,000
TOTAL INCOME		1,743,800
EXPENSE:		
Building Improvements	1530	1,800
Machinery and Equipment	1540	4,600
Office Furniture & Fixtures	1550	3,300
Marina Improvements	1560	1,200
Advertising & Promotion	4010	21,500
Repairs & Maintenance	4100	5,000
Auto Repairs	4110	1,200
Air Conditioning	4115	500
Boardwalk Maintenance	4125	400
Mooring Maintenance	4135	5,000
Computer Repair	4150	1,000
Carpentry	4160	900
Cleaning Service	4170	2,000
Exterminating	4175	600
Electric Work	4180	5,000
Floor covering - Carpet	4190	300
Heating Repair	4200	400
Janitor Service	4210	3,500
Janitor Supplies	4215	1,500
Key/Lock Repair	4220	1,800
Landscaping	4230	500
Masonry	4240	1,000
Office Equipment Repair	4250	200
Office Equipment Contract	4255	8,000
Parking Lot Repairs	4260	750
Plumbing	4270	2,500
Redecorate Building	4280	0
Roof Repair	4290	0
Rubbish Removal	4310	9,500
Snow Removal	4320	1,500
Window Accessories	4330	0
Window Washing	4340	400
Boat Repair Operations	4410	350,000
Boat Supplies	4420	50,000
Building Paint	4430	150
Dock Repairs	4440	10,000
Dock Supplies	4445	500
Ice	4450	4,500
Marine Equipment Repair	4460	3,000
Pier Repair	4470	0
Power Post Repair	4480	500
Marina Supplies	4490	4,500
Uniforms	4495	1,500
Marina Miscellaneous	4500	3,500
Operating Expense Other	4600	0
Electricity	4620	78,000
Fuel - Heating	4630	4,500
Lease Expense	4700	0

(continued)

Table 2-2. *Continued*

General Ledger Account Name	General Ledger Account No.	Budget Amount
Insurance - Building	5000	60,000
Employee Insurance	5100	10,000
Payroll - Non Administration	5200	60,000
Benefits	5210	25,000
Payroll Taxes	5230	10,000
Printed Supplies	5310	500
Security	5400	90,000
Water and Sewer	5500	6,500
Miscellaneous	6010	1,500
Employee Recruitment	6020	2,500
Auditing & Accounting	6100	10,000
Computer Supplies	6110	500
Consulting	6200	5,000
Medical Insurance	6310	25,000
Pension Plan	6320	10,000
Life Insurance	6330	0
Disability Insurance	6340	0
Insurance Other	6350	0
Legal Fees	6410	1,500
Office Supplies	6510	2,000
Payroll Administration	6610	78,000
Benefits Administration	6620	0
Payroll Taxes	6630	0
Postage	6710	800
Subscriptions/Memberships	6720	150
Telephone	6730	5,500
Travel Expense/Meals	6740	2,500
Taxes Real Estate	6800	35,000
Miscellaneous Administration	6900	1,500
OPERATING COSTS		961,950
Principal Payments		150,000
Interest Payments		250,000
TOTAL OPERATING COSTS		1,361,950

3
Regulatory Considerations

Virtually any marina and small craft harbor construction will require some form of regulatory approval. The scope of the project will to some degree impact the complexity of the regulatory process, however, even minor maintenance work may require a significant permitting review. In today's environmentally sensitive climate, it is prudent to ascertain the regulatory approvals required before attempting any coastal construction to avoid costly construction delays or litigation and fines. The regulatory process should not be considered, by the marina developer, to have only negative aspects. During the review process better ways are often found to accomplish development goals than were initially envisioned. The properly permitted facility is also often safer and more attractive. It is also usually true, however, that the regulatory process not only adds to the cost of a project by preparation of the permitting documents, but also increases the project cost by implementation of the regulatory requirements, or results in a reduction in potential revenue by downsizing or otherwise altering the project.

A major premise of regulatory review is protection of the public's interest in access to the nation's waterways and coastal areas. The so-called public trust doctrine, whereby the states hold certain lands in trust for all the public benefit, is continually being challenged by private interests, but generally the findings support the public benefit premise. We must not lose sight of the fact that although marinas may be developed by a private interest, they provide a necessary access that allows the public to enjoy the waters held in public trust. There should therefore be a partnership between marina development and regulatory agencies to foster the creation and construction of good marina facilities that ultimately will be occupied and used by the public.

Regulatory agencies may be generalized into three categories: federal, state (provincial), and local. This may vary in some countries but the concepts and often the policies are similar. The federal government is most concerned with major issues such as: water quality, fish and wildlife, protection of wetlands, flood hazard protection, impacts of development on maritime commerce, effects of dredging and filling, and consistency with federal coastal management policy. State agencies are concerned with public access, traffic and transportation issues, wetland protection, fish and wildlife, structural adequacy, and local property impacts. Local or regional agencies are concerned with issues of: coastal and wetlands protection, impacts to abutters and neighborhoods, conformity with municipal master planning, traffic and parking issues, and economic and social benefits to the community.

Approvals may be categorized in three groups: licenses, permits and certificates of compliance or consistency. Licenses are usually associated with the granting of permission to use land or water that does not belong to the petitioner, i.e., governments may license the occupation of tidal flowed lands for marina use. Permits are generally associated with some form of structure: building permits, sanitary system construction permits, pile or dock placement permits, etc. Certificates of compliance or consistency may be issued when the project demonstrates that its construction will not violate a certain set of criteria. A water quality certificate is often issued after a proposed project has been reviewed and the regulatory agency finds that the project will not compromise water quality standards for the site area. Consistency certification may be required to assure that the project is in conformance with a coastal development master plan.

The approvals system is often structured in levels that require one set of approvals to be obtained before another level will issue an approval. By the nature of politics, the local approval is usually the first approval issued, then a state or provincial approval, and last will be the federal approval. The time frame to acquire all the approvals will vary greatly, but is usually not less than six months and often as long as three to five years.

It is impossible to explore all the rules and regulations that have been formulated to regulate waterfront development. It is of interest, however, to consider the three major categories of regulation: federal, state, and local, and to look more closely at their general posture on regulatory issues.

3.1 REGULATORY AGENCIES

Federal

Federal regulatory approval is vested in different types of agencies for any particular country. Although specific policies will vary, the primary interest is often similar from country to country. A look at legislation and policy in the United States provides an overview of federal regulatory requirements for marinas that will generally apply in other countries.

The United States Army Corps of Engineers is the lead federal agency regulating the use of the navigable waters of the United States. The federal authorization for Corps of Engineers regulations was originally contained in the River and Harbor Act of 1899. This act provided for regulation to protect navigation and the navigable capacity of the nation's waters. Many activities associated with marina and small craft harbor development continue to fall within the jurisdiction of this act. National concern for environmental issues resulted in a revision to the act in 1968. The revision addresses concerns of impacts to national resources including fish and wildlife, conservation, pollution, aesthetics, ecology and other public interest issues. The 1968 revision was designed to provide a "public interest review" as opposed to generally commercial interest concerns of navigability associated with the original act of 1899.

A greater Corps of Engineers role in the protection of national resources was created in the Federal Water Pollution Control Act Amendments of 1972. Under Section 404 of the act, the Corps of Engineers was mandated to establish a permit program to regulate discharges of dredged or fill material into the waters of the United States. The initial interpretation of Section 404 regulated only those navigable waters of the United States traditionally considered under the language of the River and Harbor Act of 1899. Public concern was expressed that the Corps of Engineers should regulate the nation's wetlands as well as navigable waters. A challenge was made to the Corps of Engineers area of jurisdiction. The challenge was heard in the United States District Court for the District of Columbia. On March 27, 1975 the court ruled that the Corps of Engineers jurisdiction, under Section 404, extends to regulate all "waters of the United States," including wetlands. Further definition of the Corps' role in wetland protection was made in the Clean Water Act of 1977.

Congress has specified that "waters of the United States" shall be construed in the broadest constitutional interpretation. The definition includes: "waters of the United States that are subject to the ebb and flow of the tide shoreward to the mean high water mark and/or are presently used, or have been used in the past, or may be susceptible for use to transport interstate or foreign commerce. A determination of navigability, once made, applies laterally over the entire surface of the waterbody and is not extinguished by later actions or events which impede or destroy navigable capacity. The term includes coastal and inland waters, lakes, rivers, and streams that are navigable and the oceans." Local Corps of Engineers offices should be consulted for specific interpretations of local jurisdiction.

The definition of wetlands includes areas adjacent to "waters of the United States" that are inundated or saturated by surface or ground water at a frequency and duration sufficient to support, and under normal circumstances do support, a prevalence of vegetation typically adapted for life in saturated soil conditions. Adjacent wetlands include those areas that are separated from other "waters of the United States" by man-made dikes or barriers, natural river berms, beach dunes and the like.

Tributaries to "navigable waters of the United States" include adjacent wetlands, lakes and ponds. Included also in Corps of Engineer jurisdiction are all other "waters of the United States" such as isolated wetlands and lakes, intermittent streams, and other waters that are not part of a tributary system to interstate waters or to navigable waters of the United States, when use, degradation or destruction of these waters could affect interstate or foreign commerce.

The River and Harbor Act of 1899 contains two sections of primary interest to marina development. Section 9 of the River and Harbor Act creates the authority for requirement of authorization from the Secretary of the Army, acting through the Corps of Engineers, for the construction of any dam or dike in a navigable water of the United States. The act originally included the permitting authority associated with construction of bridges and causeways in navigable waters, but this authority was transferred in 1966 to the United States Coast Guard when it came under the newly created Department of Transportation. However, any discharge of dredge or fill material associated with bridge or dike construction still requires a Corps of Engineers permit under Section 404 of the Clean Waters Act.

Section 10 of the River and Harbor Act of 1899 is the section that most generally impacts marina development. Section 10 requires authorization from the Secretary of the Army, acting through the Corps of Engineers, for the construction of any structure in or over navigable water of the United States. This includes deposition or excavation of material in federal waters or any obstruction or alteration in a navigable water. A Section 10 permit will also be required if the construction of a structure or other work affects the course, location or condition of a waterbody under federal jurisdiction. Representative work included under the act includes: commercial or recreational fixed or floating docks, piers, wharfs, dolphins, weirs, breakwaters, groins, jetties, bulkheads, rip rap slopes, revetments, pilings, permanently moored vessels, aids to navigation, pipelines, tunnels, canals, boat ramps, aerial or subaqueous transmission lines or any semi-permanent or permanent obstacle or obstruction. In other words, any structure or work in a defined navigable water will probably require a Corps of Engineers authorization.

Section 404 of the Clean Water Act is again administered by the Corps of Engineers on behalf of the Secretary of the Army, for the discharge of dredge or fill material into all waters of the United States. This broad definition includes wetlands both adjacent to and isolated from a navigable water. The act regulates activities such as placement of fill required for construction of any structure, whether temporary or permanent, including: site development fills for industrial, commercial, residential or recreational projects, dams and dikes, shore protection, bank stabilization, beach nourishment, causeways, roads, breakwaters, pipeline or outfall structures, ponds, etc. Temporary access roads, construction cofferdams and discharge of dredge material dewatering effluents also require authorization.

In addition to authorization under Section 404 of the Clean Water Act, transportation of dredge material to be deposited in ocean waters requires approval under Section 103 of the Marine Protection Research and Sanctuaries Act of 1972.

The Corps of Engineers permit review process considers a number of factors including: general environmental concerns, conservation, water quality, safety, energy needs, navigation, land use, historic assessment, fish and wildlife value, flood damage prevention, economics, aesthetics, water supply, recreation, food production and health and welfare of the general public. The purpose of the permit program is designed to:

1. Insure that the nation's water resources are protected
2. Assure that use of the nation's water resources are in the best interest of the public
3. Provide full consideration for the environmental, social and economic concerns of the public.

To comply with the public interest review policy for permit authorization, the Corps of Engineers issues "Public Notices" advising all interested parties of a proposed activity for which an individual Corps of Engineers permit is sought. Public Notices may be received by any interested party upon written request to the appropriate Corps of Engineers office. Comments received from the Public Notices or through other sources will be reviewed and considered in the Corps' project review process. Public hearings on specific projects may be held if appropriately petitioned by the public, at the request of other interested federal agencies or if the Corps deems a public hearing of import to the review process.

The Corps review process will include interagency meetings with other Federal agencies such as the Environmental Protection Agency, Department of the Interior (Fish and Wildlife Service), Department of Commerce (NOAA, National Marine Fisheries Service), Department of Transportation (United States Coast Guard) and other appropriate agencies. Coordination is also instituted with the cognizant state department responsible for permitting the project on the state level.

Federal Coastal Zone Management Act. Marina development projects proposed along the coastal United States will also be involved with justification of the project within a set of coastal zone management standards developed by each state under guidelines established by the federal government. Projects outside of the coastal zone will not be subject to the program requirements for consistency determination.

A continuing interest in protecting and preserving the nation's irreplaceable coastal and inland wetlands during the 1960s and early 1970s precipitated the enactment of the federal Coastal Zone Management Act of 1972 (P.L. 92-583). This act recognized the importance of coastal and inland wetlands and the potential adverse affects that were being imposed on the resource by uncontrolled development along the nation's waterbodies. Section 302(a) of the act states "there is a national interest in the effective management, beneficial use, protection, and development of the coastal zone."

The act established a policy administered by the Secretary of Commerce through the National Oceanic and Atmospheric Administration (NOAA) and created the Office of Coastal Zone Management. Substantial modifications were incorporated into the act by P.L. 94-370, Coastal Zone Management Act Amendments of 1976. A key component of the act is the delegation of implementation of the coastal zone policy making to the states. Section 302(h) states in part, "the key to more effective protection and use of the land and water resources of the coastal zone is to encourage states to exercise their full authority over the land and water use programs . . . for dealing with coastal land and water use decisions of more than local significance."

In order to foster state involvement and creation of state coastal zone policy, a system of federal grants was established to provide federal funding for implementation of the coastal zone program on the state level. Specific guidelines for preparation of state programs is outlined in 15 CFR Part 923, *Federal Register 40* (6):1683-1695. The first step, by a state, is to develop a management program which is submitted to the Office of Coastal Zone Management (OCZM) for review, correction of deficiencies, and ultimately approval as consistent with federal policy. Upon approval the federal government will provide financial assistance for initial implementation and continuing program development. The federal government will periodically review the entire coastal zone management program and determine its continuing validity and provide or withdraw funding as appropriate.

The state requesting participation must demonstrate that its coastal zone management program is comprehensive and addresses the primary concerns of the federal act. Some areas of importance that the federal government feels have been adversely affected by uncontrolled growth and economic development include: wildlife habitats, open space for resource development, nutrient rich feeding and breeding areas, and aesthetic, historic, cultural and ecological values of marine resources.

State programs must clearly present policies, standards, objectives, and define the criteria upon which program interpretation will be made. The program must have a clear sense of direction and provide an enforcement mechanism to insure compliance with the program.

The end result of the coastal zone management program is to assure that any applicant for a federal license or permit affecting water and land uses within the coastal zone must demonstrate to the state, which in turn certifies to the federal government, that the proposed project is consistent with the federal coastal zone management policy. This is the so-called federal consistency requirement of coastal zone permitting. A fairly wide latitude is offered to each state in developing a program that is compatible with its own coastal requirements. The acquiescence of the state federal consistency on behalf of the proposed project is a necessary component for an applicant to secure an Army Corps of Engineers permit.

The general procedure for securing a federally consistent project is to obtain the appropriate application and policy statement from the state coastal zone management agency and prepare a consistency statement that addresses each policy in terms of the proposed project and how the project may or may not impact the policy objectives. This written document will be reviewed by the state coastal zone management agency and it will issue a letter of consistency compliance or indicate the areas of deficiency that must be mitigated before a consistency certificate will be issued.

State Regulatory Agencies

Each state has developed its own policies regarding waterfront development. The marina developer will be required to submit a formal application to the regulatory agency along with plans and other information to describe the proposed work. The application will be reviewed by the agency, and if the project meets the agency's criteria some form of approval such as an assent, license, or permit will be issued to construct the proposed project. The review process is usually vested in a state agency such as a Department of Environmental Protection or a similar agency.

Rules and regulations vary widely between states, however, in general, a minimum of the following must be supplied with a formal application for a proposed project.

Formal application—Completion of a standard form providing information identifying the applicant, indicating adjacent property ownership, zoning of the property, describing the project, a listing of other approvals required and specific questions on impacts of the project.

Description of the property and its location—A location plan is generally required identifying the site on a topographic map, street map, nautical chart or other vicinity map.

Plans—Usually 8.5" × 11" plans showing the site, the extent of the proposed work, the basic design of structures, topography and/or bathymetry, land and water elevations, drainage, transportation access, parking, boat sewage pumpout provisions and other pertinent information to adequately describe the work. More complete plans including definitive design drawings may be required for complex projects. Some jurisdictions may require that plans be prepared and certified by a registered professional engineer, architect, or land surveyor.

Other approvals—Copies of approvals required and obtained to date should be submitted to the state agency. Other approvals may be: zoning commission acceptance of the proposed project as conforming to existing zoning, planning board review and approval, conservation commission approval, board of health compliance certificate, building department permits, redevelopment authority approval, port and harbor commission approval, etc.

Testing—Copies of testing performed on the site to satisfy specific issues such as: traffic assessments, dredge material analysis, soil boring and geotechnical analysis, wind/wave climate analysis, hazardous waste site testing, wildlife inventory, biogeographical analysis of the littoral area to determine the aquatic life quality and quantity, etc.

Consistency determination—A written analysis addressing each of the coastal management act policies on coastal resources and the impacts on the resources by the proposed project.

Except on all but the smallest noncontroversial projects, it will probably be necessary to retain the guidance of legal, scientific, and engineering professionals to assist in compiling data and responding to issues raised in the application process. The applicant can often assist the process by participating in data collection, agency contact, and obtaining site surveys.

Upon receipt of all required information, the approval agency will generally assign a file number to the project and place the application into the review process. Most projects will be reviewed in a sequence based on the date of receipt by the agency. Public hearings may be required if the project has controversial aspects or is of sufficient scope to trigger an agency threshold requirement for a public hearing. Often a group of interested citizens may petition the agency to allow a public hearing, whatever the size and scope of the project. The agency's in-house technical reviewers will review the project content for compliance with rules and regulations and administrative and legal staff may review the project for procedural and legal aspects. Usually the state will require prior approval by local agencies before issuing the state approval. If the project is judged to meet all the pertinent criteria, a formal approval will be issued.

Local or Regional Agencies

On the local level a number of municipal boards or agencies may be involved in reviewing a proposed project. Typically, municipal agencies include: a conservation commission, a planning commission, harbormaster, shellfish commission, zoning board, board of health, public works commission and building department. Hopefully there is one central agency that accepts and is responsible for processing an application. In many communities, however, the applicant will have to apply to each agency separately and often obtain approvals in a sequence. Some boards will not sign off on their area of approval until another board has acted. This is particularly true in obtaining building permits, generally for upland work, where the building official will want to see satisfactory evidence that the project is in compliance with zoning, has approved sewerage and water supply plans and conforms to coastal policy regulations.

The local approvals process is also where the greatest voice of the community will be heard. Most local meetings will be advertised locally and you can expect to find a large turn out of the public, which may often be quite vocal, if the project is at all controversial. It is definitely a positive step if the project can obtain initial unofficial municipal approval prior to the submitting of formal applications. It is a mistake to file approval applications with federal and state authority before having briefed the local community and developed a base of support at the community level. None of us likes to be ignored, especially in our own backyard. It is also important to note that both state and federal authority will heavily weigh the feeling of the community toward the project.

3.2 PUBLIC TRUST DOCTRINE

As waterfront development has blossomed, especially with the intense interest in recreational facilities and public demand for more access to the water, the question of who owns the land under water and what the public rights are in this area has become a central issue. In most developed countries the jurisdiction of water rights falls into a category often called "public trust doctrine." Public trust doctrine evolved from ancient Roman Law. It has been refined through the ages and takes many forms in each jurisdiction. The common thread, however, is that the public enjoys certain rights to the waters extending from the shore and that these rights are expressed in law and held by the political body for use by all the inhabitants.

During the heyday of the Roman Empire, the Romans realized that their existence relied heavily on waterborne commerce and transportation. They instituted an early legal code that provided for the common ownership of essential waterfront property. The body of law was formulated in the Institutes of Justinian. The demise of the Roman Empire and the ensuing period of history known as the Dark Ages caused a loss of the notion of ownership for the common good and restructured property ownership into private hands often under the feudal lord system. In England, the king claimed title to all the coastal lands and disbursed land and maritime resources to a favored few under exclusive private ownership. Eventually the public renounced the feudal land ownership policies and in 1215 with the institution of the Magna Charta, some public rights in coastal tidelands were restored. Although private rights were maintained, the concept of public interest lands was developed in which the king held certain lands in trust for the benefit of all the public. This then became the basis for what is now considered public trust lands. Early settlers in North America brought with them the English Law containing the public trust doctrine. Some colonial states modified the public trust doctrine by extending ownership of private land to the intertidal zone, between high water and low water, ostensibly to promote the development of private wharf structures and the enhancement of waterborne commerce. However, even with the resumption of private ownership of flowed tidelands, the colonists realized the need to protect the public interest and restricted the ownership of land below high water with the premise that this land would remain available to the public for fishing, fowling and navigation.

Continued interest to foster private development in the flowed tidelands caused various states to grant special permission to develop the intertidal zone under so-called wharfing statutes. These statutes allowed for the construction of private piers and wharfs seaward of the mean low mark and into undisputed public trust waters. The proliferation of structures in this area prompted a reconsideration of the impacts of privatization of the waterfront and the loss of public access to this land and water. The formation of governmental boards of harbor commissioners and a decision by the United States Supreme Court in *Illinois Central Railroad* v. *Illinois* (1892) began a reversal of the exclusive use of the waterfront by private interests. In *Illinois Central Railroad* v. *Illinois,* the Court ruled that a large tract of tidelands granted by the Illinois legislature to the railroad violated the public trust doctrine and the grant was ruled invalid. The case set the stage for the current policy that states have the requirement to protect and preserve the tidelands for public interest. A number of other legal decisions over the years have sought to clarify and expand upon the original arguments. Suffice it to say that in the United States, a well documented body of law prevails to establish the right of public interest in tidelands. Some states along the east coast of the United States may have coastal laws based on historic colonial law which provided private ownership of land to the low water line, with limited public use of the intertidal area. The specifics for each jurisdiction should be investigated on a local basis to assure that any proposed waterfront development respects the common law and does not violate the tenants of public trust. Generally today, private use of the public trust lands must be accompanied by implementation of public amenities and overall the project must be able to demonstrate that the public benefit outweighs the public detriment.

3.3 LITTORAL AND RIPARIAN RIGHTS ISSUES

The right of use or occupation of waterfront property beyond high or low water is a complex issue. Generally some degree of private rights are associated with upland ownership of waterfront property in the waters adjacent to the property. The waterbody may be a river, lake or ocean. Riparian rights are those associated with land adjacent to a river or stream whereas littoral rights pertain to land adjacent to a lake or the sea. Frequently the terms are used with some interchangeability although riparian rights terminology is often used, incorrectly, when addressing land along the sea.

The issue is not so much one of land underwater ownership since in most jurisdictions the local political body, the states in the United States, own the land under water between high and low water out to some territorial limit. This is not always true, however, as several New England states have land grants that evolve from colonial times bestowing on the land owner, ownership of submerged land out to an historic low water line. However, these lands grants (bestowed by the King of England), did reserve certain public rights, namely, the rights of fishing, fowling and navigation within the so-called private tidelands between high and low water. Other bodies of legislation may also control other aspects of coastal development in the submerged tidelands or along the coastal zone such as the Federal Coastal Zone Management Act and certain activities under the jurisdiction of the U.S. Army Corps of Engineers.

Whatever the jurisdictional considerations, there is usually an implied right to wharf out in the littoral area by the waterfront land owner. The delineation of the littoral or riparian boundaries is often an issue which is usually determined by a court. In a number of states a court called a land court will be the judicial body that determines the extension of property lines in the littoral zone. It is not necessarily a valid presumption to assume that the littoral extension of property lines will follow the natural extension of the upland property. This may be the case, but more often the court will use a set of standards, accepted by tradition, to determine boundaries based on geographic features of the surrounding shoreline. For many marina projects it will be unnecessary to involve the court in property line determination. Historic plans or traditional usage of waterfront land may have established rational property line extension into the water area. The proposed project may also be located well within property boundaries so that explicit definition of the property lines by a court is unnecessary. If the scope of the project layout will require structures in close proximity to perceived property line extension in the water, it may be desirable to have the property lines determined by the court to prevent future challenges on the validity of the placement of the structures in the water.

A typical court determination of property or littoral rights will begin with a close examination of the property deed. The deed may specifically define the water rights associated with the upland, which may present a clear and concise definition for the court to follow in establishing or confirming the lines. More often however, the deed may vaguely reference littoral rights by statements such as "together with the beach, shore and flats in front of, adjacent or appurtenant to the piece or parcel of land above described."

If the deed examination does not provide a clear definition of property rights in the littoral area, the court may then attempt to establish if there has been some "custom, conduct or usage" that demonstrates a past definition of a property line that has been recognized and agreed to as the littoral boundary for the property. Documentation of past permitted structures or licenses and permits defining property lines are a good source for documentation of past custom, conduct and usage. In many cases the court will accept past usage and custom as sufficient to continue acceptance of the prior littoral boundaries. Of course, as in all judicial deliberations the court may find otherwise.

If the property deed is insufficient to determine littoral boundaries and a case for past "custom, conduct or usage" can not be made, then the court may rely on a formula for geographic distribution of littoral boundaries based on certain coastline configurations. In general the coastline will be viewed in plan (bird's-eye view) and theoretical baselines established between natural headlands or around promontories (see Fig. 3-1). From the baselines created, perpendicular lines may be drawn between the baseline and the property parcels. The lines developed in the littoral zone will then be construed to be the court accepted property line extension from the upland property. In creation of the baseline, man-made structures and placement of filled areas may not be considered in the baseline generation. Often research will be performed to establish the historic high water or low water lines as far back as written and recorded records will allow. In Boston, Massachusetts,

REGULATORY CONSIDERATIONS 57

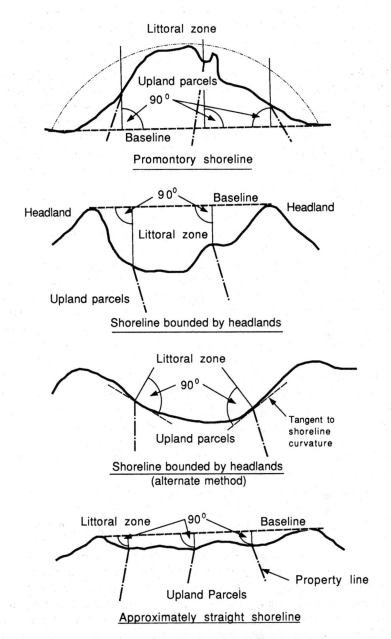

Figure 3-1. The question of the extension of upland property lines into the water area is important to determine the usable water area associated with a parcel of property. These diagrams present several typical methods of extending property lines based on the configuration of the surrounding coastline.

the authors have reviewed and used documents dating to the 1600s to establish historic water level locations.

In using the geographic configuration method of littoral line definition the court will look at each specific case and develop a scenario for that area of land, but in general may use the following basis of interpretation. When the coastline is relatively straight, a theoretical baseline will be created that averages the straightness of the coast or perhaps runs from minor headland to minor headland. From the baseline a series of perpendicular lines will be created that begin at the base line and terminate at each upland property line intersection with the high water line or other water line. If the area under study is concave, such as an embayment, the baseline may be created between two major natural headlands. Again from the baseline, perpendicular lines will be constructed to intersect the upland property lines at the appropriate water line. An alternative for embayment property line extension may be to construct a tangent to the curvature of the shoreline at the intersection of the shoreline with the property line, and extend the property line seaward as a line perpendicular to this tangent. When the upland forms a promontory or projection into the waterbody, a baseline may be constructed through a curve circumscribing the promontory. This baseline will then cut across the upland, and perpendicular lines will emanate from the baseline intersecting the appropriate waterline along the shore and extending outward to the circumscribed curve.

Each court or jurisdiction may use different methods of creating littoral boundaries so it is prudent to seek knowledgeable counsel to prepare documentation before beginning the planning of structure placement in littoral or riparian areas and to establish the background for an abutter challenge to littoral or riparian usage. Most courts will adopt a perspective that the land owner accepts his littoral rights as he finds them, that is, the court will not easily alter accepted littoral boundary determination for the convenience of the land owner.

3.4 LENGTH OF APPROVAL TERM

In days gone by, the length of approval term was simple. A permitted structure or facility, once constructed, was permitted virtually forever. In some states, permits or licenses were considered irrevocable and if rescinded by the authority, the action was considered a taking of private property and could result in monetary damages payable to the structure owner. Today, however, it is a new ballgame. Permits and other approvals generally have two finite lives. One is the time frame after approval issuance in which to implement the approval. The other time frame may be a definite term of years for which the granted approval is valid.

The construction or implementation time frame is usually in the order of three to five years and generally may be extended by the issuing authority if so petitioned in a timely manner prior to expiration. The purpose of this requirement is to effect timely implementation and not have a project under construction for an indefinite period. Most regulatory agencies are sympathetic to unavoidable delays caused by economic conditions, changing demands and even pure laziness, if time extensions are requested in an appropriate and timely manner. Projects involving dredging may have further restrictions specifying that dredging may be carried out only during certain months to protect local marine resources. Because of the construction time restriction, it is important that any project continuously monitor the expiration dates on approvals and file for extension consideration in a timely manner. Many, many facilities have allowed approvals to lapse only to find that completion of the project or routine maintenance requires a completely new filing, often under new and more stringent regulations. The regulatory process has become so unwieldy in some areas that initial approvals will often reach expiration before other necessary approvals have been issued. This is particularly true with local approvals that may be annually renewable or have other short terms. Prudent marina developers or owners will have "tickle" files to remind them of the expiration date of each approval, so they may file the necessary papers for renewal before an approval lapses.

The length of the life of approvals will vary from area to area. Concern about implementation of public trust doctrine and protection of future public needs will often result in an approval having a finite life. Many jurisdictions are opting for terms of between 10 and 50 years. A major problem with time restrictions on approvals is the inherent lack of confidence, by investors, in a project having a finite life. Generally lenders want a life that will assure adequate protection for the investment and enough time to generate a fair rate of return on the investment. Terms of 30 or 40 years are desirable, terms of less than 20 years appear to be very difficult to finance. Irrevocable licenses or approvals are very difficult to obtain and if obtainable will probably require an act of the legislature or other similar governmental body. In marina development today, marina life expectancy is usually considered to be 25 years. In this period of time the initial structural facilities will be near the end of their useful life, if that long, and probably the market demands will have changed to the point where alterations and modifications to the facility are in order. Under this hypothesis, term approvals may be tolerable if renewable options are available to allow a continuity of operation if so desired. Regulators rarely consider the economic plight of the marina entrepreneur and often fail to realize the lack of economic value associated with a facility that is nearing the end of its permittable time. It is therefore important that term approvals be accompanied by a mechanism to allow valuable water dependent resources to continue through means of reasonable renewal processes. Of course, the regulatory agencies must maintain the right and opportunity to adjust regulatory policy and controls to suit changing needs and demands. An example of valid regulatory term control is in areas of port activity where national security or commerce may require the use of a waterfront area that is currently occupied by recreational boating interests. New deep water ports are difficult to develop and therefore the capability to restructure existing ports must be maintained. A marina developer embarking on development in a port area should be fully appreciative of these potentials and be prepared to accept some form of reasonable restriction on project development.

Term approvals have caused considerable discomfort among those marina developers who have opted for dockominium or long term lease of slips. Certainly the sale or long term lease of a slip whose very existence is governed by a finite regulatory life is highly speculative. From an investment perspective, what is the value of a slip nearing the end of its permitted life? These questions continue to be debated and yet many people are willing to enter in such risky investments to "own a piece of the waterfront."

3.5 PREAPPLICATION MEETINGS

Preapplication meetings are an important part of the regulatory process. At the initiation of a project, copies of all applicable regulatory rules and regulations should be obtained and reviewed before commencement of design development. It is imperative that the initial marina planning be performed with regulatory constraints in mind.

During the design development process many issues will arise that are not fully addressed in the rules and regulations. It is unwise to complete a design concept and submit a regulatory application without first obtaining some input from the regulatory agencies and the local community. It is usually prudent to consult with local community officials early in the project to determine community reaction to the proposed project and hopefully to establish a positive working relationship with the local officials. Neglecting local community reaction and input will frequently result in the community adopting an adversarial position, even for meritorious projects. To address these issues before a design concept is completed, a preapplication meeting can be requested at which the basic design goals and content can be presented and discussed. The reason to have basic design goals established at the time of the preapplication meeting is to prevent the regulatory agency from designing the project for you, which if allowed to happen, may seriously compromise the design intent. It is far better to present a well prepared program of design goals that meet the intent of the regulations and which therefore may limit scrutiny to specific areas. This also helps the regulators by defining the project scope and allowing them to offer constructive (hopefully) criticism on the merits of the project as it relates to their jurisdiction. If properly prepared and presented, a preapplication meeting can provide very valuable guidance from the regulatory agency.

To maintain some degree of confidentiality, it is recommended that no written materials be left with the regulators and that the regulators be asked that the project confidentiality be maintained until such time as a formal application is submitted. Most agencies will respect this request and only discuss the merits of a project in-house. Of course you must be prepared for the eventuality that the agency may go public with the project but hopefully the design is far enough along that public scrutiny will not impact the process at this stage.

The meeting should be conducted in a friendly, nonadversarial climate. The project proponent should not have a large number of attendees that will make the agency reviewers feel overwhelmed and on the defensive. Hopefully, the agency will not arrive with a dozen people as well, but they might if it appears the project may be controversial. The meeting is a time to listen and not to argue about issues or details. Absorb what the agency representative has to say and be thinking of the ramifications to the project. Politely challenge appropriate issues but refrain from direct confrontation. During the formal application review there will be plenty of time to dig in your heels or challenge policy. The goal of the meeting is to receive a cursory, noncommittal assessment by the reviewer of the potential for the project to make it through the application process and to flag areas that require further investigation or that will be challenged.

A successful preapplication meeting can speed the formal review process and create the feeling that the regulators have been included in your decision making process. The knowledge that you care about the regulatory policy and issues and are trying to address these items can be very beneficial as the project moves through the regulatory review process.

Keep the meeting short, one to one and one half hours and be a good host if the meeting is on your turf. Arrive on time if the meeting is at the agency and do not keep the agency representative waiting if the meeting is at your office. Be prepared to conduct a brief site visit to acquaint the reviewer with the field conditions. Be open and responsive but do not volunteer more than is required. Be a good listener. Have a good time!

3.6 SCIENTIFIC STUDIES

Today's regulatory environment requires the applicant to provide a great deal of scientific and engineering information. The acquisition of this data may be time consuming and costly. It is imperative, however, that appropriate data be developed both to satisfy regulatory requirements and to allow the designer to prepare a design that is responsive to the site conditions. The usual design development process consists of developing a conceptual design which in a general way addresses environmental concerns. The conceptual design is massaged until the design basically fits the site constraints and the economics of the project. At this point only very basic scientific studies may have been carried out. The next logical step is to fill in the information voids by initiating the necessary scientific studies. However, all too often the cost of the appropriate studies is great enough that the developer pushes to move directly into the approvals phase on the assumption that if approvals can be secured the scientific studies can always be performed prior to final design. If the project is able to move through the approvals process then it often proceeds directly to construction without the scientific studies and often without a final design. There is a very real danger in executing a project in this manner. Often the permitted design does not have sufficient detail to discover major design problems, problems that only become apparent in the field during construction.

Examples of the pitfalls include lack of definitive geotechnical information (soil borings and interpretation of the soil data). Without benefit of detailed soil investigation, field conditions may radically alter mooring systems which appeared appropriate during the approvals period. Bedrock may be discovered to be high enough to preclude conventional pile embedment. Soil capacity may be so low that pile embedment must be greatly lengthened to acquire sufficient pile capacity. In projects with a dredging component, high bedrock may either preclude reaching design depth or require extraordinary rock excavation methods, at very high additional cost. If appropriate testing of proposed dredge material is not performed, and subsequent dredging reveals a highly toxic or otherwise unsuitable dredge material, the project costs may escalate dramatically. Without accurate bathymetry, it is impossible to calculate a rational estimate of the quantity of dredge material to be removed. Bottom contouring may also affect basin flushing, sedimentation and littoral drift.

A proper investigation of the ambient wind and wave climate is most important and should be performed during the conceptual design phase. Without benefit of this information, strategic errors can easily be made affecting marina slip orientation, perimeter protection and design load analysis. The basic wind and wave work, together with comprehensive bathymetry will allow consideration of marina flushing characteristics, sedimentation effects and impacts to littoral drift. The investigation should also include an analysis of transiting boat wake characteristics and the information developed should be input into the conceptual design process.

Other scientific studies that are often appropriate include traffic analysis, marketing studies, biogeographical studies to assess the quantity and quality of marine biology, fin fish spawning and migratory investigations, water quality parameters, wildlife population and impacts, wetland vegetation investigations and delineation of wetlands, and economic studies to support the economic value of the project to the community.

Early attention to securing the appropriate scientific studies will assist in the approvals process by having definitive information upon which to make knowledgeable decisions or refute challenges to the project.

PART 1—INFORMATION SOURCES

1. Adie, D. W. 1984. *Marinas A Working Guide to Their Development and Design,* 3rd edition. New York: Nichols Publishing Company.
2. American Society of Civil Engineers. 1969. *Report on Small Craft Harbors, ASCE-Manuals and Reports on Engineering Practice-No. 50.* New York: American Society of Civil Engineers.
3. Barnes, H. 1982. *The Backyard Boatyard.* Camden, ME: International Marine Publishing Company.
4. Blain, W. R. and Webber, N. B., eds. 1989. *Marinas: Planning and Feasibility, Proceedings of the International Conference on Marinas, Southampton, UK 1989.* Southampton, UK: Computational Mechanics Publications.
5. Chamberlain, C. J. 1983. *Marinas, Recommendations for Design, Construction and Management,* Vol. 1, 3rd edition. Chicago, IL: National Marine Manufacturers Association.
6. Comerford, R. A. 1987. *Marina and Boatyard Industry Financial Performance.* Wickford, RI: International Marina Institute.
7. Dunham, J. W., and Finn, A. A. 1974. *Small-Craft Harbors: Design, Construction, and Operation, Special Report No. 2.* Vicksburg, MS: U.S. Army Corps of Engineers, Coastal Engineering Research Center, Waterways Experiment Station.
8. Fitzgerald, A. R. ed. 1986. *Waterfront Planning and Development.* New York: American Society of Civil Engineers.
9. Lahey, W. L., Zurier, L. S., and Salinger, K. W. 1990. Expanding public access by codifying the public trust doctrine: the Massachusetts experience. *Maine Law Review* 42 (1): 65-93.
10. Naranjo, R. 1988. *Boatyards and Marinas.* Camden, ME: International Marine Publishing Company.
11. Norvell, D. G. and Egler, D. G. 1987. *Financial Profiles of Ten Marinas.* Macomb, IL: Center for Business and Economic Research, Western Illinois University.
12. Norvell, D. G. 1986. *How to Prepare a Business Plan for a Marina.* Macomb, IL: Center for Business and Economic Research, Western Illinois University.
13. Phillips, P. L., ed. 1986. *Developing with Recreational Amenities: Golf, Tennis, Skiing, Marinas.* Washington, DC.: The Urban Land Institute.
14. Polhemus, V. D. and Keyes, G. S. 1988. *Strategy Handbook for Recreational Small Boat Harbor Financing, IWR Report 88-R-1.* Ft. Belvoir, VA: Water Resources Support Center, Institute for Water Resources, U.S. Army Corps of Engineers.
15. Ross, N. W. ed. 1989. *Marina Dictionary.* Wickford, RI: International Marina Institute.
16. Ross, N. W. and Dodson, P. E. eds. 1988. *Dockominium: Opportunities and Problems, Proceedings of the 1987 National Dockominium Conference.* Wickford, RI: International Marina Institute.
17. Ross, N. W. ed. 1983. *A Special Literature Search for the Marine Boating Industry.* Narragansett, RI: University of Rhode Island Marine Advisory Service, University of Rhode Island.
18. Torre, L. A. 1989. *Waterfront Development.* New York: Van Nostrand Reinhold.
19. Webber, N. B., ed. 1973. *Marinas and Small-Craft Harbors. Proceedings of the University of Southampton Conference, April 1972.* Southampton, UK: Southampton University Press.
20. Wortley, C. A. 1989. *Docks and Marinas Bibliography.* Madison, WI: University of Wisconsin Sea Grant Advisory Services.

PART 2
SITE EVALUATION AND ASSESSMENT

4
Site Selection

The selection of an appropriate site for new marina development has become extremely difficult in the last two decades. The escalating demand for housing, a strong economy and the desire to live "on the waterfront" has led to massive landside development of prime waterfront land. In effect most, if not all, "good" sites for marina development have been developed.

The evaluation of a potential site for marina development should consider the following basic characteristics:

- Appropriate zoning and in conformance with public plans for the area
- Sufficient water depth (no dredging required)
- Adequate upland area (no fill required)
- Adequate water frontage
- Protected exposure
- Outside of a wetland or resource protection area
- Outside a designated port area
- Outside an area of restricted historic preservation
- Not adjacent to a public beach
- Near a metropolitan area or other market
- Appropriate land elevation (above flood hazard areas)
- Access to utilities
- Adequate transportation infrastructure

A potential marina site should be zoned for marina use or have the ability to be rezoned without great difficulty and should be compatible with local master plan and community development goals. The site should have sufficient water depth for marina requirements without the need for dredging, have adequate land area to avoid the need for filling, and be protected from significant effects of wind, wave, wake and water level change. The site should not be in or immediately adjacent to a wetland or resource protection area, nor should it be in an area of significant historical protection. Any form of development that negatively impacts a wetland or resource protection

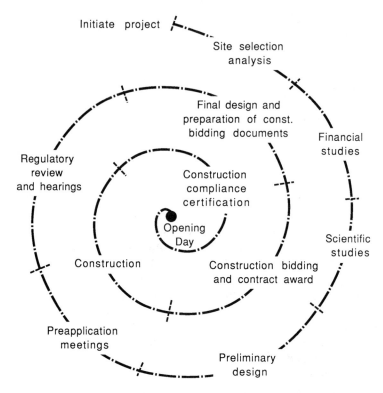

Figure 4-1. Marina development spiral showing the processes involved between project initiation and opening day.

area is highly unlikely to be permitted. Areas of significant historical protection may be subjected to restrictions on the type, size, and character of developments within the protected area. The project should not be in an area of designated port development or proximate to a public beach or other area where public health or safety may be allegedly compromised by marina siting. A new marina site should be within 50 miles (80 km) of a major metropolitan area and have a demonstrated demand for marina facilities. An adequate transportation system servicing the site such as major highways, bus or rail service is a positive feature. Very few potential marina sites will meet all these criteria so it will be necessary to weigh the impacts of the deviation with the desirable criteria and assess the impact on the resource and the project viability. If an existing marina site is contemplated for acquisition or expansion, the same parameters should be evaluated to assess impacts to continued marina viability and potential for future expansion and improvement.

Once a desirable site is identified, a knowledgeable plan must be developed to maximize marina density and landside features while reasonably conforming to the myriad rules and regulations affecting waterfront development. The development process may be considered to follow the form of an ever tightening spiral as demonstrated in Figure 4-1. The outer rim of the spiral represents the site selection process. The path to the center of the spiral is the route to follow to ultimate marina construction and operation. If the site selection is compromised by introducing many obstacles to marina development, the length of the spiral path to the center point will be lengthy and costly. By optimizing the site selection process the spiral path will be shortened and marina development hastened with fewer challenges to development strewn along the path.

Perhaps the greatest impediment to successful marina development and subsequent operation is the selection of a site that offers inadequate protection from external environmental exposure.

4.1 EXPOSURE

The ideal site will be protected from exposure from all directions, provide protected access from the major water body and sustain a comfortable climate in the marina for the berthed boats. Rarely, if ever, can all these conditions be met. It is therefore important to assess the impact of the exposure conditions, then orient the marina to mitigate the exposure and investigate the possibility of incorporating engineered structures to ameliorate the most significant exposure conditions.

Most locations will have a prevailing wind direction during the predominant boating season. The wind direction should be used to orient the marina slips to provide the most comfortable riding conditions for peak winds from this direction. Often the prevailing wind direction will change with seasons. The impact of off season wind directions should be assessed. In the northeast United States the prevailing summer winds are from the southwest and during the winter months the prevailing wind is from the northwest with wind often out of the northeast. With this spread of wind direction it would be difficult to orient the marina toward a single favorable wind direction except that during the winter very few boats are in the water. Therefore a marina orientation considerate of the summer wind direction is usually a suitable design choice. Assessment of the magnitude of the wind speed is also important and the frequency with which it blows. A wind rose should be consulted to obtain a general idea of the magnitude, frequency and direction of wind at a given site. Further discussion on wind effects on design parameters is covered in Chapter 5.

Sea state conditions relate to wind effects and also to local effects along the shoreline. A well protected site that can only be accessed by a narrow inlet that develops breaking waves at its entrance may not be a good choice for marina development. Difficult access from seaward is often very hard to mitigate and if such a condition exists it may be sufficient cause to reject a potential marina site or existing marina acquisition. In the past, it was often possible to obtain federal government assistance in providing major engineered structures such as breakwaters or channel entrance protection. Today, however, it is generally unlikely that the government will create major harbor improvement structures solely for recreational boating interests. If the capital project can be associated with some important aspect of maritime commerce, then assistance from the federal government may be possible. An example might be the blending of a marina project with a water transportation terminal (water shuttle, commuter boats) such that the navigation channel and channel entrance protection structures could be government sponsored to provide a necessary public service in the form of water transportation. The marina might benefit from the dredged channel and channel entrance protection structures.

Current is another major exposure problem to be reckoned with in marina siting. **Currents of more than one knot (1 kt) are usually too swift to sustain a safe and viable marina.** Nearly all docking maneuvers in a marina require a 90 degree turn somewhere in the docking process. As a result there will generally be some time when a vessel will have to turn broadside to a current that exists. Boats tend to become less maneuverable when forced to turn into, out of, or broadside to any current flow. If the maneuver occurs in a confined fairway between rows of berthed boats, considerable difficulty can be expected in executing a safe docking maneuver. If there is a choice, the difficult maneuver, such as entering the confined slip, should be made with the boat heading up into the flow of the current. Under these conditions the boat can maintain the most control. There are very few ways to mitigate a strong current other than to construct massive diverting structures. If no option exists to mitigating current velocity and the marina will be constructed, then it is prudent to properly orient the berthing layout and to allow extra room in channels and fairways to accommodate a margin of error in boathandling.

Exposure is often thought of in terms of weather events, sea states or current flow; however, exposure could also relate to marina security if the marina were located in a neighborhood with a high crime rate, to pollution if a marina were located under a major highway bridge with resulting roadway salts, dust, and petroleum products being deposited in the boats beneath, or next to a major industry where emissions from smoke stacks would be deposited on boats.

Large river or lake water level fluctuations are an obvious exposure consideration. Issues that may be important to address include: accessibility to and from the marina during periods of water level change, water depth at the marina during low water levels, suitability of the mooring system to accommodate large water level fluctuations, and water quality during water level fluctuation.

If issues of exposure can satisfactorily be answered or solutions engineered to mitigate the conditions, then the optimum marina size and the ratio of land to water needed to develop a viable marina may be considered.

4.2 MARINA DEVELOPMENT GOALS

If an appropriate site is obtained or an existing site desired to be upgraded, the first item on the planning agenda should be to establish the marina development goals. A representative checklist follows:

- Year around in-water berthing
- Seasonal in-water berthing
- Dry stack storage
- Seasonal land storage
- Transient dockage
- Boat yard services
    Boat haul-out facilities
    Engine and mechanical services
    Carpentry
    Painting
    Electronics
    Fiberglass repair
- Boat fueling facility
- Ship store
- Food services
    Restaurant
    Dining room
    Snack bar/fast food
    Take-out service
- Clubhouse
    Swimming pool
    Tennis courts
    Health club
- Lease or sale of slips
    Seasonal lease
    Dockominium
    Cooperative
- Associated with upland residences or offices

Is the marina to have year around wet berthing; is dry stack storage desired; will ancillary services such as boat repair, fuel, boat sales, restaurants, ship chandlery or yacht club facilities be provided? Will the marina be associated with upland residences, townhouses or condominiums? Is it desirable to develop the marina for long term slip leasing, seasonal rental or to "dockominiumize" the slips? Is transient dockage a desirable feature, and if so, how should dock space be allocated for this use? To answer many of these questions it is helpful to prepare a market survey and marina pro forma (see Chapter 2). These two documents will reveal the profitable direction that the marina should pursue and determine if the venture is economically justifiable. Identifying the desired services and form of ownership will allow the question of marina size to be quantified. Every item on the marina goals agenda requires some form of space to be fully integrated into the marina plan. Using paper and pencil, or a computer, various components may be sized and roughly located to ascertain general requirements for land and water space occupation. If a specific size site is under consideration, the planning goals may have to be adjusted to suit the available land and water area.

Sometimes an inexperienced marina developer will make the assumption that a certain number of boat slips are necessary for a marina to be profitable. Developing a marina on a preconceived number of slips without benefit of appropriate studies is a poor way to begin a project. **While there is a general perception, in the marina industry, that a minimum of 250 slips is needed for a marina to be financially viable, the validity of this argument has yet to be established.** Every marina is site specific and should be evaluated on its own merits. Many successful marinas operate with a minimum number of slips. Size alone is not a guarantee of success! A number of profitable marinas exist with as few as 35 or fewer slips. The name of the game is to provide those facilities that can generate the maximum rate of return on minimum investment while addressing the current and projected demand. In the case of major urban developments associated with upland projects, the number of slips provided is often irrelevant since no number of slips capable of fitting the available site can justify the investment if the marina must stand alone on its revenue producing capability. Some of the reasons for this situation are the high cost of land in urban areas, additional expense associated with aesthetic treatment to the marina to complement the upland development (i.e., pier surface treatment, promenades, landscaping, etc.), the high cost of providing landside amenities for the marina that must occupy what might otherwise be rented at a high per square foot cost, and the incompatibility of such services as boat haul-outs, boat storage and boat servicing with many upland developments. Once a rational number of boat slips has been established for a specific marina development, generalizations can be made about the amount of land versus water area that is necessary to sustain a viable marina.

A traditional rule of thumb for marina development has been to provide a land area of approximately one to one and one-quarter times the water area. This ratio appears to have changed and the more appropriate ratio is closer to the reverse: a water area of one to one-and-one-quarter the land area, (see Fig. 4-2). Since the mid 1970s, the average size of recreational boats, in service, has increased both in length and beam. As the vessel size increases so does the required area for berthing, fairways between berths, and entrance and maneuvering channels. In many areas there is a trend to provide berthing for large 65 foot to 150 foot yachts, requiring large berthing and maneuvering areas. However, the primary upland facilities to service the larger boat size (heads, showers, laundry, etc.) remains relatively constant. Also many boats in cold regions transit south (or north in the southern hemisphere) during winter, and there the need to provide winter storage space on the land is eliminated. In other areas wet winter storage is popular. All of these changes have resulted in a need for a greater water to land area ratio.

The question of automobile parking has also come under considerable study and debate in the last decade. In the past, zoning codes have generally required automobile parking ratios of 1.5 or 2.0 parking spaces per recreational boat berth. Recent (1988) studies indicate that marinas rarely require

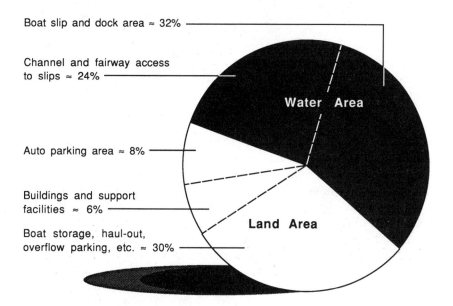

Figure 4-2. Rule of thumb pie chart for land area versus water area utilization.

such generous amounts of parking space to adequately service marina users. The current recommended ratio is 0.6 to 0.8 parking spaces per recreational boat berth. Although not universally accepted, the 0.6 to 0.8 ratio is becoming acknowledged by communities and zoning codes as a rational provision for parking space requirements in marinas. Except on major holidays, such as the Fourth of July in the United States, this ratio will adequately accommodate the marina needs. Overflow parking can often be found in unused land storage areas or in urban areas at adjacent parking garages or office parking lots. Peak marina use days generally coincide with work holidays or weekends when office and commercial space may be vacant.

4.3 WATER DEPENDENCY

Maintaining water dependent use along the waterfront is the hue and cry of concerned citizens and regulatory agencies worldwide. It is important that the limited waterfront be used for those purposes that can only be performed by direct access to the water. Fortunately, marinas are a water dependent use. However, ancillary uses within a marina might not meet the test for water dependency, but rather may be termed as water enhanced activities. Such an example is a restaurant associated with a marina development. Although the restaurant's ambience may be enhanced by its location on the waterfront, it does not have to be located on the waterfront to fulfill its mission of preparing and serving food and/or entertainment. Even if the financial contribution of the restaurant is necessary to the success of the marina, a regulatory agency could find such a facility to be nonwater dependent and request more extensive review or actually deny its approval for construction. It is therefore important to maximize water dependent content, especially if seeking the introduction of nonwater dependent ancillary activities.

4.4 PUBLIC ACCESS

Public trust doctrine was discussed in Chapter 3 as a regulatory issue in many jurisdictions. The public trust doctrine establishes the public's rights along waterfront property. During the site selection process it is wise to keep the public trust issues in mind. Somewhere in the regulatory process, there will be concern for implementing a public benefit into the project, if public trust lands are involved.

Typical public access features that may qualify for satisfying the public benefit issue include: provision of public walkways or promenades, boat launching ramps open to the general public, hosting a public institution (university marine biology station, aquarium, museum, berthing for government vessels, etc.), providing scenic overlooks, providing a public sailing program, providing dockage for public water transportation, providing dockage for otherwise displaced commercial fishermen, and constructing public fishing piers.

The public access issues should not be viewed as a major development stumbling block. In many cases the public access requirements will be quite simple to implement, and concerned regulators will appreciate the needs for safety and security at a marina and not force unreasonable public access demands. In one project where there is a public beach on each side of the marina property, the marina developer was required to provide limited access to a waterfront boardwalk (sunrise to sunset) for public access from the beach area. The haul-out facility was deemed to be a safety hazard and the marina was allowed to divert people away from the haul-out area during hauling operations. No other public access requirements were made. As a benefit for providing public access, the marina was allowed to construct approximately 30 slips in state controlled waters.

Another project in an urban waterfront was required to provide berthing space for an 85 foot and a 45 foot municipal fireboat, provide a dinghy landing for anyone anchoring or mooring in the general vicinity, provide a 120 foot long by 8 foot wide public landing float, provide full public access around two rebuilt pier structures and provide a water shuttle landing float. The remaining waterfrontage allowed the construction of 13 marina slips. In this case the public access demands were quite substantial and were required because the primary project, located on prime waterfront, was a residential condominium and, therefore, classified as nonwater dependent.

In some cases public access requirements may increase the liability for a marina. The requirements of any permit or license should be carefully reviewed before accepting it and assurance obtained from competent legal counsel that implementing the permits will not create an unreasonable liability for the marina. The cost of providing the public access and an associated liability will, of course, be borne by the marina developer, not the agency requiring the access.

Rather than adopt an attitude that resists the implementation of public access provision, we recommend developing your own program for providing some form of public access accommodation before the challenge is issued by the regulatory agency. In this fashion, the developer shows a desire to comply with the regulations and thus can develop public access that is compatible with the site and development goals. If left to the regulatory agency, the specified access may seriously compromise the development goals.

4.5 ENVIRONMENTAL CONSIDERATIONS

A number of site related environmental considerations must be considered in the selection of a site. These are summarized here and in detail in subsequent chapters. Among the "environmental factors" that can affect a potential marina site are: wind, wave and boat wake, water quality, seasonal temperature (ice or sustained heat), quality of available municipal resources (potable water, trash and garbage collection, fire protection, police protection, hospitals, etc.), proximity to sensitive community facilities (hospitals, schools, churches), access to transportation networks, hazardous waste contamination, suitable topography for development goals, susceptibility of the site to storm surge and flooding and suitable ambience to attract the target market.

4.6 SITE SURVEYS

To determine and confirm the suitability of a site, a series of site surveys may be needed. The requirements of regulatory agencies should be kept in mind when performing site surveys so that all necessary data is acquired the first time.

Paper Work Survey

A first survey is not really a site survey but is a paper work survey to determine the validity of property ownership. This survey would be the title search. The title search should not only trace the record of property ownership such as through deeds, but also determine any past permits issued to the property and any encumbrances associated with the land. Local and state agencies such as a Board of Health, Conservation Commission, or Department of Environmental Protection should be consulted to determine if any complaints have been filed against the property which may reveal the potential of hazardous wastes on the site or other conditions that may compromise a "clean" development. The local assessors maps should be consulted and confirmed with the deeded property ownership. Zoning maps should be consulted to determine zoning on the site and any restrictions associated with the zoning.

Property Survey

Once the paper work survey is in hand a local qualified land surveyor should be retained to perform a property survey and prepare a plan of landownership. **Depending on the size of the land and water parcel, suitable plan scales might be $1'' = 40'$, $1'' = 100'$, or $1'' = 200'$.** The property survey should also indicate any water rights or property line extensions into the water area that have been determined for the site. Legal easements should be included on the plan as well as locations of water related channel lines, if any. It is also helpful to relate the property lines to an appropriate coordinate grid system, either a state system or a federal grid system.

Topographic Survey

Along with the property boundary survey, the surveyor should perform a topographic survey of the land at contour intervals of 1′, 2′, or 5′. The existing contour of the land will govern contour intervals. Also, it is advisable to have the surveyor establish a benchmark for ground elevation control and relate the benchmark to the local standard water level. Continuity of elevation always seems to be a problem in waterfront development. Many regulatory agencies will use different elevation controls such as the National Geodetic Vertical Datum (NGVD), mean low water, mean lower low water, or mean high water. Meanwhile many communities have their own elevation base which may or may not directly relate to the permitting agencies elevation base datum. Check this carefully and be sure all consultants and contractors work from the same base, or understand that different elevation bases are being used for specific aspects of the project. Once water level elevations have been established, it is a good idea to set a tide board or water level gage somewhere on the site. The water level gage should be protected and monitored to confirm local water level conditions. The details of a typical tide board are included in Appendix II, Useful Information, Figure A 2-1.

The surveyors topographic plan should show all the prominent land features as well as land elevations. Some typical items include: structures, shorelines (high and low water), utilities to the extent recoverable, overhead obstructions (power lines, gates, structures, etc.), roadway widths, trees and major vegetation, wetland delineation, and any other features that describe existing site conditions.

Bathymetric Survey

Accompanying the topographic survey should be a bathymetric survey, if possible, which provides information on the below water bottom contours. Typically a series of lines projecting from shore (transects) are established by using ranges on shore for reference. The ranges may be as simple as lining up two structures or driving stakes in the ground to allow offshore delineation of the theoretical range line. Soundings are taken at prescribed intervals on the range lines either directly by rod (ruler) reading, by a tape with a weight on it or by depth sounder. **The sounding data is usually taken on a grid of 25′ × 25′, 50′ × 50′ or 100′ × 100′ (see Fig. 4-3) depending on the rate of change of the bottom slope.** The depths may be shown on the plan as individual water depths or translated to smooth curves (contours). **The datum used, (e.g., mean low water, NGVD, etc.), should be clearly labelled on the plan.** The water depth data will be extremely helpful in determining a marina slip layout, assessing dredging requirements, and developing a mooring design.

78 PART 2/SITE EVALUATION AND ASSESSMENT

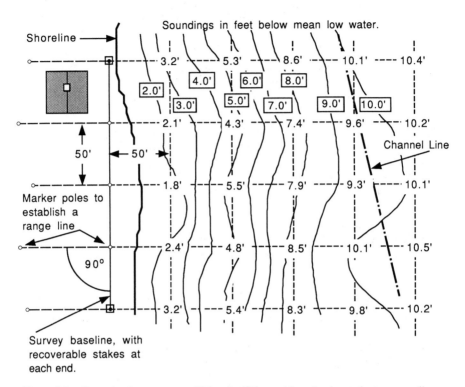

Figure 4-3. Example of a survey on a 50 foot by 50 foot grid used to locate bottom soundings. Contours are roughly drawn in between sounding data points. If possible tie the survey grid into the appropriate federal or state coordinate grid system.

If physical topographic and bathymetric data are unavailable, the usual case when first considering a site, and as an alternative to having surveys performed, it may be possible to develop enough basic data to assess site potential from local nautical charts, such as those prepared by the National Oceanic and Atmospheric Agency (NOAA) or the Defense Mapping Agency (DMA), in the United States, and the land features estimated from topographic maps prepared by the United States Geological Service (U.S.G.S.) or similar agencies in other countries. A good source for topographic maps is local sporting stores that may carry such maps for hunters and fishermen.

Additional Surveys

A number of additional surveys may be required, depending on the specific site. If the site has any history of commercial operation that may have disposed of hazardous waste on site, a survey may be required to assess the condition of the site soil for hazardous waste. Test pits or corings should be made and the recovered soil samples should be analyzed for hazardous waste content. If the ground is found to be contaminated, a difficult and costly process may be required to remove or mitigate the hazardous waste material. Prior to a site acquisition, it is advisable to ascertain the condition of the site for hazardous waste and, if it exists, either not to acquire the site or define the responsibility of the current or past owners for making the site acceptable for development.

If there are wetlands on the site and it is necessary to impact the wetlands to develop the project, a detailed survey of the wetlands, by competent biologists, should be made to regulatory standards. The survey may include delineation of the wetlands, cataloging of the wetlands species inhabiting the area, and assessment of the value of the wetland as an important ecological resource. If the wetlands are to be impacted, mitigation may be required such as compensatory wetland development, or it may be necessary to minimize the impacts by selective structure or engineering design. Since the United States currently has a "no-net-wetlands-loss" policy, any consideration of wetland development must be well investigated and mitigation planned to obtain regulatory approval.

If the site contains undeveloped land, it may be necessary to perform a wildlife inventory to ascertain quality and quantity of the indigenous wildlife species. Bird counts may also be required if the area is frequented by rare or endangered bird species or the area is a nesting or migratory bird resting area. The same concerns may be made for fin fish in the site waters. It is particularly important to study fin fish impacts if dredging is involved in the project development or if major fixed structures are contemplated (dams, breakwaters, wave barriers, filled piers, etc.) that may impede fin fish migration or spawning. In addition a biogeographical survey may be required that assesses quality and quantity of biota dwelling on or in the bottom sediments, (shellfish, sea worms, plant species, etc.).

Some marina developments may effect sedimentation transport, local flushing characteristics or tidal exchange. Additional site surveys may be required to determine existing conditions and develop data to allow design of an engineered solution to any impact. In addition to the above mentioned surveys, dye tests may be required to determine water flow characteristics and water level monitoring stations may have to be established to record changes in tidal flow and elevation changes over time.

4.7 COMMERCIAL MARITIME CONFLICTS

Many marinas are constructed in close proximity to traffic lanes or in harbors frequented by commercial maritime traffic. The effects of commercial maritime traffic that passes in the vicinity of the marina site such as vessel wake, vessel suction, visibility, and propeller wash must be considered and procedures or design features incorporated to mitigate the effects. In dealing with large commercial shipping, the old adage "discretion is the better part of valor" should prevail. In confined waterways most commercial traffic is constrained by draft and maneuverability. These constraints may create an undesirable effect on the marina location. Many larger harbors have defined channel lines which are reserved for use by all maritime traffic but which should not be encroached upon by marina or any other waterfront construction. Early in the marina design phase, the location, extent and restrictions of any adjacent channel lines should be thoroughly investigated and defined on the plans.

In the United States, the U. S. Army Corps of Engineers establishes federal channel lines and has a detailed record of their locations. In most cases, channel lines are referenced to state coordinate grid systems so that the lines may be referenced on project site plans with relative ease. Many states may also have channel lines or state harbor lines which govern the extent to which local construction into the waterway may be permitted. Even local communities may have special channel lines, so a comprehensive search into applicable documents is recommended. In other countries channel lines may be established by various military or domestic departments.

The recent intense development along many United States coastlines has caused the Army Corps of Engineers to scrutinize the extent of development toward channel lines and deny permits for structures in proximity to the channel lines that might compromise vessel navigation or channel maintenance operations. One Corps of Engineer Division has issued guidelines for the permitting of structures adjacent to federal channel lines. **These guidelines recommend a setback from federal channel lines of three times the channel control depth.** Therefore for a 35 foot control depth channel, the closest structure would be approximately 105 feet from the channel line. The Corps of Engineers has recognized some historic use of structures closer than three times the water depth and has agreed to look at specific cases in rendering a decision on the suitability of the proposed structure layout. The three times water depth criteria appears to have evolved from a generalized dredge material side slope of three horizontal to one vertical (3:1), therefore a setback of 3:1 will provide adequate, unobstructed area to provide channel side slopes during channel maintenance projects. The three times water depth setback argument fails when the bottom bathymetry adjacent to the channel is less than zero such that the channel side slopes can be created in less horizontal distance than three times the water depth (see Fig. 4-4). As an example: if the water depth 50 feet from a 35 foot control depth channel is 10 feet, then the depth of soil associated with the channel side slope is 35 feet minus 10 feet or 25 feet. A 3:1 side slope for this condition will need to be

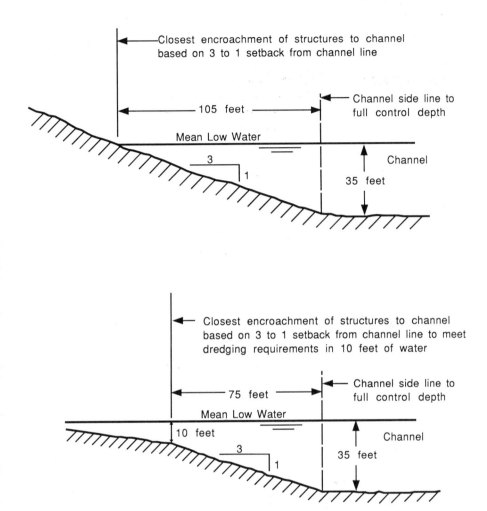

Figure 4-4. A cross section of a channel showing 3 to 1 setback of structures to allow dredging from a channel with differing side slope conditions.

only 3 × 25 feet or a 75 foot setback rather than 105 feet as suggested by the Corps of Engineers guideline. It is therefore prudent to become aware of the intent of guidelines and understand how they may be challenged. The need for a setback from federal channel lines itself raises some question in that the channel lines were created to provide a defined area of operation for traffic control and maintenance. However, since vessels are more difficult to maneuver than say automobiles, it is wise to provide some buffer area between a

82 PART 2/SITE EVALUATION AND ASSESSMENT

channel line and a fixed structure to accommodate the occasional wandering of vessels off the channel. Channels are also usually only intermittently buoyed so the actual channel line is harder to define than the pavement of an auto highway.

Primary commercial/recreational maritime conflicts result from the vessel wake wave generated by passing commercial traffic. If the commercial traffic is passing a site in a designated channel or other historically used channel and is complying with appropriate rules of the road it may be incumbent upon the marina to provide wake wave attenuation. Commercial traffic is often constrained by draft and maneuverability and must maintain certain positions and speed to transit an area. Maritime law generally allows for the exclusion of liability on vessels restricted in maneuvering if they must create a wake wave in order to complete the maneuver or safely transit an area. Tug boats, for example, may generate a short term large wake or propeller thrust if maneuvering a vessel around a bend or otherwise adjusting the position of the vessel (see Fig. 4-5). Tug boats or any other vessel not attending to the assistance of a maneuvering commercial vessel are subject to the rules of the road and other applicable regulations like any other vessel underway. The provision of wake wave attenuation is covered in more detail in Chapter 7. It is however prudent to provide such protection if the marina or other vessel berthing site is subject to this type of action. Due to numerous complaints by recreational boating activities on the effects and damage caused by commercial maritime vessel wake, the United States Coast Guard

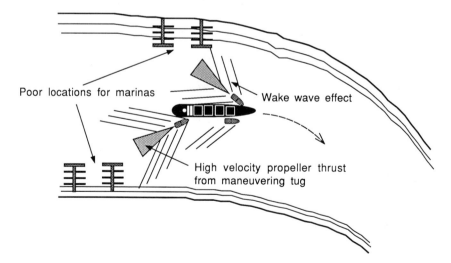

Figure 4-5. Effect of maneuvering vessels in a marina located in a narrow waterway.

has become more interested in reviewing marina permit applications and advising applicants that construction of a recreational berthing facility adjacent to a commercial navigation channel might result in unacceptable wake wave and or maneuvering vessel propeller thrust and the attenuation of this action is the responsibility of the marina developer. The Coast Guard does not want to arbitrate complaints from newly constructed marinas on passing vessel wake if such activity was and is normal to the area.

In addition to constraints of draft and speed, many commercial vessels are constrained in their ability to see recreational boats. Ocean going vessels and river traffic with long lengths of barges being pushed may have severe line of sight restrictions that compromise their ability to observe small boat traffic especially in the area forward of the vessel. Figure 4-6 represents a large commercial vessel with the bridge aft positioned so that in conning the vessel, an observer on the bridge looking forward cannot see below a line from his eye through the highest point of cargo or ship projection ahead. This may create a significant blind area of visibility. In terms of marina location, it would be prudent to site a marina so that passing vessels do not have a blind spot in an area where vessels may be entering or leaving the marina. This is especially true where channels or waterways may bend. In very difficult situations it may be necessary to establish some form of vessel control system in the marina to alert entering or departing boats that commercial traffic is in the area. Several marinas use red and green control lights at the marina/channel entrance to warn boaters of impending traffic. Flags or semaphore could also be used as warning devices.

Area forward of ship's path which has restricted visibility due to vision obstruction created by ship structure or deck cargo.

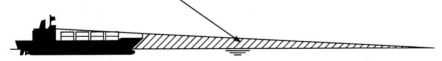

Small boats or other objects within the blind area may not be readily detected by an observer on the ship's bridge.

Figure 4-6. Line of sight and blind area for an observer on a ship's bridge without deck cargo and further impaired with deck cargo.

Location of fairway access points from main channels should also be sited only after a study of lighting and traffic patterns. If possible avoid a marina entrance in a position where a rising or setting sun may blind transiting maritime traffic to vessels entering or leaving the marina. Provide as much distance as possible and at as small an angle as possible to the main channel. The analogy of an entrance ramp to a major automobile highway and the usual long in-run to allow the cars to reach highway speed and blend easily into the flow of highway traffic is appropriate. A sharp entrance to a navigation channel may create confusion in that the ultimate direction and speed of an entering vessel may be difficult to estimate, increasing the potential for conflict. Likewise, provide the greatest visual distance possible from the navigation channel to the marina activity. This is especially a problem in urban marinas where the marina may be tucked in behind or adjacent to pier structures such that vessel movements are not discernible from the channel or active waterway. A vessel accelerating out of a blind area between piers may pose a significant navigation hazard. Often incorporating marina entrances parallel to the channel or waterway will force the boater to deaccelerate and ease out of the channel or enter the channel at a small angle. The desirable wake wave attenuation devices may be oriented to function as appropriate entrance structures.

Although often overrated, large vessels transiting narrow rivers or channels may create a so-called "bank suction" where the mass of the vessel first pushes a volume of water out of its way in the constricted channel and then the water rushes into the hole left by the passing vessel. The inrush of water may create a suction immediately aft of the vessel. This suction can cause damage to any smaller vessel in the area. The intercoastal waterway along the east coast of the United States is famous for the horror stories about the effects of bank suction on small vessels transiting the waterway when passed by larger vessels often using excessive speed in passing. For adjacent marinas these effects may cause difficulty in the maneuvering of boats within the marina during a time of large vessel passage or these effects may subject berthed vessels to excessive mooring line pulls and severe vertical and horizontal movement. It is often the sudden unexpected nature of such action that really causes the damage, especially to people. In a known storm condition, people prepare for sudden movements and take necessary precautions to prevent injury. The unexpected movement caused by a relatively instantaneous passage of an infrequent vessel may catch people off guard, resulting in serious personal injury or boat damage. Commercial traffic must move in fair and foul weather and at all times of the day, so instances of transiting effects are not limited to specific times or weather conditions.

In many commercial harbors there is a perceived animosity between commercial and recreational boating interests. Each group believes the other most responsible for the problems that have developed in the use of the public waterways. Both arguments may have merit. Recreational and commercial interests need further education as to the problems each has and how mutually they may be resolved. Most commercial operators are required to obtain a license to operate their vessel and with that license comes a responsibility to operate safely and with due concern for other interests. The commercial operator however is faced with the requirement of operating a cost effective vessel on a time/money restricted program. He is not allowed the luxury of undue delays or large deviations in route or speed, conditions often available to the recreational boater. Licensed commercial operators are concerned that in any commercial/recreational incident they stand to lose their license, often whether the operator is the negligent party or not, and the loss of license often means the loss of livelihood. These are legitimate concerns that must be recognized by the recreational boater and the marina developer. Any new structure in the waterway that places a licensed operator at greater risk will be challenged.

It is often prudent to seek out concerned local interests such as harbor pilots or tug boat dockmasters as well as shipping interests and discuss development plans at an early stage. Many times simple and easily incorporated suggestions regarding marina alignment, traffic flow patterns or seaward access may be generated by these discussions and resolved before the issues are raised in regulatory review. Many commercial operators are recreational boaters themselves and can relate to the needs of the recreational marina. Of course some commercial interests will be totally unreceptive to new recreational boating facilities and no amount of discussion will alter their perspective. In those cases the best defense is to define the opposition issues and prepare well reasoned responses based on technical facts or adjust the site design to mitigate as far as possible the concerns presented. Remember, the traditional regulatory mentality favors the commercial interests over the recreational interests. Federal agencies like the U.S. Coast Guard and the U.S. Army Corps of Engineers are mandated by law to service an unimpeded flow of commercial goods and services by waterborne transportation.

4.8 AUTOMOBILE TRAFFIC ASSESSMENT

Marinas and other waterfront developments are generally designed to accommodate large numbers of people. Unless the development is close to a major metropolitan or urban area, there probably will not be adequate, accessible, or convenient public transportation to the site and therefore a majority of people using the facility will arrive by automobile. The movement into and out of the development may result in traffic conflicts with the existing community infrastructure. It is common in many regulatory review procedures to require a traffic assessment study. The traffic assessment study will attempt to quantify the volume of traffic in the area, define the peak times of traffic movement and identify areas of potential traffic conflict. Traffic assessment study requirements will vary between agencies, but in general the study will include the following considerations.

- A description of the proposed project and area under study with supporting area maps, topographic plans, highway network plans, site plans and zoning maps.
- A description of the existing conditions, including the roadway network, traffic volumes, air quality, accident reports, other transportation modes available, and the existing level of service of the transportation network.
- An analysis of the projected number of automobile trips that will be generated by the project and the distribution of traffic, whether into and out of the site, or passing by the site.
- A description of anticipated future conditions around the site including projected traffic volumes, other transportation modes, local area development impacts, and potential future capacity of the roadway network.
- An analysis of development options and mitigation measures.
- The proposed construction schedule along with information on any aspects of construction that may impact the capacity of any of the existing transportation networks. A description of all measures to be taken to minimize impacts of noise and dust pollution.

Although the study is best performed under the direction of professional traffic planners or consultants, it is helpful for a marina developer to understand what constitutes a traffic study, and in some cases the traffic study may be undertaken directly by a knowledgeable marina developer.

5
Winds and Storms

The wind along the shore is probably the most noticed, most utilized, and the most feared feature of the boating environment. Go down to the docks sometime and count the number of wind measuring devices, anemometers, at the top of many boat masts. Of all the weather phenomena, the wind, or lack of it, is the single most important environmental aspect of boating activity and marina design. The wind is used to power boats. Many hours may be spent waiting for it to abate, or sometimes for it to "come up." It warms us up, it cools us down, it chills us, it can drive the rain hard. It blows the beach sand around, makes the surface of the boatyard parking lots airborne, it can spoil a varnish job, blow the paint off a brush, and blow objects overboard. It can create one of the most powerful forces on earth, the water wave, and it can blow structures, docks, and boats out of existence. How dock arrangements are designed, how boats are tied up, and how many lines are used are all determined by the expected wind conditions for the site. An even greater concern arises when boats are tethered by a single line on a swing mooring exposed to the wind out in open water. Is the line strong enough and will the mooring anchor perform as expected?

Wind velocities for nautical purposes are generally referred to as the air's speed in knots rather than in miles per hour or mph. A knot is a nautical mile per hour. Since a nautical mile is slightly longer than a statue or land mile, 6081 feet versus 5280 feet, a knot of wind is 15 percent faster than a mile per hour, mph, of wind. Thus 20 knots of wind is the same as 23 mph of wind. Not much of an important difference until you start to talk about 60 knots of wind which is the same as 70 mph. This points up the fact that the use of knots for wind speed can be deceiving and tends to make us think that things are calmer than they really are. Also, few wind systems are constant in their speeds. **There is a gustiness to consider which may cause a variation of 30 to 40 percent above the so-called mean or average or steadier wind.** A 30 knot maximum gust or peak wind system may range downward to a longer term low of 20 knots, for example. Thus, a "steady" or "sustained" wind yardstick of perhaps 22 to 24 knots should be used for computing the longer term effects of the wind such as wave height prediction. One good blast of wind every 10 minutes or so may create mooring line problems or call for boat cover tie down parties, but it does little for raising up a higher sea condition.

5.1 WIND SYSTEMS

Typical, onshore afternoon breezes are generally caused by the heating of the air above the land, drawing the cooler air from over the water towards the shore. The reverse is sometimes true during the nighttime hours when the water cools more slowly than the land. These daily wind events are often punctuated by the passage of small convective afternoon thunderstorms or weather fronts which contain lines of squalls and thunderstorms. These events are usually short lived with perhaps very high winds for several minutes and then they are gone. Events like these have little significant effect on raising high sea conditions over open waters, although a healthy chop can develop rather quickly. However, they can have important effects on wind loadings on moored boats and dock systems. It may only require several minutes of these high winds to effectively place some of the highest loadings on a dock system. Longer period storms with similar wind velocities do not necessarily increase the wind loading forces, but just make the forces persist for a longer period of time. So marina systems and associated structures must be able to withstand even these short period momentary blasts which may reach 40 to 60 knots and on occasion, even higher speeds.

The more extensive low pressure storm systems may have wind velocities no higher than the short lived squalls or thunderstorms. However, these high winds may persist for hours from one direction, perhaps shifting slowly around the compass. Depending on the speed of travel of these wind producing weather systems, high winds may have to be endured from one direction for many hours. Many of these storms have an associated weather front causing abrupt shifts to the winds after blowing in one persistent direction as these systems move over the site.

Tropical storms, typhoons and hurricanes bring a whole new set of problems into the boating world and marina design. It is important to understand that our world of common experience relates to wind systems generally of 40 knots, with occasional experiences up to 50 knots. The winds can reach velocities of 135 knots (155 mph) and beyond, but our experience level rarely even reaches to the halfway point. A halfway point in this wind velocity scale for example may be 64 knots. This is just the beginning point for the awareness of the potential fury of defined hurricane and typhoon conditions. The mere word hurricane then relocates our anticipated experiences into a realm of often unpredictable, but almost always unbelievable wind intensities and forces. Marina design under some of these conditions is impossible. Few man made structures can withstand the fury of an intense hurricane without damage. For example, the famous 1938 hurricane that struck the northeastern seaboard, hurricane Camille of 1969 along the Louisiana and Mississippi coast, and the 1989 hurricane Hugo as it came ashore at Charleston, South Carolina, all produced unbelievable devastation.

5.2 SOURCES OF WIND INFORMATION

A person can live in a particular area and still not be really informed about the local short and long term wind systems. Where can information about these winds be found? The answer is from the weather gathering facilities of the federal government, both civilian and military. The National Weather Service, part of the National Oceanic and Atmospheric Agency (NOAA) of the Department of Commerce is the major source for weather information, with the U.S. Navy providing additional information from their air bases and ships at sea. Most sizeable airports, for example, have had prepared for them a climatological analysis of the weather data that they have recorded over the years. These data are published in small pamphlets or booklets. The most useful for major airports is the series entitled "Climatology of the United States No. 90., Airport Climatological Summary, NOAA." These are marvelous sources of all kinds of weather data, but especially wind data. Table 5-1 shows an example of the type of wind presentation available for the month of March, from the *Airport Climatological Summary for Washington National Airport*. As can be seen, the percentage of occurrence of wind velocity and direction is summarized on a monthly basis and provides some

Table 5-1. NOAA provided Wind direction vs. Wind speed (percent frequency of observations) for the month of March at the Washington National Airport.

Wind direction	Wind Speed in Knots								
	0-3	4-6	7-10	11-16	17-21	22-27	28-33	34-40	>40
N	.5	1.3	2.8	2.4	.2	.0			
NNE	.2	.9	2.2	1.4	.0				
NE	.5	1.9	2.6	1.2	.0				
ENE	.3	1.4	1.7	.6					
E	.8	2.5	1.9	.7					
ESE	.2	1.0	.8	.0					
SE	.3	1.5	.9	.2					
SSE	.8	2.9	1.5	.2	.0				
S	.7	4.5	4.6	1.7	.1				
SSW	.4	2.5	1.6	2.2	.5	.1			
SW	.5	1.6	.9	.6	.1				
WSW	.3	.4	.7	.3	.1	.0	.0		
W	.3	1.0	1.8	1.7	.3	.1	.1		
WNW	.2	.4	1.5	2.1	.7	.2	.0		
NW	.1	1.4	3.3	5.1	1.5	.2			
NNW	.3	1.9	4.4	5.1	.9	.1			

details of the wind systems for marina sites within a reasonable distance from the airport. Most metropolitan areas have several sources of wind and weather data. New York City has three airports within a 15 mile radius and all produce airport climatological summaries. Other types of weather publications concerning local winds systems from diverse locations are also available. These are exemplified by the following samples from the greater New York area:

> National Weather Records Center, Ashville, N.C. Wind Direction vs. Wind Speed Tabulation, Battery Place (at the southern tip of Manhattan Island), New York. January 1956 to December 1960.

> Office of Chief of Naval Operations, Aerology Branch, National Weather Records Center. Floyd Bennett Field, Brooklyn, New York. Surface wind data summaries: April 1938 through October 1942, April 1945 through August 1957, and 1958 through 1962.

> National Weather Records Center, Ashville, N.C. Wind Direction vs. Wind Speed Tabulation, New York Pan Am Heliport, Manhattan, New York. May 1966 to February 1968.

Sometimes the smaller airports throughout the U.S. have had special studies performed. For limited periods of time very useful and informative wind frequency and velocity information was obtained at selected locations. As an example of this, a program conducted a number of years ago involved a wind to electrical energy conversion study. Many airports, including the smaller ones which normally do not routinely record weather data, collected wind data for over 10 years for these projects. This data is published by the National Weather Service in site specific publications such as *Percent, Frequency of Occurrence, Wind Direction versus Wind Speed, Groton-New London Airport Wind Summary, 1950 to 1961 and 1961 to 1963. National Climate Center, N.C.*

The major source for wind data anywhere in the United States is the National Oceanic and Atmospheric Administration's Environmental Data Service, National Climatic Center, Ashville, N.C. Most of the publications mentioned can be obtained from them. They are very cooperative in helping to find wind and weather data for a particular location and in providing copies of the data for a nominal fee. Still other sources involve the local electrical utility companies, especially nuclear power plants, water filtration plants at reservoirs and sewage treatment facilities along the shore. These facilities will often make their data available as a public service, and often these utilities are located closer to the proposed marina sites than are the weather gathering facilities of the larger airports.

5.3 WIND VELOCITIES OF INTEREST

For marina design, most people tend to think categorically about the wind effects from the perspective of average and extreme conditions. The first important category is the ambient, average or so-called normal wind condition for the site. This tells something about direction, frequency of occurrence and average speed and indicates whether wind and wave protection may be needed. As an example, Table 5-2 shows the annual average winds from coastal areas including Newark, New Jersey which is representative of New York Harbor and Raritan Bay; James Creek which lies at the junction of the Potomac and Anacostia Rivers in Washington, D.C.; Anacortes, Washington at the eastern end of the Straits of Juan De Fuca; and Bridgeport, Connecticut representing central Long Island Sound. The direction, average wind velocities and percent occurrence are shown for Newark, while just the direction and percent occurrence are shown for the other locations. Comparisons for many parts of the mainland United States discloses that typically average wind velocities generally range between 8 and 12 knots and therefore only Newark's velocities are shown. However, the frequency of occurrence in the wind's direction can differ considerably from site to site.

As highlighted by the asterisks, Bridgeport, Connecticut shows a definite bias for wind systems from the west and west-northwest over all other

Table 5-2. Average annual wind speeds and percent of occurrence for selected areas.

Wind direction	Average speed kts	Percent of occurrence			
	Newark NJ	Newark NJ	James Ck Wash, DC	Anacortes Wash	Bridgeport Conn
N	9.5	9.4	5.7	4.7	9.8
NNE	8.9	6.4	4.3	3.7	6.3
NE	8.7	3.6	5.9	3.2	7.7
ENE	7.6	2.2	3.4	2.6	5.4
E	7.2	3.1	4.5	3.3	3.4
ESE	7.5	3.2	1.6	2.6	2.4
SE	7.9	4.8	2.8	4.2	1.7
SSE	7.7	3.7	5.8	4.9	1.9
S	7.2	7.8	*15.4	4.1	3.1
SSW	7.7	6.8	*12.2	3.6	3.1
SW	8.6	*10.7	4.7	5.8	5.3
WSW	9.1	9.3	2.5	7.6	7.0
W	9.9	9.1	4.7	*15.7	*12.7
WNW	11.9	8.8	3.7	*10.5	*12.0
NW	12.0	6.7	8.4	7.4	9.7
NNW	10.9	3.5	*11.5	4.0	7.3

* Annual bias for wind direction

directions. Anacortes, Washington, at the other end of the United States, also has a strong bias from the same direction, with westerly winds blowing more than 25 percent of the time in each case. The Washington, D.C. area which is typical of the Chesapeake Bay region in general, exhibits average winds from the south, with a somewhat unusual secondary bias from the north-northwest. Thus, these frequently occurring average wind conditions have the unusual attribute of coming from opposite directions and may double the design problems. Newark, on the other hand, only shows a weak and general bias for winds from the western half of the compass, with little specific concentration of wind directions to note. All sites indicate the infrequent occurrence of winds from the east and southeast, however, these can be the extreme storm wind directions and eventually must be given special consideration in any wind analysis.

In general, these annual average winds tell something about the boating climate for the area, and if used for wave height prediction, something about the mean wave heights and arrival directions that may be expected. It is the first information that may be used to determine the feasibility, advisability, ease or convenience of establishing a marina at a particular site. If the selected site is generally already protected from the winds which occur most frequently on an annual basis, the potential development costs start out at a lower rate.

The next piece of information that needs to be known relates to the direction and variation of intensity of the winds by the month for the area. This information is even more useful in predicting waves than the simple averages for the site. For example, as shown in Table 5-1, NOAA provides an excellent statistical description of the winds on a monthly basis throughout a normal year. These are the more specific wind data and details that can be used to provide a knowledge of the annual wave climate for the area based on the annual winds.

Maximum Winds

The next category of winds of concern are the so-called maximum winds that can occur in the area. These are the winds that have been recorded to have occurred during the full period of historical measurements for the site, whatever that period may be. For some sites in the U.S. it may only span one or more decades. In other areas, it may go back to the latter part of the nineteenth century. It is not surprising to find that the longer the period that wind data has been recorded at a particular site, the higher are the observed maximum winds. A most important source for this maximum wind information is the publication *Comparative Climatic Data for the United States through 19(year). National Oceanic and Atmospheric Admin. Monthly Average and Maximum Wind Speeds.* This publication includes every major U.S. Airport and is updated every few years. It provides some very important

Table 5-3. Maximum winds for selected airports.

Month	Wind direction, degrees true			Highest recorded speed knots		
	Newark	La Gua	JFK	Newark	La Gua	JFK
Jan	300	045	260	45	59	45
Feb	230	045	250	40	56	40
Mar	270	315	280	37	52	38
Apr	270	270	260	44	51	38
May	320	315	160	44	45	38
Jun	070	315	290	48	47	35
Jul	180	315	340	39	51	32
Aug	090	315	300	40	55	40
Sep	050	045	020	44	61	40
Oct	110	135	260	42	57	38
Nov	090	000	050	71	59	38
Dec	320	315	060	48	49	40

information on maximum wind velocities and directions recorded on a monthly basis. Several examples abstracted from this publication are shown in Table 5-3. The data was taken over a 37 year period for Newark and 34 years for La Guardia and 22 years for JFK, all located within the greater New York City area. Measured maximum data for JFK is lower in velocity, and typically due to the shorter collection period than the data for Newark or La Guardia.

Even more valuable, if you are lucky to have a marina site near one of the selected airports, is a publication prepared a few years ago *Extreme Wind Speeds at 129 Stations in the Contiguous United States,* National Bureau of Standards Building Science Series 118, Department of Commerce. This was a one-time publication which listed a long series of maximum annual wind measurements and predicted long term wind return periods. These data provide the maximum wind that may be expected every 2, 5, or 10 years, or what ever period is desired, in some cases up to 1000 years and more. A sample of the type of information available from this publication is shown below for Wilmington, North Carolina:

Fastest mile wind speed: September 27, 1958, N at 73 knots

Next fastest recorded: August 11, 1955, NE at 60 knots

Next fastest recorded: February 25, 1956, SW at 55 knots

The 10 year return period predicts the highest wind at 57 knots

The 20 year return period predicts the highest wind at 64 knots

The 50 year return period predicts the highest wind at 77 knots

The 100 year return period predicts the highest wind at 89 knots

Also helpful is another publication that presents past high wind measurements *Historical Extreme Winds for the United States—Atlantic and Gulf of Mexico Coastlines.* National Oceanic and Atmospheric Administration, May 1982. Using all of these publications and wind sources, the historical experience for the maximum winds for the area is well documented. These are generally the winds that will be used for maximum wind loading calculations on the marina structures, docks and boats and for the prediction of the maximum storm wave heights after appropriate adjustment for the peak values have been made.

Extreme Winds

Finally, some consideration must be given to the potential for the occurrence of the extreme winds that may visit an area. Maximum winds recorded for a particular area will generally include the extreme wind experienced to date. However, in many areas some of the highest winds experienced, or which can be experienced, are not recorded because of power or instrument failures or because the data record is too short. This is quite common for many coastal airports and forces speculation on the potential extreme winds that may occur.

Tornados and extreme blasts from thunderstorms are devastating but generally short lived. Little important wave activity is generated from these events. However, considerable structural wind damage can be sustained over periods as short as several seconds. Predictions of higher wind gusts in thunderstorms, as well as the potential 200 knot winds of a tornado are not generally considered as marina design criteria since little preparation can be made to withstand these hit-and-run wind blasts. Hurricanes, typhoons and tropical storms, however, come in a variety of intensities, occurrence frequencies, and directions for many coastal and near coastal areas. Some consideration is usually made in the design of the marina depending on the specific details of the arrival potential of these weather systems in a particular area. In many cases, design for other than marginal hurricanes is not feasible or cost effective, however, the potential occurrence of these events and their possible intensity is necessary knowledge for the marina designer, developer, owner or operator.

Historical records may not have been collected long enough to include hurricane or typhoon conditions for the area. They also may not have measured peak occurrences due to equipment failure. Certainly, few recording sites have been in operation for 100 years in order to document the so-called 100 year storm. Most areas of the U.S. Atlantic seaboard and Gulf coast are vulnerable to these hurricanes, with Hawaii exposed to typhoons. This is why it is necessary to speculate on how bad, bad may be for the particular site during the design process and if anything can be done about it other than to buy insurance. **There are five different categories of hurricanes** as described by the NOAA Saffir/Simpson scale, with each successive higher level category associated with greater devastation. A Category 1 hurricane, for example, will bring sustained winds from 64 knots (74 mph) up to 83 knots (95 mph) and a storm surge of 4 to 5 feet above normal water levels. The arrival of a Category 3 hurricane will be accompanied by sustained winds of from 96 knots (110 mph) up to 113 knots (130 mph). The famous 1938 hurricane in the northeast is classified as a Category 3 hurricane. The arrival of a Category 5 hurricane will be accompanied by sustained winds of 135 knots (155 mph) and greater. This is the upper limit to the category classification and any hurricane with winds above 135 knots is considered to be in this category. There is no upper limit set for the wind velocity. Also predicted is a storm surge water level of greater than 18 feet above normal for a Category 5 hurricane. Only two Category 5 hurricanes have ever arrived at the U.S. coast since accurate records have been kept from about 1899, and both were in the Gulf of Mexico. One of the most intense hurricanes ever recorded for the continental U.S. was Camille, which struck the Louisiana and Mississippi coasts in August of 1969. The 1989 Category 4 hurricane Hugo was only the second Category 4 hurricane to ever reach the Atlantic coastline. Two important sources of information relative to these devastating storms are the publications:

Frequency and Motion of Atlantic Tropical Cyclones. NOAA Technical Report NWS 26, March 1981.

Tropical Cyclones of the North Atlantic Ocean 1871—1987. NOAA, National Weather Service.

Using these publications, one can determine the past history for a particular coastal area relative to the frequency and intensity of these storms. For example, hurricane conditions can occur in the New York, New Jersey, Sandy Hook coastal area referred to as the New York bight, with the NOAA estimated return periods of:

Winds equal to or greater than 34 kts: 5 years

Winds equal to or greater than 64 kts: 15 to 20 years

Winds equal to or greater than 100 kts: greater than 100 years

Considering only direct hits on the New York bight area, only one hurricane has arrived since 1899. That Category 1 hurricane occurred in September of 1903. A number of other hurricanes have come close to the New York bight area. Most of the hurricanes that generally affect the New England states pass this shoreline to the east. This creates high winds, but sometimes less than hurricane conditions. These higher winds start mainly from the north initially, then swing rapidly counter-clockwise to the west and southwest depending on the hurricane's speed of travel. It is interesting to note that there are no unusual high wind data recorded for the 1938 hurricane for the New York bight. For most New England areas, that hurricane is considered to be representative of the 100 year storm. Hurricanes which have passed to the east of the New York bight are shown summarized below. Some of these hurricanes created severe damage on Long Island and coastal Connecticut, only a few miles to the east of Manhattan Island, New York City.

Mo/Year	Category	Name
9/1903	1 (reached NJ and NY)	—
9/1938	3	—
9/1944	3	—
8/1954	3	CAROL
9/1960	3	DONNA
6/1972	1	AGNES
8/1976	1	BELLE
9/1985	3 (2 further east)	GLORIA

As can be seen, a marina designer can get an abundant amount of information concerning the wind conditions for most local areas from the U.S. Government. Although many of these publications seem dated, most are still available either from NOAA or the local library, and some are revised periodically. These data allow the determination or description of the expected wind conditions as a four part analysis, including the ambient or day to day wind intensities and directions during all seasons, the annual maximum conditions, the longer term maximum conditions, and the extreme conditions that may affect an area. Very often with sufficient data available, it is convenient to categorize the wind study in terms of velocity and direction with return periods of 1, 10, 20, 50, and 100 years. This then becomes an effective tool, a data set with which to determine the suitability of a potential marina site.

5.4 DATA COLLECTION AND USAGE

Surface wind observations are reported at the height or elevation of the particular anemometer installation. These heights vary considerably from place to place and may have changed several times during long term wind data reporting periods. In the early 1960s, most airport anemometer elevations were standardized at between 20 and 25 feet above the ground, depending on the site. Data collected before that time must be checked for anemometer heights, since it can significantly affect the usefulness of the data. After all, wind velocities of 40 knots measured from the top of the Pan Am building in New York City find little application for wave prediction of the surface waters of New York Harbor. There are many other less obvious examples, but the important thing is to know the anemometer height or the data is less useful. The wave height and period prediction equations from the Army Corps of Engineers, *Shore Protection Manual* are based on sustained winds as measured or corrected to 10 meters or 33 feet above the water's surface. Where possible, if the information is available, the wind data must be adjusted from its value as measured at the anemometer's height, to this 33 foot elevation value using the commonly accepted logarithmic decay relationship as given in the *Shore Protection Manual*. This is certainly not an exact process since the measuring site, for example an airport, may be several miles away from the marina site and at a different elevation.

Peak Winds

Historically, peak wind speeds were originally recorded in statute miles per hour as the defined fastest mile, and then converted to knots, or nautical miles per hour. Fastest mile wind speeds, v_f, are defined as the fastest speed in miles per hour at which wind travels one mile, or the sustained wind velocity with the air moving one mile. This term developed from early anemometers which measured the distance moved by the air. The wind was timed by stopwatch, and after the anemometer cups measured an air flow of one mile on an odometer, the velocity was computed. If this was done several times during the highest wind period, a determination could be made of the maximum wind velocity that occurred, hence the term fastest mile. Present practice followed today is that the recorded wind speeds of the anemometer record are visually observed and the extreme velocity or peak winds are reported. These are now considered to be synonymous with fastest mile wind speeds. An alternative method called the fastest observed one minute wind speed, v_m, has been used in the past, and is still recorded as one separate entry on daily surface weather observation sheets. It is determined from the recorded data, but is a specific wind average over a one minute period. This value is lower than the fastest mile type of observation due to the longer period used for its determination. Studies involving both reporting methods during the same wind episodes provide a rule of thumb involving the higher wind speeds. **It is generally accepted that the fastest observed one minute wind speed can be converted to the fastest mile, or peak wind, by adding 10 miles per hour, where $v_f = v_m + 10$.** Although these differences in definition of the recorded data may appear rather subtle and perhaps unimportant, a 10 mph maximum wind difference can make a predicted variation in wave heights of a foot or more.

It is quite important for the marina designer to understand exactly what wind data was measured and how it is being reported. Some of these velocity concepts and definitions are best illustrated by abstracting wind data and comments from the daily surface weather observation sheets of several airport control towers during the 1985 hurricane Gloria. This so-called Category 3 hurricane had winds in excess of 87 knots to the east of its landfall at Fire Island, Long Island, New York. Surface weather observation data sheets for the three local airports for September 27, 1985 show the following maximum recorded winds:

Newark Airport: winds reaching 40 knots from the east, then later peak gusts to 58 knots from the west, with the fastest observed one minute wind speed of 44 knots

La Guardia Airport: winds reaching 55 knots from the east, then later peak gusts to 56 knots from the northwest, with the fastest observed one minute wind speed of 46 knots

JFK Airport: winds reaching 45 knots from the east, then later peak gusts to 50 knots from the west, with the fastest observed one minute wind speed of 46 knots

As can be seen, the fastest observed one minute wind speed very nicely converts to the fastest mile by adding 10 miles per hour. However, they still remain peak wind speed measurements, with short duration times. These peak winds may provide an almost instantaneous force on marina structures and boats. Peak gusts may easily and quickly blow out the windows of a building, but must act on a boat or a dock system for longer than just several seconds in order to have them respond. The force of the wind, as tempered by the weight or mass of the object involved will determine how fast the object reacts to the wind. This is the famous Newtonian law of physics that acceleration (A) equals the force (F) divided by the mass (M), or the more familiar, $F = M A$. In other words, the buildup of the speed of movement of a boat tied in its slip, when acted upon by a wind gust, will be governed by the size or weight of the boat, the boat's windage area, and its underwater drag. So normally, it takes more than just a short period maximum wind gust to have important effects on boats or dock systems. A longer term, so-called mean or duration averaged wind speed must be considered in determining the wind's structural loading on an object. Unless actual time response constants are known for various floating objects, dock systems or fixed structures within a marina, an engineering judgement or estimation must be made. Once values of these time responses are determined for the important components of the marina, the duration averaged wind speeds can be calculated from the peak winds using the specific methods and equations described in the *Shore Protection Manual*. The method is quite simple and straight-forward and allows the determination of a so-called averaged fastest mile wind speed over the duration length of the event based on the measured peak wind velocity as reported in the wind data. These adjustments to the recorded and published peak wind data will then provide the proper wind velocities for application to wind loading determinations.

Peak wind velocity adjustments are also necessary for wave height predictions. A ten minute high wind gust does not produce very large waves, and a one minute gust has little or no effect on the water. The maximum wave heights to be produced from a given wind condition depend upon the average wind velocity versus the time over which it blew at that velocity. The peak measured wind data must first be converted to duration averaged wind speed over a time period commensurate with the available fetch distances and necessary wind duration for the generation of the wave fields. The process for the time averaging of the wind in wave height and period prediction involves an iterative technique in which a prediction is first made using the fastest mile and a minimum duration is determined commensurate with the fetch distance. A time averaged wind velocity is then computed for that first trial duration in accordance with the same time averaging method for wind loading from the *Shore Protection Manual* cited above. A new wave height, period and new duration is forecast and a new time averaged wind speed determined. The process is repeated for each new duration as determined until convergence occurs with a changing duration-averaged wind velocity of less than one knot.

Extreme Wind Velocity Predictions

For the extreme wind occurrences the duration averaging method finds less general applicability. This is because for the most part, the extreme wind predictions are statistically determined and partially speculative. However, NOAA provides a slightly different methodology for smoothing the hurricane wind fields into longer term sustained winds. It would be nice to believe that conditions for a marina site will not exceed those maximum wind conditions already determined from recorded historical data. However, for many coastal regions it is entirely possible that higher extreme wind conditions will occur, as was the case in the Carolinas with hurricane Hugo in 1989. Sustained extreme wind velocities above the maximums already measured can occur under hurricane conditions. Total destruction of the boating facility, and the boats that it contains is a reality for consideration. There is no way to predict when or if these excessive velocities may occur in the future or to realistically design in-the-water structures to withstand them. In some cases of exposed coastal marina sites, offshore hurricane conditions with high velocity sustained winds are a concern due to the relatively large waves and surges that will occur and travel to the vicinity of the site. A marina designer must be prepared for the inevitable by at least speculating on what the highest return period winds may be and understanding what one may be dealing with.

By definition, the peak predicted winds of an arriving hurricane are called sustained winds which NOAA defines as winds lasting at least one minute. As has already been discussed, a considerably longer time period than one minute is necessary for these extreme winds to operate on the water's surface and generate the extreme waves. Predicted hurricane wind velocities must be treated differently for wave height and period determinations than the actually measured historical peak winds discussed above. This is partly due to the way hurricane force winds are predicted and the fact that these storms have forward motions ranging up to 60 knots. Consider a hurricane without winds, as a simple mass of air moving forward at 60 knots. Automatically, this so-called wind-less hurricane has 60 knots of wind as it moves over a location. Any wind that it possesses on its own as the storm's air spirals towards its center or eye, must be adjusted for the storm's movement speed. Its forward speed may add to its sustained wind velocity on one side of the storm, where the circular winds are in the same direction as the storm's movement, or reduce their intensity when they oppose its direction of movement on its other side. With a counter-clockwise swirl to these storms in the northern hemisphere, this forward motion effect creates the strongest hurricane winds on the right hand side of their trackline as they approach the coast.

Hurricanes tend to move slower in the lower latitudes of the eastern United States and Gulf coast and faster to the north. For the Atlantic coastal area, the long term average for hurricane forward speeds has been 30 knots, with the 1938 hurricane traveling at 60 knots. For a storm moving at 30 knots, with the highest winds typically within 60 miles of its center, winds from one steady direction will rarely last more than two hours. On the other hand, at a forward speed of 15 knots, these winds may blow from one direction for twice that time. In order to generate the extreme waves, the wind must sustain its highest velocities from one direction for a period of time. Basically, the storm's direction of passage, as well as its speed as it approaches and passes an area, determines its ability to produce the largest extreme waves. Unfortunately, the direction of passage is the most difficult factor to predict. However, for the extreme winds of a hurricane, sustained wind durations of one-half hour is all that is necessary to produce fully developed sea conditions in most sheltered or partially sheltered areas with shortened wind fetches.

How sustained or steady the winds will be for the production of waves depends on the forward speed of the hurricane. The faster the forward speed of the storm, the steadier the higher winds will be. In accordance with the procedures outlined in the *Shore Protection Manual,* these longer term sustained winds are generally calculated, as being equal to 0.865 times the one minute maximum winds, defined sustained winds, for a stationary hurricane. It can generally be assumed that the measured one minute maximum wind values also include the effects of the storm's forward motion. Therefore, if the hurricane is moving, an average factor of 0.5 times the forward speed of the storm is assumed to be included. This is obviously a compromise since it cannot be known which side of the hurricane will reach the site, i.e., should the full value of the hurricane's forward speed be added or subtracted? Following accepted procedure, for a storm movement speed of 30 knots, a marginal Category 3 hurricane with defined one minute sustained winds of 96 knots may produce the longer term sustained winds of 85 knots for developing the sea conditions associated with it ((96 kts − 15 kts) × 0.865 + 15 kts). A fast forward speed of the storm tends to smooth out the lulls in the wind pattern but at the same time it tends to shorten its duration from one direction. Higher gusts of less than one minute may also occur. Since the extreme wave buildup duration for most restricted marina sites is relatively short, the sustained winds, as determined by the method described above, should be used for predicting the extreme wave heights and periods without further alteration or averaging.

To some extent, this manipulation of the hurricane wind predictions is very simplistic and naive. However, it does provide some mathematical tools and methods to help deal with and give order to some of nature's most awesome phenomena. Hurricane wind conditions are based on statistical classifications and speculative wind data from the National Weather Service. Within any hurricane category, maximum wind gusts, sustained winds and sustained directions cannot be predicted with any certainty. The wind velocities to be used for design purposes can only be based on the best available information. No advice is inferred or offered as to what the extreme wind conditions may be during any hurricane event and there are certainly no guarantees. Historical data in terms of hurricane wind velocities and landfall locations are available. There is also some data to produce statistics relating to the hurricane's probability of return with various maximum wind intensities. For those extreme wind and wave height predictions, the question is always what should be protected against and which hurricane return period is to be used? These questions are not easily answered.

5.5 USES FOR WIND DATA

Relative to the construction of a marina, the greatest interest generally lies in the so-called normal annual maximums and the historical maximum wind conditions that have visited the area. This information tells what to expect during a normal year, and tells what the maximum conditions may be for the design process. Knowledge of the extreme conditions that are possible is also useful for the selection of design intensity levels. Extreme conditions may simply set the upper limit of the protection offered by the marina's perimeter or may be totally ignored. These concepts are very subjective because the historical maximum conditions may be too severe to design for without even considering the potential extreme conditions. Acceptable wave limits within the marina and wind damage failure percentages must be established with each level of wind and wave return period prediction to allow the marina developer to decide the level of protection that the marina will be warranted for. It may be for simply annual maximum conditions, or with the proper natural protection, perhaps for the 100 year return period storm. With the proper use of the available data, generally, wind predictions can be made that describe the marina site from daily occurrences to conditions that will occur once a century. Once the full spectrum of wind conditions has been carefully delineated, the environmental effects of these winds, such as boat and dock wind loadings and the wave fields that will visit the marina are the next items to be established.

6
Wave Climate

Water waves are often difficult to deal with, discuss or understand without invoking rather esoteric mathematical equations. The object of this chapter is to provide basic word descriptions of waves and their associated phenomena, as well as to expand some relatively simple wave concepts into more useful rules of thumb and relationships for the marina designer or engineer.

When we look out over the water and see waves, our eyes view a surface manifestation of ridges and valleys. The rowing chant of the old New England whaleman stated it slightly differently in the refrain "hill and gully riders—hill and gully" as they pulled in unison on their oars in pursuit of the whale. The waves being viewed may have been created by the wind, but what caused them doesn't really matter. Water waves are basically the same whether they are locally created by the wind, a tide rip, or a boat wake, or are swells arriving from a distant storm. Characteristically there is a hill part, called the crest and a gully part called the trough. Seldom if ever does one exist in the absence of the other, and seldom does only one wave exist. Generally there will be a series of them, called a train of waves. Sometimes these trains are made up of many waves having identical characteristics, sometimes the trains will consist of a mixture of different sized waves. There are specific names applied to different parts of waves to help communicate better and explain and discuss them more clearly. Figure 6-1 shows a series or train of several waves along with the wave part names which apply. For now, the discussion will remain focussed on wind produced waves, with perhaps an occasional reference to boat wake waves, as appropriate. Wakes and wake waves require separate treatment due to their importance to marinas and are covered in greater detail in Chapter 7. However, the discussion of the physical aspect of waves in this chapter generally applies to wake waves also.

6.1 THE ANATOMY OF WATER WAVES

This section explores the fundamentals of water waves, the names of their parts, how they develop, their behavior and ultimate fate. To start with, wave height is probably the most important element of the wave's anatomy. This

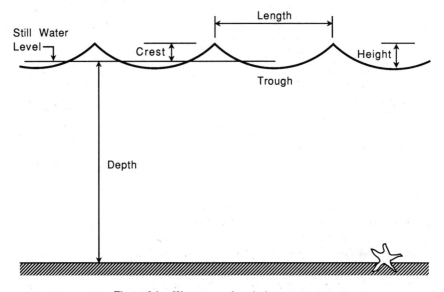

Figure 6-1. Water wave descriptive part names.

height of the wave is the distance from the top of the wave to the deepest or lowest part of the trough as measured vertically and is the distance usually given in answer to, "How big are the waves out there?" **Within bays, estuaries and river environments (the most probable sites for marinas), wind produced wave heights for fair weather conditions may be 1 to 2 feet. Moderate storms may produce waves with heights of 3 to 5 feet, while extreme conditions or exposed locations may have waves 6 to 8 feet in height. Open ocean areas, far from land will produce 30 to 50 foot high waves, with the highest ocean wave to have been reported measured at 112 feet.** Wind produced waves exist in a random or chaotic manner. The sea surface shows a mixture of many waves, varying in height, including large and small waves. This mixture of waves is referred to as a random sea, where if it is desirable to talk about the average wave height, it would be necessary to systematically measure all the waves and compute an average height statistically. Generally we speak of significant wave heights. This is the wave height that one usually reports when asked to judge how high the waves are. It is the wave height that a person's eyes and brain working together, observe and conclude when looking over the sea. It is the wave height or sea condition logged by mariners since records have been kept in the form of ships logs or diaries. It is what any observer sees as they look out over the water, regardless of their technical training. Scientific study of waves has shown that this observed significant height turns out to be the average height of the highest one-third of the waves that are observed. An observer tends to see the larger waves because they are intimidating. Putting it more simply, an observer's mind is already starting to exaggerate how big the waves are before they even have an opportunity to write it down or tell someone.

Another vertical distance which is commonly used is the crest height. The wave crest is that part of the wave which is above an imaginary line representing the still water level, in the absence of waves. It is the so-called ridge that is observed in a wave field as opposed to the trough or valley portion. The crest height is measured up from this imaginary still water level, while the depth of the trough is measured downward.

The width of the crest and the width of the trough are measured horizontally along this imaginary still water level line. From Figure 6-1 it is apparent that the wave crest is higher than its trough is deep. In addition, the crest width is shorter than its trough. Thus the wave's appearance is peaked. The wave is separated from its neighbor by a flattened and somewhat elongated trough. The distance of this separation is referred to as its wave length. Since the wave can travel, there is a factor of speed, or how fast it can go. Wave travel can be determined and quantified in feet per second or knots or any velocity units. When dealing with waves however, the most common practice is to talk about how long it takes for a wave to pass-by or how long it takes for one wave crest to be replaced by a following wave crest. This length of time is called the wave period and is generally expressed in seconds. **Fair weather, moderate wind conditions in bays and estuaries commonly produce 2 to 3 second wave periods. Storm conditions may produce 4 to 5 second waves. Boat wake waves will vary, ranging upward to 4 seconds and sometimes beyond. Deeper water conditions along the open coastline may produce 6 to 8 second waves, while deep ocean waves fall into the period category of 9 to 15 seconds, with the longest extreme storm produced waves reaching 20 second periods.**

The precise shape of the wave is open to some controversy. However, it is generally accepted that the crest width is shorter than its trough width, the crest height is greater than its trough depth relative to still water level and that the crest has more of a point at its top with the trough being more rounded. The crest sometimes tumbles over, generally in the direction of the wave's travel. That is called breaking. If it only occurs at the very top of the crest it is referred to it as a whitecap. If whitecaps are observed on the water, two things are known. First, the winds are at least 12 to 15 knots, and second, the waves are in a building mode. They will become higher as time passes.

Thus far the focus has been on the water's surface to describe the wave. This is natural since that is generally all that is ever seen of a wave. However, a wave has a complicated motion structure within the water that produces it. What is seen on the surface is visible because that is where the top of the water is, and where the observer generally is. The wave that is seen on the surface exists below the surface also, and generally with the same characteristics. Below the water surface the waves are difficult to see, but they can be felt. Here the water moves in unison with the surface wave, at least down to some limited depth. Furthermore, the wave length continues to exist, the period as seen on the surface, and the direction of travel remains the same. It is part of the same wave that is seen on the surface, except that with increased depth in the water, its height diminishes. The water movement below a passing wave crest has a circular path and moves in the same direction as the wave's crest on the surface. As the trough passes, the water movement reverses itself and moves backwards. Putting these motions together, the water below a wave moves in a cyclic manner with each passing crest and trough, but never quite returning to its starting point, as indicated in Figure 6-2. If you follow the path of a water particle underwater, its net movement is down wind. This motion tends to produce a slow wave current in the water moving in the direction of the wave.

How deep do these waves extend or at what depth do their motions become negligible and effectively die out? The simple answer is: at a depth which is one-half of their wave length. So if you can measure the distance between each crest, the waves effectively don't exist at a depth which is greater than half that distance. If that distance turns out to be greater than

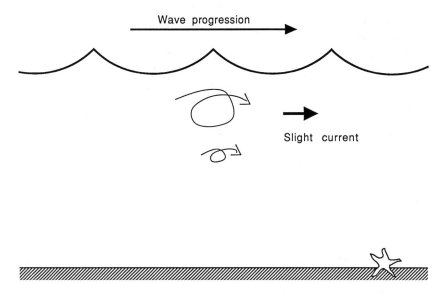

Figure 6-2. Cyclic water movement under a passing wave showing the production of a wave produced water current.

the water's depth, then the wave can feel the bottom. Effectively, a part of the wave is now being dragged along the bottom and the water reacts to this. Subtle changes occur in the wave's underwater structure. The subsurface water motion changes from a mostly circular pattern to a somewhat flattened ellipse as the water depth becomes increasingly shallow.

The important point to remember is that the wave does exist at depth. It is just not a surface phenomenon. Wave effects on marinas must therefore be addressed from the water's surface down to a depth where the wave dies out enough so as to not be a problem. This may mean wave protection may be needed for a marina which extends down into the water for a distance of perhaps one-third of the wave length of the important waves. This could mean protection that extends all the way down to the bottom of the marina, if the water is shallow.

Wave Development Process

Wind waves are produced as the wind blows over the surface of the water. Basically there is one fluid, the air, rubbing against another fluid, the water. The result of this relative movement is the formation of waves on (and in) the water. The stronger the wind blows, the longer it blows, the higher will be the developing waves. The process is quite complicated and not completely understood. Descriptions of the process tend to be over simplified. However, they are based on observations of and experimentation on what appears to happen in the wave production process. In relatively simple terms, starting at one end of a bay or similar water body, waves will form as a wind starts to blow over the water. Since the newly formed waves move downwind, they remain under the influence of the wind that created them. The wind may freshen, i.e., increase in strength, or remain steady. As the developing wave remains under the wind's influence, more energy is added to the wave by the wind, and the wave's height increases. The developing wave's period and wave length remain fixed while the height increases due to the energy added to the water's surface by the wind. As long as the wind continues to blow and the wave remains under its influence, the wave's height increases. Eventually the wave grows to a height which cannot be supported by the water and it collapses or breaks. Stating it another way, it becomes too steep and falls over. This steepness is measured by the wave's height (H) relative to the wave's length (L). **When that ratio, H to L becomes greater than 1 to 7, the wave tumbles, producing foam which is referred to as a whitecap.** That is not the end of the wave, however. Nature simply says that the best way to overcome being too steep is to reform the wave with a longer wave length (and slightly lowered height). Now the reformed wave is stable, and has some room to continue to grow higher, since its longer wave length can now support it. As this reformed wave travels down wind, it continues to grow, becomes unstable and breaks and reforms again. The process is repeated many times producing many whitecaps and a building sea.

The higher the winds and the longer they can blow on these developing waves, the larger the waves will become, both in terms of height and wave length. This is an important concept. The stronger the winds are, the higher the waves may grow over a short distance and a short period of time. Conversely, lower winds may form the same high waves if allowed to blow over longer water distances. Since these waves are moving downwind, their travel distance is increasing from the point of origin. The greater this travel distance, the higher and longer the waves will be. This travel distance is referred to as the fetch distance. Remember, the longer the wind fetch distance between your marina site and the far shore, the greater is the wave height potential. A long fetch distance under low wind conditions can create some medium sized waves. A short fetch distance under high wind conditions can create some medium sized waves. However, a long fetch distance under high wind conditions can create some very troublesome waves.

Water depth can play an important role in the development of the waves. In deep water, the wind's energy can be distributed down into the water for many feet. In shallow water, the wind's energy gets trapped; the energy is concentrated between the surface and the bottom. Shallow water allows more rapid wave growth while limiting the wave's ultimate height. It also will cause the waves to have shorter wave lengths, thus producing steeper waves more quickly in the building process. This quickly turns a shallow water body into a foam covered sea of rapidly building and steep breaking waves.

Wave Behavior and Fate

One of the best ways to gain some knowledge and useful information on wind waves is to examine what factors are necessary to produce waves of various sizes. Table 6-1 shows some wave heights, wave lengths and periods that will develop under a number of fetch distances, starting with just one-half of a nautical mile. Shown for the selected wind velocities in knots are the wave's significant height in feet, its period in seconds, the approximate time in minutes that it takes to build that height over the fetch distances shown, and the resulting wave lengths. The wind velocities shown are the persistent wind velocities, acting uniformly and lasting at least as long as the duration shown to be needed. Remember, these wind velocities are not simply the peak velocities measured, but are the longer term persistent average winds discussed in Chapter 5. The table was prepared for water depths of 15 to 20 feet over the wind fetch distance and can be further expanded and extended by adding 10 percent to the wave heights for average water depths of 25 feet, and subtracting 20 percent from the wave heights for average water depths of 10 feet. No adjustment for depth need be made to the wave periods shown. The

Table 6-1. Generalized wave characteristics based on wind fetch distance.

Wind fetch distance in nautical miles

Wind	0.5nm	1nm	2nm	4nm	6nm	10nm
10kts						
Height	0.3ft	0.4ft	0.5ft	0.8ft	0.9ft	1.2ft
Period	1.1sec	1.3sec	1.7sec	2.1sec	2.4sec	2.9sec
Duration	<30.min	46.min	73.min	115.min	152.min	180.min
20kts						
Ht.	0.6ft	0.9ft	1.3ft	1.8ft	1.9ft	2.3ft
Per.	1.4sec	1.8sec	2.2sec	2.8sec	2.8sec	3.2sec
Dur.	<25.min	35.min	50.min	60.min	70.min	90.min
WL	10.ft	16.ft	22.ft	34.ft	40.ft	50.ft
30kts						
Ht.	1.1ft	1.5ft	2.1ft	2.6ft	3.0ft	3.5ft
Per.	1.7sec	2.1sec	2.6sec	3.0sec	3.3sec	3.8sec
Dur.	<20.min	30.min	45.min	50.min	55.min	75.min
WL	14.ft	22.ft	32.ft	44.ft	56.ft	68.ft
40kts						
Ht.	1.5ft	2.1ft	2.7ft	3.5ft	4.0ft	4.5ft
Per.	1.9sec	2.4sec	2.8sec	3.4sec	3.7sec	4.2sec
Dur.	<15.min	25.min	30.min	35.min	45.min	60.min
WL	18.ft	26.ft	40.ft	56.ft	68.ft	80.ft
50kts						
Ht.	2.0ft	2.8ft	3.5ft	4.4ft	4.9ft	5.4ft
Per.	2.0sec	2.6sec	3.0sec	3.7sec	4.1sec	4.6sec
Dur.	<15.min	25.min	30.min	35.min	40.min	50.min
WL	20.ft	32.ft	46.ft	66.ft	78.ft	92.ft

significant wave heights which are given are a statistical determination of the highest one-third of the expected waves. This height may be further adjusted to provide an estimate of some of the highest random waves to be expected. If the heights as shown are increased by 30 percent, an approximate value for the highest 10 percent of the arriving waves can be determined, and for the approximate value of the highest 1 percent of the arriving waves, increase the heights by 65 percent.

If the waves are to continue to grow during their travel of the fetch distance, the winds must continue to blow during that time. The wind must blow for a sufficient time to allow the waves to remain under its influence over the entire fetch distance. Included in Table 6-1 is the minimum duration time necessary for these waves to build to the heights shown. Looking at the 10 knot wind produced waves of Table 6-1, after a fetch travel distance of 0.5 nautical mile, the waves have reached a height of 0.3 feet. If the fetch distance allows, they will continue to grow as they travel downwind and become 0.5 feet in height with 1.7 second periods after 2 nautical miles, taking 73 minutes to reach that state, and so on up to the 10 nautical miles shown. Note the fact that the wave growth rate slows with an increasing fetch distance. For the 10 knot wind example used above, the wave's height after 6 nautical miles is within 25 percent of its final height after the travel of an additional 4 nautical miles. Since the waves arrive at a waterfront site having traveled from the far shore under a steady wind, even if the wind continues to blow for a longer period than shown, the waves will become no higher. Any wave height increase must now be caused by an increase in the wind's speed. The waves are considered to be fetch limited in height.

Once waves are removed from the wind conditions that produced them, they will continue to travel basically in the same direction. They will slowly decay in height. The distance that they may ultimately travel will depend on their period or wave length. The longer the period, the further the waves will travel before dying out. They no longer are building and therefore are not breaking and reforming. They tend to form more uniform wave trains, with more rounded crest tops, and are referred to as swell. The deep ocean waves with very long periods will last and travel as swells for many miles, sometimes thousands of miles from where they were produced. In these cases, these long period swells may arrive in a bay or harbor marina site and cause what is called surging. Shorter period waves die out more quickly, and the locally produced swells are noticed for only brief periods as the wind dies out.

Winds may shift during the course of a storm event. When this happens, the waves produced from one direction will die out, while a new series of waves builds from the wind's new direction. Often, this results in what is referred to as cross sea conditions. The marina site may experience several wave trains arriving simultaneously, but from different directions. The physics of water waves dictates that these arriving cross seas will reinforce each other. This means that when two crests coincide, meeting from different directions, their heights will add together. Thus two wave trains crossing, one with a height of 2 feet, another with a height of 3 feet will produce a momentary wave height of 5 feet.

6.2 ESTIMATING WAVE ACTIVITY AT THE SITE

A critical design element for a marina or waterfront site is the determination of expected wave activity and wave behavior on arrival at the site. This is important from a first impression viewpoint as well as from a more detailed determination which may be made later. In this section, the emphasis is on making estimates of the arriving wave activity at a site, in order to allow a relatively quick determination of the feasibility or development problems that may have to be faced. In order to gain some knowledge of potential wave problems at a site, there first must be a fundamental understanding of how waves behave once they are formed, what happens as they reach the more shallow water of the site, and from what direction will they be arriving.

Table 6-1 illustrates an interesting feature about wind waves in a coastal or harbor environment. **If you can measure or estimate the height of the wave at a particular site, the height of the wave in feet is approximately equal to the period of the wave in seconds.** This is particularly true for the winds of 20 knots and greater. Conversely, if the period is measured, some knowledge of wave height is obtained. If these observations can be made under average wind and storm conditions, one can begin to gain wave height and period information that is useful when examining the potential use of various wave attenuation devices. Some basic information on the height of the storm waves for a particular location can also be obtained from the base flood elevations determined for all coastal areas by the United States Federal Emergency Management Agency (FEMA) flood insurance studies. The study pamphlet published for each waterfront community generally lists the 10, 50, and 100 year return period storm produced still water flood elevations. It also lists the elevation of the top of the wave crests as the base flood elevations upon which the insurance rates are established. **The height difference between the still water and base flood elevations represents the wave crest height, which by rule of thumb is 70 percent of the actual maximum wave height expected to arrive at the shoreline.** So if the 100 year return period storm wave crest elevation is 4 feet above the still water flood level for the same storm, then waves 5.7 feet in height (H = 4 feet /0.7) are the largest waves expected. Furthermore using the period and height relationship in this case, these waves may have periods of 5 to 6 seconds. Thus, with some simple, but admittedly opportune observations, or a visit to the local town planner's office, the estimated wave picture and potential problems can begin to take shape.

Further information can be gained from the relationship of the wave length (L), the period (T), and the speed (C) of the wave's travel across the surface of the water. The period of any wave, once formed, will not change. If the wave breaks and reforms, that is a reformed wave with a new period. However, if that doesn't happen, the period remains constant. **The relationship $C = L/T$ tells all. Since the period (T) must remain unchanged as a wave travels, then if it slows down, that is if (C) becomes smaller for some reason, then the wave length (L) must get smaller.** As a wave enters shallow water, it slows down, its (C) becomes smaller. The more shallow the water, the slower the wave travels and the closer the waves get to each other. Shallow is a relative term in this case. Waves exist not only on the surface, but as previously described, extend down into the water a distance which is equal to one half their wave length. So if a wave is travelling in water whose depth is half a wave length deep or less, then the wave will be affected by the bottom.

Its forward speed will be slowed, and it may also have its direction altered even if it is still under the building action of the wind. **A rule of thumb equation that is helpful in determining the wave length from a knowledge of period is $L = 5.12\ T \times T$. Furthermore, if this wave length is divided by 3, the expression tells how deep the important aspects of the wave's influence may extend in the water column. To determine the wave's speed in knots under general conditions, the equation $C = 3 \times T$ can be used.**

So given a situation where the wave's period is 3 seconds, its approximate wave length will be 46 feet and its speed of travel as an individual wave will be 9 knots. However, the physics of water waves in relatively deep water dictates that the overall wave energy travels at only half this speed. Thus, if a lake or bay is 9 miles across, 3 second waves produced out in its center will take at least one hour to subside after the wind dies out. This 3 second wave energy will have left the center of the lake or bay and travelled the 4.5 nautical miles to the shoreline in that length of time.

Carrying these approximations further, a wave will break, or become unstable when the depth of the water in which it is travelling becomes more shallow than 1.3 times the wave height. **Putting that a different way, the wave will break when the wave's height is 78 percent of the water's depth.** Thus a wave 6 feet in height cannot exist in water as shallow as 8 feet. In that case the wave will break, spill, and reform to a lower height. This is a useful rule of thumb which is often overlooked in determining the maximum wave that may approach a marina site. Many large water bodies, fresh or salt, can produce waves which head towards a lee shore with considerable heights. However, the depth of water surrounding the marina site may be shoal enough to provide a filter for wave heights and thus limit the maximum wave height that can arrive regardless of how hard the winds may blow.

An additional piece of information is the crest length of the wave. This crest length is measured as a distance along the top of the wave's crest, at a right angle to its travel direction. **For wind produced waves, the length of the wave crest is generally equal to a distance which is approximately 3 times the length of its wave length.** This does not apply to wake waves which obviously have different origins. But in the case of wind produced waves, it means that in some applications, it can be assumed that an arriving wave crest has a limited length and may not simultaneously strike a breakwater, bulkhead, seawall, or floating attenuator along the structure's entire length. **Another rule of thumb linked with the wave length is that the on-coming wave cannot sense the presence of an object in its path which is smaller than one-quarter of its wave length.** In other words, a one foot diameter pile will sense, feel or be impacted by a 20 foot long wave length wave. The pile may be moved, bent or broken depending on the height of the wave. However, the wave will not sense the presence of the pile, and passage of it will not affect the characteristics or power contained in that wave by any measurable amount. This has important implications for what can be proposed for blocking waves around the perimeter of a marina. In the example given, an element 5 feet across would be the minimum effective element size that would be needed to reflect significant portions of this wave which has a wave length of 20 feet.

Any object that fulfills the one-quarter wave length criteria will create a reflected wave, or at least a portion of a reflected wave. If the object extends to the bottom and intercepts the wave totally, a wave of equal height will be reflected back off the object. This means that a 5 foot high wave, 100 percent reflected, will produce waves in front of that reflecting surface which will reach 10 feet in height, and be made up of the 5 foot high arriving wave combined with the 5 foot high reflected wave. This reinforced wave will take the form of a series of standing waves, non-traveling, but creating a series of waves which will reach out several wave lengths from the object, as shown in Figure 6-3. This reflection may occur with waves arriving at an angle with the reflecting surface and create chaotic conditions as a pattern of isolated wave peaks called clapoti. The arriving wave crests are reflected off a surface at an angle equal to, but in the opposite direction from their arrival angle. These

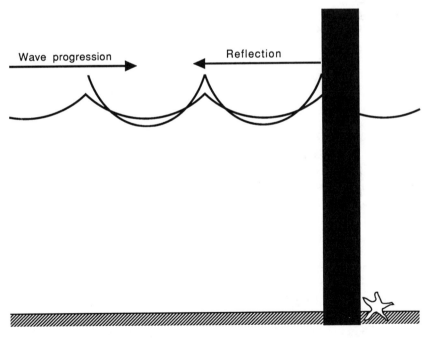

Figure 6-3. Wave reflection off a vertical wall showing the additive effect of the incident and reflected wave crests.

peaks are superposed wave crests, adding together as the wave crests cross. This reflected energy, and its patterns are predictable and controllable through design. However, such waves must be recognized as a formidable problem, especially where a marina is designed with smooth vertical bulkheads or walls within its perimeter. Some semi enclosed marinas with limited openings and vertical walls on four sides have big problems when wave energy enters the marina and then batters about, reflecting from one wall surface to another until dissipation occurs, many minutes later.

Wave Directional Changes — Refraction

Waves which sense the bottom of shoaling water will be steered by the changing depth of the water in which they are travelling. This steering of the waves may cause a concentration or focussing of the wave's energy, or conversely, its spread over a larger section of the shoreline, depending on the shape of the bottom. This process of the waves being steered by the bottom depths is called wave refraction and results from the wave's propensity to turn towards shoaling water. Relatively simple wave transmission physics, borrowed from the theories of light and acoustics, allows predictions to be made of how waves will be altered in their travel directions. This is done by following the path of a wave orthogonal, a line perpendicular to the crest of the wave in the direction of its travel, from the wave's deep water source to its final breaking point with a continuous monitoring of wave height and wave length alterations along its travel path. The wave's height can be altered by the amount of refraction that occurs, and by bottom friction, with some height increases being possible, depending on the bottom shoaling conditions. In general, the longer their refracted travel paths and the greater their directional refraction away from their down-wind direction, the lower will be their arrival heights.

The actual arrival direction and heights of these waves have important influences on the orientation, type, and structural support of any wave attenuation devices or protective structures which are to be used for a marina or for a revetment along a shore. In many cases waves may arrive at a marina site from directions which differ greatly from the directions of the winds that produced them. They may arrive at marina sites from around what would appear to be protective headlands or other obstructions. The process of refraction can steer the waves into some very unexpected directions. Figure 6-4 shows an example of a wave refraction chart for a potential marina site in Raritan Bay, New Jersey. The west side of the headland was expected to be well protected from long period storm waves arriving from the east. As can be seen in Figure 6-4, these waves effectively refract completely around the headland and into the western bay area of the site, reaching the shoreline with considerable wave heights. After the completion of a full refraction

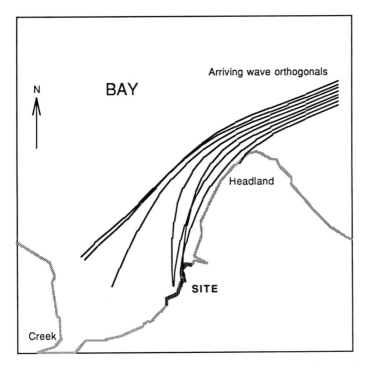

Figure 6-4. Wave arrival refraction diagram showing easterly waves managing to arrive at the western side of a so-called protecting headland.

analysis, the marina designer was alerted to this unexpected perimeter protection problem. Following the wave height and period predictions, a wave refraction analysis is the next important study to be made. Without an examination of the potential refraction of waves, their arrival directions and ultimate heights at a proposed marina location, the fact that they may arrive at all may remain undetected. This could be a disaster after the waterfront development is completed and the storm waves start arriving.

The water depth plays an important role in the amount of refraction and wave height alteration that will occur as the waves travel over the bottom. The object of a wave refraction study is to determine the possibility of refracted waves and to investigate the height alterations that may take place as these waves travel and refract towards the marina site. As can be seen from Figure 6-4, the orthogonal lines represent the direction of travel of the wave crests. The spreading of these orthogonal lines indicates wave crest divergence and commensurate wave height diminishment. Convergence of these orthogonals indicates wave height increases, with the arriving waves actually larger than the original offshore waves. As the waves travel downwind towards the shore, refraction and bottom friction will turn the waves towards shoal water causing their direction to alter from being directly downwind. This refraction and bottom friction can cause their arrival direction at the shore to be different from the wind direction, and their heights may sometimes be lowered or increased from the straight line of travel predicted heights. Estimates of these wave height changes caused by refraction can be made if the patterns of the converging or diverging wave orthogonals can be determined. The increase or decrease in refracted wave heights can be estimated from simple proportional measurements between the orthogonals. If the distance, for example, between arriving wave orthogonals decreases by 50 percent, the wave heights may be increased by that percentage over their deeper water or unrefracted heights. If the divergence opens the spacing up by 50 percent, the wave heights may be halved. Such information is seldom easily obtained and must be left to the more sophisticated techniques of computer programs and models. However, some information about arrival angles can be easily determined with the use of a navigational chart or other bathymetric mapping of the site and its surrounding bottom depths.

One assumption that is necessary here is that the wave feels the bottom at a depth of half its wave length. Another assumption is that the wave will arrive at the shore with its orthogonal nearly perpendicular to the shoreline. Figure 6-5 is a chart which gives a general idea of what happens to refracted waves as

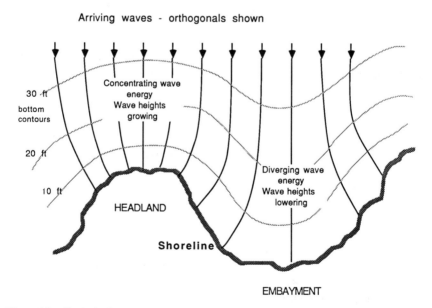

Figure 6-5. Typical refraction patterns for arriving waves at either a projecting headland or an embayment indentation along the shore.

they arrive from offshore at either a headland or an embayment. Note here that off a headland the waves turn towards the closer shoal water and converge, creating higher waves, while within an embayed area the same wave tendency to turn towards shoal water causes divergence of the orthogonals which will result in a decrease of the wave height. The better location for a marina becomes obvious. Sketches of arriving wave directions and inferred refraction can be made on a navigational chart of the bottom depths of the site, and using these basic assumptions and refraction principles. These sketch diagrams can provide some gross information about the arrival angle of the longer period waves, perhaps even some information on converging or diverging wave orthogonals. If a proposed wave barrier can be located on the schematic drawing, some idea can also be obtained of the impact angle of the arriving storm waves.

Summary of Wave Estimates to be Made

Most of the information that has been discussed thus far can provide a somewhat reliable picture of the wave climate that may affect a marina or waterfront development site. Armed with a navigation chart of the area in which the site is located, along with some local knowledge of the normal and storm wind systems, and perhaps the Federal Emergency Management Agency flood levels, it is possible to make an estimate of the normal wave heights, periods, wave lengths, crest heights and crest lengths that may affect the site. This can include the storm waves as well as the extreme waves that may have to be contended with, their speed of buildup and decay, wave height breaking limits at the site, and the depth down into the water column to which it may be necessary to have wave protection extend. In addition, it is also possible to estimate the arrival direction of the waves as well as the refracted arrival direction with a possible indication of wave height changes due to converging or diverging wave energy. All of this general information can now be used to start to obtain an idea as to what will be necessary at the site as far as wave protection is concerned. With this information, coupled with further general analysis on wave protection devices, some back of the envelope estimates may be made relative to the potential methods and gross costs for protecting the marina.

6.3 DETERMINATION OF DETAILED WAVE ACTIVITY FOR DESIGN

As discussed, a good deal of general information relating to an estimated wave climate near a potential marina or waterfront development site can be easily obtained. At the very least, this information will suggest how much wave protection may be needed for normal and extreme weather conditions, what kind of protection may be considered and what the extrapolated costs may be. Further work must be carried out by professionals, educated and experienced in these engineering sciences. Eventually, detailed information will be necessary relative to the forces on any wave protection structures, the effectiveness of these structures, and the degrees of protection that they may provide. In order to make these more detailed determinations, design calculations, computer simulations and professional judgements must be made. Rules of thumb must then be abandoned for the more precise determinations that will allow the marina to be constructed properly and to withstand the expected storm environment at reasonable costs. Furthermore, fairly accurate predictions will have to be made relative to the wave climate that will exist within the finished marina. In order to determine that, accurate determinations of the area's wave climate must first be made.

Wave Climate Determination

The object of any marina siting study or design analysis usually relates to the wind produced wave fields. The approach is to determine, based on historical wind data, the temporal wave climate that is expected to normally exist on an annual basis in the vicinity of the proposed site. In addition, the extreme waves expected to visit the site within a 1, 10, 20, 50, and 100 year return period, including extreme hurricane wind conditions, may also be examined. These are important factors for proposing alternative solutions for solving any existing wave problems at a marina site and properly designing the system to withstand the expected extreme waves. The wave heights, periods, their arrival directions, and especially how often they may occur, are important economic considerations of the problem. Ultimately, a cost/benefit ratio must be established in order to decide which wave regimes may be ameliorated within the financial constraints and which ones may be too expensive to block entirely, but must be attenuated to some extent. The approach usually taken is to examine the worst case to be expected, based on the best available wind data and predicted weather parameters. Often, the worst case produces design conditions which are not feasible or logical criteria on which to base a marina project. Expense, space availability, or a desire to protect the planned facility up to an arbitrary storm intensity limit may cause a reduction in the level of an acceptable design. In that case, shorter storm return periods with their commensurate lower wind velocities, wave heights, and flood water levels should be contrasted against their probability of occurrence and the damage level that may be acceptable. This will lower initial capital costs while increasing potential future costs for damage repairs and insurance.

The maximum annual wave climate defines the waves that will have to be either accepted or ameliorated in order to provide a normally safe and comfortable marina basin, while the extreme wave conditions represent the structural design criteria needed to successfully design and install the marina structures for the target storm intensity. These are the waves that the marina is expected to survive by definition.

The procedure to be followed carries the analysis from the determination of the frequency of the common waves that the location will experience, through the determination of some of the maximum waves normally expected, to the extreme wave potential, especially in hurricane prone areas. The analysis is then completed with an examination of the wave spectra. This examination tends to highlight the relationship between those waves that the site will be protected from and those waves that must be survived.

Wave vulnerability for many sites stems from two sources referred to as the near field source and the far field source. The near field source consists of those waves which can be locally produced within the environs of the estuary, bay or harbor which contains the marina site. The far field source consists of those waves produced by winds over the greater distances of the connecting ocean or other larger body of water. These far field waves would be expected to travel from the direction of their generation, potentially into the local body of water which contains the marina site, through the process of shallow water refraction. These far field waves are expected to be long period, possibly surge type waves affecting the site. The possibility of their arrival at the site is a function of their wave lengths, their access to the bay or harbor and their direction of initial approach from the far field sector. These far field waves may arrive at a marina site simultaneously with the locally produced near field waves. Their combined effects must be included in any wave climate study.

The wave data is developed in two parts. The first set of data, based on historical data, pertains to the normal annual wind conditions expected at the site. They represent the so-called normal wave climate to be expected. A second set of wave data, determined for the site, is from the historical maximum wind conditions. These wave data represent the maximum waves expected for the site based on 1, 10, 20, 50, and 100 year return maximum wind periods, as desired, from any exposed direction regardless of any previous experience. All wave heights should be generally determined for mean high water conditions near the proposed site. This generally provides for slightly higher waves than will be determined for the more shallow, mean low water depths.

Histograms for Annual Wave Climate Conditions

The first set of wave data to be analyzed, based on historical data, pertains to the normal annual wind conditions at the site. They represent the so-called normal annual wave climate to be expected. The percentage of occurrence of wind velocities from each 22.5 degree sector (i.e., N, NNE, NE, etc.)

affecting the site, for January through December should be determined from the available wind summaries. Wind velocity categories may include: 0-6 knots, 7-10 knots, 11-16 knots, 17-21 knots, 22-27 knots, 28-33 knots, 34-40 knots or the maximum wind velocities for each month if known to be higher. These data are further combined to produce wind data for trimonthly or seasonal periods, representing the expected annual conditions as recorded. These trimonthly wind summaries can be used to produce average wave climates for the site for each of the periods starting with January-February-March or any starting combination desired. Thus, four, trimonthly periods of data can be prepared in order to produce histograms of locally produced wind wave conditions at a site. Histograms may be prepared on a monthly or multimonthly basis, depending on weather variability at the site and the detail desired. However, experience has shown that trimonthly periods are usually adequate and annual histograms are sometimes appropriate. The data used in preparing these histograms include the wave height and period predictions and the frequency of occurrence of the specific wind conditions described above.

Examples of the trimonthly histograms are shown as Figure 6-6. These histograms show the percentage of time for each trimonthly period that wave heights are normally expected to be less than given values. The wave heights used for the histograms are the significant waves. The highest 10 percent of

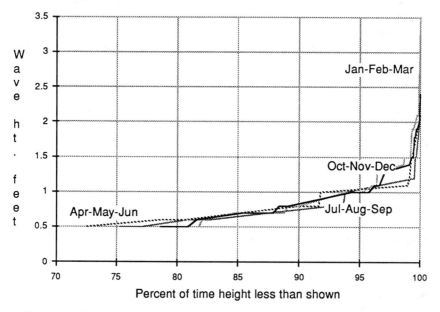

Figure 6-6. Wave height histograms for trimonthly periods of the annual wave climate.

the predicted waves may be slightly higher than shown. Since these waves are predicted from long term wind data, they must be considered the norm for any particular year. Anomalous years will occur where the waves may be greater or smaller. The histograms provide information on the frequency of occurrence of wave heights and periods during an average year in the vicinity of the marina site. In general, the curves of Figure 6-6 show that there is little difference at this particular site in wave heights expected throughout the different seasons. In this case an annual histogram would be sufficient. Wave heights during the winter months generally reach 3.5 feet. For approximately 70 to 75 percent of the year, wind waves with heights no greater than 0.5 feet generally occur. The site is apparently not a particularly rough site relative to the wind waves from its exposed directions. On balance, the site does not experience large waves very often, and with the exception of perhaps a few storm episodes each normal year, only moderate wave protection would be considered necessary.

Figure 6-7 shows the annual wave period histogram. This histogram combines all seasons and shows that for approximately 98 percent of the time, wave periods are less than 3.0 seconds. The periods drop off rapidly becoming less than 2.0 seconds for approximately 90 percent of the time. Here again, the data indicates that only modest protection would be necessary for these annual wave episodes most of the time.

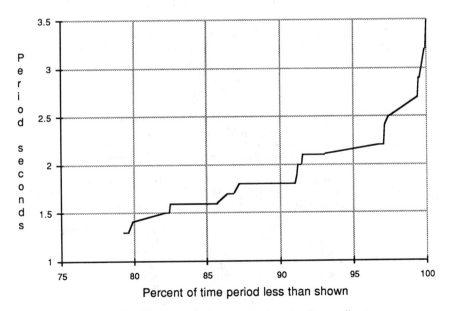

Figure 6-7. Wave period histogram for the annual wave climate.

Maximum Waves

The second set of wave data is developed from the historical and predicted maximum wind conditions which have been determined for the site. These wave data represent the maximum waves expected for the site based on the desired 1, 10, 20, 50, and 100 year return maximum wind periods from any exposed direction. This is a conservative assumption since these wind velocities will generally not have been observed from all of the site's exposed directions. However, for the establishment of the maximum waves for the various return periods, along with the selection of a design wave for the project, this is considered the best procedure.

In order to obtain more precise information on these waves, particularly their arrival directions at the site and any potential convergence of their orthogonal paths, a refraction analysis is also carried out. The normally produced near field wind waves may have periods too short to be significantly affected by bathymetric refraction caused by the short fetch distances and their direct arrival at the site. However, the longer period far field maximum waves will generally be examined for the effects of bathymetric refraction. A manual method is clearly described in the *Shore Protection Manual*. Unfortunately this method is very laborious. The fastest way to carry this process out is through the use of a wave refraction computer program, several of which are also referenced in the *Shore Protection Manual*. The end result of the wave prediction and refraction work is the development of a table or series of tables showing the details of the maximum expected waves and their arrival directions. These tables may take several forms depending on the particular site. One type of maximum wave table is shown as Table 6-2. It is based on two storm return periods, 20 and 100 years, and on the potential simultaneous arrival of both the far field waves and the

Table 6-2. Maximum waves generated from simultaneous arrival of far field and near field wave regimes.

	Maximum Waves Arriving at the Marina Site Perimeter			
	20 year return period		100 year return period	
Wind Dir degrees T.	Ht in ft Hcomb/Hs	Signif period, surge in sec.	Ht in ft Hcomb/Hs	Signif period, surge in sec.
090	<1.0	6.3 surge	1.0	7.3 surge
113	2.9/1.9	2.0, 5.5 surge	4.0/3.1	2.4, 6.0 surge
135	3.3/2.3	2.3, 4.7 surge	5.1/3.6	2.7, 5.4 surge
157	4.7/2.5	2.4, 4.4 surge	7.4/3.9	2.8, 5.1 surge
180	4.8/2.6	2.5, 4.7 surge	7.6/4.1	2.9, 5.4 surge
202	5.5/2.8	2.6, 4.9 surge	7.8/4.1	2.8, 5.4 surge
225	<1.0	5.4 surge	1.0	5.9 surge

near field waves, shown listed as the significant wave height in feet (Hs) and the significant period in seconds, and their occasional combination (Hcomb) heights as they may superimpose together momentarily. Shown also are the periods for the near field waves, generally the shortest, and the far field waves are shown as a surge.

Another maximum wave table may result in only the use of the near field waves because of the marina's somewhat isolated location. Table 6-3 summarizes the maximum near field waves arriving at a proposed marina site, including the highest 10 percent waves in feet (H_{10}), the significant wave heights in feet (Hs), and the significant period in seconds (T), all shown as $H_{10}/Hs/T$, for the given return periods.

Following the preparation of these return period maximum wave tables, some information on the level of site protection that is needed or desired can begin to be addressed relative to appropriate cost considerations.

Wave Energy Spectra

A wave energy spectra graph, showing wave height squared versus wave period, can be very helpful in determining potential protection devices for the site. The spectra should be prepared for the annual wave climate as well as for the higher return period waves. It represents the highest waves expected at the site for each wave period determined from the histograms and the maximum storm return period waves. The spectra provide information on the period associated with the higher waves that the site should be protected from. At the same time it can indicate that it may be impractical to protect the site from the longer return period extreme waves. Figure 6-8 shows a sample spectra presenting wave conditions for a marina site. The annual waves have an energy peak at 3.5 seconds which also corresponds with the highest ambient waves as well as the longest. At many sites this may not

Table 6-3. Maximum waves from near field wave regime.

Direction degrees T.	Maximum Near Field Waves			
	1 yr return $H_{10}/Hs/T$	20 yr return $H_{10}/Hs/T$	50 yr return $H_{10}/Hs/T$	100 yr return $H_{10}/Hs/T$
022 (NNE)	4.2/3.3/3.2	5.6/4.4/3.5	6.0/4.7/3.6	6.4/5.1/3.6
045 (NE)	3.1/2.4/2.5	3.7/2.9/2.6	4.1/3.2/2.7	4.3/3.4/2.8
067 (ENE)	2.5/2.0/2.2	3.0/2.4/2.3	3.4/2.7/2.4	3.6/2.8/2.4
090 (E)	2.1/1.6/2.0	2.8/2.2/2.2	3.1/2.5/2.3	3.3/2.6/2.3
112 (ESE)	2.1/1.6/2.0	2.8/2.2/2.2	3.1/2.5/2.3	3.3/2.6/2.3
135 (SE)	3.1/2.4/2.6	4.1/3.3/2.8	4.5/3.6/2.9	4.7/3.7/3.0
157 (SSE)	3.1/2.4/2.6	4.1/3.3/2.8	4.5/3.6/2.9	4.7/3.7/3.0
180 (S)	5.4/4.2/3.7	7.2/5.7/4.0	7.4/5.8/4.1	7.6/6.0/4.2

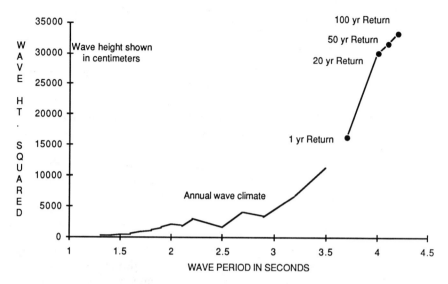

Figure 6-8. Arriving wave energy spectra for a proposed marina site.

always be the case since longer period, but lower height waves may also arrive. For the longer storm return periods the energy peak shifts from approximately 3.7 seconds for the assumed 1 year return period storm to only 4.2 seconds for the 100 year return period storm with only a slight increase in height. These relatively short period storm waves are attributable to the fact that the available fetch distance for the site is limited by the harbor configuration. In this case, the wave spectra results suggest that protecting the marina site from both the 20 and 100 year return period waves may differ little in cost and should probably be considered.

6.4 A DISCUSSION OF WAVE PREDICTION DIFFICULTIES

The wave climate development and analysis as discussed in the previous section should be carried out in accordance with accepted practices and guidance as delineated in publications produced by the U.S. Army Corps of Engineers, U.S. Naval Facilities Engineering, NOAA, and various other federal publications as well as numerous journal publications of the American Society of Civil Engineers. Experience has shown that there are many problems and pitfalls to be avoided in preparing a complete wave climate analysis for a waterfront site. The discussion that follows, attempts to highlight some of these prediction difficulties and suggests some methods that may be followed while using the predictive framework and data provided by these government and professional references.

Using the published monthly wind values for an area by direction and frequency, along with the available effective fetch distance, air stability factor and water depth, the wave climate can be determined. All wave heights, wave lengths and periods may be predicted by using the equations given in the *Shore Protection Manual* for transitional depth waves.

The significant wave height and period, in the form of Hs and Ts or the equivalent Hmo and $0.95 Tm$ as discussed in the *Shore Protection Manual*, are the basic results from the commonly accepted predictive methods. **The significant wave height represents the average of the highest one-third of all the waves for a given set of prediction assumptions. The expected height of the highest one out of ten waves, H_{10}, is statistically determined to be 1.27 times the significant wave height, with the highest one out of one hundred waves, H_1, determined to be 1.69 times the significant wave height.** These relationships are the commonly used wave height distributions, accepted by the public sector, specifically the National Oceanic and Atmospheric Agency, the Army Corps of Engineers, the U.S. Navy and the Federal Emergency Management Agency, as well as by the private scientific and engineering sector. These wave predictions can all be made using the so-called hand method with the nomograms provided in the *Shore Protection Manual,* but in general, it is more precisely done using a digital computer program of the equations.

The next level of wave analysis relates to the extreme wave conditions, and is carried out in much the same way. The problem in this case is that the prediction of the wave heights and periods produced from extreme winds of hurricane velocities, blowing over shallow water with reduced fetch distances remains to be fully substantiated. Hurricane wind conditions are based on statistical classifications and wind data from the National Weather Service. Within any hurricane category, maximum wind gusts, sustained winds and sustained directions cannot be predicted with any certainty. Few if any verification comparisons during intense hurricanes have been carried out in the field for the commonly accepted wave equations given in the *Shore Protection Manual*. The determination of the proper fetch distances, water depths, as well as the value of the air stability factor to be used can greatly influence the wave height predictions with quite large and perplexing variations. There are other more advanced and sophisticated wave prediction computer models, however, the *Shore Protection Manual* equations remain the best predictors generally available for the purposes of engineering design along the shore.

Certain information must be obtained in order to construct a proper wave analysis for a marina site. The categories of information are limited in number, but unlimited in the volume of information within each category, and the interpretation that must accompany it. In order to carry out the wave height and period predictions it is necessary to have knowledge of:

Time averaged wind velocities and duration near the site

Fetch distances over which the winds will generate the waves

Water depths at the site and over the fetch distance

Air stability factors as they may vary seasonally

Each one of these elements affects the prediction of the waves in terms of height and period. Each one possesses some uncertainty in its determination, and in its application, and collectively they can result in wave predictions that must be carefully interpreted and evaluated. It cannot be over emphasized that wave height prediction, especially extreme wave height prediction remains, unfortunately, more of an art than a science. Each one of these elements is briefly examined in the following paragraphs in order to show how misuse or misinterpretation of the predictive process may lead to either disastrously weak structures or costly overbuilt ones.

Time Averaged Wind Velocity

For use with the *Shore Protection Manual* wave prediction equations, the fastest mile wind speed, or peak values usually obtained from short time observation periods as reported by the National Weather Service must be converted to time averaged wind speeds, at 10 meters of elevation (33 feet) over a time period commensurate with the available fetch distances and necessary wind duration for the generation of the wave fields. These procedures are outlined in the *Shore Protection Manual,* but what is not included is how to interrelate the time averaged wind, the duration over which it is to be averaged and the resulting wave height and period. The process for the time averaging of the wind for wave height and period prediction involves an iterative technique in which a wave height and period prediction is first made using the fastest mile. A minimum duration is determined in the process commensurate with the fetch distance, and a time averaged wind velocity is then computed for that duration. A new wave height, period and new duration is then forecast. The process is repeated for each new duration as determined until convergence occurs for the changing wind velocity average.

The final result produces the corrected wave producing wind velocities, adjusted for the required duration and fetch distance, along with the prediction of the wave fields.

Wind Fetch Distance

Some method is generally used to determine the fetch distance over which the wind will blow to produce the wave field. For the extreme wave, most marina sites are fetch limited, which tends to negate consideration for the duration of the winds. For example, with a wind fetch distance of three-quarters of a nautical mile, an 80 knot wind needs to blow for less than half an hour in order to develop a fully arisen sea condition. A fully arisen sea is one which can get no higher no matter how long the wind continues to blow at that speed. Many marina locations are sheltered from several different directions and often have only one or two major fetch distances to worry about. Often this major fetch direction faces down a restricted or narrow harbor configuration where the actual wind fetch distance that will generate the waves is less than the available straight line fetch. This actual or useable wind fetch distance must be determined using some acceptable method.

The procedures of the *Shore Protection Manual* prescribes either the straight line fetch distance to the opposite shore or the use of a nine small angle radial method. In the nine small angle method, a series of radial straight line distances upwind from the site are averaged and the result becomes the useable fetch distance from that direction. The size of the angle to be used between the radials is not standardized but is usually assumed to be 3 degrees. Previous editions of the *Shore Protection Manual* gave a procedure described as the summed cosine method and coined the phrase effective fetch. Other scientific literature also describe this procedure or a variation of it which is called the cosine-square method. The predictive equations in the more recent edition of the *Shore Protection Manual* under predict the waves if the effective fetch method of the previous edition is used and testing has shown that the use of straight line fetch was a better predictor. Some of the testing was carried out in a narrow but long water body indicated to be Lake Erie. This tends to reduce the validity of the test results for small water bodies such as bays and harbors. From experience, there remains a strong suggestion that for narrow coastal water bodies of limited size, the straight line fetch assumption may be overly conservative. The presently accepted procedures for long narrow water bodies can best be described as blurred. By way of example, the following wave height predictions are made for 70 knots of wind over a 3.5 mile straight line distance for a long narrow harbor and compared to the nine radial alternate method for fetch determination. Both are suggested as acceptable methods in the *Shore Protection Manual:*

Method	Fetch distance nautical miles	Wave height in feet H_{10}	Wave period in seconds
Straight line distance	3.5	9.5	4.4
3 degree radials	1.8	7.3	3.5

As can be seen above, how the fetch distance is interpreted can increase the predicted wave height by 30 percent and the period by one second. These are significant differences, especially when considering the forces that they will produce. The forces will vary as the square of the wave height and thus the computed forces may differ by 70 percent.

Water Depths

The sensitivity to depth for the wave prediction equations indicates that the science is somewhat more secure in this case. The effects of bottom friction and water column depths on waves appears to be better understood and documented. What generally is lacking here is the proper determination of the average water depths over the fetch distance and appropriate application of what the water depths will be during hurricane conditions. Since the strongest Gulf and east coast winds are associated with hurricanes, the level of hurricane severity also sets the expected water level rise, or tidal enhancement. For a Category 1 hurricane, water depths may increase over the norm by 4 to 5 feet, and with a Category 3 hurricane by 9 to 12 feet. If a marina site has wind fetches over mean low water depths averaging 10 feet, in an area that has a 5 foot mean tide, and if a Category 1 or 3 hurricane were to arrive at the worst time, that of high tide, the following wave height predictions for 70 knots of wind over a 2 mile fetch distance will result:

Water depth in feet	Wave height in feet H_{10}	Wave period in seconds
10. MLW	5.7	3.5
15. MHW	6.6	3.6
19.(+4' tide)	7.1	3.6
24.(+9' tide)	7.4	3.7

Here can be seen a marked difference in predicted wave height for given water depths. The differences get greater as the tidal height enhancement is increased by the storm intensity. The interesting aspect of the prediction in this case will relate to the force calculations where, although the waves are smaller during a low tide storm episode, the resulting wave forces in the shallow water may be the largest, along with the potential for breaking waves adjacent to a structure.

One last comment is in order here. Most sites have varying depths over their wind fetch distances. As a wave grows in height, water depth plays an increasing role in its development towards the end of its fetch distance of travel. **A common practice is to use an average depth for wave height determination which is representative of the last 25 percent of the wave's travel and growth distance as it arrives at the site.** This later travel distance is at a point where the waves have grown and nearly fully developed and the depth of water will become more influential than when it first starts out.

Air Stability

Air stability relates to the efficiency of momentum transfer from the air moving over the water to form waves. This energy transfer is greatly enhanced by a coupling mechanism called turbulence. If the water body is cold relative to the moving air above it, air turbulence and the efficiency of energy transfer to form waves is somewhat suppressed. If the water is warm compared to the air, turbulent interchange of momentum is more easily effected across the boundary layer between the two fluids. What results is either larger or smaller waves for a given wind condition. If the water is warm and the air is cold, larger waves will result than if the reverse is true. The air stability factor, as specified in the *Shore Protection Manual* can be easily computed and then applied as a multiplier for the wind velocity. The result can either lower or raise the design wind velocity and therefore the wave height and period predictions. Again by way of example, using 70 knots of wind over a 3.5 mile fetch and a 10 degree F. spread between the water and air temperature:

Condition		Air Stability factor	Wave height in feet H_{10}	Wave period in seconds
Warm air/Cold water	~10 F	0.9	7.7	4.0
Cold air/Cold water		1.0	8.6	4.2
Cold air/Warm water	~10 F	1.1	9.5	4.4

This often neglected aspect of wave prediction can create enormous differences in the end result. The air stability factor of 1.0 assumes the air and surface water are nearly the same temperature as may occur during the summer along the east coast or most of the time in the Gulf coast area. A factor of 1.1 assumes the air to be colder than the water as in the fall and winter, with a 0.9 factor for warmer air springtime conditions or typical for the west coast of the U.S. most of the time. It is difficult to predict what the air or water temperatures may be during these high wind events. The practice suggested by the *Shore Protection Manual* is that if actual data are not available, then the more conservative air stability factor of 1.1 should be used. As can be seen above, this may cause the predicted wave to increase by almost two feet in height and one half second in period. Careful evaluations must be made for the proper air stability factors used in the preparation of a wave climate.

6.5 DESIGN WAVE DETERMINATIONS

Once the complete wave climate has been carefully prepared, the project's design wave must be determined. This is the wave that sets the standards for what the project is expected to withstand while sustaining minimal damage. Waves greater than the design wave may occur and eventually destroy the project, but it is the design wave that the structures are being designed to withstand and thus it becomes a value judgement that must be carefully considered by all involved. The process of deciding on the design wave involves a lot more than a simple scientific evaluation. The first question to be asked is which return period storm is the design storm, then which statistical wave height should be used, followed by how does that affect the economics of the project, and finally what does that give for protection? The process becomes an iterative decision loop between the scientist/engineer and the waterfront developer. Most waterfront projects cannot be economically built to withstand the 100 year return period winds and waves. Usually, some compromise storm intensity level, determined by the return period, must be decided on and the risks evaluated. Once the storm intensity level is decided, at least for the first iteration, then the design wave is determined.

The design wave criteria from the *Shore Protection Manual,* suggests that the highest 10 percent of the waves from the design storm should be used as the design waves. However, the Navy's *Coastal Protection Design Manual,* suggests that the significant wave heights could form the design waves if at least 20 years of wind data are available. The procedures followed by the Federal Emergency Management Agency uses a wave height which is just slightly less than the highest 1 percent wave. The possible mix and interchange of the criteria, proper wind velocities, fetch distances, water depths and air stability data further exemplify the judgmental aspects of the problem and the decision must be cast in the light of local knowledge and experience.

The design wave selection guidance offered by accepted practice does not consider catastrophic wave episodes. For example, within any hurricane or high wind produced wave train, the statistically determined one hundredth wave has a prodigious height and the one in one thousandth is even higher. It is in general, not economically feasible to design for this contingency, because these statistically predicted waves may not occur during any particular wind wave event. On the other hand, the actual arrival of the one hundredth highest wave or possibly the one out of one thousandth wave, with their potential for coastal destruction must be accepted as part of the natural hazards for the waterfront developer or engineer. From a statistical standpoint, these extreme waves may be the first two waves to arrive or the last two waves to arrive or they may not arrive at all. The end result of any predictions relating to wave conditions cannot be considered as an absolute prediction of what will happen. These predictions, based on commonly accepted practice, can only be considered as a guide, and to aid in the formulation of decisions as they may relate to management or engineering calculations.

Putting aside the other uncertainties, what about the determination of a design wave? Once the design storm has been established then the question of "what design wave?" must be decided. The question implies that there may be a choice even after all of the other problems have been faced. Deciding on which hurricane intensity to use is perplexing enough. However, upon comparing two hurricane produced wave intensities, the choice of direction sometimes becomes clearer. For example, compare the set of design conditions shown below using a 3.5 mile wind fetch and a depth of 30 feet at mean low water:

Wave heights in feet	Category 1 20 year return 70 knots peak +4 ft tide	Category 3 100 year return 105 knots peak +9 ft tide
Most frequent wave	3.8	6.1
Average wave	4.8	7.6
Significant wave	7.7	12.2
Highest 10%	9.8	15.4
Highest 1%	12.9	20.4

As can be seen, designing a marina for a Category 3 hurricane may be quite a challenge. Should the design be for the 100 year storm? If the Category 1 storm is selected, the question of which design wave to use remains open for the structural problem. Should a 7.7, 9.8, or a 12.9 foot design wave be used? Obviously, a realistic approach is required when assessing the degree of protection to be afforded and the amount of damage that may have to be tolerated. The analysis process can be iterative until a final compromise is attained. Basically what all this boils down to is that the design is determined by the economics of the project, and is not necessarily dependent on to the worst conditions expected. This is the difference between coastal or offshore oil rig work where nonfailure of the structures is required, and marina design where it is a question of how much failure can be accepted. The uncertainties discussed above require expert evaluation based on experience, local knowledge, and to a degree, intuition. Once the design wave has been determined after weighing all the facts, a written rationale for the selection must be carefully documented along with an assessment of the risks.

6.6 FEDERAL EMERGENCY MANAGEMENT AGENCY MAXIMUM WAVES

Once the project's design wave array has been determined, it is also instructive to carefully examine what the Federal Emergency Management Agency (FEMA) has had to say about the wave and flood levels for the site. General experience has shown that the wave heights predicted by FEMA are not all that conservative. In fact in many cases wave heights which have been independently predicted are often higher than FEMA's predictions, but seldom if ever lower.

Studies carried out by FEMA determine the base flood elevations which are shown on the rate maps and published in their summary studies for a particular municipality. Base flood elevations are determined from a datum of 0.0 National Geodetic Vertical Datum (NGVD), and represent the top of the maximum wave crest for a particular return period. Very often this means the 100 year flood level, since many FEMA studies do not examine more than one return period, particularly the older studies. The wave heights used by FEMA to establish these base flood elevations are determined in accordance with the publications: *Methodology for Calculating Wave Action Effects Associated with Storm Surges,* which has several revisions, and *Users Manual for Wave Height Analysis,* also with revisions. Both of the above publications were produced for the National Academy of Sciences, Washington, D.C. by the Panel on Wave Action Effects Associated with Storm Surges of the Science and Engineering Program on the Prevention and Mitigation of Flood Losses, Building Research Advisory Board, Commission on Sociotechnical Systems, National Research Council. The procedures in these publications have been fully adopted by FEMA for use in their determination of the rate maps.

The procedure used for the determination of the flood level elevations involves the superposition of a storm wave crest atop the predicted still water elevation. If the predicted wave has a height of 3.0 feet or greater, the Zone is designated a Velocity, or V-Zone. Within this V-Zone, the lowest horizontal structural member of a habitable floor must be above the highest reach of the waves, or above the V-Zone elevation as designated. It appears that if any ancillary appurtenances are attached to these horizontal structural members, they must not pose a threat to the structural integrity of those members should they be damaged or torn away from those members by wave activity.

The predictions for the wave heights used to establish this superposed wave employs an assumed 100 year return period storm with its associated waves and storm tide surge. The wave heights associated with this 100 year storm are predicted by using a series of empirical equations which standardize the method throughout the Atlantic and Gulf coastal regions. The method fundamentally assumes 80 mph (70 knot) winds, blowing in a direction which will bring the largest waves ashore at any particular site. The method is expected to produce a predicted wave height which represents a wave which is slightly below (approximately 5 percent below) the average height of the highest one percent of the waves. It is referred to as the controlling wave height, Hc. The actual height is taken as 1.60 times the height of the maximum significant wave expected at the site. Its determination is not very site specific since the method ignores potential extreme wind conditions, ignores the updated procedures of the *Shore Protection Manual* including wave refraction, and it also uses a generic selection of water depths (fixed at 26 feet) and fetch distances. FEMA does not publish the period of these waves.

FEMA assumes that the wave crest portion of a calculated wave height extends above the Still Water Storm Tide Level (SWL), for a distance of 70 percent of the wave height. Thus, a designated base flood of Elevation 15 with a still water level (SWL) flood Elevation at 11.6 means that the crest of the wave will be at a distance of 3.4 feet above SWL, giving a predicted controlling wave height, Hc, of 4.86 feet or a significant wave height, Hs, of 3.04 feet for the particular return period storm. The analysis provides a good check on the independent wave work as previously described. Needless to say, the crest elevation of a design wave height that is determined below the level of the FEMA base flood elevations should be carefully re-evaluated.

Local base flood levels are generally determined by running one or two coastal flood computer models using historical data for winter storms and for hurricanes. These computer programs determine still water flood level variations over a 500 year time period; the period includes the statistical occurrence of the tidal cycle. In general, winter storms and hurricanes are the cause of the storm surges determined from the computer models. The 10, 50, 100, and 500 year still water flood levels are obtained from these computer runs and they include the probability of flooding occurring during all stages of the tidal cycle. So for example, the probability of the 100 year flood level reoccurring during the 500 year run was determined from the combined effects of both the predicted storm surges and their frequency, but not necessarily using the highest possible surge, and the state of the tide, again not necessarily high water, to give a summed value of the still water flood level that shows up every 100 years. Thus, either a winter storm or hurricane may have contributed to the highest flood levels for the 10, 50, 100 or 500 year return period, depending on how the computer model statistically combined their frequency of occurrence, intensity and duration with the astronomical tidal cycle. So, basically the model does not just assume that the maximum storm surge is just perched atop high tide. Stating it in another way, a maximum storm surge occurring at a lower tide may be the most probable combination for the 100 year flood level, or perhaps it was a lower storm surge added to a higher tidal state. What ever it was, it resulted in the highest still water flood level expected every 100 years. It may take 500 years to cause the maximum combination of the greatest possible storm surge with the occurrence of high tide.

Going one step further, storm devastation also includes the wave fields superimposed upon the tidal flood elevations. The assumed 100 year return period wind remains as stipulated by the National Academy of Sciences as 80 mph or 70 knots, and is treated for wave height prediction as being a continuously sustained wind over the period of wave growth duration. This is a generic wind, set by the National Academy of Sciences, and applies to every location where the FEMA studies were carried out. It is not a site specific wind as should be used. FEMA uses a probabilistic combination of storm surge, tidal cycle and wind wave crest heights that produces the 10, 50, 100, and 500 year base flood elevations which represent the top of the wave crest. Individually, the base flood elevations may not result from either the highest potential floods combined with the highest potential tides combined with the highest possible waves. They represent the highest occurring statistical combination of these physical phenomena. The point is that separately, the waves, winds and flood levels at any site may be individually greater than those used by FEMA to statistically determine the base flood elevations and must therefore be individually considered in the design of a structure placed by the shore. For the purposes of establishing flood insurance rates, these wave height and flood predictions are adequate for their intended use. However, caution should be exercised when using these FEMA wave height determinations for scientific purposes or engineering design.

7
Water Related Site Conditions

Wind and wave activity at a site is a primary consideration, but there are a number of other physical phenomena that may present even more formidable problems that must also be addressed. Basic water site conditions are defined as those existing conditions or situations that a marina designer must often "live with" due to the site's unique location. These basic site conditions may range from man made circumstances to the expected natural situations. Planning a marina begins with understanding and evaluating these existing water site conditions, conditions which may also include boat traffic and the ensuing wakes along with the natural conditions of tides, tidal currents, lake levels, river flows, floods, and ice. The following sections discuss and explain these various phenomena in an effort to alert the marina designer to problems that may arise from unexpected sources that sometimes tend to be neglected.

7.1 SHIP AND BOAT WAKES

A very important site consideration relates to the problem of the arrival of ship and boat wakes. Many commercial interests, as well as state and federal regulatory bodies consider wake production by commercial vessels as the price of doing business on the water. Slow moving craft don't generally cause wakes, but at the same time don't generally make any money either. Commercial craft are usually moving at significant speeds, attempting to get somewhere or deliver a product to some point in the least amount of time. The wakes that they may generate are of little concern and the philosophy is that a marina will either have to live with an already existing wake problem or construct its perimeters so as to eliminate the problem. Recreational craft can generate wakes also, but in their case their speed can be slowed or regulated to the point that they will not create problem wakes. All regions and water bodies differ in terms of the commercial and recreational boat activities and must be analyzed as to the extent of the wake problem for a proposed marina site.

If the proposed marina site is in a busy commercial harbor, then the thrust of the analysis will be to determine the periods and heights of the wake waves which are usually produced. This is then followed by a determination of how best to protect the boat basin from them. If the proposed site is mainly in a recreational boating location, then the analysis must determine the severity of the problem, whether the marina can be designed to co-exist with these wakes, or if some regulation of the speed of passing vessels can be implemented. Many people feel that arriving boat wakes are an infringement on their shoreline ownership rights. To some extent this may be true, but boat wakes are waves, and waves affect the shore from natural conditions also. If boat wakes are generated from vessels going about their business or pleasure at reasonable speeds for the area, the occurrence of boat and ship wakes simply becomes one of these natural conditions that shoreline sites must contend with.

The increased usage of our waterways is a fact. Water taxis, excursion and sightseeing boats, along with a myriad of recreational craft are appearing in greater numbers. Wake production can only be expected to increase in the future. Planning on that increase at the time of marina siting, orientation and design is just good business and may save considerable amounts of money in the future as well as producing an attractive and desirable location for potential marina patrons.

Except for the channel buoy system, maintained by the U.S. Coast Guard, recreational and commercial ship and boat traffic patterns are seldom regulated, and boats travel in all parts of harbors and rivers where water depth permits. There are few boat or ship speed limits generally enforced. Commercial vessels including oil tankers and barges, ferries, high speed water taxis, and excursion vessels of all sizes may use the waterway along with tugboats, fireboats and Coast Guard cutters. They all contribute to the wake problem and are very difficult to regulate just for the benefit of wake reduction at a marina site. The basic concept here is to understand such problems and create a marina design which copes with them.

Vessel Wake Production

If commercial or recreational vessels must travel past a potential marina site, boat and ship wakes will be created that must be examined for arrival height, frequency and direction in an effort to evaluate potential problems that must be considered in the marina's perimeter design. Small power boats of 20 feet and less are quite common in most river, bay and inlet locations. High speed operation is frequently the case, however these small craft generally are not capable of producing very sizeable wakes, particularly when they are in a planing mode of operation. Larger recreational power boats can present problems when operated at their higher speeds. The medium to large sized power boats of the planing hull variety are designed for efficiency at the higher speeds. When these boats are in a full planing mode, only moderate wakes are generally produced. When they are traveling below their planing speed, the hull configurations are such that very significant wakes are produced when traveling at speeds of between 6 and 12 knots. The medium to large sailboats, considered to be displacement type hulls, are usually capable of motoring at 6 knots and in some cases at slightly higher speeds. Their hulls are designed for efficiency at these speeds and sizeable wakes are usually not created as they travel through the water. The trawler type displacement hulls are generally designed to be most efficient between 6 and 8 knots, which places them in the category of sailboats as far as wake wave production is concerned. There is a notable exception to this when these hulls are driven at speeds of 10 knots and greater. Their efficiency drops significantly and their wake waves can become quite large.

Vessels such as ferries, tugs and fireboats tend to produce a considerable wake. For the most part these vessels have displacement type hulls and are often seen moving at close to full, or what is called hull speeds. At these speeds, some hull designs can produce incredibly high wakes, and in some shallow water depths they can produce bizarre single crested wake waves which give the appearance of being walls of water traveling on the surface. These wakes can be very dangerous for other small craft transiting the area as well as a problem for a marina sited nearby. The larger, 250 to 550 foot long freighters, tankers and barges, generally move along channels at moderate speeds of 8 knots and less, usually creating only a slight wake.

Table 7-1 shows the potential wake production for displacement type hulls as a function of vessel speed. The actual wave height produced depends on the shape of the hull, with the more blunt, bathtub hull shapes producing some of the largest waves. In addition, many of the larger recreational planing boats in the 35 to 50 foot range produce extremely large wakes when moving at less than their usual minimum planing speeds of around 10 to 12 knots. Table 7-1 presents a summary of wake periods produced, versus boat speed in channel depths of 25 feet. Boat speeds from 3 knots up to 15 knots are shown. Wake periods produced are shown to range from 0.8 seconds to 5.1 seconds for high speed boats traveling in shallow water.

Figure 7-1(a) shows the anatomy of a displacement hull boat wake in terms of its makeup and travel direction relative to the boat's courseline. For a full displacement hull, traveling at or below its hull speed, the wake produced by the combination of its bow wave and stern or transverse wave travels off to the left and right of the vessel's heading or courseline at an angle of approximately 35 degrees. **An approximate value of the hull speed of a vessel acting**

Table 7-1. Displacement vessel wake production, depths and lengths in feet, wake period in seconds and vessel speed in knots.

Boat Speed	Water Depth	Wake Angle With Course	Wake Wave Length	Wake Wave Period
3.0	25.0	35.3	3.7	0.8
4.0	25.0	35.3	6.0	1.1
5.0	25.0	35.3	9.3	1.3
6.0	25.0	35.3	13.7	1.6
7.0	25.0	35.3	18.3	1.9
8.0	25.0	35.2	24.0	2.2
9.0	25.0	35.2	30.3	2.4
10.0	25.0	35.0	37.7	2.7
11.0	25.0	34.7	46.0	3.0
12.0	25.0	34.2	56.3	3.3
13.0	25.0	33.0	69.3	3.7
14.0	25.0	30.5	87.3	4.2
15.0	25.0	25.5	117.0	5.1

as a displacement craft, not planing through the water, can be determined by taking the square root of its waterline length and multiplying it by 1.3. Using this general rule, a 40 foot boat has a hull speed of approximately 8.2 knots. If it is driven any faster than that speed it is considered to be approaching its critical speed and depending on the depth of water, the 35 degrees of wake travel to the right or left of its courseline decreases towards 0 degrees. It is a very complicated phenomena and difficult to simplify in a few words. Figure 7-1(b) shows an example of this type of wake produced at this critical point. The wake wave is now traveling parallel to the boat's courseline, or at an angle of 0 degrees difference from it as shown. It is about as high a wake wave as a vessel can produce. Within water depths generally found in our rivers and harbors, all vessels which attempt to reach a planing mode of travel, momentarily go through this critical speed and produce their maximum wake. Once a planing mode is reached, the angle of the wake wave with the courseline quickly increases from 0 towards 80 degrees and tends to travel almost sideways from the hull as shown in Figure 7-1(c). At the same time, the

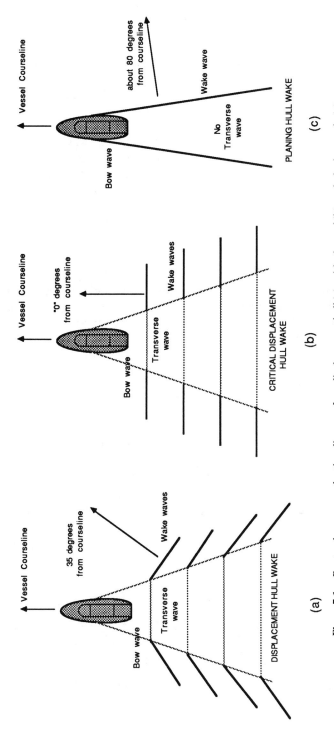

Figure 7-1. Boat wake wave production diagrams for a displacement hull (a), critical speed (b), and planing mode (c).

height of the wake drops dramatically and the stern or transverse portion of the hull wave disappears. The reverse process happens as the boat slows or comes off its plane. Many operators of fast moving planing boats producing a relatively small wake, mistakenly slow down in an attempt to make even a smaller wake while passing other boats or shore facilities. It becomes quite disconcerting to the operator when a very large wake wave suddenly forms astern of his vessel, seemingly out of nowhere, and heads towards the shore. The actual wake wave heights that may be produced relate to specific hull designs, vessel speeds and water depths. The vessel's efficiency of movement, its fuel consumption versus its speed, is often indicated by the size of its wake, or how much energy it looses to the water by producing these wake waves.

Wake Travel and Behavior

The wakes start out next to the boat's hull in their highest form, reducing to half this height at a distance of approximately one wave length from the hull. Their heights are reduced slightly for every wave length of travel, but the number of wake waves in each wave train increases. So as these wake waves travel, they become lower in height but more numerous. They represent the spreading of wave energy over the surface of the water, rather than the dying out of a wake wave as would appear to be the case. Figure 7-2 indicates the

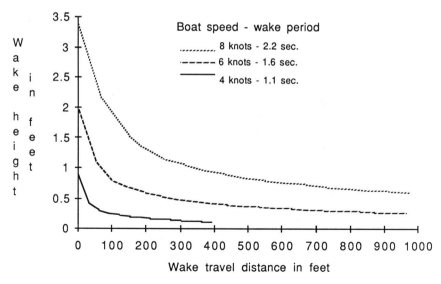

Figure 7-2. Wake wave height in feet versus distance travelled for several boat speeds producing the wales and the wake wave's period.

WATER RELATED SITE CONDITIONS 141

wake height changes of several wake waves as they travel over increasing distances from their source. The greater the wake's period, the further it may travel with significant height. The wake wave periods remain unchanged as they travel, however, as the water depth shoals, their wave lengths will shorten. Furthermore they are subject to bathymetric refraction if they travel in water depths which are less than half of their wave lengths. For the larger wakes with 4 second periods, in 25 feet of water their wave lengths are 85 feet. As they spread towards a shore, departing a channel perhaps and entering water which shoals, a considerable amount of refraction can occur just as it does for arriving wind waves. Obviously, the further the marina is away from the channel, or the slower the speed of the passing ship or boat traffic, the fewer the potential problems.

In order to place all of the above information into perspective, wake wave refraction diagrams Figures 7-3 through 7-5 were constructed for a typical marina site using a representative wake wave. That wake is assumed to be

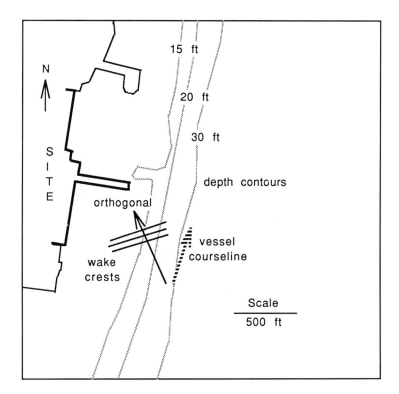

Figure 7-3. Base wake wave refraction chart for a proposed marina site.

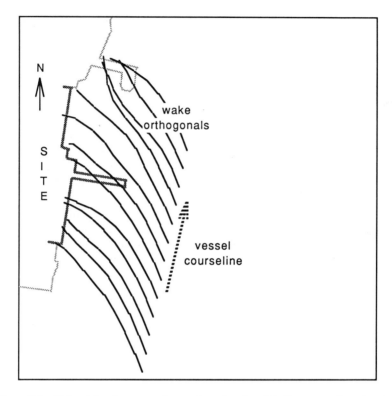

Figure 7-4. Refracted wake wave orthogonals as they travel to the proposed marina site.

produced from a displacement or semi-displacement hull vessel, approximately 40 feet in length, traveling at a speed of 10 to 12 knots in 25 to 30 feet of water. This will produce a wake wave with a characteristic height of approximately 4 feet near the boat's hull and approximately 2 feet in height 100 feet away from its hull. The wake wave will have a period of 3 seconds and will be refracted by the bottom as it travels into shallow water. The purpose of these refraction diagrams is to determine what the characteristics of the arriving wakes at the marina site will be in terms of residual wakes heights and arrival directions. That information can then be used to determine any needs for wake amelioration, or the success of any attenuation devices that may be proposed. For example, two scenarios pose the situation for a northbound and eastbound boat passing the marina site. As their wakes reach the shoal water near the marina they are refracted strongly towards the shore. The wake heights will diminish as they travel due to the distance, refractive effects and the friction of the shoaling bottom.

Figure 7-3 shows a plan view of a potential marina location. A traveling boat's courseline is shown, along with a series of wake waves traveling off at

WATER RELATED SITE CONDITIONS 143

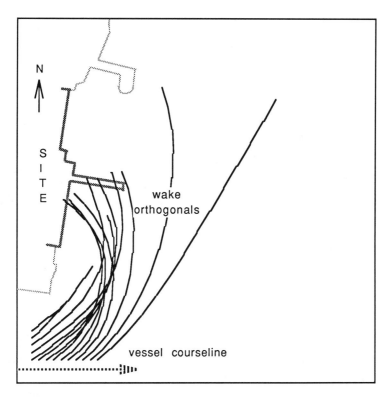

Figure 7-5. Refracted wake wave orthogonals as they travel to the proposed marina site.

an angle of 35 degrees to the left of the boat's heading. The wake wave's orthogonal, a line perpendicular to the wave crest, is also shown indicating the wake wave's initial travel path. Figure 7-4 shows a computer prepared, wake wave refraction chart, depicting the travel direction and fate of the wakes produced from a northbound vessel as displayed by the wake orthogonals. In this case they tend to provide wake wave disturbances which travel along and into most of the potential marina site. Figure 7-5 repeats this scenario for boats heading to the east, emerging from a channel located just to the south of the site. As these wake waves reach the edge of the channel, the shoaling water refractive effects tend to direct most of these waves towards and into the marina area with some serious concentration of wake energy occurring within the marina site. There appears to be a considerable focussing of the arriving wake waves at this marina site that will have to be addressed in the perimeter design. With the distances indicated from the channel, the wakes produced by these passing vessels may decay in height by 50 to 75 percent. Unfortunately, even reduced-in-height wakes with periods on the order of 3 seconds will produce annoying rolling and pitching conditions for the docked boats. This is because most recreational boats are sensitive to these periods, and motion is easily induced. These wake arrival patterns are often neglected by marina designers, but as demonstrated, these patterns can create some unexpected and unwelcome wave conditions at marina entrances and within the marina's perimeter. Unfortunately, this is often discovered after construction has taken place.

7.2 COASTAL TIDES

The sea and its connected arms rise and fall with a rhythm that can best be described as the "heartbeat of the ocean." Generally, the tide is thought of in the context of some rhythmic movement of the sea which is tied to the effects of an astronomical body such as the sun or moon. However, tidal excursions can also be caused by the effects of the wind, atmospheric pressure differences associated with a storm, and even seismic or earthquake generated waves called Tsunamis. Sometimes the different causes of tidal excursions happen at the same time, producing extraordinarily high or low water levels, that are many times the normal ranges associated with the astronomical tides.

Tides exist in other nonmarine bodies of water also. Any body of water, a lake, pond and even the so-called solid earth experience tides. Of course these tides may be only a fraction of an inch, but they do exist. The forces creating tides are pervasive indeed. Every object on the earth responds to these forces, including people. Our weight changes a few fractions of an ounce as we experience these semidiurnal, or twice daily, tide producing forces.

Expectation and an understanding of these occasional high or low stands of water is necessary before any activity should take place along the shore. A knowledge of the highest historical tide levels certainly aids in decisions as to where shoreside support facilities are to be built. A pier or dock that gets submerged monthly, or during storms is not only useless but hazardous to the boats tied up to it. Even more important, if a marina basin must be dredged to a certain depth, or an access channel or canal created, the depth of that dredging must be determined from a good understanding of the extreme tides and their frequency of occurrence. If this is not done, there may be times when a docked boat may be temporarily resting on the bottom or a transiting boat may be unable to maneuver, or leave the marina. Additionally, during an extra high tide, a swing mooring that is too short can be picked up by the boat it tethers and carried off to a new location, which could be a rocky shoal or the open sea.

In some coastal regions where tides are large, pleasure boating is often regulated by the tides. Low tides can allow shallow water barriers to be temporarily placed in a boat's path and prevent access to the sea or prevent return home. So for the safety of boats, marina structures, and even life, it is advisable to know the tides at the proposed marina site. Know what makes them operate and know what their ranges will be under both normal and abnormal conditions.

Ordinarily, the sea defines its own boundaries by a process of erosion or carving of the edges of the land. The normal rise and fall of the tide, produced by the effects of the sun and moon, generally determine where these boundaries will be. From time to time the tidal excursion is augmented by additional happenings such as storms, resulting in the breeching of the normal ocean boundaries and an ensuing coastal flood. When this happens, additional sculpturing of the land takes place, homes are washed away, sand bars relocated, piers and docks submerged and boats lift their moorings and drift off. Such higher than normal tides, as well as lower stands, have happened in the past and should be an expected part of our design criteria.

The normal rhythm of oceanic tides varies from place to place along the coastal U.S. **Although tides usually exhibit what is called a semidiurnal variation of 12 hours and 25 minutes, some areas have only a diurnal variation of twice that or 24 hours and 50 minutes.** These are generally found in the Gulf of Mexico. In still other regions a combination of these tides occur and they are referred to as mixed tides. In this case, major variations occur on a diurnal basis and minor ones occur in between. Figure 7-6 shows typical tidal variations around the U.S. coast. The eastern seaboard is characterized by a semidiurnal tide with tidal excursions differing greatly from one location to another. The Gulf region displays both diurnal tides as found at Pensacola, Florida or a mixed variety at Key West and Galveston. Tides along the west coast of the U.S. are again of the semidiurnal variety with large ranges typical of the northern coastline.

The tidal heights shown in Figure 7-6 are relative to what is called Mean Lower Low Water (MLLW). It does not mean that it is the lowest the tide can get, but only the average low tide stand as determined over a long period of time. Since the twice daily tides are generally not equal, charted depths published by NOAA are referenced to Mean Lower Low Water. That is the statistically determined low water stand using only the lowest of the two daily tides. Of course there is a mean higher high water also, where the highest of the two daily tides has been used to determine the high tide level. The mean range of the tide is defined as the vertical distance between a mean low water and a mean high water. Mean ranges around the U.S. vary from less than 2 feet in the Gulf of Mexico to almost 19 feet in northern Maine and up to 30 feet near Anchorage, Alaska.

There exists an array of tide elevation terms that are important to define and understand. Starting from the highest mean water levels, there is Mean Higher High Water (MHHW), Mean High Water (MHW), Mean Low Water (MLW) and Mean Lower Low Water (MLLW). These terms are simply

146 PART 2/SITE EVALUATION

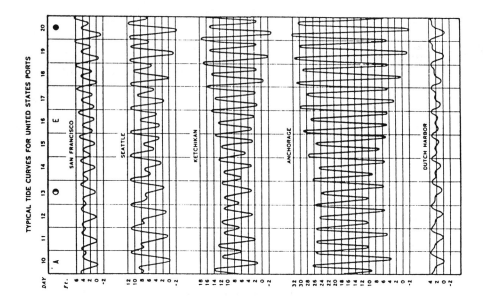

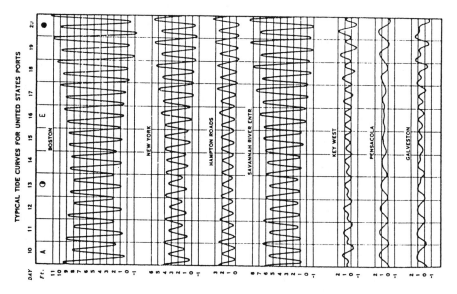

Figure 7-6. Tidal height ranges for the east and west coasts of the United States. After the NOAA Tide Tables.

statistical determinations for a particular locale. For example, the MHHW level of the tide is a long term average of the higher stand of one of the two daily high tides, while MHW is simply the long term average of both of the daily high tides. There is also a correction factor for elevation of the Land Surveyor's datum of the National Geodetic Vertical Datum (NGVD). Back in 1929 when this datum was established around the coastal U.S., the midpoint of the tide, Mean Sea Level (MSL) was used. Since then, the oceans have risen more than half a foot, and the location of NGVD 0.0 datum has become closer to mean low water with the passage of time. It is always best to check with NOAA for the latest correction factor between these two. One must be extra careful in comparing water soundings or land elevations using NGVD datum with those based on MLW or MLLW datum. More than one marina basin has been designed and dredged to minus 8 feet deep NGVD only later to be discovered as having only MLW depths of 4 feet.

How the Tides are Produced

Imagine that the earth is completely covered with an envelope of water. That envelope would assume a shape determined by the earth's gravity and, in the absence of other forces, would form a concentric and uniformly deep covering of water. A uniform sea level would be experienced everywhere on the surface and no rhythmic variations would be expected to occur. There are other forces that will influence the shape of this water envelope and to a much lesser extent, even the shape of the rigid earth. The most important of these other forces are the mutual gravitational attractions between the earth and the moon and to a smaller degree, between the earth and the sun. Because of the sun's great distance from the earth, in comparison with the closeness of the moon, the sun's gravitational influence on the earth is about half that of the moon. These gravitational attractions cause a bulge to occur in the water envelope on the side of the earth facing the sun or the moon. This can be understood as a simple attraction or pull exerted on the water envelope which quickly conforms to the force. A similar bulge occurs on the opposite side of the earth. The explanation for this lies in the realm of astrophysics and is not dealt with here. Every point on the earth is affected by these two water bulges and usually experiences two tides daily. These two bulges are caused by the attractive force of either the sun or the moon. Each of these bodies produces a separate bulge pair and these pairs are independent of each other. Since the moon's gravitational attraction is a little more than twice that of the sun's attraction, the moon's bulge system dominates. It is easy to imagine that when the sun and the moon line up with each other and the earth, both independent bulge systems augment each other and the total bulge which results is at a maximum. This alignment of the earth, sun and moon occurs twice monthly: first, when the sun and the moon are on the same side of the earth, which gives us a new moon, and again two weeks later when the moon is on the opposite side of the earth, which results in a full moon. The tides during this time are referred to as spring tides and they are generally the largest ranging tides for the month. The word spring refers to the fact that the tides spring or rise up to their highest, and has nothing to do with the season. On the other hand, twice monthly the sun's and moon's gravitational attractions are at right angles to each other. At those times the bulges are displaced 90 degrees from each other and the net resulting bulge is at its minimum level because the water is more evenly distributed. The tides during this time are referred to as neap tides. These have a considerably smaller vertical range.

Since the earth turns on its axis every 24 hours and since these bulges are formed and controlled by the sun and the moon, they will change their shape and their arrival times as the relative positions of the earth, moon and sun change. Experiencing all of these movements, the sea level goes from a high stand to a low stand and back to a high stand again, forever repeating itself. For an observer at a fixed point on the earth, the bulge system would appear to be a gigantically long wave traveling around the water covered earth at a speed that brings us under the crest or high point in the wave every 12 hours and 25 minutes on average and under the trough or low point of the wave in between times. The tidal height range of these bulges is only several feet. The height will vary slightly as well as will the time between the tides. These variations are caused by a number of cyclic changes in such things as the distances of the sun and moon from the earth, and the seasonal variation of the sun to the north or south of the earth's equator. One of the more important variations is the moon's monthly close approach to the earth. This is called perigee and occurs each lunar month, or every 27.3 days. The moon's tide creating force is then at its strongest, and what are called perigean (or perigee) tides are experienced. These movements and their combinations cause the tides to be closely related to the seasonal solstices and equinoxes. **There can be many combinations of the moon, sun and earth positions so that basically the tides will take as long as 18.6 years to repeat themselves. That is why when tides are recorded for purposes of making statistical predictions and for the preparation of the tide tables for publication, the recording must be done over a 19 year period, which is referred to as a tidal epoch.**

The "Real" Tides

The earth is not a water covered planet although 71 percent of its surface is water. With the exception of the circumpolar sea just north of the Antarctic continent, there is no place where the oceans completely encircle the earth.

The main tidal effects result from the complete and continuous tidal bulges or tide waves produced in the circumpolar ocean which surrounds the Antarctic continent. Here, as the earth rotates, a tide wave with two crests travels uninterrupted around the earth twice daily.

As this wave travels the circumpolar ocean it disturbs the water and sets up other waves which travel northward into the Atlantic, Pacific and Indian Oceans. These waves have the same period as the wave that set them off on their journey northward. Traveling northward, these tide waves meet up with and interact with the smaller independent tide waves created in each of the ocean basins. This interaction sets up an interesting pattern of tide waves moving across our oceans with heights of only several feet but wave lengths of thousands of miles and periods equal to the periods of the semidiurnal forces that created them. For example, when the crest of the tide wave approaches the eastern seaboard of the U.S. it is almost parallel to the coast with minimal variations from north to south. But in the eastern Pacific the tide wave touches the west coast with a wave that moves from Mexico to Alaska. Table 7-2 shows the similarity of the times of high tide along the east coast in contrast to the south to north sequence of high tide on the west coast.

These traveling tide waves establish the tidal ranges and times of high and low water observed at open coastal areas. Generally, open coastal tides have small ranges and are reflective of the ranges of the tide waves set off initially in the circumpolar ocean. Notable exceptions to this occur in semi-isolated water bodies like the Gulf of Alaska, the Gulf of Maine, and the North Sea, to name a few. In these areas the arriving waves from the southern oceans react with the local independent tide to produce 20 to 30 foot tide ranges along the coasts of Alaska, Maine, Great Britain and northern Europe. No matter what the range may be, or how fast these tide waves travel up from the southern

Table 7-2. Variations of times of tidal occurrence along east and west coasts U.S.

Times of high tide along the east coast		Times of high tide along the west coast	
Sandy Hook, N.J.	1220	La Union, El Salvador	0812
Delaware Breakwater	1302	San Diego, Ca.	0917
Charleston, S.C.	1225	San Franciso, Ca.	1152
Savannah River, Ga.	1228	Halfmoon Bay, Ca.	1212
Mayport, Fl.	1255	Astoria, Or.	1346
Miami, Fl.	1228	Victoria, BC	1450

oceans, their periods are fixed and their heights are determined by the astronomical forces that produced them. That is to say, the semidiurnal period is the same on a world wide basis, as is the occurrence of the ever changing height variations with minor alterations caused by shallow water near our coasts. These same tide waves will now enter our bays, harbors, lagoons, inlets and sounds and will undergo some transformation. These transformations or changes will account for the quite diverse variations in tidal heights and behavior observable around the U.S. and is at least one reason for the marina designer to be fully aware of how these transformations affect the proposed site.

Coastal and Embayment Tides

As the deep ocean tide waves travel towards the land they must move over the more shallow Continental Shelf regions bordering our continents. When this happens the waves tend to slow down and some alteration of the wave form begins to take place. Although the period of the semidiurnal tide is fixed by astronomical forces, most coastal areas experience a shorter interval from low tide to high tide and a longer time between high tide and low tide. For example, in the vicinity of New York, Raritan Bay exhibits a time interval between low and high water of around 6 hours and 5 minutes while between high and low water the time averages 6 hours and 20 minutes. The speed of the tide wave is determined by water depth. The trough of the tide wave has less water under it than the crest and therefore travels slower in shoal water. Thus the crest, or highest point of the wave, can catch up somewhat, resulting in a shorter interval of rising tide and a longer time for a falling tide, while the total period remains 12 hours and 25 minutes. Not only does the wave length of the tide wave shorten due to the shallow water, but the wave height begins to grow. Sometimes the height can grow to several times its deep water size as it reaches the coastline. It will all depend on how far it has had to travel in the shallow water before it reaches the coast.

If the tide wave enters a large embayment along the coast and is allowed to travel unimpeded up that embayment, such as the mouth of the Mississippi River, it will continue to travel as a progressive wave. An observer along the banks sees the wave pass by as the rise and fall of the tide. The crest goes by the observer, heading up river and high tide is noted, when the trough goes by, low tide is observed. An observer at the mouth of the river experiences high tide first, with locations further up river experiencing high tide progressively later as the wave moves up the river. As the river gradually shoals, this progressive tide wave is acted upon by friction causing it to lose energy and height. At some distance up the river, no tide can be detected at all because the tide wave has been completely dissipated.

If the embayment ends abruptly, that is without a river of any significant size at its head, the tide wave is reflected off the far end and sends a wave back down the embayment. Effectively the wave has been turned around and sent back out to meet with the next incoming tide wave. Long Island Sound, San Francisco Bay, Puget Sound, and the Bay of Fundy are all examples of this kind of embayment. Many smaller embayments can also produce reflections of their incoming tide.

When reflection occurs, the embayment no longer has a progressive type of tide with the time of high tide becoming later further up the embayment. The reflected tide wave now interacts with the next incoming tide wave and produces a standing or stationary wave in the embayment. This resultant standing wave will now be twice the height of either the incoming or reflected wave with the two waves simply adding their heights together. No coastal embayment exists which is long enough to support anything other than a small segment of the wave length of the incoming tide wave. This means that with a standing wave, high tide will occur almost simultaneously at every point in the embayment. The same applies for the time of occurrence of low tide. Figure 7-7 shows a typical embayment, Long Island Sound, with a standing wave in it. Two points should be noted here. High tide and low tide all along the shores of Long Island Sound should occur close to the same time. In addition, it appears that at the head or the almost closed

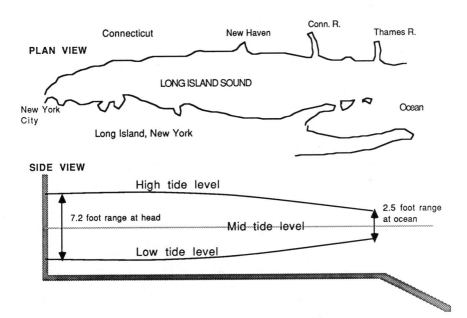

Figure 7-7. A combination top view or schematic plan and side view of Long Island Sound. The side view indicates the high ranging tide at the head of the Sound, near New York City, in comparison to the relatively small ocean tide near its mouth. The effect is similar to the Bay of Fundy tides.

western end, some 80 miles from the sea, the tidal range is greater than at its mouth. The greater range shown is more than a simple doubling of height of either the incoming or reflected wave that produces the standing wave. The tides at the head end range upwards to 7.2 feet while at the mouth they are less than 2.5 feet.

A very important feature of an enclosed or semienclosed body of water is that if disturbed they will oscillate or slosh back and forth with their own distinctive period of oscillation. Tilt a dishpan or tub of water and watch it slosh back and forth. Watch the tides change at the dishpan's rim from high to low and back again. Also note that in the center, no tides exist at all. The period of oscillation of that dishpan is determined by its dimensions. The larger the dishpan, the longer the period. A shallow pan will oscillate more slowly than a deeper pan of the same length or diameter. What you are observing is a free oscillation, much the same as a pendulum would display as it swings back and forth.

Coastal or open ended embayments are not dishpans. But open ended embayments can still have their own free oscillation whose period is determined by the dimension of the embayment. When an open ended embayment tries to oscillate it finds that one end, the open end, is clamped firmly to the ocean. This means that the embayment can oscillate up and down to a lesser degree towards its open end compared to the head or free end as shown in Figure 7-7. Due to the size of some embayments, their natural period of oscillation, if disturbed, can be in terms of hours. The longer the embayment, the greater will be its natural period just as in the case of the dishpan.

Now what can disturb or excite an embayment to oscillate? The answer is the rhythmic tide wave itself. If an embayment has a small natural period compared to the semidiurnal tidal period, little effect takes place. If the natural period of oscillation is near that of the tidal period, very profound effects occur. The natural oscillations of the bay are greatly enhanced by the incoming energy of the tide wave and extreme tide level variations result within the bay. For example, the classic Bay of Fundy tides. Here tides range upward to 30 feet near its closed end. Its natural period of oscillation is approximately 12 hours and 20 minutes. Almost identical to the 12 hours and 25 minutes of the semidiurnal tide.

A less dramatic example of this response is Long Island Sound, located between New York and Connecticut. Here the natural period is 11 hours and 30 minutes. But as shown in Figure 7-7, the head end experiences almost three times the tidal range as the mouth. Both the north and south bays of San Francisco have higher tidal ranges at their heads than near their common entrance at the Golden Gate Bridge. Narragansett Bay in Rhode Island is another example of a standing tide wave. Chesapeake Bay, on the other hand, is bordered by so many rivers that a standing wave never gets set up. These rivers tend to drink up the incoming tidal energy without providing any significant reflection. Its tides are representative of a progressive wave and little or no amplification takes place between head and mouth. Except in those embayments which are dominated by large rivers, some amplification of the tide is to be expected. No matter what the degree of amplification, the period of the standing wave remains that of the semidiurnal tide. Table 7-3 lists some of the major embayments around the U.S. where amplification occurs.

Sometimes embayments have narrow entrances which restrict the incoming tide wave. Regardless of the type of tide that exists inside, tidal ranges outside are sometimes greater than what exists inside. A choking effect occurs because the water level inside the embayment cannot rise fast enough to keep pace with the level outside. The opposite is true for low tide. This choking or throttling effect results in smaller tidal ranges occurring within the embayment compared with just outside. To a degree, this choking phenomena provides some sheltering from extraordinary tidal ranges occurring just outside the entrance. On the other hand such conditions also cause extremely high velocity currents at the entrance constriction due to tide level differences. These currents tend to flow rapidly and reverse abruptly as the tide changes. However, once inside, a mooring or anchorage is better protected from the ravages of abnormal tidal ranges. Generally these choking characteristics are most pronounced in small embayments with restrictive sand bars at their entrances. They can be found all along the coastline.

With a little investigation, one can determine the particular type of tide wave to be expected at a marina site, either standing or progressive, and what amplification or choking of the tide may occur. The type of tide wave then determines when maximum tidal water currents will occur, either at high

Table 7-3. Mean tidal ranges at selected locations for entrance and head conditions.

	Mean tidal ranges in feet	
Location	Entrance	Head
Narragansett Bay	3.1	4.6
Long Island Sound	2.5	7.2
Boston Harbor	9.0	9.6
Delaware Bay	4.1	6.0
Puget Sound	6.7	10.5
San Francisco South Bay	4.1	6.1
San Francisco North Bay	4.1	4.6

water or midtide levels. This has important structural loading considerations for the marina designer. This is further discussed in Section 7.3, Water Currents.

The Prediction of the Tides

Because of hydrodynamic considerations there is no theoretical way to tie together the predictable astronomical forces with the actual response of the ocean in any given area. The approach taken for tide prediction, therefore, uses an empirical method, i.e., the history of tides at any given location. Each moment of the day, all points on the earth are subjected to the forces that produce tides. The times and strengths of the maximum forces are highly predictable. What is not easily predicted is how the water in a particular area will react to these forces. The time and intensity of the twice daily maximum tide force can be calculated for San Francisco Harbor, California for example, but one must look to observations or measurements of the tide in the harbor to see what the water's response will be. Thus through these tide measurements it can be determined how long after the maximum force occurs that high tide will occur and to what height the water will actually rise.

So to find out the characteristics of a certain harbor or coastal region, tidal records must be made and compared with the astronomical forces which cause those tides. Since most of the important force variations repeat themselves on an annual cycle, at least one year's worth of tide data must be gathered. Analysis is performed on the data and the response of the water correlated with the magnitude and times of the tide producing forces. Once that has been accomplished for a particular locale, future tidal predictions can be made based on the expected astronomical forces. Some tidal forces produced by the moon only repeat themselves every 18.6 years, so the longer the tidal record obtained, the better will be the predictions. Generally, to produce tidal predictions a 19 year cycle of data collection, the tidal epoch, is considered standard.

Tide predictions for most areas are published yearly by the National Ocean Survey, National Oceanic and Atmospheric Administration (NOAA) of the U.S. Department of Commerce. These predictions are published in a book called *Tide Tables, High and Low Water Predictions*. One edition covers the east coast of North and South America, including Greenland. Another edition covers the west coast. Most other cruising guides sold by private publishers, as well as information given in the daily newspaper, use these publications as their source.

The *Tide Tables* as this publication is generally referred to, is divided into several parts. One part lists the major stations for which detailed data have been gathered enabling daily predictions to be made. These major stations cover important regions along the coast. A second part of the *Tide Tables* lists substations that provide corrections of tidal times and heights for thousands of specific coastal points and embayments. These corrections are then applied to the daily prediction at the major stations to provide detailed information. Detailed instructions as well as examples are given in these publications, making for easy use of the tables.

Special notes about tidal idiosyncrasies are footnoted on the substation pages of the *Tide Tables* and provide valuable information. These *Tide Tables* also include information on how to determine the height of the tide to be expected at any time, not just high or low water, and the tables provide the information for the calculation of sunrise, sunset, moonrise and moonset. The NOAA *Tide Tables* can be purchased at most large marinas, book stores or it can be purchased directly from the U.S. National Ocean Survey, Rockville, MD.

Meteorologic Tides

Although the normal astronomical tides are quite predictable, there are occasions when the weather plays a role. During periods of calm wind and in the absence of storms, the tidal excursions follow reliable patterns. These patterns can be disturbed, however, and tides can exceed their normal ranges by many feet. Furthermore, the predicted time of occurrence can be changed, causing either early or late arrival times of high and low water. These abnormal tides can be caused by two meteorologic conditions: wind and barometric effects.

Winds blowing from the water towards the land, onshore winds, can effectively pack the water into an embayment or elevate its level along the coast and create coastal floods. These onshore winds will hasten the attainment of high tide and make it last longer than expected by holding the water in place. Conversely, offshore winds tend to empty embayments and prevent the tides from coming back in again. Thus, the water level can be extraordinarily low and persist prior to and after astronomical low tide has occurred.

The real meteorological tide crunch comes when a storm occurs. All coastal storms have a low pressure center. This low pressure, lower than the atmosphere surrounding the storm, causes what is called a barometric effect to occur. The water over which the storm travels, mounds or humps up under the low pressure center. The situation is similar to a drink rising up in a soda straw which has in part been evacuated by the drinker. This hump of water can elevate up to three feet or more under intense coastal storms and can be many miles across. This increased water level travels with the storm's center as a wave crest. As the storm comes ashore, this wave, now called a storm surge, responds to the shoaling water and gets even higher. So as the storm passes over or along the coast, many feet of water are added to the normal tide level. If the storm occurs at the time of astronomical high tide, extensive flooding can occur.

With an unusually intense, rapidly moving storm, the storm surge arrives abruptly and has the characteristics of a large steeply sloped wave at its leading edge. This is the phenomena that does most of the water damage during hurricanes. During the 1938 hurricane in southern New England, the storm surge brought a rapidly rising high tide some 8 to 10 feet above normal, while hurricane Hugo brought tides reaching 20 feet above normal to the Carolinas. These extraordinary tides must be considered in the design of pile top elevations and mooring chain scope. Floating docks built to withstand large wind and wave forces are of little value if they can float off over the tops of their restraining piles, or lift their anchors or perhaps even be pulled under.

Measuring the Tide

For many marinas, the state of the tide is a useful piece of information for both the marina management and the marina patrons. The posting of the times of high and low tide can be useful, especially in areas where low tide may mean difficulties for returning vessels. Even more useful can be a visual tide level mark or gauge so that the departing boater gets an instant idea as to water levels.

A simple direct reading tide gauge can be established in any marina to provide the marina patrons and operators with tide information at a glance. A simple board with one foot marks painted on it will suffice. Intermediate interval marks can be included for greater reading accuracy. One-tenth of a foot is as close as anyone can predict the tides at best and with any waves present it certainly is as close as you can read the tide board. The vertically mounted board, attached to a seawall, fixed dock or pile, should be marked with a zero point representing normal low water conditions and at least one or two foot marks below this level for those abnormal low water conditions. The number of foot marks above the zero point will depend on the marina's location. Refer to the *Tide Tables* to determine the appropriate range.

Once you have the board made, mount it at the desired location with its zero mark set at MLW. This can be determined by a land surveyor from the elevation of the National Geodetic Vertical Datum (NGVD) and the correction factor to MLW. However, the easiest and least expensive procedure to follow is to look up the time of low tide from the *Tide Tables,* properly corrected for your substation if necessary, then when low water occurs put a mark at the waterline of the pile, dock or pier on which the tide board is to be mounted. This low tide mark will probably be at some point a number of inches above or below the MLW mark unless you chose a low tide which was at zero feet in the *Tide Tables*. In either case, you know then that MLW is so many inches above or below your original mark. Measure to the actual MLW level with a ruler and put in a nail or thumb tack. Remember that the winds, locally or offshore, may disturb the water level and it won't be where its predicted to be. That's why measuring and marking the location of MLW several times is important in order to smooth out the error the winds may introduce during the process. Its best to do the marking only on those days when the wind is calm and no storms are nearby. Early mornings and evenings are usually the best times for quiet sea conditions. Repeat this process as the opportunity presents itself. After about 8 to 10 marks, your tacks or nails should cluster within about 1 to 2 inches of each other. This establishes MLW for the site and the tide board's zero mark can now be located and the board mounted in place. Others can now be placed throughout the marina as desired, using this first tide board's zero elevation as a guide. Once set up throughout the marina, the boaters can use this information for trip planning, dock line and other umbilical cord adjustments, as well as for watching the occurrence of storm tides or unusual low water conditions.

Unusual Tides

Site evaluation as well as marina design must also consider some of the unusual tidal occurrences. If marina basin or access channel dredging is part of the design, a knowledge of the frequency of very low tides is essential to keep the boats off the bottom. Some extraordinary tides result due to the close occurrence in time of the twice monthly spring tides and the once monthly perigean tides. Closeness in time would refer to their occurrence within 24 hours of each other, thereby causing an extreme augmentation of the two effects and resulting in very large tidal ranges, possibly 40 percent greater than the simple spring range alone. Fortunately this occurs only a few times a year. These particularity high tides become even more dangerous when they occur during severe weather disturbances. It is interesting to note that when statistically considered, a more than random number of major tidal flooding episodes have occurred which involved the coincidence between perigean spring tides and strong coastal storms. What is even more remarkable is that these episodic situations have occurred simultaneously on both the east and west U.S. coasts whose weather patterns are supposed to be unrelated.

It is difficult to find any direct cause and effect relationship between intense storms arriving at the same time that these unusually high tides occur. There is no theoretical concept to support a relationship, but there is some physical evidence from past weather and tidal records that such a relationship does exist. The caution seems clear: perigean-spring tides and bad weather can combine to cause coastal flooding all to often. The prudent marina operator or manager, as well as the marina's patrons should take note that if very high astronomical tides are expected, it is not a bad idea to take added precautions for the possibility of high winds.

7.3 WATER CURRENTS

The oceans and associated coastal waters by their physical nature want to remain at rest. Motion is resisted by the water because of its viscosity or internal friction and its own inertia. Some force must be present to cause it to move. Once set into motion, if the force ceases, friction gradually robs the water of its energy and it comes to rest again. The motion imparted to the water is localized in the sense that only water being acted upon by the force gets set into motion. Away from the force, the water remains motionless. This is best illustrated by the wind blowing over a body of water. The surface, that part being acted upon directly by the wind will move. A few feet down, perhaps no wind effect will be felt. Some distance away from a particular wind field, little or no current exists. A current produced by the tides however, acts from the top to the bottom of the water column because the tidal current force caused by the tide waves acts uniformly in the vertical direction.

Water movement or currents affecting marinas can be categorized as rivers or streams, tide produced, wind produced or density produced. However, the last type of current, density produced is usually of little concern to most marinas. It refers to the current effects caused by horizontal differences in the water density or weight. It produces systems like the Gulf Stream and aids in the general circulation of the oceans but is of little significance in relation to local or coastal current velocities.

Currents are usually referred to have a set and drift. The set refers to the direction in which the water is moving and the drift refers to its speed. Unfortunately, this is just the opposite to how the wind is described. For the wind, the direction from which it is coming is stated. A north wind blows from the north while a north setting current is going in that direction. Current set is usually given in terms of compass headings and drift in terms of knots. Remember, a knot is a velocity of one nautical mile (6080 feet) per hour. The often heard term of a "knot per hour" is an incorrect usage of the word knot. A knot is a little more than a land mile (5280 feet) per hour, about 15 percent more. Most water areas don't move very fast in terms of current drift. Tidal currents are generally the fastest category of currents with which a marina may have to contend. In some coastal regions, tidal currents may run as high as 6 to 8 knots, with most other currents having a drift of a fraction of that.

Local Wind Driven Currents

Significant water movement caused by the wind acting on the water's surface can occur through a marina. Local or small scale wind systems push the surface waters around. If the winds are persistent in terms of velocity and direction for periods exceeding several minutes, movement of the surface water becomes detectable. As the wind commences to blow over the surface of the water, waves start to form, small at first and then growing larger with time. The wind's grip on the water improves as these small waves form. This grip starts the very uppermost layer of water moving. Gradually, the upper layer begins to influence the layers below through a process of vertical mixing and turbulence and begins to set them into motion also. This wind affected layer grows from a thickness of several inches, and with time, reaches some ultimate depth depending on the strength of the wind, how long the wind has been blowing and the water's tendency to mix downward. The ultimate thickness of the layer is difficult to predict without knowing something about the water properties such as temperature and salinity, if it is in the marine environment, and how they change in the vertical direction. It may reach a thickness of 20 feet and sometimes even more in the deep ocean. If the water gets colder as you go deeper, or more saline, there is a resistance to deeper mixing or thickening of the wind affected layer. The intensity of the temperature or salinity change with depth plays an important role in limiting the ultimate thickness of the layer to be pushed around by the wind. One can imagine this layer which responds to the wind as simply a slab of water moving around on top of a fixed or stationary reservoir of deeper water.

For any given wind velocity, several hours are required for full development of this slab of water and its movement. When the current or wind drift is fully developed, it can be 2 percent of the wind's velocity and generally downwind in most shallow water areas. In areas which experience persistant strong winds this wind current may have to be considered in determining the force loadings on structures in the water. A wind of 25 knots can produce a water current on the order of 0.5 knots. The downwind movement direction may be modified by the shape and proximity of the shoreline. If this wind driven layer thickens, it can develop its own structure and internal circulation. On the surface windrows appear which are the result of the development of a rather interesting, vertically oriented cellular structure in the water's motion. Windrows are long strings or lines of flotsam. Debris of all kinds is collected as surface strings brought about by a convergence or coming together of the surface waters. These windrows are closely aligned with the wind and are an aid in determining the wind's true direction.

Wave Currents

A more important and predictable effect of the wind is the development of wave currents. To the eye, a wave's motion appears as though it is traveling across the surface of the water. Indeed its form is doing just that, but the water under this wave form tends to move only slightly from its original position as the wave passes. This is discussed in greater detail in Chapter 6 which deals with water waves. As a series of waves pass a given point, the water particle motion is induced below the surface and is continuous in the form of looping particle orbits. The diameter of the orbit of the topmost particle is equal to the surface height of the passing wave. These particle orbital paths do not quite form perfect circles, and end up making a series of loops with a net movement in the direction of wave travel.

Thus each water particle moves slowly in the down wave or wind direction, creating a wave current. With an increase in depth, away from the surface, this movement decreases, quite rapidly until it ceases at a depth equal to one-half of the wave length of the wave creating the motion. So the surface has the greatest horizontal wave current with less movement at depth. Since the waves are produced by the winds and travel down wind, the wave current also tracks downwind. It is roughly equal to 2 percent of the wind velocity, just as in the case for pure wind drift. The wave drift is a localized phenomena and if waves are present, a down wind or down wave movement will be experienced.

Remember, this is a wave produced current. There can be a wave or swell field arriving at the marina site and at the same time little or no wind. Waves produced elsewhere can travel into an area and a wave current will exist. In this situation the wind's velocity cannot be used as a guide to the wave current's drift or speed. Its direction is still established by the waves direction, but in the absence of the wind that produced it, computation of the current speed requires a knowledge of the wave's height, wave length and speed. **As a rough estimate, one-tenth of a knot of wave drift can be assumed for each foot of wave height. So if a swell system is estimated at 2 to 3 feet high, it may produce a current of 0.25 knots at the water's surface.** The important point here is that when the wind blows persistently from one direction over several hours, there will be surface water currents set up which may combine both wave and water drift effects and have total velocities on the order of 4 percent of the wind. A 20 knot persistent afternoon wind may produce water surface flows of 0.8 knots through a marina.

If the winds are blowing directly onshore, the water tends to pile up against the shore as a result of all this surface movement. Some of the excess water will move parallel to the shore and some of it will sink vertically downward and move back offshore at a depth away from the wind's effects. This accumulation or pile up of water creates a nice thick layer of surface water against the shoreline. In the summer, it makes the beach goers happy with all its warmth. On the other hand, it also carries the surface flotsam and debris to the beach or into a marina, providing warm but dirty water. This pile up along the downwind shoreline is not detectable to the eye since the surface is elevated only inches above the normal level. With an offshore breeze, the surface waters are transported away from the shore. A reverse circulation is set up and colder water from offshore moves in at depth and rises to replace the surface waters which have been blown off. Now the water near the shore is colder, but usually clear and free of debris.

Tidal Currents

With the rise and fall of the tides, water must be supplied to elevate the sea surface and must be taken away when the surface falls. This water is supplied and removed in the horizontal direction and produces what are called tidal currents. A flood current refers to water moving in response to a rising tide and an ebb current refers to a falling tide. Slack water occurs at some point in between, while the current switches from ebb to flood or back to ebb again. It is that moment in time when the water motion effectively stops.

Since tides are predictable, it stands to reason that tidal currents should also be predictable. They are characterized as not only the most predictable of the current systems but generally have the highest velocities. An embayment with a tidal range of 10 feet must move a considerable volume of water in and out over its tidal period with understandably large velocities. Five and six knot tidal current velocities are not unusual, particularly at restricted embayment entrances. Tidal currents under the Golden Gate Bridge in San Francisco, for example, run up to 6 knots under spring tide conditions. The Strait of Juan De Fuca off Washington State and Long Island Sound can have 4 knot currents in some parts. More benign areas such as San Diego Bay, California or Miami Harbor entrance in Florida possess currents which approach 3 knots. Few coastal locations are free of the effects of strong tidal currents. Generally the greater the tide ranges, the higher will be the tidal current velocities, and these in turn can be modified into even higher velocities due to restricted or narrow channel entrances, either natural or man made.

Characteristically, tidal currents are reasonably uniform from top to bottom except where modified by bottom friction or fresh water river flows on the surface. In the latter case, an incoming flood current in a river mouth tries to move uniformly up the river while on the surface the fresh water moves in the opposite direction.

Tide waves come in two different forms: progressive and standing waves. The relationship between the height of the tides and the tidal currents of flood, ebb and slack will differ with these two different wave forms. For example, as a progressive tide wave moves up a river, the maximum flood current occurs at the crest of the wave, high tide, and the maximum ebb current occurs in the trough or low tide. Slack water occurs at the mean water level position between those two extremes as the wave passes by. Table 7-4 gives high tide and maximum current times along the Hudson River up to

Table 7-4. Hudson River, New York—Progressive Wave Tide

	High Tide Time	Max. Flood Time
The Battery	1946	1946
George Washington Bridge	2032	2011
Tarrytown, NY	2131	2106
Peekskill, NY	2210	2146

Peekskill, New York, to illustrate this point. Most rivers will cause the times of maximum tidal current to differ somewhat from the times of high tide because of the strength of their natural flow toward the sea. A river that flows rapidly will delay the time of maximum flood and reduce its intensity. The Columbia River for example has no tidal flood current phase due to its strong flow. Slack water and strong ebb currents are all that occur. In the case of the Hudson River, it can be seen from Table 7-4 that the time of high tide progresses up river in step with the time of maximum flood current.

The Chesapeake Bay is one of the few embayments in the U.S. which has a progressive tide wave instead of a standing wave. Here, the embayment is bordered by so many rivers and streams that the tidal wave's energy is absorbed and not reflected. For the most part within the bay, high tide and maximum flood current occur almost simultaneously, as they should for a progressive tide wave.

The typical standing tide wave produces tidal currents which are slack at the approximate times of high and low tide and flow fastest in between these times. This is just the opposite to the progressive type of tide wave. Thus at either high or low water, everything comes to a halt, the rising tide as well as the currents. Table 7-5 shows that the time of occurrence of high tide in Long Island Sound is almost the same as for the occurrence of the slack current for the entire length of the Sound. The correlation between the time of occurrence between high and low water and slack current is not always perfect or according to theory. In some embayments which exhibit a standing wave, there is a slow progression of the times of the tidal occurrences towards the head similar to a progressive wave type of tide wave.

Current velocities at the entrance of the standing wave type of embayment are usually much faster than those at the head. Long Island Sound is a good example. Here currents at the eastern or open end of the Sound typically run at 4 to 5 knots while near Greenwich, Connecticut, near the head, they are almost negligible, regardless of the state of the tide. Tidal currents are strongest under the Golden Gate Bridge (which is at the mouth end of San

Table 7-5. Long Island Sound — Standing Wave Tide

	High Tide Time	Slack Water Time
Westbrook, Ct.	0257	0303
Sachem Head (Madison, Ct.)	0310	0312
Bridgeport, Ct.	0320	0347
Greenwich (Cos Cob)	0326	0357

Francisco Bay), as well as at the entrance to Puget Sound, Washington, to cite additional examples.

A number of complicating effects come into play which cause tidal currents to behave in unexpected ways. River flows, bottom shape, and configuration of the embayment all play a role in setting up the complicated tidal current patterns that are observed. Eddies around the backside of islands and peninsulas, all help to produce irregular high velocity flows that do not seem to be related to anything the tides are doing. Where Long Island Sound joins the East River at the northern end of Manhattan Island, New York City, some very interesting and treacherous water exists. Here the standing tide wave in Long Island Sound is mismatched in time by over an hour with the progressive tide wave coming up the East River. That is why the location is called Hell Gate.

As with tidal prediction, the regular and predictable periods of recurrence of the tidal currents exist. A semidiurnal tide produces semidiurnal tidal currents and spring range tides produce higher currents than do neap tides. But with all the velocity and direction alterations cited above, the only way tidal currents can be predicted is to form a data base by measuring the currents under as many differing tidal conditions and at as many locations as possible. This has been done for many areas and is published by the Department of Commerce, NOAA in the form of the *Tidal Current Tables*. These tables are divided into two parts: the daily pages which show times of maximum flood, ebb and slack water at selected measurement stations around the U.S., and a second part which gives correction factors for many substations in the vicinity of the major measurement stations. This is very similar to the way the *Tide Tables* are presented.

Two publications are prepared each year, one for the Pacific Coast of North America and Asia, and another for the Atlantic Coast of North America. They are quite easy to use for specific locations of interest. NOAA also prepares detailed tidal current information in chart form for selected areas. The northeast coast is well covered by these charts for areas including Chesapeake Bay, Greater New York Harbor, Long Island Sound, and most of the coast of Rhode Island and Massachusetts. To the south of Charleston Harbor, South Carolina, no charts have been published and only Puget Sound and San Francisco Bay are published for the west coast.

7.4 LAKE SURFACE ELEVATION CHANGES

If the site being evaluated is on a lake, a change in water elevation must still be considered. Most lake levels will rise and fall with either the short term or long term rainfall and land runoff. These inland basins of water are the temporary storage locations for sometimes vast quantities of fresh water. They also form areas for recreational boating, and depending on their sizes, form sites for marina installations. Lakes come in all sizes and depths. The elevation of their surface may change significantly during an overnight rainstorm, or within a matter of an hour or less due to wind setup from one end of the lake to another. These short term elevation changes may be on the order of several feet, perhaps 6 feet or more as in the case of Lake Erie at Buffalo, or as much as 20 feet in some of the western reservoir lakes. Long term changes for most large lakes are generally regulated, where possible, through inlet and outlet engineering structures by interstate or international commissions, and these elevation changes are kept as small as possible. Generally, the Army Corps of Engineers, the U.S. Geological Survey or a state environmental office can provide historical information on lake elevation changes. Reservoir lakes however, may change elevation by hundreds of feet from season to season. Often, information on these changes can be obtained from the public or municipal water authorities regulating these reservoirs.

Although most lake elevation changes may seem small by comparison with tidal swings on the ocean shores, many marina sites have not considered these elevation changes in their designs and find themselves either high and dry during some years, or with submerged fixed docks during others. It is obviously a solvable problem with just a small amount of necessary research and design. Ignore it and it can mean total marina shutdown for one or more seasons.

Great Lakes Surface Elevation Data

Surface elevation changes present special design problems to marina developers and operators. The Great Lakes of the U.S. and Canada are an important example of this. Their sizes and interconnection prevents any significant control of lake levels beyond certain small adjustments. Seasonal variations of all the lakes may fluctuate on average of 1 to 2 feet, with longer term records showing variations of from just over 1 foot below chart datum to just over plus 5 feet. The largest swings in elevation occur in Lake Erie and Lake Ontario. The Army Corps of Engineers, Detroit District, and Environment Canada combine to publish a *Monthly Bulletin of Lake Levels for the Great Lakes.* This two page document contains a wealth of information on the present levels, past levels and 6 months of predicted (probable) lake levels for all 6 of the Great Lakes: Superior, Michigan, Huron, St. Clair, Erie and Ontario. These data are very useful for the marina designer and engineer, and also the marina operator who must confront these changing levels.

An important modifier to these longer lake level data is the occurrence of wind setup elevation differences from the up wind end of the lake to the down wind end. The longer the axis of the lake over which these winds blow, coupled with the strength and duration of the wind, the greater will be the end to end differences in lake elevation. Wind setup is caused by the extreme winds pushing the water of the lake in a downwind direction and creating higher levels along the lee shorelines. Winds from the west-southwest in Lake Erie for example, create lake level elevations at Buffalo which are 4 to 5 feet higher than normal levels while at Toledo they may be 3 to 4 feet below normal on an annual basis. These temporary higher water levels along a lakeshore create an environment in which higher waves may approach the shore due to the deeper water. In addition, the lowered elevations may significantly alter the temporary wave refraction patterns of the incoming storm waves. Information on wind induced setup is published by the National Weather Service and curves are produced by the Army Corps of Engineers showing the percent occurrence of lake level fluctuations.

The NOAA Technical Memorandum NWS TDL-54, *Climatology of Lake Erie Storm Surges at Buffalo and Toledo,* dated 1975 provides prediction data on lake setup which was determined from 33 years of collected information. Lake Erie has the unique distinction of having its longest axis oriented from west-southwest to east-northeast, which also happens to be the prevalent direction of the ambient and storm winds on the lake. Wind setup creates a condition of surface tilt for the lake. Considering the enhanced elevations at Buffalo for example, commensurate lake level depressions are experienced at the same time at Toledo, with a nodal point of no elevation changes existing approximately half way in between. During the period of data analysis from 1940 to 1972, a period of 33 years, the greatest setup experienced in Buffalo was on February 16, 1967, and resulted in a value of 7.79 feet above the mean lake elevations. A similar elevation occurred in December of 1985, when the Buffalo lake level reached 7.8 feet above the mean. With a lake level depression of 5 to 6 feet at Toledo during these episodes, the water body sets itself up for still another problem called seiching.

Lake Seiching

As discussed above, with the water surface of Lake Erie poised with an elevation difference from Toledo to Buffalo of possibly 10 to 12 feet or more, once the wind setup releases the water's surface, a rather large slosh of water will occur back towards Toledo. The situation is much the same as if a dishpan full of water was tilted or tipped and released. The water tries to relevel itself, but in doing so goes back and forth in the dishpan, setting up what is referred to as a standing wave. Both ends of the pan experience a periodic rise and fall of water level, which gradually damps itself out until the water in the dishpan is at rest. This sloshing phenomena is referred to as a seiche (pronounced saysh). Much the same happens in Lake Erie, only more slowly, with periods on the order of several hours or more, giving the effect of a short period tide change at either end of the lake. Observers along the lakeshore's center would see little effect, if any. That is the location of the node or fixed point around which the lake's surface oscillates.

Most lakes will produce seiches to a greater or lesser extent depending on their size, depth and characteristic winds. Most lakes are small by comparison with Lake Erie, especially when compared along its longest axis, aligned with the wind. The other Great Lakes do not have seiches of the size or regularity produced in Lake Erie. However, seiches which produce 2 to 3 foot elevation changes can result in unexpected flooding. Even the long and narrow Finger Lakes in New York State are known to produce problems when strong winds blow along their long axes. Before doing any marina design work in lakes, especially as it pertains to fixed or floating docks, mooring scopes, and wave barrier heights, it is prudent to thoroughly investigate the historical occurrence of these phenomena.

7.5 RIVER FLOODS

Siting a marina along the banks of a river does not remove the problem of water surface elevation changes that occur in the sea and lakes. If the river empties into an embayment or coastal area where tides exist, these tides may very well also exist for many miles up river. This will depend on the volume flow rates of the river, as well as the surface elevation of the river caused by rising land topography. During high river flow rates, the effects of the tide are pushed seaward, down river. During low flow rates, the tides may be just as great 50 miles or more up river. The thing that seems to amaze many people is the fact that although the tides may be experienced many miles up river from a marine tidal source, there may not be any salt water reaching that point. In other words, the tides consist of fresh water from the river. Effectively, the ocean pushes on the lower, connecting end of the river and the surface elevation increase is transmitted many miles upstream, but the salt water does not move up the river. The tidal pulse creates a changing seasonal flow situation in combination with the speed of the river current. During times of high river flow, a rising tide may produce few or no upstream currents. There may even be a reduced flow in the downstream direction at that time, while on a falling tide the downstream current may flow with great vigor.

Another condition that should be considered in evaluating a potential river bank site for a marina is springtime floods. This problem comes from either melting snows further upstream or spring rain runoff, or both. In some areas it can cause floods with elevation changes of 20 to 30 feet and sometimes more. A river generally flows within its channel banks. When it crests above these banks it floods the surrounding country side. The area of flood is called the floodplain and a marina sited within a floodplain will require a supporting infrastructure that is kept above the highest levels of the floodplain. Marinas placed in a floodplain where several feet of water will flow every spring produce problems for dock and building structures. In many areas, flood elevations are also accompanied by high velocity flows of the river. Most states have tight controlling regulations concerning structures within floodplains, with site restrictions within municipalities specifically set forth by the Federal Emergency Management Agency (FEMA). Before serious consideration is given for site development along a river, a thorough investigation must be made. The FEMA floodplain and flood elevation data as well as any building restrictions for a proposed site can be obtained from the local town or city planning office. If they cannot provide this information, most states maintain this information within their environmental regulatory departments, or if that fails, FEMA has a number of regional offices which will provide assistance. There are marinas which have been successfully built and maintained in flood plains. However, these tend to be the exception, "grandfathered in" before FEMA. There are now regulatory, design and operational problems which all interrelate that must be first explored with the local and state goverments before significant amounts of money are spent on a potentially nonfeasible project.

7.6 ICE

Ice engineering research along with resulting techniques have made great strides in providing the marina designer with the tools for coping with ice conditions. It remains a seasonal problem and only for certain locales, affecting both the boats and the structures of a marina. If such conditions are improperly dealt with, the results can be fatal to a poorly prepared or designed marina. Ice and its interaction with marinas has been well studied, with many problems solved, ameliorated, or at least identified. An excellent source for detailed information on ice and marinas is the book *Ice Engineering Manual for Design of Small-Craft Harbors and Structures,* by C. Allen Wortley, University of Wisconsin Sea Grant Institute, 1984. Because that book is available the problem is only briefly discussed here. Before working in an ice prone area, the marina designer must have a working knowledge of the basic physics of ice. The following paragraphs are intended to give only a cursory survey of the field.

Local Ice Conditions

Specific ice reconnaissance for a area should always be carried out by either direct observation during the winter or by obtaining reliable local knowledge. For the purposes of obtaining local knowledge of past ice conditions a Harbormaster, the Coast Guard, local marine contractors or other nearby marinas are often good sources of information. Sometimes several sources should be used since one particular bad year often becomes the imagined norm for the area. Information to be obtained includes the extent of freezing from shore to shore, the ability to walk on the ice to other shoreline points, ice thickness, when the ice begins to break up, the amount of dock and pile damage and the extent to which bubbler systems are used with good success.

In the absence of documented year to year data on the occurrence of ice, one method of establishing some information on ice coverage and ice thickness is to obtain data on the number of freezing degree days (FDD) that normally occur in the area. This data can be obtained from the weather data reported at local airports, just as the wind data is obtained for forecasting wave heights. It generally includes reports on the average freezing degree days on a monthly basis. **Ice thickness can be estimated to a reasonably accurate degree by utilizing the relationship of expected ice thickness in inches, which is equal to the product of a locality factor, times the square root of the sum of the FDD during the winter.** This equation is commonly used for still water lakes of all sizes but it can also be adapted to marine tidal embayments through the modification of the value for the FDD. For moderately vigorous tidal movement in marine areas, the ice thickness computed from the above relationship correlates well if half the value of what is given for the FDD is used. The value of the locality factor varies with snow cover. Due to the snow cover, the excessive thickness expected from the low water temperatures are sometimes never realized. Ice growth becomes limited due to the insulating blanket of snow that produces a snow ice cover which structurally has much less crushing strength than does clear ice. This phenomenon is included in the equation through the so-called locality factor. **This locality factor ranges from 0.33 to 1.00 where the lowest value relates to heavy snow cover on the ice and 1.00 relates to bare ice.** Snow fall data is also collected and reported in conjunction with average freezing degree days. Only moderate snow amounts would be expected to accumulate on the floating ice in most U.S. marine areas regardless of what may be reported at the local airport.

Bays, harbors and other open water areas, including rivers, follow a generally similar scenario when subjected to prolonged winter conditions. These areas may freeze from shore to shore with a coherent ice sheet during parts of a moderate to severe winter season, depending on the site's location. The local extremes of the climate, be it either marine or inland lake, coupled with storm action, seiche and tidal currents can break up parts of the ice sheet into various sized floes several times during a severe winter. These drifting ice pieces act as thin floating lenses which are subject to wind stress forces on their surfaces. They can be set in motion by the wind and wind currents and driven on top of each other, resulting in what is referred to as rafted ice. Individual floes are often driven on shore by the winds with great force, freezing in place for the winter. These ice formations can at times cause considerable damage locally, but their net effects are largely beneficial. Spray from winds and waves freezes on the banks and structures along the shore, often covering them with a protective layer of ice. Ice piled on the shore by wind and wave action does not, in general, cause serious damage to beaches, bulkheads, or protective riprap, but it provides additional protection against severe winter waves. A severe winter often progresses with the breakup, rafting, melting process repeating itself several times before the season ends by February or March, with many intervals when loose floe ice is present. This is a more dangerous time of the year for the shorelines and structures in the water. They are now subject to impact by moving ice floes. The speed of movement will be dictated by the water currents and wind stress.

Ice Forces

Ice structural damage can be caused by friction and adhesion of ice, vertical forces from ice sheets, and horizontal static and dynamic loads. Ice adheres tenaciously to most construction materials including the surfaces of wooden piles. Even if the ice sheet frozen around a structure's surface breaks away, a collar or band of rough ice generally remains. This remnant ice allows the rapid rejoining of an ice sheet to the structure. Many kinds of low friction coatings, jackets, and other mechanical and even electrical devices have been tried in order to reduce this ice adhesion. The adherent ice sets the stage for the structural problems by enabling the floating ice to become joined to the structure. This means that the forces operating on the floating ice are transmitted to the structures.

Once the ice is joined to a structure, especially a cylindrical pile, vertical lifting forces are directly transmitted to the pile. In a tidal or seiche area, these vertical lifting forces can physically jack the piles upward during a rising water level. With sufficient ice formation and the refreezing process, the piles can be pulled from the bottom, and their attached structures broken. This is the most common form of failure in ice conditions. This lifting action does not occur uniformly and the structure may be left partially elevated and badly distorted. The end result is loss of a usable facility and a requirement for rebuilding, perhaps on a yearly basis. The tendency of the ice to attach itself to a pile can be reduced by the use of a bubbler system.

These systems utilize the fact that the unfrozen water near the bottom of the water column possesses enough heat so that if it can be brought or circulated to the surface, localized formation of the ice near and around the pilings and associated structures can be reduced and even prevented in thin ice areas. These systems are subject to mechanical failure and cannot be relied on as the only form of ice protection.

Horizontal static forces can do considerable damage. These types of forces arise when the ice sheet is coherent from shore to shore or has a horizontal floe size of several miles. Thermal expansion of the ice due to rising diurnal temperatures in the early spring can create enormous static pressures and are considered to be the greatest cause of damage in an ice prone area. With ice sheets adhering to the structure's surface, or merely pushing against it, structural damage can occur. Free standing piles can be snapped off at their base, and entire dock systems can be pushed over from the ice pressure. **Thick clear ice tends to crush and raft when subjected to pressures greater than 400 pounds per inch, with snow ice, or recently formed salt water ice crushing and rafting at 100 to 200 psi. Ice forces from thermal expansion are on the order of 5,000 to 10,000 lbs/ft, for an ice sheet several feet thick. Ice thicknesses of 3 inches or less may not create a thermal expansion problem because the ice tends to buckle, break and raft easily, thus relieving the pressures.**

A second type of static ice loading can occur when the wind blows over the surface of a large ice field, which is either broken into floes or completely coherent. In this case the wind acts in a shearing mode and couples itself to the floe producing a horizontal pressure in the down wind direction. The force produced can be determined from the standard tangential or surface drag equation found in most fluid physics books. The force is equal to the product of the surface area, a coefficient of coupling, the air density, and the wind velocity squared. For example, it would take a 70 mph wind over an ice fetch distance of 2.5 nautical miles to produce a force of 1000 pounds per linear foot of bearing surface.

Horizontal dynamic loads are another potential ice force on structures. These loads arise when floating pieces of ice are propelled by wind and current thus generating impact loads on the structure. Small pieces of ice can also be propelled by the waves and thrown against the structure. However, this is less of a problem since the mere presence of the ice field, broken or coherent, tends to act as a floating breakwater, quelling or ameliorating the waves considerably. Potentially, wind propelled ice floes would pose the most important impact problem. If there is a potential for these types of collisions or impacts at a proposed site, further analysis will be necessary and calculations made of the magnitude of the impact force from moving ice. For example, it is assumed that an irregularly shaped ice floe measuring 10,000 feet square (approximately 100 by 100 feet), 8 inches thick, is set into motion by the wind and impacts along a 20 foot section of a dock. The maximum velocity attained by the ice floe can be estimated at 3 percent of the wind speed. Using winds of 70 mph, the ice floe could be moving at a speed of 1.8 knots. Using an ice strength of 200 psi results in an impact force of approximately 1,100 pounds per linear foot. However, if this ice floe were propelled by a five knot tidal flow or river current, the resulting force will be greater than 10,000 pounds per linear foot since the impact force is dependent on the square of the speed of the moving ice.

Ice Protection Perimeter

Regardless of the particular design of a marina dock system, using either floating or fixed docks, piles or moorings, if the system is to remain in the water during the winter, some form of perimeter protection should be considered. This protection may simply take the form of a bubbler system or it may be an elaborate and substantial barrier. If the marina is already within a barrier perimeter which protects it from the crushing and drifting ice forces, then a reliable bubbling system may be all that is necessary. If the marina is more exposed, then perhaps a combination of permanent and temporary structures may be necessary. The idea in this case is to isolate as well as possible, the sensitive marina structures from the potential damage of the expanding or moving ice, as discussed above. A knowledge of the physics of ice: how it forms, expands, contracts, and breaks up, allows the marina designer to provide for these phenomena in ice prone areas. It must be accepted that ice may cause damage to the marina structure in spite of all the precautions taken. When structures are built in ice prone areas, some degree of risk must be anticipated and economically accounted for in any long range maintenance program.

8
Perimeter Protection

After the completion of a wave climate study and a basic site condition analysis which may include wake, current and ice evaluations, most sites will be found to need some form of perimeter protection. The goal of this chapter is to offer the marina designer the tools and concepts with which to fashion the protection of a marina site. The storm wave climate studies as well as the boat wake wave analysis will generally indicate that some form of wave attenuation will be required for the marina to be considered a comfortable yacht haven during the maximum annual wave conditions. The physical conditions of a proposed site may allow the site to be relatively calm most of the time due to natural protection from near field wind waves, while at the same time, the site could be a moderately rough area under storm conditions due to the arrival of far field waves. Under those conditions significant protection may be required. In addition, studies, as already outlined, may disclose the history of ice occurrence throughout the area or strong river or tidal currents which must also be addressed. In some cases, a protection perimeter is established for on-site conditions related solely to these peripheral concerns because of natural wave and wake protection that exists. Unfortunately, few sites have this luxury, and lucky is the marina owner who may have to contend with only one or two of these problems. Generally the case is that a marina site needs wave and wake protection of some sort. In addition, there are usually secondary protection considerations which, although taken care of by proposed wave protection, may have a deciding effect on the form that that wave protection may take. Stopping the waves from entering the marina may be a laudable goal, but if the technique used creates a dredging problem, or high current conditions at the entrance, then perhaps an alternate wave protection method must be sought.

Most of the protection solutions for a marina site start by addressing the wave or wake problem. Following a logical procedure, primary protection from the annual wave regime, along with the potential for extreme waves must first be developed. Any structures suggested for wave protection must then be analyzed for their ability to withstand any projected current and ice conditions. The marina designer must have a basic understanding of how these perimeter protection devices work, how effective they are and their ability to survive. In addition, a value judgement must be made as to how much or how little protection is needed or desired, and then the appropriate protection device or structure is selected for further analysis. As necessary and appropriate, the protection concepts discussed in the following sections will also allude to these other phenomena which may be ameliorated or exacerbated by the protective structures. In addition, these engineered structures may have some effects on the sediment transport within the area, and therefore must be examined as to the short and long term impact on the marina, surrounding shoreline properties and existing channels.

Competent perimeter protection may take the form of stone barriers, wave screens or vertical barriers which are either solid or semipermeable such as the wave board type, or floating wave attenuators. The final suggested system may be a combination of these engineered structures. They each have a specific function and their own particular problems in use. Stone barriers for example, are costly structures, which are quite permanent, and tend to use a significant portion of the bottom unless the water is extremely shallow. Closely spaced steel or concrete circular interlocking piles can be used in place of a stone wave barrier. They also form a wave blocking function, but may cause other problems such as unsightliness, the need for long term maintenance, and wave reflection. In addition, the cost of the product and installation may prove to be prohibitive. Pile and timber wave screens or wave boards are often used structures. However, a potential high wave regime during storm conditions would require careful design and analysis for their use in terms of orientation, construction and longevity. High tide water depth coupled with the potential water rise during storm conditions require all of these structures to be quite large in the vertical direction for effective storm protection, and in some cases, batter piled in such a way that they may take up considerable water space along their inboard sides. In a marine environment, the unsightliness of these structures at low tide, when they may be many feet above the water's surface, must be accepted if they are to be used for marina protection.

A floating attenuator may appear to be a good choice for wave suppression during most weather conditions. It can offer protection around a marina's perimeter while remaining relatively inconspicuous when compared to wave boards or other vertical extent barriers. Furthermore, water movement and marina basin flushing are unimpeded by its presence. Unless carefully constructed and positioned, ice may cause damage, so be careful when using them where ice conditions occur. However, they generally do not form a barrier to fish movement or interfere with sediment transport along the bottom. Most floating wave attenuation devices are sensitive to wave period.

The device used to provide for wave comfort on board a docked boat must be carefully analyzed for its effectiveness within the range of wave periods that will arrive at the marina site. Extreme storm conditions may bring larger waves than can be properly ameliorated by a floating attenuator. In this case, survivability of the attenuator and its mooring system is also a priority consideration along with the docks and boats present.

8.1 WAVE PROTECTION CONCEPTS

The concepts of wave protection, wave blocking and attenuation, although many times foremost in the minds of marina owners, operators, and developers, are somewhat elusive due to a general lack of basic information. The object of this chapter is to try to remedy this situation, and nail down some basic ideas. Once armed with the fundamental concepts, designers and engineers can use variations on these ideas to match specific site conditions with the appropriate wave blocking or attenuation system. The best way to accomplish this is to start from scratch in terms of wave protection, keep things simple without going to second order effect complications, limit the mathematics, and generate ideas to help solve individual marina wave problems. No marina design book will have a tear out page applicable to your marina. It will take some translation and interpretation of the ideas and principles offered here to solve the site specific problems of any marina.

The term breakwater has probably seen its better days considering that the term is archaic, and really fails to describe specifically what is meant. Perhaps its demise should be hurried along. Basically the choice comes down to two fundamental systems, that of wave blocking or wave attenuation. We may break a wave, but that may not prevent it from picking itself up again and doing damage to the facility. Fundamentally we tend to attempt to block the wave energy entirely, by using some form of fixed structure, strong enough to totally block any and all on-coming waves. Yes, a rubble mound or stone wave barrier certainly falls into that category as well as do a variety of vertical bulkhead type barriers. A wave blocking device will definitely be massive, strong, and with its top well elevated above the mean level of the water's surface. This means that it may take up a lot of bottom space, be costly in terms of materials or labor or both, and present, perhaps, some visual pollution by not only blocking the waves, but blocking the scenic view also. On the other hand, wave attenuators tend to fall into a less well defined grouping. Somehow, they must attenuate waves, that is to say, reduce the size or height of the wave by some partial blocking or perhaps some cleverness on the part of the wave attenuator designer that robs a wave of a portion of its power.

The ultimate wave attenuator is a wide, low sloping shoreline beach made up of coarse sands and gravels. With the proper absorption and drainage characteristics, along with the room for reshaping its profile, this attenuator may be able to withstand any wave system. An effective wave blocker must similarly be able to receive all waves and withstand their forces without significant failure. As an example, a rubble mound wave barrier is generally designed to allow some small amount of progressive failure. Since it is made up of loose pieces, some movement of these pieces can be allowed, and may occur as the rubble mound adjusts itself to better withstand the waves, much the way a beach does. Other types of wave blockers may not be able to gradually fail in this mode, and catastrophic collapse is the penalty for exceeding the design wave. In some cases, wave blocking devices may be designed to protect up to a certain storm condition, allow overtopping during others, and in certain areas allow submergence below the storm tides as a survival method. Still other systems may provide for good to adequate wave protection for the normal conditions, become less protective as the waves increase in size, but fulfill that primary requirement of at least surviving through the storm. Certain chain or cable moored floating systems have this potential capability. Full protection for any wave blocking system can be very expensive and certainly is not realistic. The regulatory and environmental problems of obtaining a permit for such devices generally precludes them from consideration as do their construction costs.

Probably the most important point when dealing with wave protection is to understand that no system designed or permitted by the regulatory agencies today will offer complete protection. All permitted systems, blocking or attenuating, will have their limitations relative to wave intensity, and the site specific aspect must always be considered in the design. Furthermore, once it is accepted that at some wave intensity the protection system will offer much reduced protection, then it must next be accepted that at some level of wave intensity the system may also fail.

8.2 HOW SYSTEMS WORK

Wave protection systems categorically operate on one or more of the following physical principles: wave reflection, wave breaking, frictional dissipation, and turbulent disorientation. In examining any of the generic wave protection systems, it is important to understand exactly how and why they work. This becomes even more important if you plan to purchase a wave protection system. There are no mystical methods for wave protection, just variations on relatively simple physical principles that should be easy to explain in simple terms. The questions to be asked are, How am I going to keep wave energy out of my marina, and what is going to happen to the wave energy that I keep out? Wave energy is awesome. If you prevent some or all of it from entering your territory, in whose territory may it end up?

The general anatomy of a water wave has already been discussed. As described, a wave's influence extends down into the water to depths which approximate one-half of its wave length. For the important waves affecting boats and marinas, this usually means water depths of 20 to 30 feet. So it can be assumed that for most marina sites, any important waves that must be dealt with, extend well down into the water column, basically to the bottom. That also means that it cannot be simply assumed that if the surface portion of the wave is handled, the problems have been solved.

As a wave passes a fixed point, the entire water column under the crest moves forward in the direction of wave travel. As the trough passes, the direction of water movement reverses itself and moves backwards. A small floating object appears to make an elliptical orbit as a wave passes. No closed loop is formed because the crest movement is generally greater than the trough movement, resulting in a small amount of net water transport, or wave current, in the direction of wave travel.

The power, or rate of energy transport, is what must be dealt with or attenuated. This rate of energy transport can be measured and referred to as the horsepower of the wave as is done for automobiles. Another familiar unit of power is Watts, a term which is often used in electrical power, and yet another unit is the scientific term Joules per second. Whatever it is called, it is what you feel as a wave strikes you. It is a useful term because if some of this power is successfully blocked with a wave attenuator or wave barrier, the power that flows past the blocking structure or device can be used to determine how successful the design has been in quelling the waves. In other words, a wave can be described mathematically in terms of power relationships. Conversely, the wave that produces that power can be described in terms of size and shape. This allows examination of the effects of certain structures on a wave system by looking at the structure's effect on the wave's power. First a wave is described in terms of pure power, then it is characterized by what happens to that power as it interacts with the structures or devices and finally by the behavior of the new wave from the remaining power that gets past the attenuator.

As already discussed, the wave's power varies from the top of the crest to the water's bottom. The greatest power is located at the still water level and it diminishes rapidly with depth in a geometric manner. Thus the water near the surface contains the greatest wave power, but there remains significant wave power deeper in the water column. Figure 8-1 shows the anatomy of a wave as defined by the power of a 3 foot high, 46 foot long wave length, 3 second wave in 23 feet of water. The vertical distribution of the wave's power is indicated in a relative sense, down to the bottom. As can be seen the power increases with depth to a maximum at what would be the still water level in the absence of a wave, then decreasing rapidly at first with depth, gradually becoming more uniform towards the bottom. At a bottom depth of 23 feet, little power remains, hence the rule of thumb about one-half of the wave length for the depth of influence of a wave. As indicated in Figure 8-1 blocking this wave down to a depth of 4 feet, eliminates 73 percent of its

PERIMETER PROTECTION

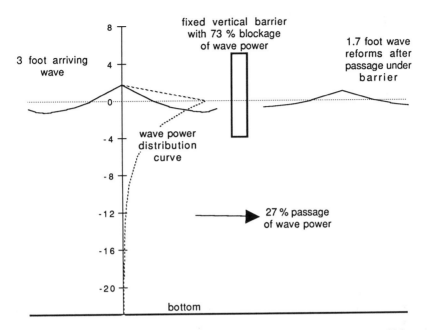

Figure 8-1. The effect of a 4 foot draft barrier on a 3 foot high, 3 second wave, in 23 feet of water. The vertical distribution of the wave power is also indicated.

power. Thus only 27 percent of this 3 foot wave's power is transmitted beneath this blocking barrier, but if that remaining power is reconstituted back into a wave behind the barrier, as typically will happen, the wave will be 1.7 feet in height. In general, approximately one-quarter of the original power of the wave has been allowed to pass the barrier, but a wave that is more than half as high as the original remains. It doesn't seem fair, but it is correct physics and something that has surprised, and disappointed, a lot of designers and builders of wave attenuation devices. A general rule of thumb here is: block one-half of the waves power and reduce the wave's height by one-quarter. Basically, all this stems from the fundamental fact that by comparison, a wave twice as large has four times the power. Stating it more formally, the power of the wave increases with the square of the wave's height.

Wave Reflection

Wave protection through the process of reflection means you have simply turned the energy and redirected it elsewhere. It is a commonly used method. That may be fine if it is turned back out to sea, but it will not be so fine if it ends up on a neighbor's shoreline. The physics of wave reflection is well understood. The wave's reflected angle will be a mirror image of its incident arrival angle. If its arrival angle is 90 degrees, it reflects back out at right angles to your barrier. At some smaller angle of incidence you intercept the wave's natural arrival course and send it off in an unnatural direction. In any event, wave reflection, simply moves the problem from your area to some other area and therefore reflection must always be given careful consideration.

Remember from the wave discussions of Chapter 6, for an object to affect an on-coming wave, that object must be at least as large across as one-quarter of the on-coming wave's length. For a common 3 second wave in 20 feet of water, the object must be approximately 10 to 12 feet across for the wave to take notice of it. If the object is sufficiently large, like a wall, the wave is stopped and turned around by the object. This reflected wave, which can be the same height as the original or incident wave, travels in the reverse direction off the wall and meets the on-coming wave. The two crests momentarily meet and create a new wave twice as high as the original wave. Not necessarily a good structure or location to moor your boat next to. Walls angled to the arrival direction of an on-coming wave, bounce this reflected wave energy off at an angle equal to the arrival angle of the wave with the wall. This reflected wave energy, travels off, crossing the wave crests of the original wave, creating double sized waves at the crossing points. It makes for messy conditions, especially if it occurs within a marina.

Wave Breaking

Wave breaking in the sense of creating conditions so that all or part of a wave is caused to break, i.e., tumble or trip over itself is also commonly used. Sea walls and perhaps some low height rubble mound or stone barriers work this way. The philosophy being, what ever wave part cannot be blocked, will break or tumble over the top of the structure, perhaps disturbing the water on the other side of the structure and creating a significantly smaller wave than the original wave. In some cases the structure can be made wide enough in comparison to the wave that it is designed for so as to allow all of the breaking energy to expend itself on the surface of the structure, leaving little more than a rush of water off its surface to the other side. With waves breaking over a structure, particularly if this occurs in a random manner without a set periodicity, the waves created in the water space behind the structure may differ considerably from those which initially arrived. The volume of water carried over the top or size of the overtopping wave and the water depths will determine what wave form this impulse will have after passing over. Wave tank experiments as well as actual measurements in the field have indicated that the resulting waves, after reforming will be lower in height and have a shorter period. The shorter period aspect can be an important consideration in developing a wave protection system. A higher, longer period wave may be caused to trip or break by some device or structure, resulting in a lower and shorter period wave that can be further attenuated by still another device that may not work on the initial wave. In a sense, what you are doing here is creating a specific site for an otherwise site specific wave attenuation device.

Turbulent Disorientation and Dissipation

Frictional dissipation of the wave's energy is quite an ambitious endeavor. A possible plan is to take the energy of a wave which, as an example, is 10 feet high, with a wave length of 150 feet, and with energy which reaches down into a depth of 75 feet below the surface, and turn it ultimately into heat, the final product of friction. Obviously no one has figured out how to do this yet, and when they do they will have invented something far more important than a wave attenuator. It is realized, however, that some form of turbulent, frictional effects will play a role in many attenuation processes. Simply slowing flowing water along some roughened surface will result in energy dissipation through heat. The problem is that the losses are too small to be used exclusively in attenuator designs. One step back from this process is the concept of turbulent disorientation. Using this concept it is possible to envision some well organized fluid movement under the action of a water wave being subverted into ever smaller chaotic motions which first destroy the organization of the fluid under a surface wave, and therefore the wave itself, and then further break down into turbulent motions of a scale small enough to either do no harm or be ignored. There is an old saying in hydrodynamics which may better describe the process of breaking down the organized wave, "big swirls have little swirls that pray on their velocity, little swirls have smaller swirls, and so on to viscosity." The idea is to get the larger threatening wave form to cascade its energy down into more numerous, but smaller physical systems that can easily be dealt with.

Engineers and scientists, well removed from the marina business, but very much concerned with waves, often work with experimental wave tanks. These wave tanks are elongated boxes which contain water and which have a mechanical device at one end to generate water waves. The wave tanks are useful devices for testing and studying all aspects of water waves and their effects on fixed and floating structures. They may be 10 feet in length or 1000 feet long. One thing that these wave tanks do not need are for waves to be reflected off the far end, returning back and interfering with the studies going on in the middle of the tank. Generally this is handled very nicely through the use of what is best described as a steel wool beach at the far end. This usually consists of debris and scraps from metal turnings, both large and small, piled up to form a sloping mound and sometimes wired in place. This produces a structure that absorbs the wave energy by effectively producing the turbulent disorientation that totally destroys the wave over a short distance, and feels the effects of only limited organized wave power. Perhaps this principle offers some innovative attenuator design possibilities.

8.3 WAVE PROTECTION DEVICES

From the framework of basic concepts and understanding discussed above, let us now apply some of these thoughts to the more common wave protection devices. There will always be new methods to solve old problems. Ideas and inventions are always sought and welcomed. Many that have already been proposed are being examined and the ones with promising futures will emerge. For the time being, this book will address only those commonly accepted attenuation and blocking methods. In many respects and for many audiences, this is more than enough. If designers can "get these right," the marina industry will have a firm base from which to further experiment and explore.

Rubble Mound Wave Barriers

The stone, rubble mound wave barrier offers advantages in the form of excellent storm protection, depending on the height, length and positioning selected. Wave overtopping and runup is greatly reduced by the sloped, rough rock surface that it presents to the oncoming wave field. Furthermore, wave and boat wake reflection off its inside and outside face are relatively small because of the energy absorbing ability of the pore spaces between the stones. It offers the best perimeter protection for use in protecting a site from the most severe waves. These wave barriers must be carefully engineered and designed for the best form, shape, elevation and position. Very careful thought must be given to them because, once constructed, very few are ever removed. They become a permanent part of the landscape and any environmental damage they may cause must either be accepted or the barrier must be removed. This may be a very expensive penalty for a mistake.

Any proposed stone wave barrier must be able to withstand storm wave conditions commensurate with the most severe design return period storm. They are effective wave barriers, being one of the few true breakwaters. Their sloping sides allow the oncoming wave to break, in the true sense of the word, similar to the way in which a wave would behave on a sloping sandy beach. They consist of small and large stones. There is a considerable amount of experience and published literature associated with rubble mound barriers. For this discussion, it is sufficient to state that they are organized piles of rocks with a sloped surface, consisting of stones which are large enough to prevent or limit movement under most wave conditions. The size of the stone used is very important, as is the ability of the stones to interlock with one another. They can be designed as either permeable or nonpermeable barriers to the oncoming waves. A rubble mound barrier designed for a not-so-severe storm, may be spread out all over the bottom when a really big storm comes along. Because of their permanency, it is necessary to design them with the worst storm in mind. Having to retrieve the pieces and reconstruct a rubble mound barrier can ruin your whole day.

Construction costs may be reduced by reducing the size and elevation of the barrier, but the stone sizes for a traditional rubble mound barrier should be determined by considering the most severe waves arriving in the area. As an example, a site with the 100 year return period wave of 7 feet with storm tide elevations flooding approximately 9 feet above the normal tide levels may require a height of approximately 15 to 16 feet above mean lower low water if no wave carry over is desired. By comparison, the 20 year return period storm may result in flood elevations of approximately 6 feet above normal tide levels with the potential for 5 foot high waves. Using the 20 year return period storm as the design storm means that if full protection were desired, the stone wave barrier would have to have an elevation of approximately 12 to 13 feet above mean low water, if no significant wave carry over is desired. However, for the 100 year storm, wave energy will overtop the barrier and enter its protective space. Remember, the wave may overtop, but make sure the stone sizes and side slopes take account of this larger wave. There is obviously a compromise in terms of what storm return period is economical to design for and how much wave carry over can be tolerated. These decisions may be made after a comparison of wave protection results that may ultimately be realized within the marina perimeter. The barrier must be high enough to provide reasonable protection under most storm flood level conditions. If a stone wave barrier were to be built at a lower level, its effectiveness could be severely reduced as compared to the potential all weather protection such a device has the ability to provide.

Because of their very effective wave blocking ability, stone barriers will not allow the transport of sediment along the shoreline, regardless of their orientation. If they are positioned at right angles to the shore and adjacent to it, they form a barrier for the littoral drift. If they are positioned parallel to the shoreline, their wave shadow prevents the continuation of the transport of the littoral drift and thus the area behind the wave barrier tends to fill in with the drift material. By their very nature, the shape of a rubble mound barrier which is basically a pile of rocks is such that it has a broad base and a narrow top or crest, a pyramid shape in cross section. The deeper the water, the larger the base or footprint will be. This footprint represents bottom loss for other uses, space, or plant and animal habitat.

An additional unfortunate aspect is that such a structure can close off a significant portion of a waterway or entrance channel, thereby causing a faster river or tidal flow in the vicinity, as well as potentially trapping ice and debris, and creating unacceptable sedimentation problems. On balance, a rubble mound barrier must not only be carefully engineered, but also very carefully analyzed for its effects on the physical system in which it is to be placed. Once constructed, it remains for a long time. Most state and federal regulatory agencies will also require a careful examination of environmental considerations, including aquatic habitats and the quality of the water trapped behind the barrier.

Solid Vertical Wave Barriers

Solid wave barriers may take the form of interlocking sheet piles, stone filled cribs, closely spaced concrete or metal cylindrical piles, or in some cases wood piles and timber barriers. They must be constructed to elevations comparable to those cited for the stone barriers and in most cases slightly higher because of their tendency to allow vertical wave runup and carryover. However, the amount of bottom that is lost is generally significantly less than with the stone, pyramid structures. Considerable bracing or batter pile support may be required for them to withstand the extreme waves. Some available commercial products have the form of cylindrical, interlocking piles, which if properly installed to sufficient depth may be used without batter support.

In most cases, because of their vertical orientation, wave carryover is more severe because the wave surges vertically along their faces and can be thrown above the top of the barrier. Riprap placed along their fronts, on their wave active surfaces, helps to reduce wave carryover and supplys some additional wave force absorbing characteristics. Solid wave barriers suffer from the same problems as do rubble mound barriers relative to their ability to interfere with any existing littoral transport or other current regimes.

Wave energy absorption by the wall or barrier can be increased, and therefore the reflection and resulting wave height reinforcement can be reduced. Different techniques are available for absorption, including stone riprap faces, or fancy design absorption elements on the face of the wall. Just remember, any design element that is expected to reflect waves in a more desirable direction must be larger than a quarter wave length of the oncoming wave or it will be invisible to the wave, and have no effect.

There are a number of disadvantages to these vertical wave barriers. In many areas when the tide is out, marina patrons may wind up sitting on their boats looking at a relatively ugly, high, blocking vertical wall with its odoriferous plant and animal life waiting for submergence by the next tide. Obviously, to be effective, the wall must be high and it will seem even higher than necessary when the tide is out. Another disadvantage is the vertical wall's ability to promote bottom scour along its base. These wall revetments create zones of extreme wave turbulence where waves breaking against these walls create vertical hydraulic jets which drive water upward and downward into the bottom substrate that supports them. This sets the sediments into

PERIMETER PROTECTION   183

suspension and enables the wave currents to carry them off. The base or toe of these vertical barriers may be the site of sufficient wave turbulence to cause actual excavation of the bottom. Solutions to this situation, if it is considered a problem, consist of placing stone or very coarse gravel riprap at the base of the barrier, or terminating the end of the barrier somewhat above the bottom.

Wave Boards and Wave Fences

Wave boards and wave fences, as they are sometimes called, can make effective wave screens. The idea is generally not to build a solid vertical wall to quell the waves, but to leave some spaces between the planks to allow for water circulation and reduced forces. Wave board spacing is critical here, and its size must be a compromise between the goals of wave attenuation and other considerations.

Wave boards are similar to solid barriers with one important exception. They take the form of either horizontally or vertically erected planks, attached to support walers, with a pre-determined spacing between the planks, (see Fig. 8-2). Their function is to intercept the wave energy in part

Figure 8-2. A vertical, timber pile and plank wave board array, protecting a marina at Stonington, Connecticut. Because of the potential wave forces, this wave attenuator is designed to overtop in extreme storm conditions.

and reflect it seaward. They will allow a certain portion of the wave energy to enter the marina area. This portion of the wave energy is dependent on the proportion of the space between the boards, and on the width of the boards. The spacing between the wave boards, and their termination distance from the bottom allows for water circulation into and through a marina, thereby providing for flushing action and water renewal as the tide level changes.

Two considerations are generally given to the functioning of wave board arrays. The first relates to the board width and spacing to be used for the installation, and the second, to the draft of these boards or how deep they are to be set. This is important because of the wave energy that can pass below the boards at different distances from the bottom and for differing wave periods. The longer the period, the greater the energy that passes below the wave boards, if they do not extend to the bottom. At the same time, if the wave boards need not extend to the bottom, flushing will be enhanced, as will be passage for bottom dwelling and migrating animals. Such considerations must be evaluated against the depth of installation required for their effectiveness. Unfortunately, in the case of shallow water the wave boards must extend to the bottom to be effective and their only difference from a solid barrier is the board spacing and the benefits that this will afford the environment.

Figure 8-3 demonstrates wave board effectiveness for 4 foot high waves. It plots the variation in the percent of board spacing allowed versus transmitted wave height. The wave boards are assumed to extend to the bottom and no energy can pass underneath to add to the transmitted wave heights. As

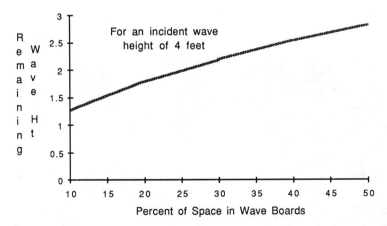

Figure 8-3. Diminishing wave board effectiveness as the spacing between the boards is increased. These wave boards extend to the bottom and allow only the passage of wave energy through the space between the boards.

can be seen, an initial 4 foot wave is attenuated down to 2 feet using 25 percent spacing, and down to 1.3 feet using 10 percent spacing. Obviously, saving money by increasing the space or gap between the boards is not cost effective for wave attenuation. However, wave board spacing is often a regulatory problem in terms of the spaces desired for flushing and biological considerations. Shorter sections of wave boards may be permissible with very close spacing, providing other large openings exist around the perimeter of the marina. The effectiveness of these devices is strictly dependent on the allowable spacing.

Consideration must also be given to the depth or draft of the wave boards. That is, how close to the bottom they will extend. In the shallow water, and if permitted by the regulatory agencies, they may extend to the bottom, with no gap underneath. In that case the wave board analysis discussed above will apply. If they do not extend to the bottom, then wave energy can pass underneath the wave boards and provide an additive effect to those wave heights transmitted through the boards themselves. This effect of passing under the wave boards will vary with the period of the wave, the depth of water, and the space allowed under the wave boards. Longer period waves have their energies distributed more uniformly in the water column. Any space between the ends of the wave boards and the bottom allows energy to pass underneath and reform waves on the other side, adding to those wave heights passed through the wave boards themselves. As previously discussed, the principle involved is power transmission through the space provided below the ends of the wave boards. The wave heights that emerge on the other side of the boards, after passing underneath, are related to the square root of the power. Even though half the power is blocked, the half that gets through underneath the wave boards, reforms waves with heights only a little smaller than the incident wave itself. For example, using a wave height of 4 feet with a 4 second period, and 15 foot water depths the following data set was prepared. The listing shows the wave heights that will be transmitted below the several board drafts given.

Board Drafts	Water space below, ft.	Incident wave height	Transmitted wave height
9	6	4.0	1.9
12	3	4.0	1.4
15	0	4.0	0.0

An important point to remember is that when energy is allowed to pass under the wave boards, the resulting wave heights must be combined with those wave heights that may pass through the wave board spaces. If the wave boards extend to the bottom, then the only concern is with the wave energy that they will pass through the spacing. As their distance above the bottom increases, additional wave energy passes underneath and the resulting wave heights within the marina become a combination of the waves entering between the wave boards and the wave energy that passes beneath them. For example, a 4.0 foot wave will become a 1.3 foot wave after passing through wave boards with a 10 percent spacing. If the wave boards are 3 feet from the bottom, an additional 1.4 foot of wave height will occur, potentially producing a combined 2.7 foot wave inside the marina, behind the wave boards. Thus, this wave board array effectiveness on a 4.0 foot storm wave is only a 33 percent reduction in height.

Partial walls, rising from the bottom but not reaching the surface have application in some situations. Because of the distribution of the wave's power vertically, walls extending upward from the bottom are not as effective as those extending down from the surface. The fact remains that a considerable amount of wave energy gets either underneath or over the top of the wave board. The wave period and water depth can make significant differences in the approach and use of wave boards. Wave boards are frequently seen attached to the outside of a fixed dock and not extending to the bottom. As already discussed, not only do these installations allow wave energy through the spaces between the boards, but they allow energy to pass beneath them. There are many marina operators who are puzzled and disappointed by the remaining wave activity within their marinas after the installation of wave boards.

Again, on balance, wave boards are not significantly more desirable or less effective than a solid barrier. They still possess most of the undesirable elevations and sediment blocking characteristics of the solid vertical barrier. They do represent a compromise with regulatory requirements for water passage, while being able to provide some wave protection for the site. As an interesting aside, water currents caused by rivers and tides can flow quite easily through the spaces between the wave boards. Most wave board arrays, consisting of closely spaced square edged planking, or perhaps even closely spaced round piles, will interfere very little with either the velocity or volume flow of a current system. The blocking provided by the boards or piles does slow the water's velocity as it approaches the barriers, however, the water pressure at the barrier increases sharply and forces the flow through the spaces or gaps in the wave boards at high velocity, resuming a more placid course on the other side. Its volume flow becomes comparable to its volume flow prior its passage of the barrier. This fluid flow principle was well described by the physicist Daniel Bernoulli (*Hydrodynamica,* 1738) and is, in part, why a hose nozzle works and airplanes can fly.

Floating Wave Attenuators

Floating wave attenuators are available in a variety of sizes, designs and materials. These attenuators can provide a degree of marina protection from the ambient summer waves and boat wakes, and withstand and provide some protection from the extreme waves, (see Fig. 8-4). From an environmental standpoint, floating attenuators are attractive because they do not interfere with the bottom ecology or aesthetic view. The consideration of the aesthetic view is especially true if the floating attenuator is moored to the bottom rather than being pile supported. They are not considered fill and can be removed without leaving any permanent damage to the environment. They allow the free circulation of water and do not form barriers for suspended and bed load sediment transport. Floating attenuators are available in the form of wooden decks with plastic or metal floats, or concrete covered foam billets. Their effectiveness is usually dependent on their mass. Their

Figure 8-4. Newport Yacht Club and Marina, Jersey City, New Jersey. A composite fiberglass and structural wood core floating wave attenuator, on the Hudson River opposite Manhattan Island, New York. The structure was designed to survive storm waves and attenuate passing boat wake, a major problem in busy New York Harbor. This innovative attenuator uses floodable pontoons to increase structure mass and achieve a 4 foot design draft.

mass is related to their volume and water displacement. The greater the horizontal and draft dimensions, the greater is their displacement and effectiveness. Some floating attenuators are created out of ready made barges, ballasted deeper into the water with sand, rock or even with water. Others may simply simulate the features of a barge and are perhaps more aesthetically acceptable. The concrete systems are widely used products. They are produced in various lengths and usually have widths of approximately 8 to 10 feet. Multiple sections can be attached in a parallel manner to provide for even wider sections, which generally improve their performance. Other versions come as very wide but shallow draft and narrow modules strung together with chain or wire (see Fig. 8-5). These modular models appear to have a potential as good survivors during extreme storm conditions, but they take up a large amount of water surface and cannot be used to moor boats, even on a temporary basis.

Most, but not all, floating attenuators must be installed in water depths that will not allow them to come in contact with the bottom even under the most extreme wave conditions, because if such contact is made cracking and

Figure 8-5. Spinnaker Island Yacht Club, Hull, Massachusetts. A floating concrete, modular wave attenuator using a chain bottom anchoring system. *(Photo courtesy: Walcon Barnegat Co., Inc.)*

structural failure may result. Specific floating attenuators may have some limitations on their use in extreme wave fields, and the manufacturer must be consulted in each case. Some attenuator products may tolerate greater wave fields, but they may also be less effective or very costly. Attenuator drafts may range from 1 to 8 feet and more, with widths and lengths variable, as desired. It is a growing field with new developments occurring relatively quickly (see Fig. 8-6). The products available today have site specific limitations. Not all floating attenuators will work effectively at all sites and under all wave conditions. Specific site protection must be carefully analyzed in the light of reasonable generic requirements for an attenuator, with an actual product contrasted with these requirements after product suitability and availability have been carefully surveyed. Unless a detailed analysis is carried out, and a floating attenuation device is either designed specifically for the site or a manufactured one carefully selected, there cannot be any expectations of significant attenuation success.

Floating attenuators operate using a variety of wave suppressing techniques. Fundamentally, constraining the water's surface with a large fixed-in-place object reflects a wave as it attempts to pass beneath it. The longer or wider the space restrained on the water's surface, the better a reflector it makes,

Figure 8-6. Offset, reflecting wave attenuator at the Philadelphia Marine Center, Philadelphia, Pennsylvania. Note accommodation for boat berthing and use of the attenuator deck surface for patio type facilities. *(Photo courtesy: Waveguard International, E. Douglas Sethness, Jr.)*

resulting in a reduced wave height once the wave passes underneath it. There can be many variations on the configuration of the system, including where the fixed-in-place object is thin in the vertical dimension and where it allows the wave to break over it, expending the upper part of the wave's energy across its surface. Additional variations or combinations can be made using both the principles of reflection, i.e., due to suppression of the sea surface, and wave blocking by having a vertical surface hanging downward. Now wave attenuation from both concepts are employed for the greatest gains.

From a theoretical standpoint the amount of pure wave blocking that a floating attenuator will provide can be easily be envisioned. With a draft of 4 feet below the water's surface and a 1.5 foot freeboard, it can be assumed, to a first order, that the wave's power which is intercepted by the floating attenuator draft and freeboard will be reflected back. The wave's crest power which passes above the floating attenuator's freeboard will, in part, be expended as the crest is forced to break on and over the floating attenuator. Depending on the width of the floating attenuator, this breaking action may simply result in spray and sheets of water passing over the top of the floating attenuator. At worst, remnants of this breaking wave may reach the lee side of the floating attenuator and create a new wave disturbance in the water on the opposite side. This new disturbance is totally disconnected from the wave that produced it and if it develops into a series of waves or disturbances, its period will be determined by the interaction of the disturbance wave's height, water depth and potential speed of movement.

As previously mentioned, a feature of a floating attenuator is its ability to place a constraint on the surface of the water. If it were rigid, or a large flat plate compared to the oncoming wave's wave length the vertical motion of the water particles under a passing wave form would be restricted. This restriction is greatest close to the surface and becomes weaker with depth. The energy that would otherwise be expended in the vertical movement of the water must now be redirected somehow. We are well aware of the wave calming effects of certain naturally occurring oils, including certain grain produced oils and sperm oil when they are spread on the sea surface. An elastic sheet can be used to best describe the effects of these oils. The sea surface, here again, is restrained and the waves reduced in height. Even the effectiveness of the ubiquitous floating tire attenuator, especially when very wide blankets of tires are used in their construction, can be explained, to a certain extent, by this restraint of the sea surface. These effects have been demonstrated in part, by studying the effects that plastic sheeting has on waves when such sheeting is spread over the water's surface. One of the most amazing phenomena of wave attenuation that can be observed occurs when large, 10 foot high waves with 5 to 8 second periods travel from an open sea condition into an area of floating ice. Even if the ice is not coherent, for example not large solid sheets, but simply slush and broken pieces floating on the surface, these large waves all but disappear within several wave lengths of travel. A technique often employed by ships with strong hulls when traveling in arctic or subarctic waters is to enter the edge of these ice fields to escape from the storm waves and ride out the storm in relatively calm waters.

Some of the wave's energy is expended and attenuated in the work of lifting this constrained sea surface. Some of the energy is also lost due to the turbulence induced, and the resulting chaotic friction that is created by the effects of the surface constraint on an otherwise well organized wave. However, these losses of energy are small in comparison with the net effects that the surface constraint has on the incoming wave fields. The most significant effect is the reflection of the wave's energy back out of the water space which is enclosed between the constrained upper surface and the bottom. If the bottom is remote, deep water, the wave's energy can, in part, be redistributed in the vertical direction, absorbed, so to speak, in the deeper column of water where the effects of the surface constraint are less. This allows more of the energy to continue on its journey and to pass beneath the floating attenuator. If the bottom is shallow relative to the wave length, this redistribution and absorption is weaker and the result seems to be the reflection back out of some of the wave's energy.

Figure 8-7 is the result of a theoretical analysis of the performance of a generic floating surface attenuator section. The analysis was performed for widths of 12, 18 and 24 feet, in order to show the improvement in attenuation characteristics as the width is increased. It was assumed that the attenuator

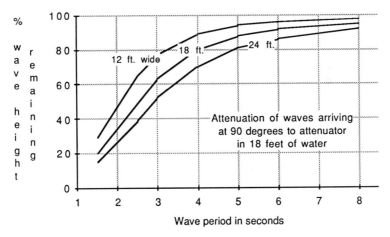

Figure 8-7. Theoretical floating attenuator performance for various attenuator widths as shown, but with no draft considered. Wave arrival angle is 90 degrees to the face of the attenuator.

was a surface constraint on the water and did not have any draft. It was further assumed that the attenuator was restrained, fixed in place, and of sufficient mass to prevent movement in either the vertical or horizontal directions. The performance of floating attenuators is directly related to the wave periods that they will be exposed to. The longer the wave period, the less effective will be the wave attenuation. Their application is best realized under normal annual wave and wake conditions, with the expectation of survival under the extreme conditions. Figure 8-7 also shows the performance characteristics in terms of the percentage of the original wave height remaining, after passing the attenuator, versus the wave period for waves arriving at a right angle to the face of the attenuator. As can be seen, the attenuator's performance improves with increasing width, but falls off rapidly as the wave periods increase to 4.0 seconds and beyond. As can be seen from Figure 8-7, for a 3.0 second wave a 24 foot wide attenuator with no draft will reduce that wave's height by 50 percent.

If the waves were to arrive at an angle of 45 degrees with the face of the attenuator, attenuation improves. Figure 8-8 shows the attenuation effects with the attenuator receiving the waves at an angle of 45 degrees. In this situation the 24 foot wide attenuator reduces the 3 second wave by 60 percent. The greater the arrival angle to the face of the attenuator, the greater will be the wave attenuation. This means that if a floating device is to be considered for use at a particular site, it should be oriented to receive the extreme storm waves at the smallest angle possible, thereby providing the greatest effective protection for the narrowest attenuator width.

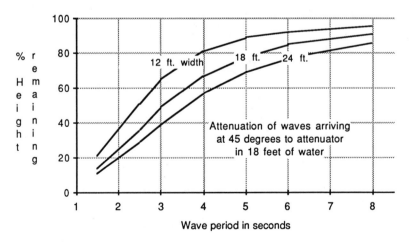

Figure 8-8. Theoretical floating attenuator performance for various attenuator widths as shown, but with no draft considered. Wave arrival angle is 45 degrees to the face of the attenuator.

A variation in attenuator draft also has a profound effect on wave attenuation. Figure 8-9 shows the theoretical effects of a constrained barrier with from 2 to 6 foot drafts, but without any width. Improvement in wave attenuation is considerable for the shorter, 1 to 3 second waves as the draft is increased, while improvement is less dramatic for the 4 second to 8 second waves. With a 4 foot draft, a 3 second wave is attenuated by 43 percent of its original height, while the 8 second wave is attenuated by only 15 percent. No attenuation improvement can be expected from decreasing the arrival angle, in contrast to the case of attenuator width, since the principle in effect in the attenuator draft case is simple wave blocking.

If the effects of width and draft are combined, the attenuation results are significantly increased. Figure 8-10 shows the effects of combining a 24 foot wide structure with increasing drafts, at 45 degrees of wave incidence. For the 4 foot draft case, attenuation of nearly 70 percent of their original heights for the 3 second waves occurs. Even better attenuation can be expected if these waves will be arriving at an angle of less than 45 degrees.

The effectiveness of the floating attenuator configurations shown are the most optimistic possible. They represent theoretical computations along with the assumption that the attenuators are fixed in place, without movement. This will not be true for most installations where the attenuators may be moored or pile supported and movement will occur. Unfortunately, most floating objects placed on the surface of the water are difficult to fix-in-place. The particle orbits of the waves produce movement and their wave attenuation effectiveness is drastically reduced. There are several remedies

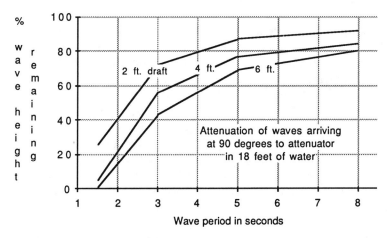

Figure 8-9. Theoretical floating attenuator performance for various attenuator drafts as shown, but with no width considered. Wave arrival angle is 90 degrees to the face of the attenuator.

194 PART 2/SITE EVALUATION

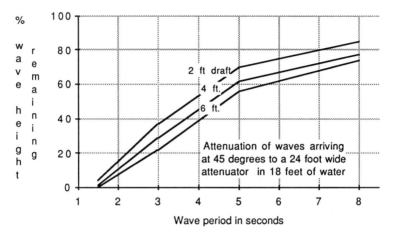

Figure 8-10. Theoretical floating attenuator performance for various attenuator drafts and with a 24 foot width. Wave arrival angle is 45 degrees to the face of the attenuator.

to this situation. First, make the floating surface in contact with the water as wide as possible. Make it greater than half the wave length, or better still, equal to it. This helps to significantly reduce the movement because different parts of a wave will be acting on it simultaneously. Another method is to create a large or massive floating object through water displacement, with the resulting natural period of the floating object being greater, and preferably much greater, than the period of the wave that it is supposed to attenuate. Still another method involves using horizontal restraining devices, such as pile supports, which restrict the horizontal movement, and using as massive and long and stiff an attenuator as can be practically placed at the site. The length and stiffness then allows those parts of the floating attenuator which are not being lifted by the wave to provide a rigid beam support to resist the uplift movement. If the floating attenuator is too light, wave lifting of one end or side lifts the rest of the floating attenuator and its effectiveness is greatly reduced.

A floating attenuator that moves up and down or back and forth has a significantly reduced attenuation ability, regardless of its width and draft. The displacement weight of the floating attenuator is very important not only from the perspective of the deepness of its draft, but also from the perspective of the tendency for attenuator uplift. As previously discussed, wind waves are short crested, particularly as they develop or grow in size. The lengths along their crests may range from 1 to 3 wave lengths. Well developed swell systems may have longer crests, but many of the waves expected to be quelled by a floating attenuator fall into the short crested category. Generally, a wave with a 3 second period in 20 feet of water may have a crest length of from 40 feet and ranging upward to 120 feet. Most wave floating attenuator systems will have lengths several times this distance which means that only a part of the floating attenuator will be acted upon, or be uplifted by the arriving wave, leaving a substantial part of it attempting to resist the movement. The stiffer the construction of the floating attenuator, the greater will be this effect since more of the floating attenuator is participating in this uplift resistance. The same concept can be applied for the longer crested wave systems, including boat wakes, which characteristically are long crested. This is a simple matter of orientation so as to cause the arrival of the boat wakes, or longer crested storm and swell waves to occur at an angle with the axis of the floating attenuator. To a greater or lesser extent, the floating attenuator is thus affected by the wave in a sequential manner, with the wave crest running along its leading edge and attempting to uplift the floating attenuator in a piecemeal fashion. The stiffer the structure, the longer and the heavier it is, the less success these short crested or angled waves will have in moving the system either vertically or horizontally. Some design consideration must also be given to avoiding having an easily moved attenuator exposed to a wave field with periods close to the natural period of the attenuator. Very unnatural gyrations may result if this occurs.

Floating wave attenuators will not produce any significant wave diffraction around their ends as will a stone wave barrier or wave boards. In addition, reflection of the larger waves will not be an important consideration because of the combination of breaking waves over the top of a low freeboard attenuator, and the amount of long period energy allowed to pass under the device. The same can be said for some shorter period wind wave and boat wake energy effectively attenuated by these devices. This makes them particularly suitable for use alongside channels or areas where frequent boat operations occur.

The use of a floating attenuator basically provides protection from most waves experienced under normal conditions. It is simply expected to survive under the extreme conditions while providing perhaps only a modicum of protection. That survival probability will be enhanced if the attenuator is chain moored rather than pile supported.

All of the principles discussed above and utilized in the various commercial and homemade attenuators have been reasonably well known for many years. For the Normandy landings during World War II, the Allies developed an extremely effective floating attenuator for protecting the landing craft on the beaches. It had a crucifix cross section and was made of concrete and steel, and employed almost every principle that has been discussed. It was called the Bombardon floating breakwater, as history records. Unfortunately, the waves of the North Sea proved too great for it and a severe storm destroyed it before it could do its job.

On balance, the use of floating wave attenuators for many sites appears to be the best solution for providing wave and wake protection while doing the least harm and providing the smallest impact on the environment. In general, they will do little for suppressing the storm waves with their longer wave periods. If a particular site is vulnerable to such storm conditions, then the floating attenuator may not be a suitable solution. However, perhaps it can be designed or selected to provide for the everyday type of wave protection and at least survive the storm condition, providing the same thing can be assumed for the marina dock installation behind it.

8.4 PROTECTION DEVICE SURVIVABILITY

The ability of a rubble mound wave barrier to progressively collapse is an advantage when it must experience extreme conditions. At least this progressive failure may mean that it may not experience catastrophic failure. Regardless of the expected intensity of the conditions, it is reasonably certain that the rubble mound at least will remain. Some damage is expected, in fact a properly designed barrier has already had determined for it the damage probability expected under progressively worse conditions. The point is that this type of structure may suffer progressive failure, piecemeal if you like, but the failure of one part of the barrier will not necessarily bring down the rest of it. Stating this another way, the load on one part of a rubble mound barrier is not distributed to other parts. Each section must stand alone, neither transmitting any significant loads along the barrier, or receiving any supporting strength from other portions of it. When dealing with vertical wall type wave barriers, or floating attenuators, their failure can be expected to be more catastrophic. For example, the vertical wall type of barrier generally does not fail progressively. In the case of the rubble mound barrier, the failure mode is usually the displacement or dislodging of the placed stone, creating a breach along sections of the mound. Those portions of the mound that remain intact, continue to function as designed and are little affected by those portions that have failed or partially failed. For a vertical wall type of barrier failure may be catastrophic resulting in almost instant loss of protection. Since these types of barriers are sometimes interlocking or are structurally connected for strength, this very feature which provides the strength may also result in the complete failure of the system. Once one part of the system begins to fail, total catastrophic failure can follow. If the vertical barrier consists of individual and isolated elements such as closely spaced piles, regardless of their diameters, the concept of progressive failure remains with those wall sections that remain intact, continuing to perform as designed.

In the case of the floating wave attenuators, catastrophic failure can also occur. These floating structures are simply moored vessels which most likely have a maximum holding capability. Whether they are cable or chain moored or pile supported, they will have their strength limitations. Many of these floating attenuators withstand very large wave impact forces as they perform their designed function. After all, they are expected to take these large waves and protect the systems in their lee. Up to a certain point the forces acting on these floating structures increase dramatically with increasing wave heights, and these forces are transmitted to the mooring systems. If the system is pile supported, reliance is placed on the pile and soil strength to withstand some significant shock loading as the waves increase in size. For most floating attenuators, a wave condition will be reached at which the attenuation begins to drop significantly and the waves simply pass the structure by. This is when the floating structure must "ride it out" as any moored vessel must do. However, for the pile supported system, "riding it out" may be considerably more difficult. Hammering on the piles by the floating structure provides for increasingly severe shock loads which must either be compensated for in the design, or eventual failure of the entire system will result. This is a case of: if one pile fails, maybe more and then many more will follow.

For a cable or chain moored system, survivability is significantly improved. As the attenuator ceases to function as an attenuator due to the size and period of the waves, the force increase with wave size is relatively small, but the movement of the structure increases greatly. This movement actually helps the structure to survive in this larger wave field. More of the structure is moved in unison by these larger waves, allowing cyclic displacement to occur and transmitting these actions to a flexible, stretchable mooring system. Until the extreme waves arrive, the attenuator remains relatively fixed in place because of its width, length or mass or all three, with the mooring system simply acting as its leash or tether to keep it in position. Under the effects of the extreme wave, this mooring system comes into play by providing the flexibility to allow for the greatly enhanced movement now required for the system to survive. Simply put, the effect of the large extreme waves changes the emphasis of design concern from a structure which must stand up to increasing forces to one which must withstand increasing movement. Flexibility and the ability to withstand cyclic stresses are now the orders of the day. A proper chain or cable moored system provides for the potential survivability of the system, with the design of the actual attenuator product and its mooring connections completing that design loop enabling the system to survive the storm. But now don't forget those systems behind this surviving attenuator.

A discussion has already been given regarding the effects of the limitation of wave crest lengths compared to the length of a floating attenuator. As pointed out, a wave crest arriving parallel to the leading edge of an attenuator may impact on only a portion of that attenuator at a given instant of time due to the finite length of the crest. That means that the remaining portions of the attenuator, not impacted directly by the crest, will provide structural support to that part of the attenuator receiving the wave's impact directly. This is an important concept. It means that a 500 foot long attenuator may receive a direct wave impact along only 150 feet of its length at any given instant of time. That also means that, providing the floating attenuator possesses sufficient structural strength, the remaining attenuator can provide significant support to that portion which was directly affected by the wave. Stating this slightly differently, it means that we are dealing with a system, rather than a simple isolated structure. If a wave is capable of providing 1000 pounds of localized impact force per linear foot of attenuator, then using a crest length of 150 feet, striking a hypothetical 500 foot long attenuator, means that the system may only need to withstand 300 pounds per linear foot of system force when distributed over the length of the attenuator. In theory, therefore, considerably less mooring restraint would be necessary in terms of pile strength, count or mooring chain configuration. As already discussed, wind waves may have crest lengths of from 1 to 3 wave lengths. Longer period swells, moving in from deeper water areas may well have crest lengths many times their wave lengths, and in their case, one should consider the wave crest striking the attenuator along its entire length. However, these swells may be much lower in height than the design wind wave, and their steepness may be considerably less. All this translates into lower forces caused by the swells on the attenuator, coupled with the fact that these longer swells and longer periods will provide more uniform uplift and rhythmic movement of the attenuator rather than a jolting impact.

The system concept for determining the loading forces caused by limited wind wave crest lengths can be optimized and the forces further reduced by angling the attenuator to the arriving crest. Obtaining the aspect of an arriving crest, rippling along the leading edge of the attenuator will reduce the localized forces and even further reduce the system forces per linear foot of the attenuator. Since the force of a wave on the surface of an attenuator manifests itself as a hydraulic pressure force at a right angle to the impact surface, the wave's force may be reduced as a function of the trigonometric sine of the incident arrival angle. These things must be carefully considered by the marina designer. It is interesting to note that in the few cases where wave impact forces have been measured on the mooring constraints of actual floating attenuator installations, the low values measured, in comparison to those calculated and expected, have surprised the researchers. Obviously more work is required here for a better understanding.

9
Planning the Marina Basin

A considerable amount of discussion and analysis of important physical planning factors have been covered in previous chapters. At this point, the information that has been developed can be integrated into a marina basin design, with suitable perimeter protection from the many natural and unnatural forces that may assault it. Once developed, the marina design must then be examined relative to the maintenance of water quality, the process of flushing, as well as the propensity for the basin to either fill with or interfere with the existing sediment transport regime. This chapter attempts to give some how to do it advice for integrating the information in the previous chapters into a successful marina perimeter design, along with explaining and pointing out some of the problems that may be encountered with the sediment regime once a trial design configuration and orientation has been established.

Some areas which are to be developed into a marina have natural configurations, shapes, or enclosures which may be integrated into the planform shape of the marina basin. Other sites may be excavated from the upland, providing the opportunity for a choice of basin shapes to the designer. Many recently planned marinas are being designed along open coastlines without any natural protection. Basically, many of the so-called good sites have been used up leaving only the marginal sites available. Most shoreline locations are usually assumed to be readily developable into marina boat basins. However, many of them should not be used for marinas. The need for extensive wave protection and bottom dredging, and the proximity of streams and river mouths, coastal currents, ice, and littoral sediment transport can all create an expensive situation for the development of marginal sites, not to mention the environmental issues and opposition that may be encountered. A thorough study and analysis of a potential marina site should be undertaken before a total commitment is made to carry out the development. Unfortunately, few approach the problem with that kind of objectivity. Usually the approach is to obtain the real estate and then do the studies and analysis to determine what is necessary to solve all of the site's problems.

What may then occur is that the site which had been anticipated as a first class marina, with significant wave protection and few problems, ends up being a marginal marina because of cost or regulatory compromises, with few boat patrons at the docks. In other words, the site may just cost too much to properly develop, or the regulatory agencies may place limitations on the use of structural solutions for solving the site's problems. It makes a lot more sense to find all of this out early in the process.

9.1 BASIN PLANFORM

In some respects, the marina developer who can excavate a marina basin out of the upland is lucky since there is usually no need to consider wave protection. On the other hand, the developer will have concerns about water quality, flushing of the basin, channel or access canal currents, basin surging and in some situations, catastrophic water contamination. Some of these water quality problems can be addressed through the proper planform design of the basin, avoiding corners, producing a circular perimeter shape with clever connections to the access channel, and even the use of mechanical water circulators or aeration systems. However, many new developments seem to end up in exposed inlets or along open coastal areas on either lakes or bays or even oceans. In these cases the primary concern is with wave protection, and the economics of the project is generally driven by this concern. The main determination involves carefully defining the wave problem.

A wave analysis for a potential marina is designed to answer three questions: What are the wave problems for the selected site? What wave protection can I enjoy within the economics of the project? What will be the wave forces on that wave protection so that it can be properly designed and remain in place after the occurrence of the design storm? In order to answer these questions, it is necessary to start out with a wind wave climate study, designed to determine the normal annual wave regime for the area and the long term maximum waves based on historical occurrences. This is especially important for the extreme wave conditions for coastal regions which are subject to hurricanes. This study provides the basic wave information, determined from available local wind data, on what normally expected wind produced waves may have to be ameliorated, and what can be expected as the maximum conditions which need to be survived. As discussed in Chapter 6, this wind wave study should also include a wave refraction analysis designed to determine the bathymetrically controlled arrival direction and the wave height alteration of the wind produced ambient and storm waves.

Depending on the site, a diffraction and reflection analyses may then be carried out, as necessary, to determine how much protection already exists at the site, and how the existing structures, shoreline shapes and revetments may interact with the arriving waves and wakes. Diffractive and reflective effects from, and around, existing structures can create waves on the marina facilities which are worse than what would exist in the absence of these structures. Plans may then have to eliminate these existing structures, or marina designs which mitigate their effects may be required. The most important point is that the designer should know what these effects are, or will be, before the marina is built, not after.

Perimeter protection determinations are then made to put into perspective the protection that is needed for the site under the various conditions of normal wind waves, boat wakes and storm waves. This is the point where some limited decisions are made about the economics of various forms, lengths and orientations of the wave protection systems which are possible and the desired storm security that the site is expected to require. The included boat wake analysis is designed to determine the potential arrival directions and frequencies of boat produced wakes in the area. The arrival direction and potential height of these wakes play an important role in the selection, position orientation and openings allowed in the perimeter. This process of perimeter site analysis is extremely important and plays a major role in determining development costs. Properly placed and angled wave attenuators or barriers, either fixed or floating, may result in smaller and less expensive structures for the site. Simply squaring off a site and placing wave attenuator devices at its boundaries may result in some very disappointing and expensive failures of wave attenuation goals.

Wave attenuation capabilities expected of the perimeter must be determined in order to match a particular device with the wave conditions. Estimates are then made about the wave climate within the marina after construction is completed. This is also the time when the detailed costs of the various levels of wave attenuation are determined and a decision is made about the type and ultimate level of protection based on the economics of the project. This part of the analysis allows the determination of the forces that the wave protection structures may have to resist, and suggests design features relating to the advisability of gravity structures, vertical barriers, or perhaps the use of chain moorings or pile supports. Once the perimeter's planform is completed and the wave attenuation or barrier structures are defined, then consideration can be given to the effects on water quality within the basin, and on flushing and siltation.

Formulation of the Perimeter

This is where the various pieces needed for a design of the marina perimeter's protection scheme begin to come together. This is where all of the time and consideration given to the understanding of wave attenuation devices begins to pay off. As a start, a wave refraction analysis should be carried out. The information from such an analysis can be used to create a composite chart of all the arriving wave orthogonals which are expected to be a problem at the site. These orthogonals should then be combined with the arriving wake wave orthogonals, examined and cast against some compromise, alternative or riparian position of the proposed perimeter shape or orientation. This process is demonstrated in the following two hypothetical examples.

Perimeter Development Example 1. Figure 9-1 shows a potential marina location at the confluence of two rivers. At this particular location, storm wave exposure is either from the south-southwest or the northeast. In addition, because of its location along a federal channel and a busy river, wake wave arrival, generated by both up and down bound river traffic, had to be considered. One or more trial perimeters can be drawn on a composite wave and wake orthogonal chart and an evaluation can be made of the wave and wake arrival angles, orthogonal spreading and energy concentration. From this composite chart some conclusions may be drawn, and perimeter protection plans formulated to provide the most suitable and cost effective protection possible.

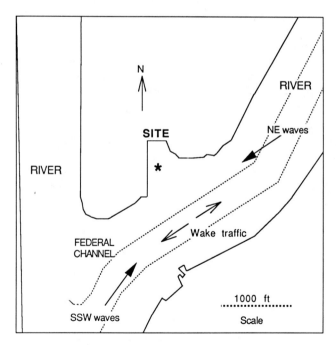

Figure 9-1. A basic site chart showing a potential marina to be located near the confluence of two rivers.

PLANNING THE MARINA BASIN 203

Table 9-1 summarizes the maximum near field waves arriving at this potential marina site. It includes the highest 10 percent waves in feet (H_{10}), the significant wave heights in feet (H_s), and the significant period in seconds (T), all shown as $H_{10}/H_s/T$, for the given return periods. These waves were then used to construct wave refraction diagrams for the site. As can be seen from Table 9-1, the wave height spread between the 20 year return period and the 100 year return period waves is relatively small, about one-half foot. This makes the design problem relatively easy since the marina can be designed to withstand the most severe waves expected by slightly strengthening the marina structures above the minimal 20 year design storm. For winds from the exposed northeast sector, wave heights on the order of 2 to 3 feet may be expected along with wave periods of slightly greater than 2 seconds. From the south-southwest the waves are a bit higher, with heights ranging 4 to 5 feet under severe conditions and with periods to 3.3 seconds. The 1 year return period waves are large enough to require some day-to-day wave protection for the marina, even if a 20 year or greater storm is not to be considered in the design. A wave with a height of 2 feet can do appreciable damage.

As a second step, a brief analysis was carried out on the potential travel of wake waves (caused by the proximity of the river channel) into the proposed site. Wake wave refraction diagrams were constructed for the site using what is considered to be a representative wake wave. That wake was assumed to be produced from a displacement or semidisplacement hull vessel, approximately 40 feet in length, traveling at a speed of 10 to 12 knots. This will produce a wake wave with a characteristic height of approximately 4 feet near the boat's hull and approximately 1 foot in height 500 feet away from its hull. The wake wave has a representative period of 3 seconds and will be refracted by the bottom as it travels out of the adjacent channel.

Figure 9-2 is the composite wind wave and wake refraction diagram showing the wave orthogonals as they arrive at the site. The marina's perimeter will receive storm wave energy from the southwest through to the northeast. This composite figure also shows the problems arising from wake arrival, with the wake waves from southbound traffic entering the marina from the

Table 9-1. Maximum near field wave data for perimeter development example 1.

Direction degrees T.	1 yr return $H_{10}/H_s/T$	20 yr return $H_{10}/H_s/T$	50 yr return $H_{10}/H_s/T$	100 yr return $H_{10}/H_s/T$
045-067	1.4/1.1/1.7	2.3/1.8/2.1	2.5/2.0/2.1	2.8/2.2/2.2
202-220	2.7/2.2/2.8	4.4/3.4/3.1	4.6/3.6/3.2	4.9/3.9/3.3

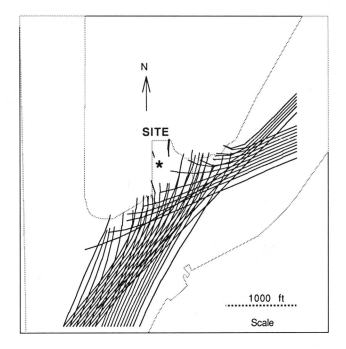

Figure 9-2. Composite wind wave and wake arrival refraction diagram.

east and wakes from northbound traffic entering from the south-southwest. The composite chart becomes a complicated mosaic, but it is very informative. There does not appear to be a natural side or end of the marina which is best suited to minimal wave protection and which is potentially the best choice for the perimeter opening for boat passage. The northeast sector will produce smaller and less frequent waves. Nevertheless, the eastern boundary of the marina must also provide some important wave blocking or attenuation. An access opening along the eastern perimeter is only slightly more desirable than one along its southern perimeter from a wave arrival perspective. An opening facing to the northeast, upstream in the river, may also allow a considerable amount of river flotsam to enter the marina.

It is readily apparent that the occurrence of wind waves, both maximum storm waves and the average chop conditions, requires complete and surrounding protection in order for the marina to offer first class, quiet and safe conditions. This is even further reinforced after examination of the arrival of the boat wake periods from the northbound and especially the southbound traffic. Northbound traffic tends to simulate maximum storm wave conditions in terms of wave periods, approximately 3 seconds, arriving from the

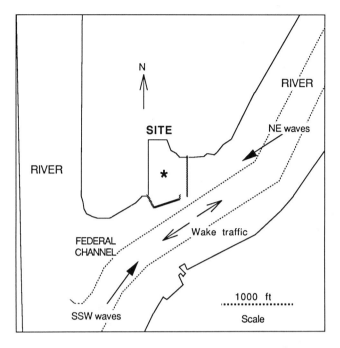

Figure 9-3. Suggested perimeter wave protection arrangement using floating attenuators.

same direction but with lower waves. Southbound traffic also simulates these maximum storm conditions with an ironic twist. In this case the wake waves have periods of 3 seconds or more which turns out to be one-third greater than the maximum storm wave periods from the easterly sector. The refraction of these longer period wake waves, coupled with their initial direction creates a situation in which their arrival tends to be slightly south of east, bringing them into the marina through the east and southeast perimeter.

Figure 9-3 shows a suggested perimeter protection for the marina. The availability of shallow water, space restrictions, nominal storm wave heights and the desires of the marina developer prompted the decision to explore the use of floating wave attenuators for the site. The perimeter layout attempts to incorporate wave and wake protection along with the exclusion of much of the floating debris from the river infrastructure. The entrance shown has a width of 75 feet. The layout addresses wave direction of arrival and the relative need for protection. It does not address the design or effectiveness of any proposed floating attenuator.

On balance, the worst waves that can arrive at the marina will come from the south-southwest. Therefore, the southern boundary must be a most effective wave attenuator and extend along this boundary as far as practicable. With the perimeter shape as shown, the western "dog leg" of this southern attenuator may be built smaller, as a less effective wave barrier because of the relatively small incident angle at which either the storm waves or the wake waves will arrive.

It also appears necessary to incorporate a wave attenuator on the eastern perimeter, perhaps as a combination attenuator/dock system. This attenuator/dock system can produce revenue at the same time that it performs its protective function. This attenuator will deal with less frequent easterly wind waves and lower and shorter period storm waves. It must still deal with wake waves from southbound traffic and their associated longer periods. This is a unique situation where instead of needing protection from annual, storm and wake waves as is required for the southern boundary, this eastern boundary needs primary protection from the wake waves. From that perspective, perhaps some costs can be saved by making this attenuator the least effective of the three discussed. If the wake situation can be made tolerable within the marina, using a minimal wave attenuator along this eastern boundary, and through proper boat orientation, the lesser storm produced maximum waves with their shorter periods will be accommodated also.

Finally, because of the potential arrival angle of the southbound traffic produced wake waves, the proposed opening as shown must be further protected from entering wake waves from a direction of slightly south of east. To do this, a stub attenuator section is added to the eastern end of the southern attenuator as shown. This stub section will intercept those wake waves attempting to gain access to the marina and disturb the southern ends of the docks. Wave and wake arrival will still occur from the south-southwest through the access opening. This is unavoidable if the entrance to the marina is desired along the channel side. However, it will produce several rougher water slips along the southern portion of the eastern attenuator/dock installation. Consideration may be given to making those rougher water slips available for larger boats which would be less affected by the waves and as a bonus have easier access to the entrance. This can work well for the marina in general where the larger boats do not have to travel any distance through the marina and thus produce less water disturbance for the smaller docked vessels.

It must be emphasized that floating attenuators are limited in their ability to quell the waves. They do not have the ability that a solid, fixed breakwater has to virtually eliminate the waves from entering the marina space. For that reason, if a floating attenuator is to be used, then the perimeter space of the marina must be closed off to the maximum extent possible with these attenuators. Stated another way, the floating attenuators let enough wave energy into the marina site by virtue of their generic limitations to attenuate waves, so that the additional wave energy that may enter through unprotected openings is certainly not welcomed and should be limited as much as possible. Less protection than has been suggested above must be weighed against the potential use and marketability of the marina. Furthermore, the amount of protection that the suggested floating attenuator arrangement may provide should be further analyzed from the perspective of product availability, effectiveness and price range.

One last comment on this example. There does not appear to be a single or unidirectional arrival pattern of the wind or wake waves displayed in the composite refraction diagram. There may be a compromise dock orientation with the boats berthed in a north-south direction to prevent a beam-on arrival angle of the wake waves produced by the northbound vessels and, in addition, during some storm conditions when the larger wind waves are expected to be produced from the south-southwest. This allows boats which are berthed in a north-south direction to receive the wake and storm waves either stern or bow-on. Wakes from southbound traffic will arrive on the beam of these north-south berthed vessels, and similar arriving storm waves from the northeast. However, these will not be as large as those from the south. This orientation may be a good compromise for the dock arrangement but some additional analysis of orientation versus benefits may be appropriate.

Perimeter Development Example 2. A second example involves a typical site along an open, exposed stretch of beach with a shallow offshore and a north facing wind fetch distance of 10 miles. Figure 9-4 shows the site orientation along an exposed shoreline. To the east, the site is separated from oceanic conditions by a long peninsula. It borders on a federal channel to its east, and has significant boat wakes generated throughout the year by fishing boat traffic. This site has serious wave problems and needs some substantial protection. Table 9-2 summarizes the maximum waves arriving at the proposed marina site, similar to the previous example, but from several more directions of exposure.

Shown on Figure 9-4 is a proposed perimeter outline for the marina site. The outline measures 1500 feet, parallel along the site's shoreline and extending 2000 feet seaward, adjacent and parallel to the federal channel to the east. The perimeter outline represents a general position for the potential use of one or more types of wave barriers or attenuation devices. The waves are assumed to arrive directly at the site and establish the wave climate for the maximum conditions as modified by the return periods. Based on an analysis of wave attenuation techniques, evaluated against the cost of protec-

208 PART 2/SITE EVALUATION

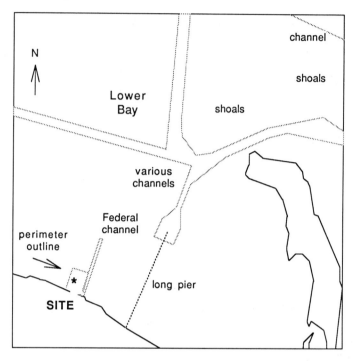

Figure 9-4. A basic site chart showing a potential marina and its proposed perimeter located along an exposed sandy beach coastline.

Table 9-2. Maximum wave arrival data for perimeter development example 2.

Direction degrees T.	1 yr return $H_{10}/H_s/T$	20 yr return $H_{10}/H_s/T$	50 yr return $H_{10}/H_s/T$	100 yr return $H_{10}/H_s/T$
315	4.0/3.2/3.4	5.8/4.6/4.0	6.2/4.9/4.2	6.2/4.9/4.2
337	4.2/3.3/3.4	6.3/5.0/4.0	6.9/5.4/4.2	6.9/5.4/4.2
000	5.1/4.0/3.9	7.5/5.9/4.5	8.1/6.3/4.7	8.4/6.6/4.8
022	5.3/4.2/4.0	7.8/6.2/4.7	8.5/6.7/4.8	8.8/6.9/4.9
045	5.3/4.2/4.0	7.8/6.2/4.7	8.5/6.7/4.8	8.8/6.9/4.9
067	3.7/2.9/3.1	5.7/4.5/3.6	6.2/4.9/3.8	6.6/5.1/3.8
090	3.5/2.8/3.0	5.4/4.2/3.6	5.9/4.6/3.7	6.1/4.8/3.8

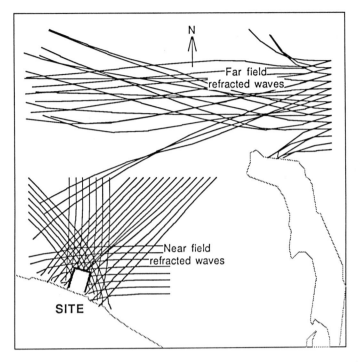

Figure 9-5. Proposed marina site and a composite wind wave arrival refraction diagram for both the far field and near field waves.

tion for each return period maximum wave regime, a decision must be made relative to the acceptable level of wave activity within the marina. This accepted wave activity must be gauged against its occurrence once annually, every 20, 50 or 100 years.

The near field and far field oceanic waves are included in the composite refraction diagram of Figure 9-5. However, as can be seen, these far field waves do not reach the site and are not used in the design process. Figure 9-5 shows that the northwest to north sector wave orthogonals progressively enter the site more directly as the wind conditions become more northerly. The northwest arriving waves are refracted by the shoaling bottom to arrive at the proposed western and northern perimeter sections at an angle of approximately 45 degrees. The northerly waves mainly impact on the northern perimeter section at right angles, while making only a small angle on the western perimeter. The north-northeast and northeast arriving waves impact the north perimeter at close to 45 to 60 degree angles, while not reaching the western perimeter at all. They arrive at only a slight angle with the east perimeter, between 15 and 30 degrees of incidence, while their orthogonals spread to some extent, lowering their wave heights. The north perimeter receives the maximum wave heights at the largest incident angle.

The composite refraction diagram leads to a suggested perimeter protection arrangement as shown in Figure 9-6. The west perimeter from the shoreline seaward to the north perimeter, must be of a solid barrier type. This is suggested because of the moderate size of the waves that can be produced from the northwest sector. The waves are sufficiently high to preclude the use of a floating attenuator. In addition, the littoral drift that exists at the site has a net movement along the shore from the west, therefore requiring a solid barrier to prevent the entrance of sediments into the marina. Wave boards will not provide either the protection desired or prevent sediments from entering and accumulating within the marina basin. The solid barrier must extend to a point on the shore above the reach of the highest storm waves in order to accomplish its task. The exact nature of the barrier is subject to design decisions, but because of the moderate wave heights expected to impact the barrier and the incident angle of their arrival, it is not necessary that it be a rubble mound wave barrier.

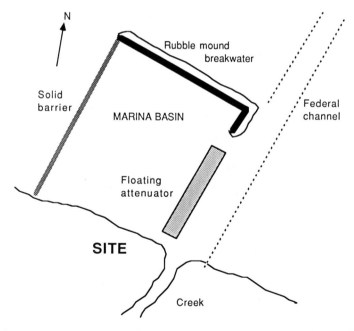

Figure 9-6. A proposed perimeter attenuation treatment for the various wave and wake intensities expected to arrive at the site from the different directions.

The north perimeter should probably be a rubble mound wave barrier. Under all wave conditions, this is the most active wave area, receiving the highest waves at almost 90 degrees of incident angle. This is the portion of the perimeter which anchors other perimeter structures and, in part, provides protection for the other wave attenuation devices along the east and west perimeter. It must be substantial and stable. Interlocking steel pile arrays can possibly be used here, however it is suggested that the seaward face of any pile structure be riprap for both additional protection and for the reduction of wave runup and carryover. In addition, because of the channel, deeper water depths and potentially higher waves at the northeast corner of this north perimeter, the wave barrier should be continued shoreward along the east perimeter until it is within approximately 1400 feet of the shoreline.

The east perimeter receives the smallest storm waves, however they still remain of considerable size. A floating attenuator of substantial construction is suggested for this perimeter. It would be placed parallel to the federal channel. A floating attenuator assures that excellent water exchange will occur within the marina perimeter. Waves from the northwest and north will be shielded from impacting on the attenuator. The extremely large waves from the north and northeast will arrive along this perimeter at a small angle, allowing a floating attenuator to be effective in ameliorating them to some extent. However, such a device must be chain moored in position in order to assure its survivability. A floating attenuator in this position will not prevent sediment from entering the marina basin from the east. This is less of a concern than is sediment entering from the west, which is the primary drift direction. Sediments from the east already contribute to shoaling and dredging problems within the federal channel. The federal channel will form a sediment trap/barrier for the marina basin, which will be similar to how it presently functions. No significant change in existing sedimentation problems for the channel from the easterly direction is expected. As a fortunate feature, the presence of the marina along more than 2000 feet of federal channel will greatly alleviate future federal dredging which will be necessitated by littoral drift from the west.

9.2 WAVE ACTIVITY WITHIN THE MARINA BASIN

The regions immediately behind any proposed wave barriers may not be entirely protected from either the ambient or the maximum waves. This somewhat reduced protection stems from a number of wave transmission sources, through, over, under and around the ends of any proposed wave barriers. For example, wave overtopping of either stone breakwaters or wave boards will occur depending on the final design height of these structures. This is generally determined commensurate with the costs and the desired level of protection under extreme conditions. In addition, the spacing of wave boards, and whether they extend down to or into the bottom, or terminate well above the bottom may allow significant wave transmission and entry into the marina.

Some wave transmission is expected even with an optimum design. Additional wave energy entering a marina area may be realized from wave windows created by the separation, gap or amount of overlap of any proposed wave barrier system. Wave diffraction around the ends of wave barriers will play an important additional role in providing the basin with still another source of imported wave energy. All of these potential wave sources must be considered in order to determine what the ultimate wave activity will be within the so-called wave protected water region of the marina basin.

The overtopping of the arriving wave fields for a rubble mound breakwater and vertical barriers is easily computed following the procedures of the *Shore Protection Manual.* The occasional combination of the far field and near field waves must be considered for the design problem, especially in terms of stone size to be used. If wave boards are used, the amount of wave energy that gets through or under the wave board array must be combined with the overtopping portion to determine the true wave climate that will exist within the marina's perimeter. If both the near field and the far field waves arrive from approximately the same direction, the potential combination of these two wave fields and the resulting increased crest heights must be considered.

The phenomena of wave diffraction around the ends of the wave barriers will also add to the wave activity within the marina. Wave diffraction is the physical occurrence of waves behind the end of a breakwater or barrier that is otherwise expected to cast a protective wave shadow in those locations. The wave energy is effectively bent around the tip of the wave barriers and into the expected wave shadow area by a physical process referred to as diffraction. As a wave train approaches the end of a solid barrier, that part of the crest that impacts on the barrier is abruptly stopped, that part of the wave crest that passes the end of the barrier can continue on its way past the barrier. However, after passing the end of the barrier as shown in Figure 9-7, the wave has had a portion of itself abruptly severed by the barrier, and although the wave continues past the barrier, there is now a shadow behind the barrier where no wave exists. The healing process for the severed wave produces a flow of wave energy along the crest and into the shadow behind the barrier. A series of circular waves are produced, still connected to the severed wave crest, but bending around behind the barrier. Nature abhors a vacuum and tries to fill the wave void or shadow behind the barrier as best it can by allowing some of the wave crest energy to spill in. That energy must come from someplace and these smaller circular waves created behind the barrier are formed at the expense of the wave crests that continue past the

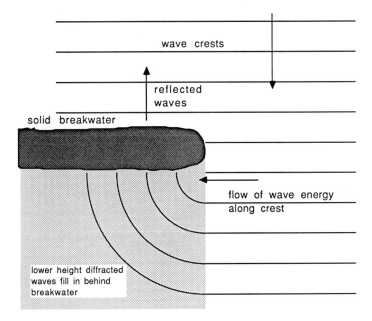

Figure 9-7. The diffraction of arriving waves behind a solid wave barrier.

barrier. That means that the portion of the wave nearest to the severed end will diminish in height as it supplies some of its energy to form the waves behind the barrier. Thus, diffraction results in two phenomena. First, the wave crests that continue past the barrier are reduced in height near their ends which have passed closest to the barrier. Second, there will be waves behind the barrier. Perhaps not high ones, but there will be waves. The further these circular waves travel and spread behind the barrier, the lower they will become.

Obviously there remains substantial wave protection just inside the end of a protective barrier, but the waves that are diffracted around the end of the barrier may still be quite high if their parent waves were high. It is a complicated mathematical exercise to determine just how high the waves will be in the shadow of the barrier. Recognizing this phenomenon is the most important point here, and a detailed determination of the wave height distributions behind a barrier should be carried out as shown in Figure 9-8. This figure shows lines of equal wave height as the waves diffract around the ends of the wave barriers forming the marina perimeter and entrance. The *Shore Protection Manual* gives the procedures for preparing these wave diffraction diagrams. General approximations can also be made from the *Shore Protection Manual's* example diffraction diagrams, but be careful

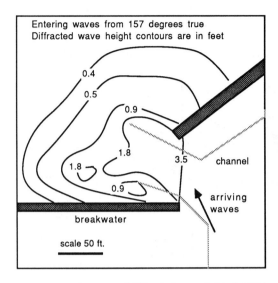

Figure 9-8. Computer generated pattern of diffracted waves behind solid barriers forming the entrance to a proposed marina basin.

here. The generic diffraction diagrams given in the *Shore Protection Manual* are for monochromatic waves, that is one period and one height for all of the diffracting waves. Generally it is assumed that arriving waves are somewhat random and a mixture of heights and periods. These *Shore Protection Manual* diagrams must be used cautiously since they show theoretical wave height increases which may not actually occur, at certain points along the wave's travel path.

An important phenomena that should be carefully examined is the wave energy reflection that will occur if vertical walls are present. Once allowed inside a marina's perimeter, and if contained by vertical surfaces, this reflected wave energy can bounce all over the marina space taking many minutes to die out and creating chaotic conditions. Wave reflection off a vertical surface will occur at an angle equal to the incident wave arrival angle, as measured from the surface towards the perpendicular. It is classically geometric and reflection diagrams are easily constructed. These refraction diagrams can also show how waves arriving nearby can be reflected into and within a marina from adjacent vertical surfaces. Remember, at each crossing of these wave crests, their heights add directly together. In addition, if a marina's perimeter consists of several vertical surfaces and if the site is subject to the arrival of long period far field waves, basin surging may result. In this case the vertical surfaces may trap and reflect these longer period waves, resulting in a continual heaving up and down of most of the water in the marina. Such happenings should be determined before the development of a complete marina protection perimeter. Few methods are available for absorbing incoming wave energy, so vertical walls and bulkheads should be carefully considered before they are specified. If they already exist in or near a proposed site, then careful analysis should be made of the potential problems they may cause and any solutions that may be available.

Once a perimeter is defined, the wave activity within that perimeter must be determined from the perspective of the allowed wave penetration. All aspects of the permeability or transmissibility of the wave barriers, the overtopping, diffraction, and internal as well as external reflection must be closely examined before the makeup and orientation of a final perimeter is proposed. If problems exist, now is the time to solve them, not after construction has taken place. People often tend to throw up their hands when dealing with water waves and say that there is little that you can do with them or to them. You simply have to live with them the way they arrive at your location of interest. That's just not true. If designers are clever enough they can refract them, diffract them, shoal them, reflect them, reinforce them, slow them down, speed them up, make them steeper, calm them and even break them. It is truly amazing what power we really have over water waves if we go about it properly.

9.3 WATER QUALITY AND MARINA BASIN WATER EXCHANGE

The water quality within the confines of a marina perimeter is often a concern of the marina operator and the local, state and federal regulatory agencies. The water that exists within the marina's perimeter came from outside of it and therefore it will be no better than what exists outside, and generally the regulators want to be sure that it will be no worse. Often, water movement within a marina will be restricted by the perimeter design and will depend largely on the amount of enclosure of the basin. The quality of the water which will then flush the area is of paramount concern because of the possibility for limited access in the marina basin. In addition, if dredging is part of the marina design plan, care should be taken to avoid the creation of a basin which is deeper than the surrounding or connecting bottom or access channel, so that bottom water will not be trapped by a sill and may be freely exchanged with the outside. Special concerns are usually related to summer water temperatures and the oxygen content as well as to the outside water that will participate in the flushing and renewal of the resident water within the marina. The quality of the water is a specifically defined condition or set of standards, set forth by each state and the federal government. These standards are generally quite similar from state to state and include such items as bacteria content, oxygen, heavy metals such as lead and copper, oil and greases, and other special chemical constituents that may be a concern locally or regionally. The water's content of these items is usually specified as not to exceed certain concentrations, or in the case of dissolved oxygen, not to fall below some fixed value.

The defining and recognizing of a potential problem is reasonably simple. If there are wave, wake or current protection barriers around the perimeter of the marina, the degree of free exchange of water between the inside and the outside of the marina is decreased by an increase in the degree of protection afforded the marina basin. This is a man made situation that did not previously exist at the site. The effect of reducing the water's flow to the outside must be given special consideration, and a series of technical studies involving the biology and chemistry of the water must generally be carried out. If a basin is to be constructed, and possibly excavated or dredged, these studies become difficult because they are predictive rather than being the relatively simple documentation of an existing condition. If the proposed marina basin is a natural, well protected basin, surrounded by land with a small entrance canal or channel, then existing water quality conditions can be directly measured under differing environmental conditions such as rainfall, changing water levels, or stream and river flows before a marina is built. These measurements may involve the installation of recording instruments to monitor a site over several days or even weeks. Often, data collection is carried out from the shore or a boat, with the measurements repeated several times over perhaps one or more tidal cycles or changing river flows. The use of dye is quite common for determining the movement of water or the flushing time required for a basin to reduce an initial volume of resident water down to some small remaining amount. This information is very useful for determining how long it may take for a marina basin to cleanse itself of some pollutant spill or contamination.

Water Quality Considerations

The concerns about the quality of the water within the confines of a proposed marina are twofold. The first concern relates to the exchange of water within the basin with the water outside the basin. How often the exchange occurs and the degree of efficiency of that exchange often indicate the chemical and biological fate of the basin's water. If the outside water has a high content of treated sewage effluent, as many rivers and bays often have, then this water should not remain trapped within the marina perimeter for any significant length of time measured in days. In other words, the water within the marina basin should have a fairly short residence time to allow it to be replaced with newer water from outside the basin. Natural basin enclosures have the advantage here because it is possible to carefully examine the so-called flushing times of the basin and measure the effects of the restricted movement of water within the basin. For a basin that is to be constructed, a designer can only speculate, contrast it with similar existing systems, or physically or mathematically model the proposed marina basin in order to predict the effects.

The second concern relates to what new or special changes may occur to the water held within the marina basin as a direct result of the marina being populated by boats. These concerns relate to sewage discharge, so-called grey water from showers and dishwashing, bottom paint leaching, fuel and lubricating oil discharges and leaks, and deck and hull washing. The concerns that may have arisen from simply reducing the water exchange with the outside and holding the ambient water in residence for a period of time, are now exacerbated by the fact that some new materials may be added to the water before it can be flushed out of the basin.

This is a very complicated problem. Discharge restrictions and the length of time a marina basin may hold the water present an array of changing and growing concerns on the part of all levels of regulatory authorities. These concerns vary by degree depending on the proposed marina's location, the type of water body involved, the potable use of that water, the marina's restrictive perimeter design as well as the potential for nearby sewage spills from municipal systems. It is a situation which is changing rapidly, often requiring the redesign of many onboard boat sewage holding and discharge systems in order to conform with a growing list of restrictive requirements.

Basin Flushing

The procedure for determining the water exchange rate or flushing time of a proposed marina basin must be site specific and is not addressed in this book. Guidelines and standards, where they exist, vary from state to state. The U.S. Environmental Protection Agency has indicated some desired goals, but there are no fixed flushing time criteria since each site is different. The methods to be used for establishing water exchange rates for marina sites must be left to the professionals and several reference sources may be found in the section on information sources. With the growing requirements for increased information by the various regulatory bodies, it has become necessary to do some computer modeling. This modeling may range from nothing more than a series of calculations based on accepted equations or determination procedures, to highly sophisticated deterministic or probabilistic computer models. In some instances, scaled, expensive physical models of the proposed marina basin may have to be constructed. Such models would include all of the water exchange mechanisms, properly calibrated, in order to simulate what will happen once the marina has been constructed.

The natural setting and surroundings of a proposed marina basin obviously play an important role in the water quality and exchange parameters of a particular site. A large tidal prism, defined as the volume of water brought in by the rising tide, along with a small low tide volume for the basin creates excellent flushing conditions and rapid water renewal. A river or stream that will flow through a marina basin may accomplish the same rapid exchange. Unfortunately, many sites are not situated in highly dynamic water movement areas. They may be in areas where the water does not naturally move at all. Often in these cases, circumstances are such that if a marina is to be built, additional engineering design must be carried out in order to improve a situation that is naturally poor. It may take the form of sculpting the shape of the marina basin to provide a planform conducive to water flow-through or circulation. Overlapping of the wave barriers arranged to act as scoops to the passing tidal or river water flow in order to direct some of the flow into or through the marina is often a good idea. Many times additional flow through pipes must be added through otherwise impermeable protection barriers in order to provide for better connections with the outside. Sometimes these flow pipes can be positioned so that they conduct well aerated surface water to the bottom of a marina basin and thus provide turbulent mixing along with direct bottom water replacement. Mechanical agitation in the form of pumps, propellers, or air bubbler systems is sometimes allowed to make up for the otherwise expected poor water quality (caused by sluggish water or the total absence of water movement) of a marina basin. Clever designs which create flow through adjacent tidal or fresh water marshes as part of the marina's resident water exchange path helps to clean the water naturally, either before it enters the marina or as it is returned to the outside water body.

Flushing calculations for a partially enclosed marina area made before its construction must usually be carried out as theoretical computations and then evaluated for feasibility. Very simple concepts or computations must give optimistic water exchange times, on the order of a few days, or there may be problems with the site. One can look at a marina most optimistically and assume that the circulation by direct water flow caused by a river or stream or tidal current will flush the basin by the simple process of water pass-through and displacement. A calculation of the marina's water volume divided by the volume flow-through rate will provide a gross yardstick for evaluation. Alternately, if no flow-through exists and the marina is located in a marine environment, then a simple tidal prism flushing calculation may also be performed. In this calculation, it is assumed that the region is well mixed and that a portion of the resident water volume is replaced by the intruding tidal prism on each incoming tide until the total resident volume is replaced. The procedure divides the marina's high tide resident water volume by the tidal prism volume brought in on each tide. The answer that results is the flushing time in terms of the number of tidal cycles. This method is generally accepted to predict the time it takes to dilute the resident water to the 50 percent level and is not very sophisticated. There are more detailed methods available which address a specific dilution factor for the flushing. In addition, they consider both the tidal and freshwater inflow, and provide for a return flow factor in which some of the resident water is partially returned to the site from the outside during a subsequent tidal cycle. Even these relatively simple computational procedures are best left to the professional, since they also require a considerable amount of interpretation and possibly calibration using dye studies of the water body. In most cases, a detailed computer model which addresses both water exchange rates, as well as oxygen or other chemical constituent reactions and replenishment rates must be included in order to provide the answers. It is not a simple task when required for a particular site.

9.4 MARINA ALTERATION OF THE SEDIMENT TRANSPORT REGIME

Vexing and costly problems that many marinas face are the problems associated with the deposition of sedimentary material within the confines of the marina's perimeter. This sedimentary material can become an expensive headache because it must be removed on a periodic basis by dredging. The problem may stem from the creation of a marina basin through dredging, with the natural siltation mechanisms simply trying to return things back to their original condition before the construction of the marina. It may also be caused by the fact that the construction of the marina required the use of engineered structures for wave, wake or current protection. These structures may provide the desired protection, but at the same time they may create quiescent conditions within their enclosures and promote the dropping out or deposition of material which would otherwise remain in suspension within the naturally turbulent or disturbed waters. In addition, much material can be transported along the bottom, near the shoreline, by the action of waves, river and tidal currents. Here again any installed structures may either be placed in the path of this moving bottom material or it may provide for an enclosed collection point where this material may now stop and be deposited rather than proceeding on along the shore.

Suspended Sediments

Most of the naturally occurring suspended sediment load within a harbor arises from either river borne sediments, sediments stirred up from the shallows of the harbor by the wind and waves, or sediments carried in from connecting coastal water bodies. Whatever the source of this suspended silt, the fact is that if a marina basin provides a quiet place for this material to settle out, siltation within the basin may proceed apace of that found outside the basin, within any maintained nearby channels. No one likes surprises in a marina basin, especially when the surprise involves the siltation of the basin, causing it to become unusable within several years. A comparison must therefore be made with the potential settlement of this suspended sediment.

The dredging history of a particular river, harbor or embayment may be available and quite useful in determining sediment fill rates for a potential marina site. Many marina sites will be located in rivers and harbors where Army Corps of Engineers federal projects exist. The location, extent and history of these generally continuing harbor and channel maintenance projects can be found in the *Army Corps of Engineers Federal Projects Book.* This book, available from the Army Corps of Engineers, not only provides specific information on the dimensions and extent of a particular anchorage, channel, turning basin or harbor deepening, but also some of the history and the frequency of the dredging. Greater detail can generally be obtained from the responsible Army Corps of Engineers District Office. The specific amounts, years and spot locations of any dredging which has been carried out can furnish information on future marina site dredging expenses. By simple calculation, a determination can be made of the siltation rate found in adjacent navigation channels. Extrapolation of that information to the marina basin then helps to estimate future dredging amounts and frequencies.

For example, the Stamford Harbor, Stamford, CT listing in the federal projects book describes a channel called the Outer Reach which connects to the next section called the Inner Reach which then branches to the west and ends in a Turning Basin, with the dimensions as shown below:

Turning Basin: 15 feet deep, various widths between 125 feet and 380 feet.

West Branch channel: 15 feet deep, 125 feet wide.

Inner Reach channel: 15 feet deep, 200 feet wide.

Outer Reach channel: 18 feet deep, 200 feet wide.

The project was completed in 1947 and the Army Corps of Engineers indicated that maintenance dredging occurred in 1964 in the West Branch and Turning Basin at its head, and again in 1979 in the East Branch. Based on a 1984 overall condition survey by the Army Corps of Engineers, the averaged across channel Outer Reach has shoaled by 1.8 feet and the Inner Reach has shoaled on average by 2.7 feet during a span of 37 years. The West Branch channel has shoaled on average by 2.6 feet and the northern portion of the Turning Basin on average by 3.8 feet over a span of 20 years. Typically, the further into the more quiet waters of the upper harbor, the greater the apparent siltation rate. Thus, from the survey data, the West Branch fill rate appears to be approximately 0.13 feet per year. If this data is extrapolated to a proposed nearby marina basin in the West Branch, it indicates that some maintenance dredging should be planned in order to balance a fill rate of approximately 1 foot every 8 to 10 years.

During more quiescent current periods, suspended sediments may be deposited within the marina basin region as the currents either stop because of the existence of obstructions, or cyclically abate as the tide changes and the silt drops from the water column. As this occurs, a relatively uniform layer of material can be plated on the bottom of the basin. With the installation of any wave or current barrier system, direct current flows may be altered in the area of the marina site, and certainly the amount of wave stirring will be reduced. However, free flushing access generally afforded a marina site provides that what suspended sediment is available outside the perimeter of the site can be brought into the marina site by the tidal prism. Information about the predicted depositional rate is necessary in order to determine what effects it may have on future maintenance dredging of the site once a marina basin is created. Towards that end, consideration of the expected sedimentation rate as well as the worst case must be examined.

A program of suspended sediment sampling should be conducted in order to determine the suspended sediment content generally existing in the water column of the area that is expected to enter the marina basin perimeter. The water samples should be obtained in a random manner, with samples taken during afternoons when the higher winds may be expected and at both flood and ebb tidal current conditions. Samples should also be obtained during periods of rain when any river runoff may be expected to contain more sediment. The sample frequency and distribution will vary for each site and it is not possible to specify the best protocol for sampling in general. However, once obtained, the water samples, generally of a quart or liter volume, should be sent to a laboratory for analysis of the suspended sediment content. Obviously, the more cloudy or turbid the water, the higher the suspended sediment content will be. Each sample will then have a sediment concentration determined for it by the laboratory in terms of the number of milligrams per liter of water. For many harbors, suspended concentrations on the order of 5 to 10 milligrams per liter (mg/l) are common with some larger commercial harbors having values of 15 to 25 mg/l. In many cases it depends on the season, the rainfall, storm activity and the size of the river that may form part of the harbor, with some larger rivers reaching 100 mg/l. Concentrations of 100 mg/l would make the water appear similar to chocolate milk.

Marina fill rates can be projected by using this suspended sediment data along with a series of assumptions which include the quantity of water borne sediment, the occurrence of the release of suspended material concentrations, and the frequency at which this happens. For example, since Stamford Harbor has already been briefly discussed, some recent suspended sediment determinations and volume calculations for the upper harbor will be used as a demonstration of the method. Assume an ambient concentration for the suspended sediments of 20 mg/l. This is the same as 20 parts per million or 10 grams per cubic meter. Convert this into a volume of sediment per volume of water by using the relationship that a cubic meter of sediment per cubic meter of water equals grams of sediment per cubic meter of water divided by the density of wet sediment in grams per cubic meter. The density of wet sediment is taken at one million grams per cubic meter of water, to be conservative. Thus, 20 grams per cubic meter divided by 1 million grams per cubic meter equals 0.00002 of a cubic meter of sediment per cubic meter of water. This volume of sediment to volume of water measurement unit is the same as 0.00002 of a cubic foot of sediment per cubic feet of water. Then, the sedimentation rate for any area where this material may settle out is given as being equal to the tidal prism volume, times the sediment concentration, divided by the bottom area. Therefore, using the potential marina basin area in upper Stamford Harbor, with a tidal prism of 33,727,000 cubic feet, times 0.00002 cubic feet of sediment per cubic foot of water, divided by 4,684,400 square feet of bottom, results in a siltation rate of 0.000144 feet per tide or 0.11 feet per year, or approximately 2.1 feet of assumed uniform shoaling every twenty years.

This calculated estimate of 0.11 feet per year is consistent with the fill rate determined from the dredging records for Stanford Harbor. It certainly provides a method of obtaining some information for a site, even in the absence of any other data. Undoubtedly, there will be isolated accumulations of sediment, since historical information has indicated that a uniform distribution of the siltation pattern does not necessarily occur. The location of these spot accumulations is generally not predictable but should be anticipated.

One last point. Since siltation occurs within the confines of a marina's perimeter structure because of the reduction of more vigorous water movement, one can assume that restoration of that vigorous movement, or stirring of the water, would tend to prevent the fine silt from dropping out of the water column and filling the basin. A cleverly distributed bubbler system or perhaps some propeller units or some other form of agitation during the lull in tidal movements can help to prevent, or at least reduce, the rate of siltation. The energy costs will easily balance the cost of physically dredging as well as the aggravation of the regulatory permit process in order to obtain permission to dredge.

Sediment Transport Interactions

Most natural shoreline sediment sizes are fine enough to be easily transported by the ambient wind generated wave fields and water currents. The water stirring and turbulent energy provided by the generation of wave fields, disturbs and picks up the sediment from the bottom and the beach or shore face and places it in temporary suspension. The water transport that exists as part of the physics of the wave fields, transports this suspended sediment in a down wave, and generally down wind direction. The direct mechanism for transport are the waves. At times, due to bathymetric refraction, these waves will be traveling at slightly different directions than is the wind. This wave induced sediment transport produces the littoral drift system along the shore. Depending on the site's particular wave conditions, the littoral drift can be bidirectional in the sense that it changes direction periodically. This means that information on the littoral drift at a potential marina site must be obtained for the short term, episodic occurrences from storm conditions as well as for the long term net effects and their interactions with the marina site.

Littoral drift, and the effects caused by its interruption both upstream and downstream for a proposed marina site can create some very important legal issues. This question of cause and effect in terms of properties disappearing as the result of a marina installation, or conversely the filling of what is naturally a deep mooring area must be carefully looked at by a professional. There is little that this book can arm the marina designer or developer with other than to emphasize that an expert in this field should be used if there is any hint of a present or future problem with the site's interaction with the shoreline.

Before the effects of a marina structure on the shoreline can be predicted, the mechanisms for maintaining the shoreline in its natural form must first be understood. The questions to be asked are: How much sediment can be moved over a prolonged time period of a year or more?, What is the net tendency of transport?, and What should be the shoreline equilibrium shape as a result of that transport?

A qualitative analysis can be carried out in order to gain some preliminary information about the potential for sediment transport in the area. The empirical equation as given in the *Shore Protection Manual* for predicting the annual transport of longshore sediments is: $Q = 200,000 \times Hb \times Hb$, where Hb is a defined term using the height of the waves at breaking, altered by the frequency of occurrence during the year. Once the average significant wave height is determined and Hb is calculated, one can make a simple determination of the net transport of sediment in cubic yards along the shore on an annual basis. The empirical equation used for these calculations assumes that a supply of material is available for this transport, and the equation gives an answer that is defined as the maximum transport value that can be expected. The process at least starts out determining the potential magnitude of the shoreline transport problem, or determining if one exists at all. If broad sandy beaches surround the potential marina site, the transport probably exists and may be a problem. If the shoreline forms a part of the pocket beach system there may be less of a problem. Generally within pocket beach systems, the total sand supply is held between the bordering headlands and the sands tend to oscillate from one end to the other depending on the wind's direction during any given season. If the shoreline forms a headland or peninsula, there may be no beach transport past that headland and therefore, perhaps, no problem.

Bathymetry shown from the aerial photographs is also a great help in initially determining a problem. Sand in the photos may be shown to be impounded in the wave shadows of islands and the rock outcrops. The bottom remains shallow behind these wave shadow features because the littoral drift is partially interrupted by their presence and sand is deposited shoreward of them. The aerial photograph may also indicate sand impoundment groins or jetties that exist along the shoreline. Some bias in sand accumulation on one side or the other helps to determine the direction of sand movement. Accumulation generally occurs on the upstream side of the groin which is interrupting littoral drift while showing starvation or erosion on the downstream or down current side of the drift.

Beaches

Many marina sites are adjacent to recreational or protective barrier beaches. The siting of a marina within or next to a beach complex requires that special design considerations be made in order to prevent the beach from being impacted by the presence of the marina. In some very large marinas, the beach may actually be contained within the marina. Such is the case in Marina Del Rey, California. If there is a beach proximate to a potential marina site, an understanding of beach processes along with a careful analysis of the site may prevent a good number of problems that may occur to these sensitive and delicately balanced coastal resources.

The trajectories of the water particles within an actively breaking wave field is a complicated process and is not well understood. However, it is known that storm waves breaking on a beach will tend to strip the beach, making it steeper, and transporting the wave suspended material seaward to form an offshore bar with a trough system just inshore. During calmer conditions this storm formed bar generally moves gradually shoreward, filling the trough and it is redeposited on the beach, creating the wider and more gradually sloping beach that was present before the storm.

When the wave activity can reach a vertical surface such as a revetment wall or bulkhead along the shore, the waves breaking against the wall create a hydraulic situation where the water is forced to move vertically upwards and downward along the face of the wall. The upward force of flowing water tends to dislodge stones which may form the face of the wall. The downward force of the water at or near the toe tends to scour the sand from in front of the wall. The presence of the wall within the wave active area creates a situation in which strong wave turbulence is produced, winnowing the sands from in front of the wall and generally carrying the sands seaward during the back rush of the wave, forming an offshore bar deposited in a less turbulent zone. The so-called storm condition is mimicked in this case by the turbulence produced against the wall, and the offshore bar system tends to become a more permanent feature. Such a wall causes the system to remain almost continuously in the storm mode condition.

Many beaches have been significantly altered by this phenomena. Broad beaches which have walls previously built far above the normal reach of the tide, eventually experience short periods of turbulence and erosion due to wave scour during extreme storm episodes. The slight lowering during these extreme occurrences simply allows more frequent approach by the waves under less severe conditions in the future. In order for the sand to rebuild to its former height, the wave activity must carry the sand up the face of the beach. The problem then becomes one where, if the wave activity is strong enough to carry the sand back up to its original elevation, the wave meets the wall that was placed there, produces turbulence and fails to deposit any sand at its former elevation. As time passes, the beaches begin to narrow, or disappear during all high tide periods, and eventually only a narrow sandy strip of beach may exist during low tide.

The marina designer must be cautious and aware of any beaches adjacent to a potential site, or within the possible influence of the site. The delicate balance of forces that creates and maintains a beach can be easily disturbed, even by activities at some distance from the beach.

PART 2—INFORMATION SOURCES

1. American Association of State Highway Officials. 1984. *A Policy on Geometric Design of Highways and Streets.* Washington, DC: American Association of State Highway Officials.
2. American Society of Civil Engineers. 1969. *Report on Small-Craft Harbors, ASCE—Manuals and Reports on Engineering Practice—No. 50.* New York: American Society of Civil Engineers.
3. Bascom, W. 1980. *Waves and Beaches: The Dynamics of the Ocean Surface,* revised updated edition. New York: Doubleday.
4. Battan, L. J. 1984. *Fundamentals of Meteorology.* Englewood Cliffs, NJ: Prentice-Hall, Inc.
5. Bigelow, H. B. and Edmonson, W. T. 1947. *Wind Waves at Sea, Breakers and Surf.,* H.O. Publication No. 602. Washington, DC: U.S. Navy Hydrographic Office.

6. Bishop, J. M. 1984. *Applied Oceanography.* New York: John Wiley and Sons.
7. Blain, W. R. and Webber, N. B., eds. 1989. *Marinas: Operation and Design, Proceedings of the International Conference on Marinas,* Southampton, UK, September 1989. Southampton, UK: Computational Mechanics Publications.
8. Bruun, P. 1981. *Port Engineering,* 3rd edition. Houston, TX: Gulf Publications.
9. Bruun., P., ed. 1985. *The Design and Construction of Mounds for Breakwaters and Coastal Protection.* New York: Elsevier.
10. Coastal Engineering Research Center (CERC). 1983. *Construction Materials for Coastal Structures,* Special Report No. 12. Vicksburg, MS: U.S. Army Engineer Waterways Experiment Station.
11. Coastal Engineering Research Center (CERC). 1984. *Shore Protection Manual,* 3rd edition. Washington, DC: Superintendent of Documents, U.S. Printing Office.
12. Dunham, J. W. and Finn, A. A. 1974. *Small-craft Harbors: Design, Construction, and Operation. Special Report No. 2.* Vicksburg, MS: U.S. Army Corps of Engineers, Coastal Engineering Research Center, Waterways Experiment Station.
13. Federal Highway Administration. 1975. *Technical Guidelines for the Control of Direct Access to Arterial Highways,* (FHWA Report FHWA-RD-76-87). Washington, DC: Federal Highway Administration.
14. Federal Highway Administration. 1982. *Access Management for Streets and Highways,* (FHWA Implementation Package FHWA-IP-82-3). Washington, DC: Federal Highway Administration.
15. Federal Highway Administration. 1985. *Site Impact Traffic Evaluation Handbook.* Washington, DC: Federal Highway Administration.
16. Federal Highway Administration. 1986. *Manual on Uniform Traffic Control Devices.* Washington, DC: Federal Highway Administration.
17. Fox, R. W. and McDonald, A. T. 1985. *Introduction to Fluid Mechanics.* New York: John Wiley & Sons.
18. Gaythwaite, J. W. 1981. *The Marine Environment and Structural Design.* New York: Van Nostrand Reinhold.
19. Gaythwaite, J. W. 1990. *Design of Marine Facilities for the Berthing, Mooring and Repair of Vessels.* New York: Van Nostrand Reinhold.
20. International Marina Institute. 1987. *Proceedings of the Marina Design and Engineering Conference,* Boston. Wickford, RI: International Marina Institute.
21. Institute of Transportation Engineers. 1987. *Trip Generation.* Washington, DC: Institute of Transportation Engineers.
22. Ippen, A. T. 1981. *Estuary and Coastline Hydrodynamics.* New York: McGraw-Hill.
23. Kinsman, B. 1965. *Wind Waves.* Englewood Cliffs, NJ: Prentice-Hall, Inc.
24. Komar, P. D. 1976. *Beach Processes and Sedimentation.* Englewood Cliffs, NJ: Prentice-Hall, Inc.
25. Kotsch, W. J. 1983. *Weather for the Mariner.* Annapolis, MD: United States Naval Institute.
26. LeMehaute, B. 1976. *An Introduction to Hydrodynamics and Water Waves.* New York: Springer-Verlag.
27. Marinacon. 1988. *Proceedings of the 3rd International Recreational Boating Conference, Singapore.* Rozelle, NSW, Australia: Marinacon.
28. McCormick, M. E. 1973. *Ocean Engineering Wave Mechanics.* New York: John Wiley and Sons.
29. Myers, J. J., Holm, C. H., McAllister, R. F. 1969. *Handbook of Ocean and Underwater Engineering.* New York: McGraw-Hill.
30. Officer, C. B. 1980. *Physical Oceanography of Estuaries and Associated Coastal Waters.* New York: John Wiley and Sons.

31. Quinn, A. D. F. 1972. *Design and Construction of Ports and Marine Structures.* New York: McGraw-Hill Book Company, Inc.
32. Sorensen, R. M. 1978. *Basic Coastal Engineering.* New York: John Wiley and Sons.
33. Tsinker, G. P. 1986. *Floating Ports, Design and Construction Practices,* Houston, TX: Gulf Publishing Company.
34. Wiegel, R. L. 1964. *Oceanographical Engineering.* Englewood Cliffs, NJ: Prentice-Hall, Inc.
35. Wortley, C. A. 1985. *Great Lakes Small-Craft Harbor and Structure Design for Ice Conditions: An Engineering Manual.* Madison, WI: University of Wisconsin Sea Grant Institute.

PART 3
ENGINEERING DESIGN

10
Vessel Considerations

To begin planning marina facilities, decisions must be made about the vessels to be serviced. Vessel sizes and configurations affect dock design, mooring requirements, fairway dimensions, basin depth, and utility criteria. In the last twenty years, a proliferation of recreational boats at affordable cost have become available to the general public. The affordability of recreational boats and the resulting entrance of large numbers of new participants into recreational boating has created a number of changes in boat design parameters. Demands placed on boat designers have resulted in recreational boats becoming more maneuverable, more spacious, and more heavily outfitted with electronic devices. Boats have increased in average beam, length and profile height dimensions to support the consumer demands for speed, comfort and livability. These changing parameters influence marina design in both berth layout considerations and service requirements.

10.1 BOAT DESIGN

Accepted principles of naval architecture dictate that certain boat characteristics are relatively fixed to allow a boat to function successfully. It appears that many of these principles have been maximized for boats in the 20 to 40 foot length range. Principles such as length-to-beam ratios, weight-to-length ratios, sail area-to-length ratios, etc., have been varied and experimented with to develop the optimum generic boat. Therefore, boat designers may continue to sculpt new appearances to small boats, but the basic parameters such as length-to-beam ratios will not change dramatically over time. During the last two decades the size of the boat used in the small boat market in coastal areas seems to have increased from an average size of 20 to 35 feet in length to 30 to 45 feet in length. The escalation in boat size has affected marina design. Boat sizes and characteristics routinely used in marina design for the last two decades are no longer valid nor is the assumption that the average boat owner is knowledgeable in proper boat handling technique.

Small boat naval architects are faced with a wide range of opposing criteria in the design of a boat. They must first develop a hull form that is suitable for the intended purpose. In general it should be stable under normal and expected operating performance conditions, it must float level and have acceptable trim, it should be streamlined to offer a reasonable horsepower to speed ratio, it must be constructible with cost effective materials, the materials should have a reasonable life and it must have amenities desired by the potential market segment. To effect a proper design a series of compromises must be made. Some of these compromises may affect the ability of the vessel to maneuver or to be handled in the marina facility. In an attempt to provide more useable living space aboard, boats are often designed with high superstructures which translates to wind load for the marina designer. Cabin configurations may provide adequate visibility forward but may restrict visibility aft, compromising ease of docking. Boat design streamlining may reduce wind load but may create access problems to deck mooring hardware. Hotel like utilities will require marina utility service of a capacity to service these needs while berthed at the dock. Modern versions of ancient boat designs such as catamarans and outriggers may cause considerable difficulty in berthing in traditional marina dock configurations. It is therefore important to understand the local boat market and what boat design trends may be offered up in the future.

10.2   BOAT LENGTH

Hard statistics on the exact changes occurring in the size increase of small boats are difficult to obtain. However, sales figures from boat dealerships, boat show activity, and marina configurations shown in permit applications indicate that the average small boat size in coastal areas is in a range from 30 to 45 feet in length. Many boating areas are also experiencing a growing demand for facilities for boats in the 50 to 85 foot range. Universally, there is also a demand for transient dockage and homeporting for megayachts in the 85 to 200 foot in length range.

The escalation in user boat length has caused many marinas to modify existing marina layouts to accommodate the larger vessels and capture that higher profit market. At the same time as average boat length has increased, the demand for berthing space for small runabout type boats has also increased due to their affordability by the public just entering boating. The explosive increase in waterfront land value, in many areas, has dictated that the land be used for the highest and best use, which is often to service the larger boats. As a result, small boats (12 to 26 feet) are often relegated to space not suitable for other boats or in modern facilities accommodated in upland rack storage. Proper marina design will provide for a range of small craft suitable for the intended market with consideration for future market trends.

For appropriate marina design, the use of the term boat length must be defined by the designer and understood by the marina developer. Does boat length refer to manufacturer's stated length-over-all (LOA) or does it refer to a boat whose total length is length-over-all plus swim platforms, bowsprits, sterndrives, bow pulpits, etc. This is important when we translate boat length to decisions on marina finger dock length, fairway dimensions and other clearance criteria. Unless otherwise stated, boat length, in marina design, should be the total extreme length of the boat including all projections for use in design considerations. This may mean that a generic 40 foot boat will not fit into a 40 foot berth if the 40 foot boat has projections that will cause intrusion into the design fairway when properly berthed in the slip. Designers should be explicit in defining their design criteria to marina owners and developers. This aspect becomes especially significant in conversion to dockominium or to a long term lease form of ownership. In this form of ownership the lease or sale documents should clearly specify the total usable length of the berth rather than specifying berths by boat length, unless boat length is clearly defined as including all projections. A prospective berth purchaser may believe that their generic 40 foot boat can occupy a 40 foot berth when in fact it may be too long when projections are included in the overall boat length. It seems reasonable to assume that in dockominium or long term lease situations, the berths will eventually become occupied with the largest boat that can be accommodated. If boats are allowed to project into the fairway between rows of berthed boats, the fairway will become narrower than designed and the potential for conflict between berthed boats and those underway will increase.

Marinas are site specific and market specific, so no definitive rules can be established on appropriate designs for boat length criteria. It is therefore important that the designer work closely with the developer to establish realistic projections for the marketability of proposed marina slips. It is also important for the designer to consider how the marina arrangement under design could be modified to accommodate other size vessels in the future should market trends change. This consideration may result in incorporating areas of alongside berthing to accommodate vessels of varying length. This arrangement also works for transient dockage, which must accommodate a wide range of vessel sizes. If sufficient water area is available, oversize fairway design may allow for future slip length increase without major changes to main walkways and utility connections. It may also be wise to develop the marina in phases, based on a master plan that would allow modification to slip sizes as the appropriate market is defined. It is often believed that with certain types of floating dock systems one can easily adjust the clear width between finger floats to accommodate a changing complement of boats. In actuality this rarely occurs because of other constraints such as utility hookups, fairway dimensions and mooring pile arrangements. It would be imprudent to select a floating dock system solely based on its ability to adjust finger float positions in the future. Adjustable finger floats sound good to the uninitiated but many other dock system criteria are much more important in dock selection, including structural strength, decking quality, flotation pontoon quality, utility access, and mooring requirements.

Marinas seem to be self generating in that a market can often be created where none exists if all other factors of marina development are considered. An example might be an urban marina where megayacht visits are nonexistent. Creation of a megayacht marina with desirable amenities and good marketing may create a destination for these yachts that did not exist until the marina offered the appropriate services. Research, creativity and design flexibility are important inputs to good marina design.

10.3 BOAT BEAM

Changing boating habits have resulted in departures from traditional boat design. Today's boater is looking for significant creature comforts and to spend more time cruising or just living aboard, than was common in the past. Family oriented boating is popular throughout the world with its inherent requirement for large living spaces. The change in the character of boating has increased the demand for hardware items such as air conditioning and heating; comfortable sleeping accommodations; functional, full service galleys and areas for just relaxing. These demands for space result in boat size increase, particularly in the beam. In order to accommodate the desire for maximum space aboard a given length of boat, naval architects have increased boat beam to near maximum limits while still respecting the principles of boat design for length to beam ratios.

In an effort to rationalize current boat beam considerations Figure 10-1 presents information on powerboat beams versus boat lengths. The boat lengths given represent manufacturer's stated boat lengths overall. Boat beams are manufacturer's reported beams and are assumed to be maximum beam dimensions. Only powerboats are presented since it is prudent to design all marinas for powerboat criteria even if a predominance of sailboats currently exists in a specific market. One reason is that the market may change and provision for the beamier boats allows for adjustment of boat mix with market changes. Another reason is that sailboats are generally less maneuverable, at slow speed than are powerboats, so the extra space between sailboats will be helpful in maneuvering in and out of the slip. The data presented in Figure 10-1 shows a dimension for the boat beam only, without

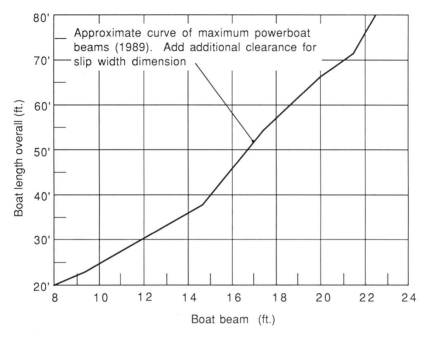

Figure 10-1. Average boat beam vs. boat length overall, based on manufacturer's product literature.

allowance for space between berthed boats. **When developing the design for clear width between finger docks, an additional 2 to 4 feet should be added to each boat beam for clearance and fendering. The amount of extra clearance provided is a design issue related to local boater experience (boathandling), type of boats to be berthed, exposure to wind and waves, water current velocity, and amount of water area available for development. Packing boats too closely may look good for revenue generation projections but might result in boat damage and unhappy tenants. If possible a minimum of 4 feet plus boat beam is recommended.** Figure 10-2 represents design criteria for clear width between finger piers. Boat beam for powerboats is usually measured at the widest point in the hull which may be a deck level flare. This may be important in cases where it appears that a boat with a manufacturer's beam will not safely fit into a dock berth, especially a floating dock. In the case of a floating dock, the interface between the float and the boat will be around 18 to 24 inches above the waterline. The boat flare may be such that boat fendering will occur beneath the flare allowing the boat to fit into the prescribed width, albeit tightly.

In marina design practice it is wise to configure a berthing layout that may have a variety of clear widths between finger docks for the same size of

236 PART 3/ENGINEERING DESIGN

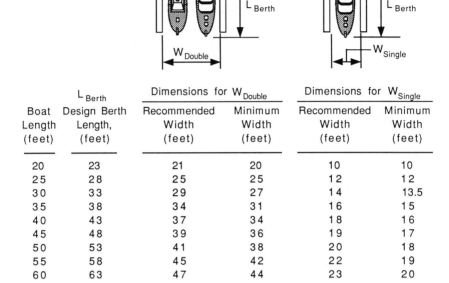

Boat Length (feet)	L_{Berth} Design Berth Length, (feet)	Dimensions for W_{Double}		Dimensions for W_{Single}	
		Recommended Width (feet)	Minimum Width (feet)	Recommended Width (feet)	Minimum Width (feet)
20	23	21	20	10	10
25	28	25	25	12	12
30	33	29	27	14	13.5
35	38	34	31	16	15
40	43	37	34	18	16
45	48	39	36	19	17
50	53	41	38	20	18
55	58	45	42	22	19
60	63	47	44	23	20

Figure 10-2. Recommended and minimum clear widths between fingers for boat beams, based on 1989 boat criteria.

vessel. In this manner, the marina operator has some flexibility in assigning slips that are most appropriate for the vessel under consideration. Two wide beam powerboats may be assigned to the widest berths while a powerboat and a narrower sailboat may occupy berths of less width. Variation is a key to success.

10.4 BOAT PROFILE HEIGHT AND WINDAGE

The marina designer usually treats the loading from wind exposure as a load generated on the average profile height of the berthed vessels. Profile height is arrived at by physically or graphically measuring the actual side or end square footage (area) of a series of boats and dividing those areas by the boat length overall to determine an average height of an equivalent rectangle. Therefore the irregular side area of a boat can be estimated by multiplying its length overall by a prescribed average height for that size and type of boat. This technique has been used for many years with varying success. The problem is that changing boat characteristics have not always been reflected

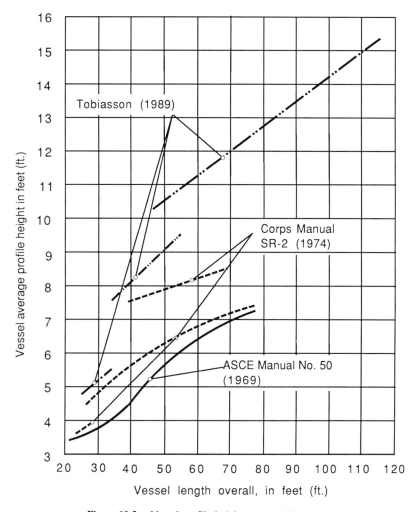

Figure 10-3. Vessel profile height vs. vessel length.

in the available reference charts used to determine boat average profile height. Figure 10-3 shows a series of average profile height curves reflecting the change in profile height over the years. One curve represents the average profile height determined and presented in the American Society of Civil Engineers, *ASCE Manual 50, Report on Small Craft Harbors,* published in 1969. A second set of curves shows an increase in profile height as developed in the U.S. Army Corps of Engineers, *Small-Craft Harbors: Design, Con-*

struction, and Operation, Manual SR-2, published in 1974. The third curve represents data generated by the author in 1983 and updated through 1989. It can be seen that boat profile height has increased dramatically and the use of older curves may result in calculations of wind loads below realistic estimates and thus result in the underdesign of dock structures and mooring devices. The curves are used to develop a square footage of an appropriate vessel area for use in determining boat loading from wind consideration. This loading can then be used to develop dock structure design resulting from vessel breasting or line pull loads and to assess the loading applied to the mooring restraint system.

The methodology in determining and applying windage loads is a subject of considerable debate. Some areas of concern that arise are: sustained velocity of the wind for the specific marina area; degradation of wind velocity due to surface terrain over which the wind passes; shape of the boat hull (streamlining) in various angles of exposure; relationship between load applied to the boat and transmission to the dock system. It is important to fully understand the implications of wind and wave exposure and how these forces influence the boats berthed in a marina. A more detailed discussion of wind and wave exposure may be found in Chapters 4 and 5. The information necessary to arrive at a comprehensive understanding of windage effects includes: wind exposure analysis for wind velocity and direction, terrain category, boat profile area, drag coefficients of the boats and shielding effects of the boats.

10.5 BOAT WEIGHT

The weight of a boat (displacement) is an important criterion for several design considerations. Certainly for boat haul-out operations it is necessary to ascertain boat weight to assure that the proposed hauling or lifting device is capable of handling the load. Boat weight may also be an important consideration in developing access roadways and boat storage areas. Straddle hoist foundations and runways must be capable of handling boat and hoist device loads. The same is true for marine railways, vertical lifts, rack storage forklifts and boat trailers. Marina designers will also use boat weights to determine impact loads on dock structures. Figure 10-4 is a plot of vessel weight (displacement) versus boat length. The input data is based on manufacturer's boat literature and does not include full fuel or stores or additional equipment that might be added by an owner. The data is an average but is suitable for rough estimates of boat weight. To determine boat weights for vessels at or near the capacity of a hauling or lifting device other methods of calculation may be required. A more detailed discussion on determining boat weights may be found in Chapter 18.

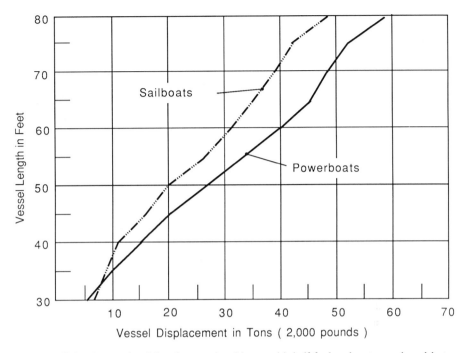

Figure 10-4. Averaged weight of recreational boats with half fuel and water on board but without owner's accessories, based on boat manufacturer's literature.

10.6 BOAT FREEBOARD

Boat freeboard is an important consideration from the perspective of ease of access to the boat from the dock system. Marinas must be able to accommodate a wide range of boat designs. Aft cockpit cruisers and sportfishing boats will usually have low freeboard stern areas that provide for easy boarding from standard freeboard floating docks or if berthed along high freeboard fixed docks, boarding may be accomplished over the higher bow profile. Today, however, the trend in recreational boat design is to provide more enclosed space, resulting in many aft cabin configurations that have high freeboard along their entire length, making boarding from low freeboard docks difficult. The higher freeboard boats will often require boarding steps placed on the dock to accommodate access to the boat. Dock freeboard versus boat freeboard therefore becomes an important consideration. It is generally felt that in areas having frequent water level changes of greater than 4 feet, a floating dock scheme is preferable to a fixed dock scheme, even if boarding steps are required. Once a boarding access scheme is developed for a particular boat, the floating dock will always remain in the same relative position to the boat irrespective of water level change. This concept must be tempered with a study of local boats that will use a facility and the boat freeboard that must be accommodated in conjunction with the range of water level change.

The proliferation of health and leisure activity associated with rowing is a special consideration. Rowing shells have a minimal freeboard, in the order of 4 to 6 inches. In addition, rowing shells have long oars (sweeps) attached to outriggers (riggers) that are not quickly detached when approaching a dock for landing. For these boats, floats must have sufficiently low freeboard (4 to 6 inches) to allow the shallow freeboard shell to lay alongside and for adequate clearance to allow the oars to sweep across the dock.

Recreational boat freeboard design has also resulted in another concern. Anyone falling off a boat, and this happens quite regularly during docking operations, may have considerable difficulty reboarding the boat. For this reason many marinas will install access ladders or footholds in floating dock systems to provide access to people in the water.

Another concern related to the often high freeboard associated with modern boat design is the angle of line pull between the boat and the dock tie-up cleats (bitts, bollards, etc.). Instead of a shallow angle of pull which generally results in a shearing action on the dock cleat, a large angle of line pull, often nearly straight up and down, may occur, which will tend to pull out the cleat. For this reason dock cleats, should always be through bolted rather than fastened with lag screws. The through bolts will act in tension with load transfer to a main dock member in bearing by the nut and washer, versus a lag bolt that is susceptible to pull out at the minimal contact between the bolt threads and the dock frame material. Cleat placement on the dock is also important to accommodate high freeboard boats by placement far enough away to develop a shallow angle between the boat and the dock cleat to reduce the pull out loading.

High freeboard boats may require boarding steps on the docks to provide access to the boat. Boarding steps may become unstable on narrow and poorly constructed floating finger docks. This is part of the reason why larger boats that may require boarding steps will usually have wider more stable finger floats.

Fendering provided between a boat and the dock will also be affected by boat freeboard. Consideration should be given in dock design to allow for a full range of boat freeboards and the necessary fendering arrangements.

10.7 SAILBOAT MAST HEIGHT

The question of vertical clearance criteria for a marina can be important in facilities where boats may have to transit through bridges, pass under overhead wires or in other conditions where height impediments to boat passage may exist. Figure 10-5 presents a plot of representative sailboat mast heights versus sailboat length. Although an attempt to use representative sailboats was made, the data should be used with caution since boat parameters are constantly changing. This data may be helpful for initial planning for limited access marina development. A more detailed local investigation should be undertaken by the marina designer before such items as bridge clearance dimensions are fixed. Note that the plot shows an added 4 foot clearance to the upper limit of actual mast heights presented. This clearance is based on general knowledge and may require additional dimension for specific cases.

A primary area where sailboat mast height becomes of critical concern is on and around boat launching ramp areas. Trailerable sailboats will be brought to the launching area with the mast unstepped for obvious, over-the-road limitations. At the launching ramp area the mast may be stepped with resulting overhead clearance concerns. If at all possible no overhead wires

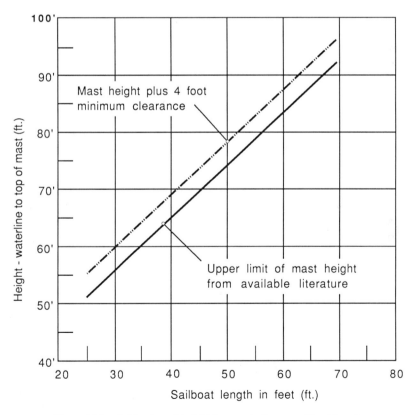

Figure 10-5. Sailboat mast height from waterline vs. sailboat length.

should be located in a boat launching area. If overhead lines cannot be avoided, the area must be well posted. When considering a potential site for boat launching ramp construction, Figure 10-5 may be consulted to ascertain the potential for use by sailboats with stepped masts. Caution should be exercised when using the data in the figure for upland consideration because the data presented relates to a boat waterline. Additional dimension of mast height must be added for the distance between the floating waterline and ground level, i.e., draft of boat plus trailer height. This dimension may add several feet to the actual height of the top of the mast.

Many marinas haul out sailboats with masts in place for boat repair work or seasonal storage. This policy may or may not be consistent with the marina's insurance coverage. Obviously, if this technique is used, the haul-out area and any area in which the boat may be moved, must be clear of overhead power lines and other obstructions that could come in contact with the boat mast. Again, caution is urged to add additional height to the dimensions in Figure 10-5 for the distance between the boat's waterline and the ground level. If sailboats are to be routinely moved around a marina, the prudent operation will have all overhead obstructions placed underground or otherwise protected from any potential contact with the boat or haul-out device.

10.8 MULTIHULLS

Multihulls create considerable difficulties in berthing at marinas. Their extra beam generally restricts the ability to berth a multihull in a conventional marina berth. If multihulls are common to the local boating area, it may be wise to provide a long run of dockage without finger projections to allow varying beam boats to lay alongside. Alternatively, a mooring area may be reserved for these special craft. Remember that multihulls are often light weight and may be very wind sensitive when moored, so allow plenty of swinging room around their moorings.

Multihulls are also difficult to haul out. Unless specifically designed for straddle hoist hauling, multihulls should not be lifted on straddle hoists. The side load compression caused by the resultant forces of hoisting sling position may damage the hulls. Being generally stable about themselves, allows many multihulls to sit firmly on a beach. This may be a suitable means of addressing haul-out needs.

If multihulls represent a significant number of the boats using the marina, extra wide berths may be designed into the facility layout, or a boat owner may opt to occupy a double berth, at the double berth price.

11

Selection of Dock Types

There are two basic categories of docks used in marina and small craft harbor design. The docks are either fixed or floating. The selection of fixed or floating docks depends on variables of water level fluctuation, use, appearance, subsoil foundation conditions, water depth, and cost. Within each category of docks are selections of materials of construction, design configurations, appearance and cost.

Fixed docks are generally constructed on site, often with prefabricated components. Deck structures and framing systems may be compatible with prefabrication techniques. The foundations for fixed docks repose on or are embedded in the bottom. The structures are generally considered permanent, although some temporary fixed docks are used for light duty and in areas where seasonal removal is necessary, such as in ice prone areas.

Floating docks may be temporary or permanent. Their permanency is generally related to the extent of their mooring system and their outfitting with utilities. A distinction is often made, however, for permitting and tax purposes, that floating structures are temporary in nature which may qualify them for special consideration. Certainly removing mooring piles or bottom anchors is more readily accomplished than removing a substantial fixed pier arrangement. The temporary nature of floating docks may not necessarily relate to the adjustability of configuration change. Many people entering the marina development industry conjure up thoughts that floating docks can easily be adjusted to accommodate changing conditions. This is generally not the case, as mooring systems, utility connections and layout fixity prevent substantial adjustment.

11.1 WATER LEVEL CHANGE EFFECTS

One of the principal considerations in dock selection is the effect of water level change. In coastal areas, tidal fluctuation will influence dock design. **As a general rule areas having tidal change in the range of 0 to 4 feet may select either fixed or floating docks. In the past fixed docks were the usual choice in this range, if water depths were not excessive. Between 4 feet and 7** feet of tidal range, both fixed and floating docks are used, but floating docks are generally a better choice because they allow a more desirable access between the dock and the floating boat. Today's boater is not interested in climbing up and down between the dock and the boat. The action is dangerous and difficult with the amount of gear and groceries carried aboard by most boaters. Areas with tidal ranges greater than 7 feet are pretty much forced to use floating docks. In addition to the great elevation difference between boat and dock at certain tidal levels, the boat mooring line arrangement and the amount of line scope necessary to accommodate the elevation change requires provision of wider than normal berth clearance widths or excessive fendering arrangements. The moored boats will tend to drift around in the berth at tide elevations where the mooring lines are slack and may be hung in the berth, not necessarily close to the dock, when the mooring lines are taut. At low water levels, the boats may be down below the dock deck level, creating an unpleasant visual climate.

River and lake level elevation changes may be quite large and occur over an extended period of time, such changes include seasonal variations. Fixed docks may or may not be desirable depending on the vertical extent of water level change. The Great Lakes area of the United States, for many years, has had a large number of fixed dock installations because of the relative constancy of lake levels. In recent years, however, the lakes have been subject to significant elevation changes, often completely inundating or exposing fixed dock installations. As a result, new installations are seriously considering floating docks to help mitigate future significant lake water level changes.

The worldwide trend in recreational boat marinas berthing craft under 100 feet in length appears to favor floating docks even in areas where fixed docks may be acceptable for the prevailing water level elevation changes.

In looking at tidal water level change, it is important to consider the spring tide elevations, or high and low water levels, rather than the mean tide levels. The spring tide levels occur on a periodic basis and are not related to seasonal changes. As an example, in Boston, Massachusetts, the spring tides may be 2 to 2.5 feet higher than the mean tide and 1 to 2 feet lower. This variation of 3 to 4 feet above a mean tidal range of 9.6 feet will significantly affect marina dock selection.

As discussed in Chapter 6, a large variation in water level may also affect the magnitude of incoming waves at varying water levels by changing the effective water column depth and possibly modifying the wave refraction characteristics of the bottom bathymetry. These effects may need to be considered in the selection of dock type with regard to structural integrity and dock orientation.

11.2 FIXED DOCK SYSTEMS

Fixed dock systems may be composed of many different components and materials. An example might be an earth filled, stone faced pier, from which a timber pile pier extends seaward. Whatever the composition and configuration, a fixed pier is associated with a permanent foundation either resting on or embedded in the bottom subsoil. The fixed pier is generally a rigid structure providing a significant resistance to imposed forces. A primary consideration for fixed pier utilization is the type of subsoil on which the structure will be founded. High bedrock formations will preclude easy pile installation and may require socketing of piles if they are selected as the foundation material. Rock makes a very good foundation, however, for filled pier structures, if adequate water depth exists where necessary. Soft, nonstructural subsoils may also inhibit pile foundations unless the piles are very long and can reach a competent soil layer. Good competent soils make a pile solution acceptable if water depths are not excessive.

Filled fixed pier structures were a major early type of dock construction. Materials such as clean backfill and rock were often available in sufficient quantity to allow cost effective construction. The filled piers are generally an extension to an existing land area and therefore make a nice transition between the shore and the water. Filled fixed docks must be designed and constructed to accommodate the wind, waves and other natural forces to which they are subjected as well as any vessel berthing loads. Filled docks or piers may have a stone face or be restrained by some form of structural bulkhead. Bulkheads are constructed of timber, steel, concrete or aluminum. Each material has it's advantages and disadvantages. Each has its own set of structural design parameters and limitations on use. A predominant type of bulkhead in use around the world is the steel sheet pile bulkhead. This type uses rolled steel sheets which are configured to interlock with each other to provide a relatively tight wall to prevent the loss of backfill materials. The sheet pile is usually driven to some specified depth by a mechanical hammer. The depth is dependant upon the height of the bulkhead, the structural quality of the subsoil material, and the surcharge load imposed on the wall. Sheet pile bulkheads may be either cantilever (free standing from the point of burial in the ground), or tied back. Tied back walls have an intermediate height structural tie to a deadman structure located in the soil behind the wall at a distance that is beyond the active influence of the soil behind the wall (see Fig. 11-1). Tie backs are usually large diameter rods connected through the face of the bulkhead to a pile or concrete structure behind the wall. Sheet piles may be of materials other than steel, such as timber, concrete or aluminum. Sheet pile bulkheads require engineering analysis and design. Improper embedment of the sheets, lack of required bending

Figure 11-1. Simple, timber, sheet pile bulkhead. System face tied back to timber pile anchors inshore by steel tie rods.

and shear resistance in the sheet pile, insufficient tie back capacity or miscalculation of active and passive soil and surcharge loads can cause a bulkhead to fail, often as a massive failure.

Filled dock structures are often difficult to permit as their configuration occupies water sheet bottom area which may contain valuable bottom dwelling benthic species. They also may interrupt littoral drift and result in corrupted flushing of the area. If large enough, filled dock structures may impede the migration of spawning fish and cause other impediments to natural aquatic life. As a land extension, filled docks often end up crossing wetland areas which today are heavily regulated to protect their role as a vital link in the natural food chain. Filled fixed docks should be avoided, if possible, in the initial facility design. If there is no alternative to filled fixed dock construction, the developer must be prepared to provide a substantial technical defense of the design.

Another typical fixed dock used in marinas and small craft harbors is a pile supported structure (see Fig. 11-2). This form of structure is relatively easy to design and construct and is often a least cost alternative. Governing design

SELECTION OF DOCK TYPES 247

Figure 11-2. Typical wood frame and timber pile, small boat, fixed pier. Note vertical, wood plank and timber pile, fixed wave attenuator in upper left corner.

parameters include: subsoil characteristics, water depth, water level change range, and sea state exposure. Pile supported docks may have piles constructed of timber, steel, concrete or a combination of these materials. Frame and deck structures may be constructed of wood, steel, concrete, aluminum, fiberglass (see Fig. 11-3), or a combination of these materials. The most common materials are timber, wood, and concrete frame and deck structures supported by steel, wood, or concrete piles.

Timber is probably the most often used piling material due to its availability, cost and constructability. Limiting factors in selection of timber piles are load carrying capacity, susceptibility to deterioration, and length availability. Steel piles can be fabricated to varying specifications to achieve desired strength characteristics. Steel can be welded or mechanically fastened to create very long piles and can be driven and handled by conventional methods. In sea water, steel piles may require protection from corrosion. Concrete piles can be cast in appropriate geometry with varying amounts of reinforcement to achieve the desired strength. Concrete piles are generally precast and often prestressed to develop better strength and driving properties.

248 PART 3/ENGINEERING DESIGN

Figure 11-3. This fixed finger pier system consists of a composite deck structure composed of a wood core encapsulated in fiberglass. The fingers are pile supported on the outshore end and attached to a concrete promenade on the inshore end. *(Photo courtesy: American Marine Systems, 13800 172nd. Ave., Grand Haven, MI 49417)*

Concrete piles are subject to various forms of cracking, which if unattended can lead to concrete spalling and reinforcement corrosion resulting in loss of load carrying capacity and ultimate failure.

Various materials are often used in combination to take advantage of the best properties of each material. Steel pipe piles are often filled with concrete, generally not for strength addition but to seal the inside of the pipe from corrosion. This practice is not as widely used as it once was because in free standing marina mooring piles the addition of concrete adds eccentric loading to the pile as lateral load application causes the pile to deflect. It is now more typical practice to leave the pipe hollow, but to provide a seal at the top and bottom to prevent water and air intrusion. A tight pipe will soon use up the available oxygen in the void space with resulting mitigation of interior corrosion.

Steel H-pile sections, which have a large surface area upon which corrosion may act, are often jacketed in concrete to protect the structural steel. The jacket, if reinforced, may also add strength and stiffness to the pile. Steel piles are also often protected by cathodic protection systems, either by impressed current or by sacrificial anodes to mitigate corrosion in aggressive waters. These systems require continuing maintenance, maintenance which is often forgotten over time, so the most common corrosion protection is the provision of good steel coating systems. Coal-tar epoxy or fusion bonded epoxy coatings provide reasonably good steel protection if properly applied and maintained.

Timber piles are usually pressure treated with preservatives, wrapped in plastic, or jacketed with a concrete shell to provide protection in highly aggressive biological and chemical areas. All types of piles are subject to mechanical damage and must be protected from such damage by their placement in the structure, by installation of fendering systems, or by sheathing in a sacrificial material such as wood planks. Ice can be a highly destructive force both by load application and by abrasion.

The design of any fixed pier must first address the intended use and site constraints. The use requirements will dictate length and width of the dock and loading considerations. Site constraints will influence dredging or filling requirements. Soil conditions will influence pile length and material selection. In areas where soil characteristics are unknown, it is prudent to have several soil borings taken and analyzed to assure achieving a foundation capable of sustaining the imposed loading.

Live load criteria is generally specified by the local regulatory agency issuing construction or building permits. Normally, fixed piers are designed for a deck live load of not less than 100 pounds per square foot of deck area. Docks having building structures or vehicle access may require increased live load considerations. Finger piers in marinas with limited access are often designed for 50 pounds per square foot of deck area.

Foundation piles are generally arranged in rows (bents) which run from one side of the dock to the opposite side. Bent spacing is usually 10 to 15 feet on centers. All the piles in a bent are connected by a pile cap. The pile cap, in timber construction, may be a single member such as a $12'' \times 12''$ member which rests directly on the trimmed pile butt and is fastened to the pile, or may be two separate members, one on each side of the pile and through bolted to the pile (see Fig. 11-4). Running perpendicular to the direction of the pile cap and fastened to it are the stringers or joists which support the deck. The stringers are closely spaced, $16''$ to $24''$, with the spacing determined by the loading, the unsupported span, and the size and type of the decking material. The deck planks are fastened across the stringers. The combined weight of piling, pile caps, stringers, decking, hardware and any other permanently attached equipment is called the dead load. The dead load in its various combinations must be added to the live load to develop the design for each component.

Figure 11-4. Detail of wood, fixed pier construction using split, pile cap. The pile is often notched, as may be noted in the photograph, to accept the pile cap timber to provide a good bearing connection between the cap and the pile. Note that the bolt securing the horizontal wood brace is recessed into the wood member to prevent possible damage to a boat that may contact the exposed member.

Other loading considerations include vessel berthing impact and line pulls on mooring hardware. Vessel impact loads can be reduced by incorporating energy absorbing fendering systems. Mooring hardware should be properly fastened and bedded. Rot in timber and corrosion of fastenings is very prevalent around mooring hardware.

The ease of maintenance, to a large degree, is a function of the design. The designer should consider a design that precludes or at least minimizes potential rot, corrosion of fastenings and deterioration of main members. Providing adequate surface water run off is essential to maintenance of timber and steel structures. Owners and users should take care to repair damaged components since damage is often the avenue for advancement of other forms of deterioration.

Although it is the designers responsibility to account for the loading considerations of a design, it is important for owners and users to appreciate the limitations that a particular dock structure may have, and not exceed design criteria.

The design life of fixed docks is difficult to predict because of the number of variables associated with life cycles. Filled fixed docks of massive construction have exhibited life spans of over 100 years. More conventional filled pier structures may provide life spans of 50 or more years without significant structural failure. Pile supported docks may have life spans of between 25 and 50 years. Marina and small craft harbor structures should have a design life of 25 years, with 50 years being a more realistic value for fixed dock facilities.

11.3 FLOATING DOCKS

Floating docks systems have become a primary component of modern marina construction. Three decades ago, floating docks were essentially do-it-your-self affairs constructed by marina staff or local contractors using readily available materials. The boom in recreational boating prompted a new industry that specializes in the manufacture of floating dock systems. The concentration of design talent and use of new material technology has resulted in the development of high strength, low maintenance, and cost effective floating dock systems. Most major marina facilities, today, will negotiate the purchase of manufactured floating dock systems rather than venture into the do-it-your-self type of construction. A major advantage to the new generation of floating dock systems is that the inherent strength of the docks allows for flexibility in design of the mooring system. It is often possible to design a mooring system that uses a small number of high capacity piles, or a few high strength mooring chains in a bottom anchored mooring design. These options provide for a stronger and more attractive marina.

There is a wide variety of choice in manufactured floating dock systems. Materials of construction include wood, metals, concrete and plastics. The ultimate selection rests on several factors. Certainly a major consideration is cost. Most dock system manufacturers cover a wide marketing area; regional, national and international. This range of area has resulted in a cost competitive market. There are three basic classes of floating docks systems. An economy class is often used for residential and small marina facilities. These systems have relatively low strength, often are lightweight for ease of handling, and reflect the economy in a low cost. They usually have minimal utilities and are not well suited for exposed or heavily loaded conditions. A middle class of floating dock system is the most common type of system, suitable for most marinas. The quality of construction is good, they provide basic utilities, and have adequate structural capabilities for normal site and use conditions. The upper end of the market offers first class construction, a full range of amenities, and significant load carrying capacity. This class is often selected to enhance a major first class upland project or to attract the up scale boater. The cost is the greatest of the three classes but this class offers high load carrying capacity for exposed sites and large boats. Within each class are a variety of subsystems that provide varying degrees of amenities and ranges of cost and strength.

A primary consideration in the selection of a floating dock system is the fact that every marina is site specific. The selected system should be suitable for the site conditions, intended use, and project budget. There are probably more than one hundred floating dock systems available worldwide. No single system is suitable for all proposed marina conditions. Before embarking on dock selection it is wise to travel around and look at various systems in operation at a number of facilities, and observe their overall appearance, maintenance requirements, cost, and suitability for the intended use. It will get confusing, as each system may offer some specific desirable features while also presenting some drawbacks for the intended use. Manufacturers will also cloud the issue by extravagant claims and derision of the competition. At some point, usually the earlier the better, professional advise should be sought to help eliminate those systems that are not suitable for the project because of structural inadequacy, quality of construction, reputation, cost, or inability to accommodate the site environmental conditions. A knowledgeable designer can usually narrow the potential manufacturer list down to a dozen or so systems that may provide an acceptable solution for the intended use. The designer and developer should work closely to arrive at a reasonable dock selection. The designer must consider layout compatibility, structural adequacy, mooring requirements, installation requirements, maintainability, and survivability during major climatic events. The developer will want to consider cost, scheduling, contract terms, life expectancy, and importantly, appearance. In any number of projects, the final selection of floating dock system has come down to appearance and perceived status of the selected dock manufacturer. This can be good or bad depending upon the ability of the selected product to meet the other important design criteria.

Floating dock systems offer a wide range of material types. Major and minor installations have been constructed in wood, glued-laminated wood, steel, aluminum, concrete, glass reinforced concrete (GRC), fiberglass and other plastic materials. Each material has it's own characteristics, advantages, and disadvantages. The choice must be made on specific site factors, intended project use, and cost considerations. From a marina designer's perspective, as opposed to upland designers who often care more about attractiveness, the most important qualities of a floating dock system are it's structural characteristics and response to the site environment. An attractive system that is not structurally sound will not be a desirable end product.

Although it is treading on dangerous ground, some comments on dock materials is in order to assist in initial material consideration. Wood in it's various forms is a primary dock construction material. It is usually readily available, can be easily handled, and can be worked with conventional tools by relatively unskilled labor. Wood products can be attractive and wood relates to marine experience as a traditional boatbuilding material. Wood products are often the least costly, unless nonstandard species or sizes are used. The structural capacity of wood products has been significantly increased by forming the wood into glued-laminated members which have desirable structural properties. Wood is durable, if properly treated and designed, is capable of easy repair, and has a reasonable life expectancy for marina structures. Many species of wood readily accept chemical treatment to enhance life. Other wood species are naturally resistant to common forms of deterioration. Disadvantages include general restriction on wood sizes based on the species and growing characteristics. Wood, as a natural material, has varying properties that may affect structural performance. It may tend to splinter, warp, and otherwise distort. It may be subject to deterioration and mechanical abrasion. It's fiber structure may cause fastenings to work loose after being subject to intense loading and load cycling. Connections should be made through substantial metal fabrications to assure appropriate load transfer (see Fig. 11-5). Length limitations result in large numbers of connections, which may become potential failure points over time. Some mitigation of this condition can be achieved by use of fabricated glued-laminated members (see Fig. 11-6 and Fig. 11-7).

Figure 11-5. This modern floating dock system is constructed of wood members and heavy duty galvanized, steel connection hardware. The wood is pressure treated with chromated copper arsenate (CCA) to assure a long wood life in a marine environment. This type of component system can be readily constructed as a do-it-yourself project or purchased from commercial dock builders as an assembled system.

Figure 11-6. A glued-laminated, wood beam and internal truss, floating dock system with individual wood deck pallets which are removeable for easy access to below deck utilities. All wood is pressure treated for durability. This photograph shows the system under construction and without utility posts and other dock amenities. *(Photo courtesy: MEECO Marinas, Inc., McAlester, OK)*

Steel is another extensively used floating dock material. It is readily available and easily worked and handled. Welding, cutting and shaping may require special equipment, but this equipment is usually readily available and may be used by semiskilled labor. Steel is obtainable in a variety of plate, sheet and rolled shapes. The material may be easily fabricated by welding and mechanical fastening to form structurally sound components (see Fig. 11-8). Steel readily accepts a variety of coating materials and surface treatments. For structural considerations steel offers the greatest flexibility of design and structural competence. Steel is subject to corrosion unless properly protected. Corrosion of steel is perhaps the greatest drawback to its use in marine structures in, on, or around a sea water environment. Steel is also a very good electrical conductor, enhancing electrolytic corrosion and possibly creating electrical danger unless properly grounded or insulated. Steel is also such a good thermal conductor, that bare decks may be extremely hot under foot. Steel floating dock systems in combination with other materials for decking and flotation have been quite successful in fresh water installations.

SELECTION OF DOCK TYPES 255

Figure 11-7. Glued-laminated, wood, monocoque box beam, floating dock system with heavy duty cleats, fender rub strip, and full service utility posts providing water, electricity (metered at the power post), telephone, cable television, and low level dock lighting.

Figure 11-8. Steel frame and pontoon floating dock in fabrication. Wood decking not yet installed. All steel members welded and protected by epoxy coating.

256 PART 3/ENGINEERING DESIGN

Concrete is a composite material made from combining cement, water and aggregate. It can be cast into an infinite variety of shapes. Its strength and density can readily be modified by altering the mix proportions and by the addition of reinforcement. Its mass tends to make concrete a desirable floating dock material because of its stability and resistance to load application, when it is cast into large blocks. Its mass may also be used, in a proper design, to quell wave energy (see Fig. 11-9). It is durable when properly proportioned and can be readily cast in the plastic state by semiskilled and unskilled labor. The raw materials in concrete may be relatively inexpensive although the casting and handling process may be labor intensive. A well designed and fabricated concrete structure will be virtually maintenance free for it's useful life (see Fig. 11-10). Concrete generally forms a very suitable walking surface for floating structures and can have a pleasing appearance. In addition to pleasant aesthetic appearance, concrete gives the impression of durability, stability and strength, all of which may be important in the marketing of marina berths. Control of the concrete mix and placement is critical to its successful function and life. It is therefore important in purchasing concrete dock systems to deal with a reputable manufacturer who has a proven track record for quality products. Concrete mix and placement errors may not be readily observable until well after product

Figure 11-9. Modular reinforced concrete floating dock with side skirt fins to enhance wave attenuation. *(Photo courtesy: SF Marina System AB, Gungalv, Sweden)*

Figure 11-10. Eight-foot long concrete floats interconnected by timber wales to create a stable floating sidewalk. Each concrete unit weighs 2.5 tons, the mass of which helps to dampen local wave energy. Modules may be connected to form lengths of 1,000 feet or greater. *(Photo courtesy: Concrete Flotation Systems, Inc., South Norwalk, CT)*

installation, when due to the stresses of use major defects or failure may occur. Concrete reinforcing may be subject to rapid corrosion and resulting concrete surface failure if inappropriate materials or placement occur. Patching of damaged concrete may be readily accomplished but the repair may not be long lasting or attractive. Major structural damage is very difficult to repair and usually requires replacement of the damaged unit. Connection between concrete flotation units must be properly designed to avoid stress concentrations and ultimate concrete failure. Do-it-yourself concrete dock fabrication frequently results in dock failure at connection locations. Even properly designed concrete systems may suffer connection failure if subjected to significant and frequent sea state loading. Utility placement and hull penetration may be a problem in concrete dock systems unless provided for in the initial design. After casting, drilling and fastening is often quite difficult and requires special tools.

Special concrete systems such as glass or fiber reinforced concrete have been gaining acceptance as a floating dock construction material. They are generally of lighter weight than conventional concrete and may be cast in thin sections. In the proper proportions the material may be sprayed on to other materials to make composite sections. The material has numerous uses for dock decking and pontoon fabrication. It must be used with care to

258 PART 3/ENGINEERING DESIGN

Figure 11-11. Darsena Di Ligano Sabbiadoro, Italy. Aluminum frame, wood deck, floating dock system exhibiting the strength of the aluminum frame and the attractiveness of a wood deck. The system is clean looking and the layout well structured. *(Photo courtesy: Walcon Marine Ltd., Fareham, Hampshire, UK)*

avoid stress cracking and impact damage. If underreinforced it will fracture off in pieces, possibly compromising the underlying material.

Aluminum dock systems enjoy a wide acceptance in marina installations (see Fig. 11-11 and Fig. 11-12). Aluminum has a high strength to weight ratio and is often a desirable structural fabrication material. It is easily worked and assembled although shaping and welding may require skilled labor. Aluminum is generally readily available, but not in the range of sizes and shapes in which steel can be found, unless specially extruded or rolled to a particular shape. It is this ability to be extruded that has made aluminum a valuable dock component. Special aluminum extrusions can be designed to achieve a desired structural property while minimizing unnecessary additional material, thus making an efficient section. The extrusion may also be formed to suit particular and varied needs. A good example is the extrusion of a shape that functions as a floating dock main longitudinal member while also having the ability to accept decking material and dock hardware without additional fastenings. The versatility of aluminum is significant for its use

SELECTION OF DOCK TYPES 259

Figure 11-12. Philadelphia Marine Center, Philadelphia, Pennsylvania. An aluminum frame, wood deck, floating dock system with adjustable finger floats. Note steel pipe pile mooring and heavy duty pile guides. *(Photo courtesy: Technomarine International Inc., Repentigny (Qc), Canada)*

as a dock manufacturing material. Certain alloys of aluminum are relatively resistant to marine corrosion. It is imperative that the proper series, usually the 5000 and 6000 series of alloys, be used for dock construction or serious and rapid corrosion may result. Aluminum is relatively soft and subject to galling when in contact with harder materials. Therefore connections made in aluminum must take account of this weakness and provide extra material or strengthening at connection locations. Aluminum may also be subject to fatigue and stress cracking. It is a very good electrical and thermal conductor so provision must be made to isolate electrical devices and provide thermal protection under foot by deck coating or covering. Without proper and detailed surface preparation, the painting of aluminum may result in premature paint failure. In general aluminum is left unpainted where possible.

Fiberglass is used extensively for boat construction but has not yet become a major dock building material. A number of years ago fiberglass was used as a dock building material but rapid failure was experienced in areas of dock connection by fiberglass cracking. Decking also suffered stress cracking and

260 PART 3/ENGINEERING DESIGN

in some cases fiberglass delamination occurred. Modern fiberglass dock systems use the fiberglass technology as one component of the product design. By combining fiberglass with a sandwich or core construction, appropriate strength and durability may be achieved. In general a solid wood core is desirable over a lightweight balsa core. Balsa core is frequently used in boat construction. Composite fiberglass has been successfully used as a structural deck system and fiberglass alone has been used for flotation pontoon units (see Fig. 11-13). The fiberglass is readily worked by experienced labor and can be easily repaired. The repairs may be as strong as or stronger than the original material. Structural backing blocks and stiffeners can easily be incorporated into the design to provide strength only where necessary, with resulting material and cost saving. Fiberglass is quite durable as witnessed by the long life sustained by recreational and commercial fiberglass boats. Fiberglass may be constructed in an infinite variety of shapes and sizes to meet specific needs. The material may be colored to achieve distinctive and attractive designs. Deck surfaces can readily be made nonskid and pitched or sloped as desired. Utility penetration can be

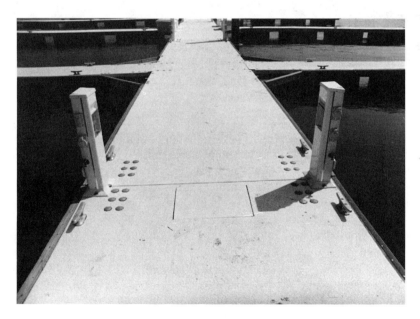

Figure 11-13. Newport Yacht Club and Marina, Jersey City, New Jersey. A composite, fiberglass and wood core floating dock system supported by fiberglass buoyancy pontoons. In addition to being a durable float construction material, the fiberglass may be colored to owner or architect specifications to provide an aesthetic architectural feature.

readily performed as needed if the penetrations are properly sealed against water intrusion.

11.4 SPECIAL CONSIDERATIONS FOR LARGE YACHT DOCKS

Dock systems used for the berthing of large yachts, often called megayachts, which we define as yachts greater than 65 feet in length, require special considerations. Obviously, the larger yachts impart greater forces to a dock system than do smaller boats and therefore the structural capacity of the dock system must be analyzed for these large forces. In addition consideration must be given to the manner in which the vessels will lay against the dock system. It may be necessary to provide substantial fendering between the dock and boat to cushion the imposed forces and protect the expensive yacht hull finish. Large yachts may also have bilge keels, spray rails or strakes, or other projections along the hull, near the waterline that might conflict with a dock structure. For some sizes of large yachts, it may be necessary to provide independent breasting dolphins for the vessel to lay against rather than having the yacht lay against the dock system itself. Dock freeboard in relation to the yacht hull side profile should also be considered. Most large yachts will be high sided and require high freeboard docks or auxiliary steps or stairs to allow reasonable access to the vessel deck. Use of steps or stairs will add load to the dock system which must be accommodated by structural capacity, and buoyancy capability must be available if the dock is a floating dock.

Mooring devices such as cleats, must be of sufficient size to accommodate the diameter of the line used by large yachts and, of course, the devices must be strong enough to resist the mooring forces. In addition to cleat strength, the cleat or other mooring device must be adequately fastened to the dock structure to prevent pull-out under load. The locations of mooring devices are also important to provide an appropriate mooring line angle between the vessel and the mooring device. Large, high sided yachts may require bow and stern mooring devices a substantial distance forward and aft of the vessel to attain the proper mooring line angle. The mooring line should not interfere with other vessels berthed in the area. Mooring a large yacht using a Mediterranean style mooring, that is with the stern fastened to a dock and the bow controlled by an offshore anchor, will require large dimension fairways to accommodate the length of line associated with the offshore mooring.

Large yachts also require significant utilities. Electric power requirements may exceed 100 amp service, with the associated large diameter and capacity feeder cables. Long distances between main power supply or step down transformers should be avoided. Typical ¾ inch diameter domestic water service may not be adequate for large yachts if water provisioning is anticipated. Larger 1.5 inch to 3 inch diameter water mains may be necessary to fill yacht water tanks in a reasonable time period. It may be possible to utilize dock fire protection standpipes for yacht water needs if fire protection water lines are approved for potable water supply and are proximate to the yacht berthing area. If sewage pumpout facilities are to be provided for large yachts, their capacity and piping should be of commercial type and suitable for the intended service.

Large yachts may also require additional upland parking space for crew and guest parking and vehicular access to or near the vessel for easy transfer of provisions and equipment. The dock system must be capable of carrying the vehicle loads and potentially large crowds of people, especially if the large yacht is used for charter purposes.

The overall requirements for large yachts require everything to be larger and stronger than the facilities provided for smaller recreational vessels. There is, however, perhaps one exception and that is that smaller recreational boats will most always be left in a marina during storm conditions whereas, large yachts with professional crews will often take the vessels to more protected locations or to sea when wind conditions are anticipated to exceed 40 to 50 knots. Therefore it may not be necessary or desirable to design a large yacht facility to accommodate a full complement of large vessels under extreme storm conditions.

11.5 SPECIAL TYPES OF FINGER FLOATS AND MOORING DEVICES

The basic type of boat mooring to a fixed or floating dock is the Mediterranean mooring system in which a boat is tied bow or stern to a float and the outshore end is held by the boat's anchor or a permanent mooring. Boats are essentially rafted together with the outshore ends positioned by a bottom anchored device. The next most popular system involves using a finger float, usually of a width equal to 10 percent of the berth length. The finger length may be shorter than the boat, the length of the boat, or longer than the boat. The last two types of finger usually provide access to the boat from anywhere along the finger length. Finger floats may be laid out such that each boat has one associated finger float (double loaded system) or where a boat has a finger float on each side (single loaded system).

A system intermediate between the Mediterranean moor and the full finger is the so-called Y boom. The Y boom consists of a structural fabrication of members shaped like the letter Y in plan view (see Fig. 11-14). Y booms are used extensively in Scandinavia and are gaining popularity worldwide. The Y boom is usually wood or metal and connects by a hinge at the forks of the Y to the main float, while the outer end has associated flotation

SELECTION OF DOCK TYPES   263

Figure 11-14.   Y boom small boat finger float, attached to concrete and wood floating dock system. Also called a Scandinavian finger. *(Photo courtesy: SF Marina System AB, Kungalv, Sweden)*

to maintain a level floating attitude in the unloaded condition. The Y booms provide a point to tie off the boat on the outshore end and act as a berth separation, helping to keep boats apart and not requiring fendering between adjacent boats, as may be required in a Mediterranean moor solution. The Y booms generally do not have sufficient flotation to allow walking down their length, but some are fitted with a small deck area in the space between the Y forks to provide some access to the main dock end of the boat. Due to their simple and minimal construction, Y booms are relatively inexpensive. Care must be taken, however, in using a Y boom, to assure that the boat loading conditions are within the structural capabilities of the selected boom configuration and material strength. The Y boom, being a slender device, allows for the creation of a relatively high density of boats within the defined water area.

Another variation for small boat berthing that allows for the maximizing of boat density is a product called Frog Hooks® (Fig. 11-15). These devices are generally constructed of wood or fiberglass and are short triangular arms that are hinged to the main float and connected to the transom of a small boat. When not in use they may be hinged up, perpendicular to the main float deck. They are used in pairs and each unit resembles an ironing board

Figure 11-15. Frog Hook® small boat mooring attachments. Note accessibility to boat over Frog Hook arm. *(Photo courtesy: Frog Hooks, Inc., South Salem, NY)*

in appearance. The outboard end is tapered and fitted with clevis pin bolts that connect to shouldered eye bolts in the boat transom. The eye bolts on the boat may also be used for attaching water skier tow lines. They are available in several sizes for boats ranging from 12 to 28 feet in length. The manufacturer claims that a space saving in berth space of more than 30% can be achieved, over conventional berthing methods, by use of the devices. They can be particularly advantageous in rack storage holding facilities where a large number of small boats may need to be temporarily berthed in a small water area. In-the-water-boat-shows have also successfully used these devices. Space between boats is kept at a minimum, generally on the order of one foot separation. The Frog Hook is strong enough to allow it to be used as a boarding step for accessing the boat. The devices are generally 42 inches long for use with boats having outboard motors and 32 inches long for boats with inboard propulsion and most stern drive powered craft. Limitations in structural capacity of these devices should be recognized if they are proposed for use in severe exposure locations. If they are used, care must be taken to assure appropriate strength and load transfer in the connection between the Frog Hook and the main float.

Figure 11-16. Tie pile mooring line attachment float, using foamed auto tire and galvanized steel ring. Float provides constant elevation between boat and pile attachment. Used extensively in New Zealand.

In situations where an offshore tie pile is used for boat mooring line attachment, a problem may result in areas which have large tidal fluctuation. The necessity of allowing enough scope on a mooring line to prevent the mooring line from hanging up the boat at the tidal extremes also allows a boat to drift around the berth during other tidal levels. A solution to this problem, noted in New Zealand, incorporated the use of a floating tire arrangement trapped by a vertical pile (see Fig. 11-16). A tire casing, preferably new, is filled with foam in the annular space normally occupied by air, and fitted with a galvanized ring that restrains the foam and provides a point of attachment for mooring lines. The tire also functions as a fender for boats getting too close to the pile. The system works quite well and is reasonably attractive. Other areas use wood and foam boxes to accomplish the same objective but without the strength and appearance afforded by the tire solution.

11.6 COST

Floating docks systems are generally less expensive to construct than are fixed dock systems. However, factors such as exposure to sea state forces, water depth, availability of local materials, and labor costs may affect the ultimate cost. There are several methods of projecting the cost for dock systems in the early stages of design development. The simplest way to assign a cost value is to consider a cost per berth. This is a coarse number, usually including the dock system, mooring and utilities. It generally does not include any support structures or utilities inshore of the gangway connection point. **Based on commercially purchased or fabricated components at 1990 values, the cost range for a berth in a medium to high class installation is between $8,000 and $15,000, both fixed and floating.** Do-it-yourself systems would generally cost less, depending on how the cost of materials and labor are allocated. Sometimes a marina manager will have his normal marina operating staff build marina dock components during slack work periods. Since the people have to be paid anyway, the slack time dock construction can dramatically lower apparent system costs. Do-it-your-self docks have a limited usability in most marinas, so it is not wise to plan a marina project totally on the basis of building it yourself and budgeting unrealistically low construction costs.

Another method of estimating dock system costs is by pricing the system on a square footage of dock surface basis. Most dock system manufacturers will provide cost estimates or bid prices on an area (square foot) cost. This method is more accurate than a per berth cost estimate, but it requires a basic marina layout plan to develop a meaningful dock area. **At 1990 values, the cost for floating dock systems ranges between $15 and $35 per square foot.** The cost is based on the dock system manufactured, delivered and installed, but without mooring system, site preparation (dredging, filling or bulkheading), and utilities. The wide spread of costs is related to the material and quality of construction of the system selected. A typical, upper class floating dock marina of 230 berths, completed in the spring of 1990, on the east coast of the United States cost approximately $35 per square foot of dock area for the complete floating dock system, installed, with high capacity steel mooring pile system, gangways and full utilities (from the shore seaward). The floating dock system alone without mooring, gangways, or utilities cost approximately $21 per square foot of dock area. These numbers must be used with caution but may provide a rough ballpark estimate for quality marina construction.

Fixed dock systems tend to cost more than floating dock systems if constructed to standard heavy duty parameters. The foundation cost is generally the area where the largest part of the fixed dock cost is generated. Pile driving may require special floating equipment and dock builder labor. **Depending on the type of construction, fixed docks may cost in the range of $25 to $100 per square foot of dock area.** The costs vary substantially because of local labor costs, materials availability, depth of water at the site, type of subsoil conditions, and use, and therefore presenting a representative example would not be indicative of costs at any other specific site.

The only proper way to determine budget costs for fixed or floating docks is to prepare an engineered plan of the marina layout and then have knowledgeable engineers or contractors prepare a budget cost estimate based on as much site information as possible. This process should begin in the early planning phases of a marina project and continually be updated throughout the planning process, right up to obtaining firm bids from competent marine contractors.

11.7 TAX AND INSURANCE ISSUES

Tax structure and insurance coverage of dock systems will be affected by local experience and tradition. There may be advantages of one system over another for tax purposes and this potential should be discussed with competent tax counsel. It may be that floating docks will be taxed as personal property and accelerated depreciation allowed. The nature of floating docks being temporary may also have an effect on the tax category.

Insurance will also vary depending on local experience and quality of construction. Generally either a new, fixed or floating dock system should enjoy a more favorable liability insurance coverage than old, dilapidated structures. It is often less costly to repair damaged docks than to purchase new replacement docks, and although insurance coverage for repaired docks may not be as favorable as a new dock system, a reduced liability should be reflected in insurance coverage cost. If the floating or fixed dock system is engineered, be sure that the insurance underwriter appreciates the quality and soundness of the design and strive for the best insurance value by stressing the quality and design of the system. Insurance agents and underwriters are generally not well versed in marina projects. It is therefore important to bring to their attention the positive aspects of the products and design. It is also wise to include your insurance representative early in the design process to assure that the marina layout and components do not present any special problems for insurance coverage. High liability items such as fuel dispensing and swimming pool inclusion will most certainly have a significant affect on insurance.

12
Facility Layout

Of all the aspects of marina design, poor facility layout garners the most criticism and is the most critical aspect of design, which when poorly executed, leads to an unsuccessful marina operation. A marina is intended for the safe and comfortable access and berthing of boats. If boats can not easily access the facility or have difficulty in maneuvering within the marina there will be damage to boats and hurt pride amongst the boat owners. Due diligence in facility layout can minimize operational conflicts and provide the degree of safety and comfort desired by both experienced and novice boaters.

Since this book is about boats, facility layout is viewed as beginning from seaward and proceeding to the upland. Each marina will have specific site conditions that may alter the recommended criteria but as long as the changes are made with an understanding of the ramifications, a functional marina should result. It helps if the marina designer is familiar with the handling characteristics of small boats of the size intended for use in the marina. The effects of a single screw powerboat maneuvering in close quarters, or the lack of backing control in sailboats are some of the input factors a designer should be familiar with when attempting a functional marina design. Effects of windage and water current on the maneuvering characteristics of boats are also important factors when deciding on a favorable slip orientation. At each stage of the layout, the designer should visualize entering the marina in a power or sail boat and effecting the maneuvering necessary to arrive at and be secured to the slip. This should be done in the context of fair weather and foul.

12.1 CHANNEL ENTRANCE DESIGN

The approach from seaward often involves maneuvering through an entrance channel. In many cases the channel will have been created by others, usually a governmental agency. In the case where an existing channel provides access, its ability to serve the needs of the marina should be assessed. If the channel is too narrow or too shallow, is it possible to widen or deepen the channel within the context of the proposed project or will the work require petitioning a governmental agency? Obtaining a governmental approval to alter a channel and getting the governmental agency to pay for the work is often a difficult and time consuming proposition, if even possible.

If an existing channel limits the viability of the proposed marina, then development of the proposed marina may not be feasible. Some developers will bite the bullet and elect to create or modify a channel. If this is to be done, be sure the bottom conditions are fully understood and that realistic expectations are made on the cost and timing of the work. Insufficient bottom profiling on a major marina project resulted in a failure to detect a high bedrock condition which, during construction, resulted in the need for an unexpectedly large quantity of rock removal at significant cost, which was necessary to obtain the required channel depth. It is also wise to perform sufficient scientific studies to have reasonable assurance that a dredged channel will not rapidly fill in and require constant maintenance dredging.

The actual dimension of entrance channels will vary with the types of boat using the entrance and with the density of use within the channel. **In general, the entrance channel should have a minimum width of 75 feet with full control depth over this width (Fig. 12-1). Dredge side slopes, where necessary should occur beyond the prescribed clear width. A 100 foot wide channel is a more preferable design criteria and should be used as the minimum, where possible. The channel control depth will be based on the deepest draft of a potential user vessel, plus a minimum of 3 additional feet of depth, plus an estimated trough depth caused by any significant wave action that may occur in the channel.** The summation of these criteria will become the design control depth and should be considered in conjunction with low water

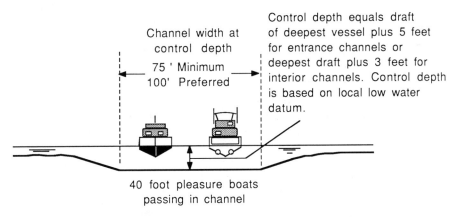

Figure 12-1. Entrance and interior channel design criteria.

information. Therefore a marina which is intended for use by boats (power and sail) up to 60 feet in length would probably require an entrance channel depth of 13 to 15 feet at low water plus additional depth if the entrance channel is subject to wave action.

Entrance channels should be properly marked by buoys, day beacons and/or ranges, as appropriate, to locate the channel from seaward and define its path to the marina. Data on how to calculate line of sight visibility from boats on the water for ranges and other structures is presented in Appendix II. If a new channel is created as part of the marina development, its characteristics should be forwarded to the governmental agency responsible for preparing nautical charts for inclusion in the next edition.

12.2 PERIMETER CONDITIONS

The perimeter protection, selected in Chapter 8, must be incorporated into the facility layout. In most cases, the perimeter protection orientation will dictate the parameters of the facility layout. The selection of dock arrangement and entrance location must be worked around the perimeter constraints. If fixed perimeter protection is to be provided, is it accessible from the shore and can it be used for boat berthing? It should always be borne in mind that perimeter protection is provided to absorb or reflect large amounts of energy so the placement of boat berths along the inside of the perimeter protection may subject the berths to some forms of severe wind wave action. An example might be the overspray off a rubble mound breakwater or fixed wave board array. A permeable wave board array may allow some wave energy to pass through the structure which may impact boats in close proximity to the structure. A floating attenuator will almost always be subject to wave spray and possibly overtopping by larger waves. If boat berths are associated with perimeter protection, it is generally best to make these berths for transients or have the provision to remove any boats from this area during storms.

Rubble mound breakwaters have a sloped inner and outer face which occupies considerable plan area. Attempting to berth boats along a rubble mound breakwater inner face may require a pier extending from the sloped side to provide adequate water depth for the boats and a reasonable access. Major fixed structures will also extend a considerable height above water to accommodate storm surge effects. This height will impede access to boats floating at water level and will create a visual barrier to the outside world.

If the preliminary site analysis indicates that perimeter protection is not required, it may still be prudent to provide wave/wake attenuation floating or fixed docks at or near the outer extremity of the layout. Floating dock systems may be constructed with extra width, mass and depth to provide a measure of wave protection. Fixed docks may be fitted with permeable wave boards to provide some protection.

FACILITY LAYOUT   271

The dock system layout should provide direct access from any entrance channel and provide adequate fairways. Setback from property lines and designated channel lines must also be considered. The setback criteria may be established by a regulating authority or based on custom in the area. Whatever the specified setback criteria or even if there is no specific criteria, the marina layout should consider the possibility of development on adjacent properties and be planned so that future adjacent development will not limit or conflict with access requirements for the marina.

12.3 FAIRWAYS

The layout of marina fairways is a topic of heated discussion. The developer wants to install as many berths as possible to adequately cover his capital costs. The designer wants a reasonable number of berths that provide only as much room as necessary to allow safe vessel maneuvering. The boat owner wants a layout that has a great amount of maneuvering room to allow for errors in maneuvering and for comfortable maneuvering in foul weather. This diversity of needs must be resolved by the designer to the satisfaction of the client, regulatory approval agency and boat owner.

Some years ago an advertisement was circulated by a marina developer offering slips for sale in a new marina facility. The advertisement showed a schematic plan view, similar to Figure 12-2, of a marina with a high density of slips. The sketch was obviously drawn without consideration for the needs of the boaters who require a fairway access to maneuver in and out of a berth. The concept often reflects the philosophy of an unknowledgeable developer who seeks to obtain the maximum rate of return on investment without due consideration for the product being marketed.

The actual dimension for the sizing of a fairway should consider several variables. The environmental climate, which we have intently addressed, is a major concern. If the site is impacted by significant winds, waves or current, then fairways should be sized above standard guidelines to allow adequate maneuvering room to accommodate these conditions. If the marina will anticipate a preponderance of less maneuverable sailboats, the fairways may be made extra wide to accommodate this lack of low speed maneuverability. Most sailboats will enter a berth bow on and can negotiate a fairly narrow fairway in this condition. Backing the boat out of the berth is where the extra width fairway comes in handy. Sailboats are notoriously bad in maneuvering control when backing down. If wind or current may also play against the backing maneuver, some sailboats will move only in the direction of the strongest force, which may not be the engine. Single screw boats will generally back to port, reducing desired directional control. The experience of the boat handler is also an important factor. If the marina serves an active boating organization, Power Squadron, Coast Guard Auxiliary, fishing club, or other group of knowledgeable boaters, perhaps some consideration can

272 PART 3/ENGINEERING DESIGN

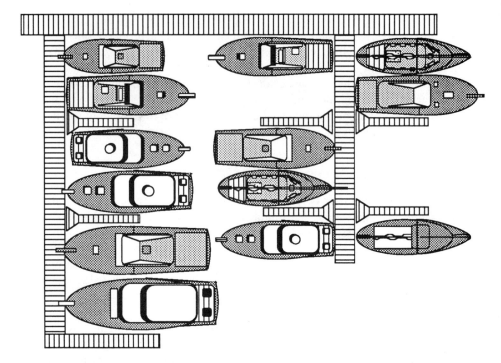

Figure 12-2. A diagram of how many inexperienced marina developers envision a marina layout to achieve maximum boat berthing capacity. The need to provide an adequate fairway to allow the boats to access the berths becomes obvious upon consideration of the proposed berth layout.

be given to reducing fairway dimensions. This must be done with care, however, since new members may not be as proficient as old hands and they might experience difficulty in maneuvering in confined areas.

The general rule of thumb for fairway sizing has been to make the clear distance between boat extremities no less than 1.5 times the longest boat length (see Fig. 12-3), and often 1.75 times boat length if maneuvering conditions warrant. In general, the greater the distance, the better for safe boat maneuvering.

A problem that has developed in upgrading existing marinas relates to traditional thoughts on the size of boats using a berth versus finger float length (see Fig. 12-4). Twenty or so years ago, finger floats were generally in the order of 20 to 24 feet in length. This design was based on the availability of wood for float construction. Wood up to 24 feet in length was readily available in the quantities and types used for dock construction. Because marinas should be able to berth boats of lengths greater than the length of a

FACILITY LAYOUT 273

Figure 12-3. Darbie Landing Marina, Salem, Massachusetts. Marina layout providing adequate fairways based on a rule of thumb that a fairway should be a minimum of 1.5 times the length of the longest boat that will be operating in that fairway. Adequate maneuvering area is also provided at interior basin spaces. *(Photo courtesy: MEECO Marinas, Inc., McAlester, OK)*

24 foot long finger, designers analyzed the finger structure and determined that boat lengths of about 1.25 times the finger length could be handled, structurally, by the finger float. Outboard tie piles were often placed beyond the end of the finger float to secure lines from the overhanging boat. As a result, it became standard practice to allow boats of 1.25 times the finger float length to be accommodated in the berth. By using modern materials and construction techniques marina dock manufacturer's have been able to provide finger float units of virtually any length. Therefore, today finger float length is considered to be equivalent to berthed boat length for structural considerations, i.e., a 40 foot long finger is structurally suitable for a 40 foot boat. The problem is that some marina operators are still applying the 1.25 times boat length rule to marinas which lack accommodation for this length extension. Two major conflicts result from applying the old rule. The first is that the marina manufacturer may not have designed the finger for the increased load associated with the larger length boat. The increase in load is significant in going from, say, a 40 foot boat to a 50 foot boat. The structural integrity of the finger may be compromised. The second condition is that the designer probably has designed the fairway to accommodate the length of boat associated with the finger length. If extra boat length is added, the

274 PART 3/ENGINEERING DESIGN

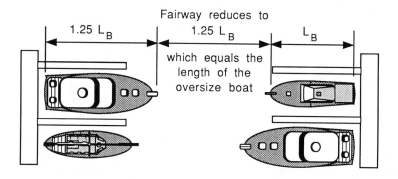

Figure 12-4. Relationship of boat and finger length to fairway dimension.

fairway clear width will be reduced, perhaps to an unacceptable dimension. In traditional marina operation, a competent marina manager may elect to stagger varying boat lengths and still have a functional marina layout. If however, the marina is "dockominiumized" or long term leased, and a provision is made that allows for boats greater than the design berth length, there will, in time, be conflict. It can be assumed that over time berth owners or their successors will berth the largest boat capable of being accommodated in the berth under the ownership or lease arrangement. If the allowed boat length is greater than the design length, the fairway width will be compromised and a lot of unhappy boaters will result. This problem is not speculative but real. We have been involved with a number of marinas where leasing agents have corrupted an otherwise proper design to create the most attractive return to the marina developer, at the expense of the lease holder or berth owner.

12.4 BERTH SIZING AND BOAT DENSITY

As discussed in Chapter 10, boat sizes have changed rather significantly in the past decade. A principal dimensional change has been in boat beam versus length. To accommodate this increase in boat beam, the marina layout must be sized accordingly. It is rare that a marina designer will have boat sizes and types specifically defined during the design layout phase. In fact the complement of vessels using the marina will be constantly changing. It therefore is prudent to design the marina layout to accommodate any reasonable size of vessels in the range of berth design. Some designers will provide a range of berth clear widths to accommodate narrower sailboats and beamier powerboats. This provision must be made with caution since the market might change and a large number of undersized berths may become unmarketable if beamier powerboats become the norm. If the marina is to be owner operated versus dockominiumized or long term leased, some flexibility in assigning slips may mitigate having a variety of berth widths. The operator can then assign berths that are appropriate for the particular size boat. Although there is rarely enough water sheet to design an extravagant berth layout, it is prudent to design the layout for the larger beam powerboats using the consideration that the opportunity exists to accept virtually any monohull that desires berthing. If sailboats occupy the full width slips, this is not necessarily a penalty, since sailboats generally require additional maneuvering room in confined spaces and will be happy with the greater maneuvering room and also with the greater spacing between berthed boats. Based on 1989 boat manufacturer statistics on boat beams and allowing for fendering between the finger float and the boat, the recommended dimensions shown in Figure 12-5 will be suitable for most monohull boat configurations. The minimum values given may be used in calm marina sites, but may create some crowding of boats. A first class marina will not skimp on berthing space. If marina berth rental or sale cost is based on square footage of occupyable space, the cost to construct to the recommended criteria will be recovered.

In marina planning it is often useful to estimate the number of boats that can be berthed in a given water sheet area. There are a number of variables

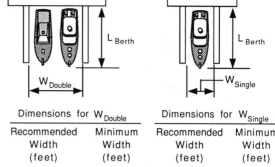

		Dimensions for W_{Double}		Dimensions for W_{Single}	
Boat Length (feet)	L_{Berth} Design Berth Length, (feet)	Recommended Width (feet)	Minimum Width (feet)	Recommended Width (feet)	Minimum Width (feet)
20	23	21	20	10	10
25	28	25	25	12	12
30	33	29	27	14	13.5
35	38	34	31	16	15
40	43	37	34	18	16
45	48	39	36	19	17
50	53	41	38	20	18
55	58	45	42	22	19
60	63	47	44	23	20

Figure 12-5. Recommended clear widths for double and single loaded floating berths.

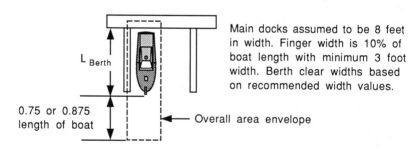

Main docks assumed to be 8 feet in width. Finger width is 10% of boat length with minimum 3 foot width. Berth clear widths based on recommended width values.

			Overall Area Envelope (sq. ft.)		Boats per Acre	
Boat Length	Design Finger Width	Dock Deck Area in Sq. Feet	Fairway Equal to 1.5 Times Boat Length	Fairway Equal to 1.75 Times Boat Length	Fairway Equal to 1.5 Times Boat Length	Fairway Equal to 1.75 Times Boat Length
20 feet	3 feet	78 sf	504 sf	534 sf	86.4	81.5
25	3	93.5	710	755	61.3	57.7
30	3	109	952	1,012	45.7	43.0
35	3.5	136	1,280	1,362	34.0	32.0
40	4	162	1,579	1,681	27.6	25.9
45	4.5	188	1,865	1.987	23.4	21.9
50	5	217	2,174	2,317	20.0	18.8
55	5.5	252	2,607	2,781	16.7	15.7
60	6	286	2,968	3,167	14.7	13.7

Figure 12-6. Estimates of the number of boats per acre for recommended berth dimensions.

that affect boat berthing density such as: main pier width, finger pier width, clear width between finger piers, and fairway dimensions. Figure 12-6 provides a range of boat density estimates for varying boat sizes and fairway dimensions. The basic layout follows the recommended main pier minimum width of 8 feet and finger pier width equal to 10 percent of the finger length with a 3 foot minimum width. The tabulation results in an estimate of the number of boats per acre of water sheet. Entrance channels, alongside dockage, turning areas or other special conditions have not been factored into these calculations and should be considered independently. The tabulation also provides a rough estimate of the square footage of dock associated with an individual boat berth. Required dock square footage may be extrapolated from this data by multiplying the square footage per boat berth times the number of the berths and adding interconnection docks and other dockage not associated directly with a berth. The resulting number will be a coarse estimate, but such a figure often is useful for first trial budget estimating.

12.5 STRUCTURE SIZING

What is the proper size for marina dock structures? This is a good question that again relates to usage, custom and structural considerations. **The minimum width of a fixed or floating main dock is considered to be 6 feet.** Floating main docks should never be less than 6 feet because they become unstable and provide a difficult walking surface. The float will also have cleats and possibly utilities that may reduce the usable space even more. **Generally an 8 foot wide main floating dock is the favored width for most marina applications.** At 8 feet, tie up hardware, power posts and other utilities will not interfere with normal walking patterns or in the use of dock carts. If golf carts or other large vehicles are contemplated then the dock width will need to be around 12 feet in width with turnabout areas provided. **Fuel docks should be 10 to 12 feet in width to provide stability, room for tying up vessels, fuel dispensers, fire protection equipment and personnel operations. Floating attenuators will be in the range of 16 to 24 feet in width, or greater.**

Fixed docks have a greater latitude in width constraints because the stability aspects of floating docks are not of concern. However, width does affect structural stability in terms of width to pile length in piled structures. A 4 foot wide fixed dock constructed on pilings that have a large free height, about 20 feet, may also be quite unstable and shaky unless battered piles are added to accept lateral load and fix the pier in the horizontal direction. Widths for fixed docks will generally be the same as for floating docks, in terms of minimum dimensions to adequately serve recreational boater requirements. Extra width can often be accommodated in fixed dock construction at minimal extra cost, so it is wise to consider additional width in areas that may benefit from increased space, such as in high pedestrian traffic or work areas.

The design width of finger floats is also somewhat arbitrary and should be based on the structural requirements of the finger float to carry the imposed boat loads, but the width should not be less than the rule of thumb of width equal to 10 percent of finger length. The 10 percent rule has been shown by experience to be an appropriate width of structural and use requirements. The 10 percent rule is modified, however, so that no finger should be less than 3 feet in width. Even at 3 or 4 feet in width a finger float by itself, before connection to the mainwalk, will generally be unstable and will require adequate connection to the mainwalk to insure its stability. A good test of connected finger stability is to walk out to the end of the finger and sway side to side. If the finger has only a minor amount of sidesway and vertical movement, it is well designed and will provide a comfortable and secure berthing structure. If the finger is very lively at the outer end, it may be flexible enough to survive significant wave action but may be uncomfortable to walk on or tie up to. Finger stability will also be affected by the type of connection to the mainwalk. If the finger is hinge connected at a location that is an extension of its normal width, it will probably be somewhat shaky at the outer end and will deflect through rotation about the hinge. If the finger is rigidly connected to the mainwalk, it will be more stable, acting like an outrigger to the mainwalk. If it is rigidly connected to the mainwalk and at locations beyond its normal width by gussets or cornerwalks, it will probably have its greatest strength and stability. Rigid connection to the mainwalk creates a system where the finger is essentially a beam cantilevering from the mainwalk. As load is applied to the finger end, the load is transmitted back to the mainwalk, which because of its mass relative to the finger, resists the effects of the applied load. In a hinge situation, the hinge relieves the stress at the expense of increased finger movement. This can be a plus, however, in high load conditions because it relieves the stress at the hinge and does not require the mainwalk connection to resist the load. The selection of finger dock and connection type must be weighed in terms of structural considerations and function. As marina float design has improved, the monolithic structure, without hinges, has found increasing favor because of its overall stability, continuity, and strength.

For generalized layout considerations it is appropriate to consider 8 foot wide mainwalks, 10 to 12 foot wide fuel or large vessel transient docks, and fingers not less than 3 feet wide or 10 percent of the finger length.

Float freeboard should also be considered. **Floating docks will generally**

have a freeboard of between 12 and 26 inches. An argument for low freeboard docks of 12 inches has been made using the consideration that a person falling overboard can often climb back up on a 12 inch high float but definitely cannot climb back up if the freeboard is 24 inches. The need for recovering people from the water is a legitimate concern but can be provided for by selective placement of short boarding ladders attached to the dock system or strategically placed safety ladders on the dock deck, perhaps adjacent to fire extinguisher locations. It is also usual to have some berthed boats in the water and many of these boats will have swim platforms or other boarding devices that can assist a person out of the water. On the other hand, it should be convenient for a person to board a berthed boat from the dock system. For boats in the 20 to 40 foot range, 24 inches of freeboard seems an acceptable height. Some boats will require boarding steps of 12 to 24 inches in height. Floating docks with freeboards much greater than 24 inches tend to become unstable and top heavy, unless they are very wide.

Fixed dock height will be based on the ambient high water level. Certainly, the dock should be above the highest reasonable water level. In areas with large tidal variation, there will be a great variation between the boat deck and the dock at some tidal levels. If the tidal variation is 4 feet or greater, floating docks should be considered over fixed docks. If fixed docks are used in areas of large tidal variation, extra berth width should be provided to account for the additional boat line scope that is required to accommodate the vertical elevation change. The boat will move around in the slip at high water when the lines are slack and be held in a more fixed position at low water, when the lines are taut. Tying off the boat lines at the midheight of water level change may help to minimize the changing line length.

12.6 WATER DEPTH

The depth of water at the marina must be suitable for the draft of the largest vessel anticipated to use the facility plus a margin for safety. Marinas used exclusively by powerboats, especially if an impediment exists that prevents use by sailboats such as a bridge or other height restrictions, may consider water depths suitable only for powerboats. Other marinas that potentially could berth sailboats should provide a depth of water suitable for either power or sail boats. Recommended minimum water depths are shown in Table 12-1 and are based on representative deepest draft vessels for the sizes listed. The table does not take into account any additional water depth that may be required to accommodate storm surge conditions where a wave trough might take a boat below the still water level. Hopefully the marina basin will not be subject to significant wave energy and the storm surge will

Table 12-1. Minimum recommended water depth in marinas

Boat Length		Minimum Water Depth			
Feet	Meters	Power		Sail	
Minimum		4 Ft	1.2 m	4 Ft	1.2 m
30	9.0	7	2.1	9	2.7
35	10.6	8	2.4	10	3.0
40	12.0	8	2.4	11	3.3
45	13.7	8	2.4	12	3.6
50	15.0	8.5	2.6	13	4.0
55	16.7	8.5	2.6	14	4.3
60	18.2	8.5	2.6	14.5	4.4
65	20.0	9	2.7	15.5	4.7

not be a problem. The water depth data is based on a mean low water or some other relevant low water level. On lakes and rivers that have fluctuating water stages a decision will have to be made as to what water level will be considered as a base low water elevation.

12.7 UPLAND BOAT STORAGE

Marina planning also often involves assessing the potential of the upland for off season boat storage. The most often used devices for hauling and placing boats on the upland are straddle hoists and/or some form of trailer such as the versatile hydraulic trailer. The actual density of boats placed on the upland will depend on the topography of the site, the proximity to the haul out facility and the desires of the marina or boat yard manager. Some rough estimates of upland boat storage density may be determined from Figure 12-7 and Figure 12-8. Figure 12-7 presents the development of a reasonable upland storage scenario using several sizes of straddle hoists. Minimal clearances are provided in the analysis. Greater room may be desirable to allow freedom of movement of people and materials around the boats. The use of a straddle hoist somewhat limits the boat density by virtue of the need to provide adequate room between boats to fit the fixed width of the straddle hoist. If boats are to be launched on a fixed schedule it may be possible to close up the spacing, but rarely are schedules completely complied with and a change in schedule might compromise launching other boats if the spacing

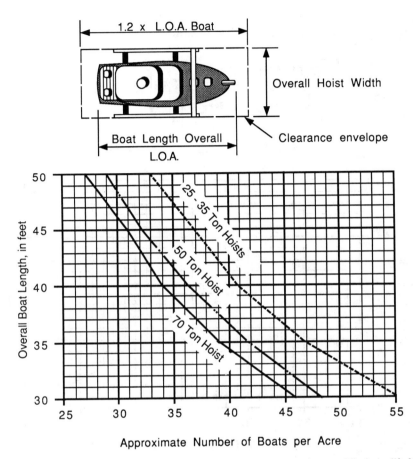

Figure 12-7. Approximate maximum number of boats per acre using straddle hoist lift for upland storage, no aisle space between boats.

between boats is too close. Also it is advisable to allow aisles between rows of boats such that any given boat may be accessed at any given time.

Note that a small boat in a large hoist is inefficient for land utilization. Straddle hoist size should be related to the predominant size of boat to be hauled and stored, or consideration should be given to augmenting the operation with a trailer to conserve space.

Figure 12-8 presents the estimated storage density based on use of a hydraulic trailer type device. Because the trailer does not require room beyond the extreme beam of the vessel, and assuming most vessels will have a beam greater than the width of the trailer, a high density of upland storage

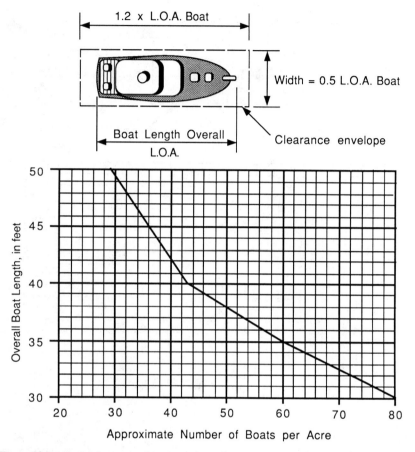

Figure 12-8. Approximate maximum number of boats per acre using hydraulic trailer for upland storage, no aisle space between boats.

can be achieved (see Fig. 12-9). As in the previous example, a number of variables may affect the actual upland storage density achieved on a specific site.

12.8 FUEL SERVICE FACILITIES

The costs and benefits of including fuel facilities in a marina development must be carefully weighed. The desirability of providing fuel will depend on several key issues. The first issue is the market for fuel services. If the marina is in an area that has numerous fueling facilities, it may be difficult to generate enough demand to cover the high capital cost of fuel tank installation and monitoring devices as well as the required insurance on such

FACILITY LAYOUT   283

Figure 12-9. Placement of a small boat by hydraulic trailer in very close proximity to other boats for maximum, on land, boat storage density. The hydraulic trailer, being of less width than the boat, allows the close placement of boats as compared to a straddle hoist that requires additional space around a boat to accommodate the frame of the straddle hoist.

facilities. The potential for fuel spills or transmission system leakage is great and the environmental clean up costs may be significant. The market for fuel sales and the resulting income must be determined to be substantial enough to carry the capital costs of fuel system installation plus the liability factor.

Another issue is the ability to site the fuel storage tank installation in a suitable area within the marina envelope. The storage tanks should be as close to the dispensing station as possible to minimize transmission line costs and potential pipe leakage or failure. The storage tanks must be designed to be environmentally safe, which may mean containment in a concrete box or dike, and resistance must be provided to prevent floating if the tanks are buried in a high ground water area or an area affected by tidal exchange. Monitoring and alarm systems will be required as well as fire protection and suppression systems.

Buried storage tanks are generally the best choice if environmental and structural requirements can be met. The buried tanks are unobtrusive and the surface area can be used for other purposes such as auto parking or boat storage. Above ground tanks will most certainly require a dike or other spill protection containment area. The above ground tanks are generally unsightly and produce an industrial atmosphere. Additional information on fuel tank installation may be found in Chapter 19.

Any dock system incorporating fuel dispensing lines must be designed for the weight of piping and full fuel weight in the pipes. In floating dock systems, this additional weight can easily unbalance the floating attitude of the system. Fuel dispensers and ancillary equipment weight must also be considered in the dock design. Additionally, a fuel facility will probably require a wet or dry pipe fire fighting system which will also require weight considerations in dock design. Fixed docks must be capable of providing adequate capacity for these piping systems when hung from deck or frame structures.

Fuel transmission pipe systems will require several local shut off valves, generally at the dispensing units, just beyond a safe distance from the dispenser (about 50 feet), at the shore end of the pipe system, and at the storage tanks. Check valves will also be required to prevent backflow and siphoning of the transmission system. Be careful to provide a relief system such that fuel in the transmission lines between shore and the dispensing units can be relieved from thermal expansion during periods of inactivity and thermal heating from normal sun related temperature rise.

Fuel docks (see Fig. 12-10), should be located in an area that facilitates access by boat. Locating a fuel dock deep inside the marina might reduce fuel product pipe line length and provide a short distance for marina personnel to travel to operate the facility, but could be disastrous if a fuel spill or fire occurs. Also the goal is to provide a rapid turnaround time for fueling boats to maximize the sale of the fuel. If the boats must maneuver inside the marina, the marina traffic pattern may be disrupted and the possibility of boat collision is increased. If a boat should catch on fire at the fuel dock, it may be desirable to release the mooring lines and move the boat away from the fuel dispensers and fuel product lines to an area more safely accessible by fire fighting personnel and equipment. The decision to move a burning boat from a dock should only be made by knowledgeable, trained personnel with a plan of attack to prevent a drifting and burning boat from increasing an already dangerous situation. A fuel dock located on the perimeter of the marina does, however, provide the opportunity to isolate the burning vessel from other areas of the marina. A recommended guide for provision of fire protection at fueling and other marina areas is the National Fire Protection Association's code and standard NFPA 303, Marinas and Boatyards.

Fuel docks should be designed to accommodate a fair amount of people and activity. A minimum width for a fuel dock is 12 feet. A greater width may be necessary to accommodate a small enclosure to provide an operating base for the fuel service personnel and a place to get out of the weather. The loading of these structures should be accounted for in the dock design. The

FACILITY LAYOUT 285

Figure 12-10. A modern fuel service installation on a floating dock system. The fuel dock is located on the outshore perimeter of the marina to facilitate easy access and maneuvering by boats approaching or leaving the dock. A long length of dock is provided to enhance fuel service during busy periods. Note also that dispenser hose reels are provided to prevent hose clutter on the dock.

length of a fuel dock will be based on the size and type of boats to be served, but it should generally be large enough to handle at least two vessels at a time. Cleats associated with a fuel dock should be substantial, easily worked, and well secured. Depending on dock orientation and wind direction, many boats may need to spring on the cleats to leave the dock after fueling. The dock should be fitted with large and substantial fendering of good quality to withstand the continual abuse of frequent dockings.

12.9 SEWAGE PUMPOUT FACILITIES

Provision of sewage pumpout facilities is becoming a mandated requirement for approvals acquisition in many areas. Inland lakes and rivers routinely must provide pumpout facilities, but coastal facilities have been lax in implementing installation due to conflicting federal and state regulations, and the ability to legally dispose of some wastes outside territorial waters. However, the marina developer should consider installation of sewage pumpout (see Fig. 12-11 and Fig. 12-12) if for no other reason than to enhance the water quality in and around the marina. The main problem in implementing the accepted use of boat holding tank systems and sewage pumpout stations

286 PART 3/ENGINEERING DESIGN

Figure 12-11. Installation of a permanent boat sewage pumpout station on a floating dock system. System is hard piped to shore and into a 1,500 gallon capacity, buried, holding tank for subsequent pickup by tank truck and disposal at an approved sewage treatment facility.

Figure 12-12. Appropriate signage to indicate a boat sewage pumpout facility within a marina.

has been their lack of availability and difficulty in use. As more and more marinas install sewage pumpout facilities there should be a greater acceptance of their use by boaters.

A common location for a sewage pumpout station is adjacent to the general area where fuel dispensers are located. Although this arrangement may be suitable from the personnel management philosophy of the marina operations, combining the two functions may be detrimental to use of the fueling area. Sailing vessels that require only infrequent fueling may tie up the fuel dock just to pumpout. On the other hand a boater will not be happy to have to move from the fuel dock to a separate pumpout dock to accomplish these functions. Pumpout facilities are also not routinely used because many boats will sit for several weeks at their berths because of weather, maintenance problems or personal desires and the owners may be reluctant to move the boat to a pumpout area just to pumpout. In foul weather this can even be dangerous.

If sewage pumpout is to really catch on, the marina should be designed for sewage pumpout capability at each berth. This is not very difficult in new construction and may be fairly easy to modify existing systems. The rigid or semirigid piping system can be incorporated into the docks with a pumpout connection along the main dock to serve two or more vessels. A portable pump can be connected between the boat and the mainline sewerage piping and the boat holding tank can be pumped out while the boat remains in the berth. If the cost of the system and service is built into the berth rental fee or association fee, people could use the system at will and with no surcharge they are likely to use the system as often as necessary with the resulting reduction in overboard discharge. For years potable water, electric power and cable television systems have been incorporated into marina services. Now is the time to include an easy to use sewage pumpout system.

As with fuel and water systems, a sewage pumpout system incorporated into a dock system must be taken into consideration in the dock design. The weight of piping, pumps and holding tanks may seriously affect dock stability and attitude. A sewage pumpout station should have electric installations that are code acceptable and should also have a separate water washdown capability. The water outlets and hoses should be clearly labelled as nonpotable water sources.

The design of a pumpout system will be determined by the vertical height between the dockside pumpout and the discharge location onshore, such as a sewerage manhole or holding tank. The length of the run between the pumpout and the discharge location must also be considered to account for line losses due to friction of the liquid moving through the pipe. Manufacturers information should be consulted for specific design. Some general guide-

Figure 12-13. A shoreside service station for emptying and washing portable boat toilets of the type found on many small boats.

lines are presented for information purposes. **Pumpout pump motors will generally be of ½ horsepower or ¾ horsepower. In general a ½ horsepower pump may be suitable for a head up to 20 feet and a ¾ horsepower motor suitable for up to 30 feet of head. Discharge pipe size and length will affect performance. Discharge pipe sizes range from a minimum of 2 inch diameter to 3 inch diameter.** To give an idea of the effect of pipe size on acceptable length of run, consider that for a system having a 10 foot head and using a ½ horsepower motor, a reasonable length for a 2 inch diameter pipe would be 125 feet, versus 333 feet for a 2.5 inch diameter pipe and 1,000 feet for a 3 inch diameter pipe. The minimum recommended size pipe is 2 inches with 2.5 inches preferred. Recommended pipe is schedule 80 PVC (polyvinylchloride). Flexible hose such as Goodyear Flexwing® is also suitable. Hose must be the same inside diameter (I.D.) as the pipe. Inline check valves in the discharge pipe may void pump manufacturers warranty, so carefully check design recommendations.

Explosion proof motors are available for hazardous locations such as those adjacent to fuel dispensers. Pumps must generally be able to pass solids up to 1 inch in diameter. The pump should also be durable enough to sustain constant passage of solids and foreign matter without unreasonable maintenance. Fixed and portable installations are available to suit particular use requirements.

In addition to providing sewage pumpout capability, a modern marina should provide a suitable facility to empty and washdown portable toilet units (see Fig. 12-13). A least one major boat sewerage equipment manufacturer offers a specially designed portable toilet service station. The unit is self standing and provides a single location for disposing of the portable toilet contents and rinsing the portable toilet unit. The service station may be connected to the main marina sewage pumpout piping or connected directly to a municipal sewerage system. The use of this type of facility prevents patrons from disposing of portable toilet contents in marina heads or other unauthorized areas.

12.10 SERVICE DOCKS

Marinas that provide dockside service or haul-out facilities may require a reserved service dock. If a haul-out facility is provided, the service dock should generally be situated adjacent to the haul-out facility and should be in a position to assist in feeding boats to the haul-out area. Other conditions may dictate that the service dock be adjacent to land based cranes for mast and engine work or close to service shops. Whatever the need, the service dock should be thoughtfully located to provide for efficient work and boat maneuvering. Many boats that may use the service dock may be without engine power, so the dock should have adequate surrounding maneuvering room and an easy approach from seaward.

Access to a service dock is best considered as a special access and should not be generally open to the public. Or if public access is necessary, the access should be constructed with extra width, minimum of 6 feet, to provide an easy flow of workers and materials while still having room for pedestrian flow. The dock itself should be extra wide, if possible, and have good stability to allow safe material handling. A recommended minimum service dock width is 8 feet, with 10 feet or 12 feet even better. Consideration may be given to outfitting the dock with special utilities and heavy duty standard utilities to service a wide range of vessels that may use the dock for short periods of time. Since the area will be used for service work, it may be advisable to provide a fire protection or suppression system as part of the service dock. The dock should also be equipped with heavy duty fendering and plenty of high capacity cleats or other tie off hardware.

The proper freeboard height for a service dock must relate to the type of vessels being serviced. The service docks may have a higher freeboard than the marina docks to provide an easy access to the larger boats that will probably require use of the service dock area. If necessary, provide portable boarding steps or several different freeboard height docks to accommodate a range of vessels.

Although the service dock is a working area, it must be kept clean and orderly. It is extremely important that the people working on boats do not track dirt, grit or oil on to the boats they are servicing. Ground in grit or oil on a teak cabin sole will not be tolerable to the recreational boater. The service dock may require special signage to indicate its function, especially for marinas that host transient boaters. An open service dock may be very attractive to tie up to by transients unfamiliar with the marina arrangement. A proper service dock area may be construed as an indicator of the quality of service offered by the marina to the boater who is not familiar with the marina's reputation for service work. A clean and well constructed service area may also reduce liability for worker injury and result in more favorable insurance premiums.

12.11 SPECIAL DOCK LAYOUT CONFIGURATIONS

In some cases, special layout configurations may be desirable to increase boat density or size and provide other marina enhancements. Any special layout must be thoroughly analyzed for suitability and cost. In most cases, deviation from standard dock components will increase cost and may result in other layout conflicts.

In locations where fairway width is limited, a method used to obtain greater berthed boat length, is the installation of angled berths (see Fig. 12-14). Instead of having the berths perpendicular to a mainwalk. the finger floats or piers will be designed at an angle. The angle creates a situation where the berth length becomes the hypotenuse of the triangle created by the angled berths. The hypotenuse being longer than the perpendicular leg allows for berthing of a boat equal to the length of the hypotenuse rather than the length of the perpendicular leg. Caution must be used however since the adjacent fairway should not also be reduced unless the boats are expected to enter and leave the fairway in only one direction. It should be assumed that a boat in an angled berth will need to turn around before heading into or out of the berth. The fairway, therefore, should be of an appropriate clear width, a minimum of 1.5 times the boat length.

Another form of special marina layout is the use of multiple berth single point moorings (see Fig. 12-15). Use of multiple berth single point moorings increase the active berthing density over one boat moorings but is not as space efficient as organized traditional marina berth layouts. The multiple berth moorings usually do not provide any utility services to the moored boats so it may be necessary to provide an area in the shoreside part of the marina to allow for the taking on of potable water, boat washdown and service needs. The multiple berth moorings will need to be accessed by small

Figure 12-14. A marina utilizing an angled berth arrangement in a narrow waterway to allow provision for longer boats than a perpendicular layout would permit while maintaining a fairway of suitable width.

Figure 12-15. A multiple berth single point mooring array. Note organized clusters of multiple berth moorings and easy boat passage between mooring clusters. *(Photo courtesy: Aqua-Matic Piers, Ltd., Milwaukee, WI)*

boat or launch service. On land dinghy storage may be required or a launch service provided. Multiple berth moorings must be designed to allow for the loading conditions of a fully occupied mooring array and provide enough searoom for array movement caused by water level fluctuation and vessel maneuvering. The cost of multiple berth moorings is generally significantly less than is the cost for a conventional marina. The use of these facilities can be very desirable in municipal harbors that have a great demand for moorings but which have limited area, and where full marina services are not necessary.

13
Materials of Construction

The selection of the proper materials for use in waterfront construction is critically important to the cost and life of the structure. The sea land interface is a harsh environment with many destructive forces working to batter, deteriorate and corrode materials. Selection of materials should be thought of in terms of suitability for the intended use, strength, resistance to degradation, cost, availability, maintainability, life, and appearance. Most applications can be accommodated by several different materials of construction. Alternative materials must be investigated and intelligent assessment made as to their suitability. Cost is most often the determining factor but it should be assessed against strength, life and appearance.

In the development of any new project it is wise to look at the types of materials that are used in the local area and the condition of those materials over time. A low first cost may be offset by high recurring maintenance costs. The appearance of a facility may be the one item that will set the project off from the competition and make the additional cost of alternative materials justifiable. Facility user perception is also important. A prospective marina customer will look at a marina system with an eye to the appearance of the system to protect his large boating investment. If the system appears weak or otherwise unattractive the potential customer may go elsewhere.

Above all, the materials selected must be adequate to perform the structural function so important in waterfront facility development. No matter how pretty or expensive the system is, if it cannot withstand the forces imposed upon it and it fails, the product or facility is a bad investment.

There are many types of materials used in waterfront construction. We will attempt to present information on a few of these materials, those that form the predominant group of materials most often encountered in marina and small craft harbor projects.

13.1 WOOD
Wood is one of man's oldest materials of construction. It's availability, strength, cost, workability and natural attractiveness have made wood the primary material of choice in coastal construction. In the United States,

there are over 100 woods available with about 60 having commercial significance. Worldwide the number of available woods increases greatly. Some woods, especially tropical hardwoods, are imported to the United States for use in marine structures because of their strength and resistance to decay and marine organism attack.

In coastal construction, wood is used for piling, fixed and floating dock structures, dolphins, bulkheads, decking, fendering and building frames. In most cases wood used in coastal environments must be treated with preservatives to enhance its resistance to biological attack. Wood has good engineering properties and can be configured to provide significant structural strength. Caution has to be exercised, however, because the extensive use of wood has reduced the availability of wood from virgin forests with some resulting loss of strength and quality. Much of today's wood comes from second growth forests and is a more rapidly grown wood, having different structural properties as compared to wood from old growth forests.

Wood is easily worked by commonly available tools and is easily repaired and altered. It accepts chemical treatment well, resists oxidation, acid and salt water exposure. In a dry condition, wood has good insulation properties against heat, sound and electricity. It can be fastened with dowels, nails, screws, bolts, and recently, adhesives. It is functional and has a pleasing aesthetic appearance.

Classification

Wood is most often designated by two general classes, "hardwoods" and "softwoods." These may be misnomers in some cases such as longleaf pine and Douglas fir which are classed as softwoods but are typically harder than hardwoods such as basswood and aspen.

Some examples of softwoods are: redwoods, pines, juniper, hemlock, cedars, firs, and spruces. The botanical name for the grouping of softwoods is Gymnosperms. Softwoods generally exhibit exposed seeds in the form of cones and have needle or scale like leaves which remain on the tree year round. Softwoods grow throughout the United States but they are limited in the Great Plains area. Woods are often labelled by geographic area of growth, such as northern softwoods, southern softwoods and western softwoods.

Hardwoods feature broad leaves and flowers with seeds that are enclosed in a fruit. Botanically they are classed as Angiosperms. Some examples of hardwoods are: beech, sycamore, cheery, walnut, maples, elm, holly, hickory, ash, and beechwood. Most hardwoods in the United States grow in areas east of the Great Plains.

Properties

Wood is fibrous in nature. It is composed of cells of organic material. The cells are long, hollow and spindle-like. The tree cross section is defined by three features: the pith, the wood and the bark. The pith is the small primary growth area at the center core of the tree. It is from the pith that most branches emanate, providing the structural shaft to support further branch growth. The wood is composed of sapwood and heartwood (see Fig. 13-1). The sapwood is the outer sheath of the tree often $1\frac{1}{2}$ to 2 inches in radial thickness. The sapwood provides the conduit for the storage of food and the transport of sap throughout the tree. The heartwood radiates outward from the pith to the sapwood. It is an area of inactive cells that formerly provided the sap transportation network. The heartwood may have a darker color than the sapwood due to deposits of various materials contained within the inactive cells. A darker heartwood is universally consistent but some woods will not show as noticeable a coloration as do others. The inclusion of materials in the cellular structure usually makes the heartwood more resistant to decay than sapwood. Both heartwood and sapwood, especially in softwoods, generally require chemical preservative treatment to be suitable for use in a coastal environment.

The bark is the outer shell of the tree which has two parts, the inner living part and the outer dead part. The outer dead part is often corky in appearance and feel. Between the bark and the wood is a layer of active, living cells

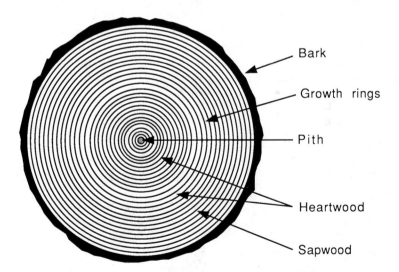

Figure 13-1. Cross section of a log showing the nomenclature for various portions of the wood.

called the cambium. This layer is not discernible by the naked eye. It is in this area that tree growth occurs. The result of continuing layers of growth are growth rings emanating from the center of the tree. In some woods, growth rings are quite discernable, while in other species they are difficult to recognize. The age of a tree may be determined by counting the numbers of rings at stump level. Some years may have more than one ring if the tree has been subjected to drought or insect attack. In some tropical woods, growth may be nearly continuous throughout the year and therefore growth rings are not well defined.

The principal constituents of dry wood are: cellulose, lignin, hemicelluloses, extractives and ash forming minerals. Cellulose is the main constituent comprising about 50 percent of the wood substance by weight. Another primary constituent, lignin, comprises 16 to 25 percent of hardwoods and 23 to 33 percent of softwoods. Hemicelluloses form about 20 to 30 percent of hardwoods and 15 to 20 percent of softwoods. Extractives may range from 5 to 30 percent and are not part of the wood but are entrapped substances such as oils, fats, resins, waxes, etc. that contribute to wood color, taste, odor, strength, density, etc.

Physical Properties

The physical characteristics of wood vary greatly with species and with other variables such as moisture content. How the wood is sawn from a log affects the way the wood will perform as lumber. The two basic ways of sawing are called plainsawed and quartersawed. Plainsawed lumber is cut from a log tangent to the growth rings. Quartersawed lumber is cut radial to the growth rings. These are not perfect definitions and in practice all lumber from a log is used so the terminology is used loosely to describe the general sawing technique. The following list presents some advantages to each of the sawing practices.

Plainsawed

- Shrinks and swells less in thickness
- Possible lower cost
- Fewer through board defects

Quartersawed

- Shrinks and swells less in width
- Less splits and surface checking
- Less raised grain at growth ring separations
- Wears more evenly
- May hold paint better

Appearance

The two terms commonly used to define appearance are grain and texture. These terms are rather nonspecific and are used only loosely to describe wood characteristics. Grain is often the term used to define the spacing of growth rings. Closely spaced growth rings would be fine grained while large growth rings spacing would be coarse grained. Grain has also been used to describe the orientation of the wood fibers as straight grain, curly grain or spiral grain. Terms such as open grain or close grain are often used to define the characteristics of the wood fiber pores. Texture is another term used to describe wood appearance and may be used synonymously with grain but most often is used to describe the wood fiber to a finer degree than by growth rings. To avoid confusion, grain or texture should be specifically defined in the context of the project to avoid misunderstanding for the intended use.

Moisture Content

The free water and absorbed water in wood has significant impact on the properties of wood. Moisture content can affect susceptibility to decay, strength, shrinkage and weight. The standard method of expressing moisture content is as a percentage of moisture compared to the weight of oven-dried wood.

The sapwood, with its living cells, generally has a higher moisture content than the adjacent heartwood, however some wood species may exhibit a significant heartwood moisture content. The terminology associated with the natural removal of water from cut green wood by evaporation is called seasoning. As seasoning progresses there is a point at which the free water in the wood cell cavities has been evaporated and the remaining moisture is associated with the capillaries of the cell walls. This point is called the fiber saturation point (fsp) and in many wood species occurs at around 30 percent moisture content. Wood with moisture content at the fiber saturation point and greater is called green wood. There is also an equilibrium moisture content where the wood has stabilized and moisture content is neither gaining or losing moisture. Wood is constantly undergoing change in moisture content due to daily and seasonal changes in temperature and relative humidity. Ideally, before use, wood should be brought to a moisture content which is consistent with the average atmospheric condition at its exposure location. Proper seasoning, handling, storage and sealing or coating of wood is desirable to minimize moisture content variation.

Shrinkage

Interestingly wood is most stable when the moisture content is high, above the fiber saturation point. Changes in moisture content below the fiber saturation point will result in shrinking and swelling of the wood. The shrinking and swelling may result in splitting, checking or warping, which may effect the performance and intended use of the wood.

Longitudinal shrinkage of wood, parallel to the grain, is generally small and usually not a major concern for coastal construction work. Transverse

298 PART 3/ENGINEERING DESIGN

and volumetric shrinkage may be a problem in some structures. Shrinkage and warping of wood decking is a common problem in coastal construction. Shrinkage is affected by a number of variables including: wood density, piece size and shape, temperature and drying rate in some woods. Most shrinkage occurs in the direction of the annual growth rings. The combination of tangential shrinkage (direction of growth rings) and radial (across the growth rings) can result in significant wood distortion. The rate of shrinkage and resulting distortion will vary considerably between woods of the same species as well as between species.

In most coastal construction projects, dimensional stability of wood will not be a major consideration. The exception as earlier mentioned is in distortion of decking. There is an age old controversy about which way to lay a deck plank as it relates to the direction of bark side or growth ring curvature. The familiar call is "bark side out" or set the plank so the growth rings look like a rainbow or the arch of a bridge. Actually this method maybe incorrect since the outer sapwood will tend to dry and shrink first and to a greater degree than the tighter growth rings near the center of the tree. The drying and shrinkage of the outer rings will cause the plank to curl in that direction causing a concavity that will hold surface water. Figure 13-2 shows the typical shrinkage patterns for lumber pieces cut from different areas of the log. In general, planks should be laid with existing convexities upward or

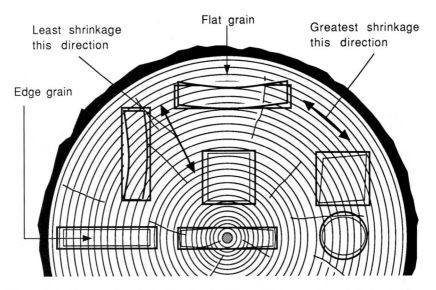

Figure 13-2. Cross section of a log showing the effects of shrinkage of wood pieces taken from different portions of the log. Note radial shrinkage is about one half as great as that of tangential shrinkage. *(Adapted from: Wood Handbook: Wood as an Engineering Material)*

if flat when installed, lay the plank to accommodate the shrinkage pattern. Better yet, use pressure treated wood and seal the wood on a regular basis to help keep the wood water resistant, this will mitigate shrinkage concerns.

Weathering

Unprotected wood will weather rapidly especially in a coastal environment. The effect of weathering may be desirable and by design or it may be undesirable. Many New England cottages are shingled with untreated white cedar which turns grayish after exposure, giving a desired weathered appearance. Weathering may cause a significant change in the wood color and/or a loss of the softer portions of the exposed wood surface, causing a ribbed or roughened surface. Weathered boards may also warp and cup. Warping or cupping stresses can be high enough to pull thin boards through their fastenings. Proper wood selection and protective coatings can help mitigate cupping and warping. The denser the board and the greater its width to thickness ratio, the greater the chance for distortion. Vertical grain boards tend to warp less than flat grain boards. A general rule of thumb is for a board or plank width not to exceed 8 times its thickness.

Color changes resulting from weathering will generally be a surface condition caused by degradation of cellulose and by biological attack. Dark colored, nonuniform or blotchy coloration may be caused by attack from fungal spores and other microorganisms. Metallic hardware and fastenings may also cause discoloration of wood surfaces, often with pronounced streaking.

Decay

Moisture and temperature are principal ingredients in the formula for decay. Constantly dry wood and wood continuously submerged in fresh water are not generally or significantly attacked by common fungi. Below fresh water, biological attack on wood may occur but is usually of a minor nature and is slow in activity. Favorable conditions for decay development occur in warm, humid areas. Higher altitudes and cool, dry areas generally retard decay. Sapwood, if left untreated, is highly susceptible to decay in virtually all species and heartwood may exhibit varying degrees of resistance to decay. For coastal construction, decay and other attack is generally so great that only pressure treated wood or certain tropical hardwoods having natural resistance to decay should be used.

Mechanical Properties

There are three principal axes in wood that influence its strength properties. The axes are longitudinal (in the direction of the fibers), radial (normal to the growth rings) and tangential (tangent to the growth rings and perpendicular to the grain). Wood exhibits its greatest strength when loaded in tension or compression in a direction parallel to the grain. Loading perpendicular to the grain may be suitable in certain conditions if strength limitations are recognized.

Compression testing has been performed to develop moduli of elasticity in selected wood species. Testing for the moduli in the radial and tangential directions is limited and only the modulus of elasticity in the longitudinal direction is well documented. Performance data is generally related to clear, straight-grained wood samples so the test data must be used with caution in wood having a lesser quality. The generally used strength properties include shear strength parallel to the grain, compression strength perpendicular to the grain, and modulus of rupture in bending. The specific values for these properties may be found in tables provided by various lumber grading associations for the species of lumber they produce or grade.

Some grading associations will provide additional information on torsion, creep, toughness, tensile strength parallel to the grain, rolling shear, and fatigue resistance. This data is also based on test samples of clear, straight-grained wood.

The quality of lumber has diminished rapidly in the past decade, as its use has accelerated for home and building construction. It is often prudent to obtain samples of the various grades of lumber that may be specified in a project to ascertain their quality prior to design and specification. It is also important to verify that the specified grade of material is, in fact, the material that arrives on site. This can be determined by reviewing the purchase orders and material invoices, specifying that certificates be provided on the wood quality by independent testing laboratories and by visual observation of grade stamps on the actual lumber.

In the design of wood structures, it also must be borne in mind that general wood properties will vary with the size, shape and thickness of the wood. The appropriate grading tables and wood properties must be consulted to ascertain the appropriate design values for the various pieces of lumber.

Preservative Treatment

Treatment of wood products to enhance their life has been part of traditional construction since at least the mid 1800s. Many naturally occurring woods lack the resistance against chemical and biological attack that would make them suitable for coastal construction use. Treatment is a viable way to provide the necessary resistance to attack by decay, insects, marine borers, and other organisms that attack and destroy wood.

Wood treatment by impregnation with toxic chemicals is effective in denying the damaging organism access to a food source. In the United States, all wood treatment involves the application of pesticides approved and registered by the Environmental Protection Agency (EPA). Depending on species, size, moisture content of the material to be treated, treatment will be effected by vacuum, pressure, elevated temperature or a combination of these methods. In general, moisture in the wood is evacuated and the preservative is introduced under pressure. The three principal preserving chemicals used for marine construction are pentachlorophenol, inorganic arsenicals and creosote. Additional information on wood preservative treatment is presented in Chapter 14, Section 14.3.

13.2 CONCRETE

Concrete is a versatile building material used for both structural and decorative purposes. It is a composite material made from three basic ingredients: cement, aggregate and water. The mixture of the cement powder and aggregate with water creates a plastic mass that can be formed to virtually any size and shape. In the hardened state concrete becomes structurally significant and with reinforcement it can be used to carry and transmit large applied loads. The material is generally durable and when properly used can have a long maintenance free life. Major advantages of concrete are the ability of relatively unskilled labor to place and work concrete and the generally favorable cost of materials.

Concrete is used in waterfront construction for piers, bulkheads, floating dock systems, launching ramps, fuel tank vaults, pipeline encasement, promenades, mooring piles and mooring anchor blocks as well as traditional sidewalks, parking lots and building structures. Marine use of concrete must consider special factors such as surface erosion from water abrasion, degradation from alternate freezing and thawing cycles, effects of sea water on reinforcement and erosion beneath concrete structures.

Cement

A primary ingredient of concrete is cement. The normal cement used in the United States is termed portland cement. Portland cement is a combination of several raw materials including limestone and clay or shale containing calcium, iron, silica and alumina. The raw materials are combined in a measured proportion and burned to form a clinker having a specific chemical composition. The clinker is ground to a fine powder and gypsum is added to control the rate of set of the cement product when water is added.

There are five basic types of portland cement produced for general use. Each type has specific characteristics making it more or less desirable for certain applications. Variations in cement fineness and chemical composition alter the characteristics of resistance to chemical attack, rate of strength gain, and the rate at which heat from the chemical reaction between the cement and water is released. The basic standard for specification of the composition of each type of cement, in the United States, is the American Society for Testing and Materials standard ASTM C 150, "Standard Specification for Portland Cement."

Type I portland cement is the most commonly specified type and is referred to as regular or normal cement type. It is used for general purpose structures which will not be under conditions that require special consideration. Although this is the generally used cement, it is not recommended for marine use because of its lack of sulfate resistance.

Type II portland cement is also a commonly used general purpose cement. It has a moderate resistance to sulfate attack and is used in marine applications as well as for sewage treatment structures and where it will be exposed to soils having a high sulfate concentration. Type II cement has a slower heat liberation (hydration) rate and generally generates less heat than Type I cement, and is therefore often used in massive or thick structures where high temperature rise may result in the cracking of the concrete.

Type III portland cement is proportioned to provide for an early high strength gain during the curing process. The early high strength gain feature allows for more rapid removal of form structures, thus speeding up the construction process. It also exhibits an advantage in cold weather construction by requiring less time to protect the concrete after placement. It should be noted, however, that Type III cement, at the same cement content as other cement types, does not provide a higher ultimate strength, only a more rapid strength gain than other types. A generalized comparison is that concrete made with Type III cement will have a slightly greater strength in one day than a Type I cement might have in three days of curing. A Type III cement with seven days of curing will approximate the twenty eight day strength of a Type I cement. Over a long period of time, the Type I cement concrete may have an ultimate strength exceeding that of the Type III cement concrete, so the type of cement specified must be appropriate for the specific use, type of construction and desired end product.

Type IV portland cement is not a commonly specified cement. It has the specific property of a low heat of hydration, and is used in very massive structures such as dams, where excessive heat liberation may cause damage to the structure. Type IV is usually only available on special order.

Type V portland cement is common to specific areas where the concrete is exposed to high concentrations of sulfate in groundwaters or soils. The southwestern part of the United States has these conditions and Type V cement may be more readily available in these areas. If either Type V or Type IV cement is to be specified, its availability should be locally confirmed.

All types of cement may be altered by various additives. The most common additive is an air entraining agent that is ground in with the basic cement to allow the formation of microscopic air bubbles in the concrete mix. The air entraining makes the concrete more workable and allows it to be more easily pumped. It also enhances the durability of the hardened concrete and is especially valuable if the concrete is subject to freezing and thawing cycles. Generally the air entraining agent is introduced into the mix by the concrete producer at the batching plant. This has eliminated the need for producing special cements which include the air entraining agent. When cements have an air entraining agent included, they are labelled as Type IA, Type IIA, Type IIIA, etc.

Aggregate

Aggregates are the bulk product that makes concrete massive. A mixture of pure cement and water has no practical structural significance. Such a mixture, in a usable quantity, would be very expensive, have excessive shrinkage and would not be very durable. The inclusion of aggregate as a filler provides an economical means of achieving mass and makes concrete a suitable construction material. The aggregate does not contribute to the chemical reaction between the cement and the water but simply occupies space. The inherent strength of the individual aggregate does provide strength enhancement and therefore selection of aggregate size can be important to a successful concrete mix.

Aggregate is generally graded by size to fall into one of two general categories called fine and coarse aggregate. The proportions of fine and coarse aggregate introduced into a concrete mix can greatly affect the properties of the fresh concrete. Too high a proportion of fine aggregate can result in greater mixing water requirements and difficulty in achieving a quality finish. An excessive amount of coarse aggregate may result in a concrete that is difficult to work and trowel finish. Aggregate size is most important if the concrete is to be pumped to the placement location. **For most work a ¾ inch or 1 inch size aggregate is adequate unless pumping is involved, then a ⅜ inch aggregate size may be specified.** In all cases, the proper grading of the aggregate and proportions of the fine and coarse aggregate will determine the suitability and workability of the concrete.

Selection of aggregate material is important. Some materials may effect the concrete by decreasing workability and durability. Such materials may require excessive addition of water and may cause surface imperfections or staining. Materials to avoid include: chert, lignite, limonite, soft sandstone, laminated sandstone, clay balls and shale. Aggregate should be specified to meet the American Society for Testing and Materials standard ASTM C 33, "Standard Specification for Concrete Aggregates" to assure proper quality.

Admixtures

It is often desirable to alter the properties of a concrete or reduce costs by introducing chemical additives into the concrete mix. The admixtures are added during or immediately prior to mixing. Air entraining additives, discussed earlier, provide a more workable and durable concrete. Other admixtures include water reducers, retarders and accelerators.

Water reducers act to reduce the amount of mixing water required to make the concrete workable. These additives can effectively reduce the amount of cement required, enhance the flow qualities of the concrete or lower the water/cement ratio. Some water reducers may be used to retard concrete setting when that characteristic is desirable. Superplasticizers, high range water reducers, can maintain a reasonable water/cement ratio while creating a very workable concrete. Water retarders may be used in combination with water reducers to offset the fast set often associated with hot weather concrete work. Retarders may reduce the concrete strength during the early stages of curing, so caution must be used to determine the specific needs of the concrete placement.

Water-Cement Ratio

Water is a basic ingredient in the making of concrete. It is the vehicle that allows the creation of a bond between the cement and the aggregate. The water mixes with the cement to form a paste that adheres to the aggregate. When the water dissipates, the cement paste hardens forming a solid mass of bonded aggregate. The quality of the hardened concrete is intimately connected to the quality of the cement paste. The higher the strength of the cement paste the generally higher the strength of the concrete, assuming a quality aggregate is used. Too much water dilutes the cement paste and weakens it. A critical relationship is the amount of water versus the amount of cement. This relationship is called the water-cement ratio (w/c ratio).

The water-cement ratio is determined by dividing the amount of water in the concrete by the amount of cement. Past practice defined the water-cement ratio as gallons of water per sack of cement. A sack of cement is defined as equal to 94 pounds of cement. A typical w/c ratio was expressed as 6 gallons per sack. Today the water-cement ratio is more conveniently expressed as a ratio of weight of water to weight of cement. The old 6 gallons per sack mix may be related to weight by multiplying 6 gallons of water by 8.33 pounds per gallon of water which is equal to 49.98 pounds, divided by 1 sack (94 pounds) this equals 49.98/94 = 0.53. **Water-cement ratios of between 0.5 and 0.55 are common.**

Quality of Mix

The quality of a concrete mix is measured both by its ultimate strength and the ability to satisfactorily place and finish the concrete. The strength properties of the concrete are governed by the mix proportions, the quality of the aggregate, the water-cement ratio, admixtures, and the quality of the placement. The quality of the placement can be affected by several conditions occurring at the time of placement. The concrete should be uniformly mixed and placed with no segregation of the fine and coarse aggregate.

Concrete should not be dumped more than several feet, as segregation will result. Use chutes or other devices or methods such as pumping to minimize unsupported dumping. The concrete should have adequate fine aggregate to permit finishing and to prevent excessive water from coming to the surface (bleeding). The finished surface should not be worked too much. Excessive working will bring water to the surface, resulting in a weak water-cement paste at the top surface (latience) which will easily crack and spall when dry. The mixed concrete should be delivered to the site in a timely fashion and placed as soon as possible. Concrete that has been mixed and unplaced for more than two hours is generally considered rejectable.

Testing

Tests of concrete are best handled by qualified testing laboratories. The certified tests will document the ultimate quality of the concrete and may be relied upon should future defects require investigation. However, there are some basic tests that can be performed in the field or prepared in the field for subsequent laboratory testing and which can be executed by a knowledgeable owner or contractor.

The simplest test of concrete that can be made during placement is the slump test. The test requires a slump cone, a flat surface, a shovel, a steel rod and a ruler. The slump test is an indication of the workability and plasticity of the concrete. The specifics of the test procedure are enumerated in the American Society for Testing and Materials Standard C 143 "Standard Test Method for Slump of Portland Cement Concrete." The slump cone is placed on a flat, clean, nonabsorbent surface and filled with fresh concrete in three layers. Each layer is rodded 25 strokes with a round, straight steel rod $\frac{5}{8}$ inch in diameter and 24 inches long. The tamping end of the rod should have a hemispherical tip of $\frac{5}{8}$ inch diameter. The filled cone should be struck level across the top with the rod in a rolling or screeding motion. The cone should immediately be lifted straight up, off the fresh concrete, taking care not to move the cone laterally. The cone is then set alongside the concrete mound, without disturbing the concrete. The tamping rod is placed across the top of the cone and over the concrete mound. The dimension from the underside of the rod (top of cone) to the displaced original center of the concrete mound is the slump of that concrete. **A slump of 1 inch represents a stiff mix and a slump of 10 inches represents a very wet mix. A normal slump will be in the order of 4 inches unless otherwise specified.**

13.3 STEEL

Steel enjoys a wide popularity as a construction material for waterfront structures. It is strong, durable, can be easily worked, accepts protective coatings, and is often cost efficient. In the United States, the quality of steel is assured by adherence to specifications of the American Society for Testing and Materials (ASTM), other countries have similar standards. There are a variety of ASTM specifications for different grades of steel suitable for a wide range of applications. Of particular interest are ASTM Specifications: ASTM A6, Specification for General Requirements for Rolled Steel Plates, Shapes, Sheet Piling, and Bars for Structural Use; ASTM A36, Specification for Structural Steel; ASTM A252, Specification for Welded and Seamless Steel Pipe Piles; ASTM A153, Specification for Zinc Coating (Hot-Dip) on Iron and Steel Hardware; and ASTM A307, Specification for Carbon Steel Bolts and Studs, 60,000 psi Tensile Strength.

Carbon steel is an alloy of iron and carbon. The chemical composition of structural steel (ASTM A36) includes approximately 0.26 percent carbon, 0.04 percent phosphorus, and 0.05 percent sulfur. Plates over 0.75 inches thick also contain small percentages of manganese, and plates over 1.5 inches thick have an additional small percentage of silicon. Carbon content is limited since content exceeding 0.35 percent may effect the weldability of the steel. Sulfur content is limited to prevent brittleness in the steel at high temperatures. Phosphorus contents greater than 0.05 percent decrease ductility and toughness. Corrosion resistance of steel may be enhanced by individual addition of copper, or nickel, or chromium, or silicon, or phosphorus. For most waterfront structures standard structural steel is utilized as the availability of corrosion resistant steels is not common and the cost may be excessive for small projects. Steel exposed to a seawater environment should be protected by an appropriate coating to reduce corrosion. Common coatings include hot-dip zinc galvanizing, coal-tar epoxy and fusion bonded epoxy. The key to successful coating life is proper steel surface preparation and application in accordance with manufacturer's recommendations for temperature and humidity.

A major benefit of steel is its weldability. Mild steels can readily be welded by standard arc welding, gas welding and thermite welding processes, which are common in virtually every area of the world. The quality of weld is very important and only qualified welders should be engaged to perform welding work. A proper weld can develop the full strength of the connected steel and will not create a weak point in the structure. If thin sections are to be joined such as steel pipe pile, it is recommended to utilize a backing plate or ring to obtain assurance of a full and complete weld. Beveling the edge of plates to be connected by butt welds can also enhance the quality and strength of the weld. Figure 13-3 shows a typical steel pipe pile butt weld with a backing ring.

Steel can also be bent in large presses to form smooth angular transitions or curved sections. Plates can be sheared to specified length to achieve straight, uniform cuts. Steel may also be flame cut, both in automated

MATERIALS OF CONSTRUCTION 307

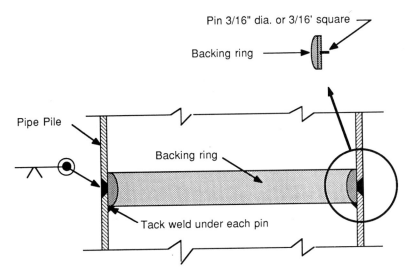

Figure 13-3. Typical pipe pile splice detail, with backing ring.

machines and in the field by use of oxyacetylene torches. Field cutting and welding is especially important in waterfront construction where dimensional criteria are often difficult to maintain due to the dynamic forces so prevalent at the waterfront.

Common steel materials are plates, from decimals of an inch thick to several inches in thickness. The size of plates is a function of the mill rolling capacity and freight requirements. Structural shapes are a most important aspect of structural steel construction. Common shapes include angles, channels, wide flange beams, H section beams (often used as piles), I beams, steel sheet pile material, pipe, and tubing (round and square). Special shapes and extrusions are also available in custom rolling. A major benefit of steel is the fact that it can be created in unusual shapes or sections by welding individual pieces. This aspect is especially important in ship construction where geometric form and weight considerations are critical to successful ship operation.

Design of complex steel structures must be performed by competent designers and often by structural engineers who appreciate the full range of stresses that may be involved in use of the structure. Structures may be subject to compressive or tensile loads, bending loads, bearing loads, shear, flexure and torsion. The analysis of these parameters requires a knowledge of structural analysis, physics, statics, and often dynamics. Although some do-it-your-self design and steel construction may be possible for minor structures and machinery repair, the prudent developer or owner will rely on professional assistance for complex structures. In the United States, the design of structural steel structures is governed by the properties of materials and design specifications presented in the *Manual of Steel Construction* published by the American Institute of Steel Construction, Inc. The manual is divided into several parts. Part 1 includes dimensions and properties of structural steel plates and shapes. Part 2 covers beam and girder design; Part 3, column design; and Part 4, connections. Also included is a section on specifications entitled "Specification for the Design, Fabrication and Erection of Structural Steel for Buildings." The manual is the bible for structural steel designers and contains a wealth of information on proper design methods and steel properties.

In marine applications, if a suitable protective coating is not applied, the prudent designer will add an extra thickness of steel material to account for anticipated corrosion. It is common to add an extra $\frac{1}{8}$ inch of material for corrosion mitigation. Another method of providing protection for anticipated corrosion is to increase the factors of safety in design. The appropriate methodology should rest with the judgement of the designer.

Since corrosion in steel structures is a constant problem, the designer should recognize this fact in the design and minimize areas which might accelerate corrosion. An example is in the placement of steel angles in a structure. The angles should be oriented so that water is not trapped by an upward pointing leg, etc. On steel sheet pile bulkheads having exposed horizontal wales composed of two channel sections, the upper channel section should have drainage holes drilled in the horizontal web to prevent water entrapment. Techniques like these can significantly lengthen structural life.

Steel is a common material used for marina piling. The most often used shapes are steel H-piles and round pipe. Of these two shapes, the round piling has distinct advantages over the H-pile section. An H-pile section may have great strength in the primary direction, line of force coincident with the web, but it is generally weaker in the side load direction. Marina mooring piles are subject to a varying load application direction which compromises the varying strength directions of the H-pile. A round pile, however, has uniform strength characteristics for any direction of load application. The H-pile also has a large surface area upon which corrosion may act. The pipe pile, if welded closed on the upper end, will generally be exposed to corrosion only on the exterior surface. However, if the pipe pile is left open or not otherwise protected on the inside, double sided corrosion may occur, effectively doubling the potential rate of corrosion deterioration. An H-pile must also be driven to extreme tolerance to allow proper placement of any pile guidance system. The round pipe pile, on the other hand, can rotate as it is driven with no subsequent difficulty in fitting the pile guidance system. In either H-piles or pipe piles, the float guidance system should incorporate an appropriate guide material that bears on the pile. Steel to steel contact should be avoided.

Steel sheet pile sections are often used in waterfront construction for bulkheads and earth retaining structures. Sheet pile bulkheads may be cantilever or tied back systems. In the cantilever arrangement, the steel sheet piles are driven into the subsoil sufficiently deep to resist the overturning moment created by the backfill material and any surcharge load on the bulkhead. The sheet piles cantilever from a point of fixity in the subsoil. There are height limitations on this type of wall due to the bending strength of the selected sheet pile and the capacity of the subsoil. The tied back system relies on embedment in the subsoil to prevent toe kick out of the sheet pile but also has a physical connection, usually around the half tide level that reduces the free height bending. The tie back system usually consists of a solid rod that connects between the wall and a deadman anchor inshore, far enough away from the wall to be out of the active soil wedge behind the wall. Obviously steel sheet piling is subject to corrosion and mechanical degradation. If affordable, the best mitigation to corrosion is to coat the sheet pile with a coal-tar or other appropriate coating. If coating is not possible, then selecting a sheet pile that has more metal than required for design load resistance is appropriate to allow a measure of corrosion to occur before the structural capacity of the wall is compromised. Some corrosion resisting steel is available for steel sheet piling but it is often difficult to obtain and it is quite expensive. If available and within budget, then using the corrosion resisting steel is advisable.

High strength steels are also available, but their use is not widespread in waterfront structures. One significant problem in using high strength steels is the need to be assured that any welding done to the steel is performed using the proper welding materials. Very frequently a mild steel welding rod will be used to weld high strength steels with a resulting failure of the weld under load. This situation has been frequently experienced with the use of used steel pipe piling. For a number of years used or uncertified oil country casing, pipe used in oil well field operations, has been purchased at a low cost for use as marina mooring piles. The pipe looks like all pipe, so it is impossible to distinguish a high strength steel from a mild structural steel. Contractors, in the field, finding it necessary to add a piece of pile to achieve the desired pile length have used an inappropriate welding rod to make the splice. The result has been significant failure of these welds, often in below water locations, making a repair extremely difficult.

Appropriately designed and used, steel is a very viable material of construction for waterfront structures.

13.4 STAINLESS STEELS

Stainless steel is a steel alloyed with nickel and chromium. The chromium content of stainless steels is generally more than 10 percent. These steels are important because of their hardness and resistance to corrosion. The corrosion resistance of most stainless steels is relatively high. Low grade stainless steels will exhibit products of corrosion, so one must be careful in selecting stainless steel hardware or fastening to be assured that the appropriate corrosion resistant series is purchased. A common high strength stainless steel is designated as 18-8, which refers to a stainless steel having 18 percent chromium and 8 percent nickel. This series and the 300 series are generally suitable for waterfront construction materials. In the 300 series, 304 and 316 stainless are common.

There are three basic categories of stainless steels. The first category contains chromium as the principal alloying element, in percentages of from 11.5 to 18.0 percent. Some alloys may contain minor amounts of nickel and will have a carbon content of between 0.08 and 1.10 percent. These steels can be softened by annealing and hardened by quenching. Alloys that occur in this category include 403, 406, 414, 416, 420, 431, 440, 501, and 502.

A second category of stainless steels contains chromium from 14 to 27 percent. Nickel content is negligible and carbon content is usually less than 0.20 percent. If annealed, the steel will become magnetic and therefore could become a problem if used around magnetic sensitive instruments like a ship's compass. Alloys in this category include 405, 430, 442, 443, and 446.

The third category of stainless steels contains chromium at 16 to 26 percent with a carbon content less than 0.20 percent. Nickel is added from 6 to 22 percent. Corrosion resistance may be enhanced by addition of columbium, molybdenum, and titanium. These steels are nonhardenable by heat treating but respond well to cold working. They are nonmagnetic but may become magnetic by excessive cold working. The alloys in this category are the 300 series stainless steels, the ones most commonly used in marine fastenings.

Corrosion resistance of all three categories of stainless steels may be enhanced by passivation. This process induces a thin film of chromium oxide on the steel surface, adding this extra degree of corrosion protection. Although generally resistant to corrosion, stainless steels do exhibit a propensity to suffer crevice corrosion and pitting. This action can be mitigated in immersed seawater applications by connection to an unalloyed (unpainted) steel surface or other cathodic protection. In these cases the lower grade steel or zinc anode will be sacrificed to the benefit of the stainless steel.

13.5 ALUMINUM

In the last forty years, there has been a significant trend to the use of aluminum in waterfront construction and especially in marina float design. Major improvement in aluminum production and design methodology developed during World War II prompted the increased use of aluminum. Aluminum is a metal renowned for its strength to weight ratio, durability, appearance, and ease of fabrication and erection. Its strength is achieved by variations in the composition of the alloying elements, therefore aluminum has a wide range of strength properties. A comparison of the specific gravity of aluminum with the specific gravity of other materials shows its relatively low weight. Copper has a specific gravity of 8.46, steel of 7.85, while aluminum has a specific gravity of 2.7. The density of aluminum ranges from about 166 to 176 pounds per cubic foot. Commercially pure aluminum has a tensile strength of about 13,000 pounds per square inch (psi) but can achieve tensile strengths of 88,000 psi in high strength alloy form. Aluminum alloys used in structural design have tensile and yield strengths comparable to structural steel.

Aluminum is naturally corrosion resistant because it forms a thin surface oxide film. If the oxide film is ruptured, the film will reform in the presence of oxygen. In the marine environment, it may be necessary to provide additional protective coating. The alloys of aluminum best suited for structural use in waterfront projects are within the 5000 and 6000 series of alloys. Usually aluminum plates will be in the 5000 series, while rolled shapes will fall into the 6000 series, with the alloy 6061-T6 being a common alloy for structural shapes. The T designator after the series designator indicates the temper of the alloy. The 5000 series, usually in the alloys of 5052, 5083 and 5086, contains magnesium which enhances their corrosion resistance, provides moderate strength and offers good welding characteristics. The 6000 series also contains magnesium and silicon which makes the series of medium strength but enhances formability for shaping and allows heat treating. The varying chemical composition of aluminum alloys requires the use of weld materials and welding techniques appropriate for the selected alloy. This is very important to assure the structural integrity of a fabricated component.

Aluminum products are available as sheet, plate, wire, rod, bar, shapes, tube, pipe and forgings. It can also be cast into special shapes and configurations for special applications. Aluminum is especially important because of its ability to be extruded. Extrusion allows for the design of special shapes to maximize design efficiency. In marinas, aluminum extrusions are frequently used as primary structural shapes for floating dock frames. The ability to be extruded into custom sections allows the use of the extrusion as a multipurpose member. An example is an extrusion that forms the side stringer of a floating dock and also functions as a guide for an adjustable finger fastening, cleat fastening, and utility tray. Although the cost per pound of the aluminum extrusion may exceed the cost of other materials, its multipurpose use may result in a net cost savings in the total structure. Its light weight makes fabrication and assembly easier than is possible with heavier materials and allows for lower cost in shipping weight.

Care must be taken when using aluminum in a marine environment to isolate the aluminum from other metals to prevent the set up of an electrolytic cell with resulting accelerated corrosion. Barriers may consist of applied coatings or thin films of isolation material. Fastenings are also a concern, with aluminum or stainless steels being the preferred fastener materials. If marine biofouling is a problem, care must be taken to apply an antifouling coating that is compatible with aluminum. Avoid antifouling paints that contain copper. Consult the coating manufacturer to assure compatibility of the product with the aluminum series to be used and with the intended application.

13.6 FIBERGLASS

Fiberglass is used extensively in the recreational boating market for boat construction and accessories. Its use as a waterfront construction material has been less enthusiastic, although a number of marina dock manufacturers have successfully incorporated fiberglass floats and deck structures into the product line. Fiberglass has been used for small boat construction since shortly after World War II, but it did not gain significant acceptance as a boat building material until the 1970s.

Fiberglass is really just one component of a two component structural composition, and therefore it is somewhat incorrect to label the product as fiberglass. The fibered glass is the textured material which is bonded together by a thermosetting liquid resin. A more proper designation is glass reinforced plastic (GRP) or fiber reinforced plastic (FRP). Both GRP and FRP are terms used more extensively outside the United States to define this material. The composite material is similar to reinforced concrete in that a reinforcement material (fiberglass or reinforcing rods) is encapsulated in a plastic capable of hardening to bond the materials together. Each material taken separately has poor structural properties, but when properly combined and hardened, they form a significant structural material. Usually fiberglass is laminated by creating a series of glass plies. These plies are often oriented in differing directions to provide near equal strength in all directions or they are oriented in a particular direction to obtain the maximum strength to weight ratio.

If properly designed, FRP products can have a good weight to strength ratio. A well designed structure in FRP can have a weight about half the weight of an equivalent steel or wood structure and about the same weight as an equivalent aluminum structure. In some applications a secondary reinforcement such as structural wood may be introduced into the FRP layup to further enhance the structural integrity of the composite product. Several marina dock manufacturers have used this technology to provide a stable, high strength floating and fixed dock deck structure. A distinct advantage of

FRP is its ability to create seamless structures, enhancing the leak resistant nature of the product, such as a boat hull or dock system flotation pontoon. FRP is also not affected by electrolysis nor by most chemical attack or seawater. From this perspective it is an ideal waterfront construction material. FRP can also be molded to fit complex curves or other special shapes. It has a low modulus of elasticity, making it a good energy absorber. The low modulus of elasticity (a ratio of the normal stress to the normal strain in the direction of applied load) can be a problem, however, in structures that require low deflections. This characteristic can be mitigated by the addition of secondary reinforcement, such as structural wood cores.

FRP is subject to cracking on impact but it can be easily repaired. It is generally cost effective and can be formed with surface gel coats to provide exceptionally smooth finishes and it can be supplied in many colors. The material has a low maintenance feature and is regarded as quite durable.

Structurally the material has a low modulus of elasticity, as previously stated, and therefore is subject to large deflection. The modulus of elasticity of FRP is in the order of 2,000,000 psi, versus steel which has a modulus of about 29,000,000 psi and aluminum which has a modulus of about 10,000,000 psi. Stiffness can be enhanced by orientation of the glass strands or by other reinforcement. The fatigue strength is lower than comparable metals so appropriate factors of safety must be introduced into structures affected by fatigue or significant vibration. Stress concentrations at changes in section or direction can also be a problem. The material, by itself, also has a low buckling strength which must be factored into any structural application. Long term loading may result in creep. FRP should be protected from significant abrasion. Its abrasion resistance is better than wood but less than metals. Usually the inclusion of fendering in waterfront structures mitigates this area of concern. If fire to the structure may be a problem, fire retardant resins may be used to prevent flame spread. Conventional resins used in FRP have a flame spread that is comparable to that of plywood.

A major problem associated with boat hulls has been the appearance of osmotic blistering. The blistering results in a pock marked hull with random fields of bubbles. A great deal of research is currently underway to determine the causes of the blistering. To date it is believed that fiberglass layup techniques and resin type may play a role in blister development. Also, continual immersion, especially in tropical waters, appears to enhance blister development. Most marina structures constructed of fiberglass are not continually immersed in water, so blistering has not occurred in these structures. Floating dock buoyancy pontoons, if constructed of fiberglass, may be subject to blistering but the presence of blistering on these units would not appear to create a problem with the intended use of the pontoons unless severe fiberglass delamination allows flooding of the pontoon. Except in cases of gross blistering, the structural integrity of a waterfront structure will probably not be compromised by minor blistering. Before undertaking the design of an FRP structure, the latest procedures and available resins should be researched to utilize the state of the art in fabrication techniques to avoid unnecessary structural compromise in the finished product.

13.7 STONE

Stone is a naturally occurring material that is used extensively in waterfront construction, principally for slope stabilization, jetties, groins, breakwaters and seawalls. For use in these applications, it should be hard, sound, durable, and free of any cleavages or laminations that might cause it to fracture or weather prematurely. Stone is available in an infinite range of sizes. It is important to select the proper size of stone for the intended use. Stone for engineered structures is generally quarried from bedrock formations and graded by type, size, and density. The type of rock selected will, in large part, be based on the available stone in the area of the project and be chosen in sizes appropriate for the intended use. If only inferior stone is available locally then the costs to transport a better quality of stone must be weighed against the possibility of premature failure of the lower cost local stone. Stone handling and placement can be labor intensive so material costs may be only a fraction of the overall project cost and the additional cost for more suitable materials may be justified.

In waterfront construction the stone must be of a type that can withstand the exposure to the water environment. Size selection is based on an engineered analysis of the forces involved in the specific use. In wave energy areas this is critical because too light a stone may rapidly be moved by the wave forces and the structure compromised. Some stone movement in rubble mound breakwaters is tolerable if the movement is a minor redistribution of materials which generally enhances their stability by further interlocking. **Generally in riprap stone structures, the greatest dimension of a stone should be no more than three times the smallest dimension.** Thin, slab type stone is subject to greater lift and drag forces and does not form as good a piece-to-piece interlock as does angular shaped stone. Angular quarry stone is the preferred stone for riprap slope armor stone. Round stone should be avoided as it is more difficult to place and does not form an interlocking mass to resist imposed wave forces. Density is important to provide the greatest weight for the least volume. Large, heavy stone will better resist the imposed wave forces associated with wave energy absorbing structures. To successfully develop a slope that can absorb energy or prevent erosion, it is first necessary to suitably grade the natural ground surface to the appropriate angles. To prevent the loss of the fine natural or core material, it is important to place a filter fabric (geotextile) on the prepared surface prior to placing any stone. This fabric is porous and allows water to migrate through the material but prevents passage of fine soil or rock materials. The filter material is generally covered with a stone size(s) suitable as a bedding material for the larger exposed armor stone. The armor stone (riprap layer) must be of sufficient size and type to resist the imposed forces. Riprap stone that is oversized may be a problem in that proper interlocking of the stone may be difficult and large void spaces between stones may allow excessive energy to reach the bedding layer and cause stone washout. Some void space is desirable to increase energy absorption of the turbulence created by the impacting water. The ideal stone gradation design will provide a gradual reduction in stone size through intermediate layers until the stone size blends with the stone size of the natural bed material.

Riprap stone is considered, for design grading purposes, to have a specific gravity of around 2.65, although stone with a considerably lower specific gravity has been successfully used in waterfront structures if other parameters are acceptable. Size may range from stone having the equivalent size of a cube 3 inches on a side and weighing about 2 pounds, to a cube 42 inches on a side and weighing about 7,500 pounds. Massive stone breakwaters may have stone as large as 25 to 30 tons. Stone larger than 30 tons are generally very difficult to handle and seldom used. Within the gradation will be a range of specific gravities to account for natural variations in stone formation. Filter bedding stone will generally range from an equivalent cube size of .375 inches on a side to 6.5 inches on a side. Stone grading requirements vary within the industry, so care must be taken to select and specify the appropriate stone for the intended use. In specifying stone, reference is often made to ASTM C-127, Standard Test Method for Specific Gravity and Absorption of Coarse Aggregate, and ASTM C-131, Standard Test Method for Resistance to Degradation of Small-Size Coarse Aggregate by Abrasion and Impact in the Los Angles Machine. In cold climates, consideration must be given to the durability of selected stone to sustain alternate freezing and thawing cycles. **A measure of resistance to weathering is water absorption which, if limited to about 2 percent, is generally acceptable.** As a practical solution to stone specification, a designer should observe the durability of in-place, local stone structures and assess their performance as it relates to use in the proposed project. If this is not possible then a site visit to the supplying quarry and inspection of undisturbed bedrock outcroppings may provide guidance as to the weathering capability of the stone.

Placement methods will vary with the project design, but in general, placement begins at the bottom of the structure and continues upward and

316 PART 3/ENGINEERING DESIGN

lengthwise to provide a proper gradation and interlocking mass with minimum voids. Bedding layers may often be randomly placed and hand spread, but surface armor stone should be controlled placed by equipment to assure a cohesive interlocking mass. It is important that armor stone be properly seated on the bedding layer to prevent, loose, tilted or unstable stones.

Stone used as a backfill for bulkheads, cells, or caisson should be clean, well graded and free of organic materials. It generally should have a high density and be well compacted to minimize settlement. In special cases, a lightweight aggregate backfill may be desired to reduce the backfill load against a wall structure. It is important to avoid a high percentage of fines in any backfill material which is directly adjacent to a wall surface as the fines may tend to wash through small voids in the wall surface with resulting settlement of the backfill material and accretion of material at the face of the wall. Many waterfront structures will require toe protection. The toe may be the base of a sheet pile bulkhead or the bottom of a sloped revetment. Toe protection will protect the structure from scour and undercutting. Stones used for toe protection are usually quite large, to prevent their movement by wave forces or currents.

Stone construction is usually considered as a means to mitigate the effects of significant forces. It is therefore important to perform the necessary environmental analysis to ascertain the magnitude and direction of the impinging forces before embarking on the design of the stone structure. Failure to properly address the environment may result in a failed structure and consequent proximate damage. Removal or modification to an in-place stone structure is considerably more difficult and costly than constructing the structure properly in the first instance.

13.8 SYNTHETIC MATERIALS

Synthetic materials of interest in waterfront construction are those of a group generally referred to as plastics. Plastics as we know them are relatively new construction materials. An early plastic used in boatbuilding was Celluloid.™ We have already discussed the primary boatbuilding plastic in the form of fiberglass reinforced plastic (FRP). Other plastics, however, have found important uses in construction products. Some examples include: dock fendering systems, cleats, flotation pontoons, tie down straps, pile caps, power posts, insulation, cable covering, nonskid mats and deck covering, and an important use in geotextile fabric to prevent soil erosion.

Plastics can be classified as either thermoplastic or thermosetting. Thermoplastic materials can be repeatedly softened by application of heat and retain the ability to reharden. Examples of thermoplastic materials are polyethylene, polystyrene, polyvinyl chlorides (pvc), and nylon. Thermosetting plastics are formed once and retain a permanent set. Examples are: silicones, epoxies, and polyesters. The different properties result from the molecular structure of the material. Thermosetting material has a cross linked molecular structure that prevents the molecules from extensive movement, essentially freezing their structure once it is formed. Each material has its specific uses. A marina power post must be rigid to support the attached hardware, whereas a float fender may need to be curved to fit the radius of the end of a finger float.

The classification of the plastic material results from the use of specific raw materials, usually synthetic resins or cellulose. Cellulose is a naturally occurring component of wood and wood related plants. Each molecule of the resin or cellulose is formed as a long, chain-like structure composed of many repeated units. The units are called monomers and when connected up form a polymer. The process is called polymerization. If the monomers are of different types, the chain-like molecule is called a copolymer. This process is fundamental to the chemistry of plastics. The process of copolymerization is much like the alloying of metals to achieve desired properties. Composite plastics may be created by adding other materials to the plastic, such as fibers and even stone aggregates. Further adjustment to the material may be made by addition of colorants, filler materials, stabilizers, and plasticizers. Each additive results in a specific desired characteristic, such as strength, increased hardness, or changes in flexibility.

Plastics have the desirable property of being formable in a number of ways to create special shapes. Forming may be accomplished by: injection molding, blow molding, rotomolding compression molding, vacuum forming, casting, laminating, and extruding. The variety of shapes and plastic compositions is virtually limitless.

In marina construction, the largest use of plastics is in the making of float pontoon units. These pontoon units have been made from expanded polystyrene, polyethylene and fiberglass. For this application the plastic must be able to resist the effects of seawater or freshwater, resist chemical attack such as is produced by gasoline and detergents, be resistant to or protected from abrasion and ice, and have a low water absorption rate. The plastics also must be resistant to thermal change associated with the in-use environment (high summer heat and possibly extreme winter cold) and not break down under constant subjection to ultraviolet rays. The material should also not be so brittle as to fracture under possible boat, debris, or ice impact. Plastics are often used as insulators due to their electrical resistance, but again they must be specified as suitable for the harsh marine environment. Cracked electrical cable insulation in proximity to the water can create an electrical hazard and potential electrolysis to surrounding metals.

Another use of plastics in waterfront construction is in the form of geotextile materials for erosion control. This material is placed on earth structures to allow passage of water but prevent the passage of fine soil material. Slope revetments and rubble mound breakwaters should have filter fabric installed between the fine and the coarse stone layers. Some factors associated with the selection of geotextiles includes: tensile strength, porosity, permeability, abrasion resistance, elasticity, puncture resistance, weight, and resistance to environmental exposure.

14

Corrosion and Material Degradation

The waterfront is a hostile environment. There are many forces at work in nature that will corrode and degrade materials that form waterfront structures. It is imperative, for the marina developer and designer, to understand these forces and to design, specify, and purchase materials that can appropriately function in the waterfront environment. Failure to do so will result in premature functional loss of the structures or worse, catastrophic failure, which may involve personal injury. Everything in this world is deteriorating with time, so it is expected that things will degrade over time. The waterfront exposure accelerates this process and can substantially reduce the useful life of structures often in a matter of months, if inappropriate materials are selected. The economic importance of proper materials selection can not be overstated. The U. S. Navy estimated in 1965, that the annual cost of damage to waterfront properties caused by marine borers and fungi alone, in the United States, exceeded $500,000,000.

Corrosion may be affected by several environmental conditions including: rainfall (acid rain); ambient humidity; proximity to water, especially sea water; temperature extremes; chemical pollution in the air; and wind, sand or snow velocity effects. The degree to which these conditions enhance corrosion will be influenced by their relative strength. In most instances it is difficult to mitigate or alter these causes of corrosion but proper structure installation and surface protection can minimize the effects.

This chapter investigates corrosion, biological attack, and mechanical degradation. The most common source of all of the above is by immersion or exposure to sea water.

14.1 SEA WATER ENVIRONMENT

The oceans cover almost three fourths of the earth's surface and contain about 85 percent of the total water on the earth's surface. Ocean or sea water is then, the primary medium in which waterfront structures are constructed.

It is also a very corrosive and degrading liquid that can rapidly shorten the useful life of an improperly designed structure.

The term seawater is used, as opposed to fresh water, to define the general characteristic of the water as being saline, that is having a salt content. Salinity varies throughout the oceans but is generally in a range of 3.3 to 3.7 percent. The usual method of expressing salinity is in parts per thousand. Average ocean salinity expressed in parts per thousand is about 35 parts per thousand. The Atlantic Ocean has an average salinity of about 37.5 parts per thousand whereas the Pacific Ocean has an average salinity of about 34.6 parts per thousand. The lower salinity of the Pacific Ocean is attributed to a lower percentage of dry wind exposure which tends to enhance evaporation of the ocean water and raises the salinity. In comparison, the Red Sea and Persian Gulf have salinities of around 42 parts per thousand, the Baltic Sea ranges between 2 and 7 parts per thousand and the Black Sea around 18 parts per thousand. Salinity will vary with depth in the ocean and with the proximity to fresh water sources such as river mouths. Generally, the higher the salinity, the more severe the corrosion will be on waterfront structures.

When exposure to seawater is considered, the first impression generally relates to visual corrosion. Corrosion is the naturally occurring conversion of metals into specific compounds. The result is a loss of the primary metal into nonstructural compounds, i.e., rust. It is generally believed that corrosion of metals in seawater occurs by an electrochemical process. The process involves both an oxidation reaction and a reduction reaction. The reactions are interdependent although separated in a molecular sense. A flow of electrons occurs between the oxidation reaction (anode) and the reduction reaction (cathode), precipitating the corrosion product. The salts in the seawater (electrolyte) provide the mechanism for electron flow. In order for corrosion to occur there must be an electrical potential difference between the anode and cathode. The electrical potential difference may result from the different composition of the metals, the condition of the metallic surfaces, or differences in the fluids in which the metals are immersed.

A less obvious but equally important form of corrosion as visual rust are pitting and crevice corrosion. Pitting is a localized corrosion often observed in aluminum alloys. It may be just a surface condition of little structural consequence or, if in an advanced state, may result in material failure. Crevice corrosion is also a localized condition occurring in areas where the protecting oxide film has failed and the component design does not allow sufficient oxygen exposure to rebuild the film. This type of corrosion is often found around fastenings, where the fastening prevents an adequate supply of oxygen from reaching the area of failed protective film. Another type of crevice corrosion occurs when the material has a high concentration of metal ions in a crevice and a low concentration of metal ions exists in the

CORROSION AND MATERIAL DEGRADATION 321

surrounding area. This forms a metal-ion concentration cell and may result in a greater amount of corrosion at the mouth of the crevice rather than in the crevice slot. This type of corrosion is often found in copper and nickel alloys.

Structural members subject to high tensile stress in a corrosive environment may also suffer stress-corrosion cracking. It is most often found in high strength steels and in other high strength alloys such as titanium. It also appears to be related to materials use in an elevated temperature environment. Welded or heat treated materials may experience stress-corrosion cracking.

Corrosion of materials is a complex subject involving chemical, electrical and mechanical considerations. Before embarking on the use of exotic materials in a marine environment it is wise to assess the corrosion potential of the materials, alone and in combination with other materials. It is also important to recognize the different potentials of materials that are to be incorporated into a marine structure. The ranking of materials by potential is called the galvanic series. The galvanic series for representative metals is shown in Figure 14-1. Most marine structures must, by necessity, use a combination of metals to form the structure. In general, the materials which form the largest area, greatest thickness, or components that are noncritical within the structure should be the anodic metal to distribute the corrosion to the largest extent possible or localize the corrosion to the least critical part. If possible the more cathodic material should be protected by painting or by other coatings. The anodic material may be left bare. By protecting the cathodic material, its effectiveness as a cathode is diminished; this may decrease the galvanic attack on the anodic material. By not coating the anodic material there is less chance for intense localized corrosion to occur in areas that are not properly coated (holidays) or bare spots. In the galvanic series, the least noble end will tend to become the corroded material, sacrificed to the benefit of the more noble material. It will be noted that the materials at the upper end of the series, magnesium and zinc, are the predominant metals used as sacrificial materials on boat hulls and other underwater structures. Materials designed to be sacrificial must not be coated. Figure 14-2 shows the representative loss of metal that may occur in unprotected steel in the above and below water marine environment.

Materials buried below the mudline will generally not be subject to significant corrosion, unless the soil material has a chemical content that will itself attack the material. Most bottom soils are not aggressive enough to be of concern in material corrosion in waterfront structures.

Biological attack is prevalent in sea water to varying degrees and may significantly degrade materials. A primary source of timber degradation in sea water is marine organisms. Marine borers are the predominant marine organism that causes serious degradation of marine structures. There are a

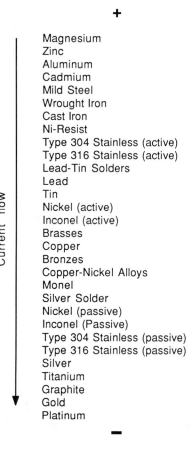

Figure 14-1. Galvanic series of common metals. The less noble metals will be sacrificed for the protection of the more noble metals.

number of marine borer species that may affect structures worldwide. Two of the more notorious species are Teredo and Limnoria. Teredinidae species have been found in waters in all inhabited areas of the world. Teredo has the general appearance of a worm and is often referred to as shipworm. The fore part of the body is equipped with cutting shells, which are used to excavate burrows or tunnels in wood materials. At the aft end of the teredo are siphons that provide water to feed the creature. In a typical infested timber pile, a cut cross section would show numerous internal tunnels excavated by the teredo.

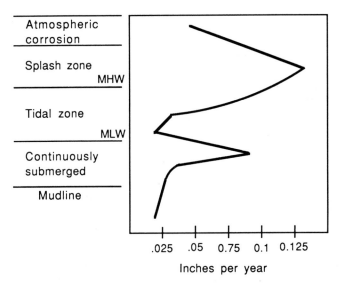

Figure 14-2. Generalized profile of steel metal loss in sea water. The intensity of corrosion will vary with atmospheric conditions, water temperature and water salinity.

From the exterior the tunnel openings would be minute and not readily visible by the human eye. The interior, however, may be nearly hollow with tunnels. Therefore a timber pile may appear to be sound on the exterior but it can be significantly weakened on the interior. Testing for the presence of teredo is difficult. A competent marine diver may be able to detect its presence by observation of the pile exterior for telltale signs of the minute tunnel entrances. Often core samples will be removed from the pile to assess the quality of the interior wood. If cores are removed they must be plugged with treated wood material to prevent an easy access for marine organism attack. Several electronic impulse detection systems have been developed to nondestructively test wood for the presence of marine borer attack. The testing is expensive and the results of the tests are subject to interpretation.

Another specie of marine borer prevalent in wood degradation in sea water is Limnoria. Limnoria is in the crustacean group of marine borers. It is also known as gribble. It is generally from three to six mm long and has a segmented, semicylindrical body. It holds on to the wood with sharp curved claws at the ends of its seven pairs of legs. Five additional pairs of legs on the posterior have blade-like terminations that function as gills to circulate water to the animal. It bores into the wood with a pair of mandibles in its mouth area. The mandibles are rasp-like, enabling it to erode away soft wood. Limnoria generally eats from the outside of the wood, which in a

324 PART 3/ENGINEERING DESIGN

timber pile leaves an hourglass like appearance. It also can hollow out a piece of treated wood if it can gain entrance to the untreated interior through any bolt hole or surface abrasion.

Preservative wood treatment may be effective against teredo and limnoria. The type and quantity of treatment will be based on local conditions and the severity of anticipated marine borer attack. Traditional preservatives are creosote, inorganic arsenicals, and pentachlorophenols. It is important that any holes, cuts or abrasions in treated wood be swabbed with preservatives or plugged to prevent marine borer intrusion to the interior of the wood. Wood may also be wrapped to protect it against borer attack. The most common wrapping materials are ship's felt, plastic and concrete. In marina mooring pile applications, wrapping is usually not appropriate because of the need to allow a pile guide mechanism to move unobstructed along the surface of the pile. Mooring piles in timber are most often treated to protect against borer attack and may be considered to be replaceable over time.

Various types and combinations of piles are extensively used in marina construction. Figure 14-3 shows some types of common deterioration often observed in piles. In addition to mechanical damage, wood piles are most seriously affected by biological attack, steel piles by corrosion, and concrete

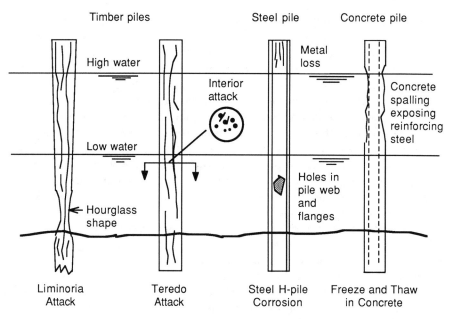

Figure 14-3. Types of deterioration that might be observed in marina piles used in a sea water environment.

CORROSION AND MATERIAL DEGRADATION   325

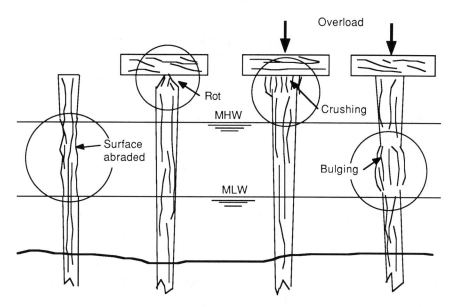

Figure 14-4. Typical forms of damage sustained by timber piles.

piles by alternate freezing and thawing or reinforcement corrosion, both of which may cause concrete cracking and spalling. Mechanical damage from overload, impact or abrasion of foreign objects may occur to all types of piles. Figure 14-4 shows some typical forms of damage to timber piles resulting from abrasion, rot and overload.

14.2 FRESH WATER AND TERRESTRIAL ENVIRONMENT

Minimal corrosion or material degradation is experienced in materials submerged in fresh water unless the water has high corrosive chemical content. If there is a significant water level fluctuation, fresh water rot may be experienced in wood at the water level change area. The most significant fresh water problem is wood rot or corrosion above the water level. Rain water or high humidity near the water surface provides a suitable environment for fungi growth with resulting rot. One form of rot attack causes wood to bleach out or lighten in color. It is commonly known as white rot. Another form is brown rot which produces a brownish residue or powder. Fungi may either stain or discolor wood or actually attack the wood, resulting in its eventual loss of structural integrity. In its most advanced stages, fungi will exhibit fruiting bodies such as mushrooms, punks, and conks. The evidence of these fruiting bodies indicates that decay has progressed to a significant degree and structural integrity is compromised. Provision of air circulation, coatings, and design to shed rain water will all help against fungi rot.

Metals exposed to fresh or rain water generally need only to have a protective coating or natural film to retard corrosion effects. The design should incorporate details that prevent standing water on or in any member.

14.3 WOOD PRESERVATIVES

The principal method of protecting wood from degradation is by adding chemicals to the wood. Wood preservation by preservative treatment has been used for thousands of years. Coating wood with pitch was an early means of extending wood life. The addition of chemical preservative is directed to mitigate biological attack by marine borers, fungi and insects. There is a wide range of chemicals that are used in wood preservation, but in all cases the effectiveness of the chemicals relates to the quantity and depth of penetration of the chemical into the wood.

The most often cited qualities of a proper wood preservative include: the preservative's toxicity to the biological attack being considered; the ability of the preservative to penetrate the wood; the ability of the preservative to sustain its function over a long period of time; reasonable cost to benefit ratio; safety in handling; lack of reactivity to metal fastenings; and noninjury to the wood itself. Environmental considerations have tempered the use of wood preservatives because of the concern of long term effects of the preservatives on benthic or terrestrial species other than those intended to be controlled. Concern is also voiced about the effect of preservatives on people who must handle treated wood either in the construction of the facility or passively by casual contact. Cost is also an important factor for consideration. Preservative treatment requires special preparation and handling of the wood products as well as the preservative material and the pressure treatment operation. In some cases alternative materials may be more cost effective.

Wood preservatives are broadly classed into two groups, waterborne preservatives and oil-borne preservatives. Waterborne preservatives commonly used in the United States include: chromated copper arsenate (CCA); ammonical copper arsenate (ACA); acid copper chromate (ACC); chromated zinc chloride (CZC) and flor-chrome arsenate phenol (FCAP). Common oil-borne preservatives include: pentachlorphenol (PCP); coal-tar creosote; creosote coat-tar solution; copper napthenate; tributyl tin oxide (TBTO) and copper-8-quinolinolate.

Waterborne preservatives have positive characteristics, providing a product that is clean, accepts paint or glue, is relatively safe to handle, is nonexplosive, has a high degree of permanence, and the solvent, water, is inexpensive. Some disadvantages are that the wood may swell upon treatment and exhibit raised grain, lack of water repellency compared to oil-borne treatments, and the need for additional drying prior to treatment.

Oil-borne preservatives are relatively permanent, provide some water repellency, are generally easy to handle and apply, have a wide range of effectiveness, are generally noncorrosive to metals, and the treatment process can be readily controlled. On the negative side, they may be toxic in both direct contact and vapor form, they may leave an oily residue on the wood surface, and they may not accept paint or other coatings.

Most effective wood treatments require pressure injection into the wood. The type of wood and the degree of pressure treatment will determine the effectiveness of the treatment. The retention of the maximum amount of preservative in the wood is accomplished by using a method called the full-cell process. This process utilizes a vacuum to remove air from the wood cells which will subsequently be filled by preservative under the application of pressure. Waterborne preservatives almost always utilize the full-cell process. Another process, called the empty cell process, is similar to the process outlined above, but it eliminates the initial vacuum step. In this case the preservative is directly injected into the wood by pressure. The amount of pressure in both processes is regulated for the species of wood being treated and the degree of treatment specified. Both processes finish with a final vacuum step to remove excess preservative and provide a relatively clean treated wood surface.

There are other nonpressure processes which tend to be simpler and less costly than the pressure injection methods. Unseasoned wood may be subjected to treatment in a preservative solution or successive solutions which allow the formation of inorganic salts in the sapwood. Preservative pastes may also be applied to the wood surface to create a preservative barrier. Another method for nonpressure treatment is to subject the wood to alternate hot and cold baths. The thermal change creates a partial vacuum in the wood structure allowing the introduction of the chemical preservative. These nonpressure treatments have been reasonably successful in providing protection against insects and decay but caution is warranted for their use in immersed marine environments.

Before embarking on the specification of a wood preservative treatment process, it is advisable to discuss the application with knowledgeable wood specialists. A number of universities have excellent wood research facilities and knowledgeable staff who can be of assistance in selecting the proper treatment for specific use areas. The Forest Products Laboratory in Madison, Wisconsin is an excellent source for information about wood products and their application.

14.4 COATINGS

Protective coatings are necessary for many metals and some wood and concrete products which are used in a marine environment. The coating functions as a barrier to isolate the base material from the corrosive elements. Proper coatings on steel surfaces are usually quite thick, with thicknesses of 15 to 25 mils not uncommon for coal-tar epoxies and other high-build products. Even with a good application, the potential for bare spots or holidays in the coating exists. The bare spots may result in the formation of an electrolytic cell and corrosion attack. It is possible to mitigate this condition by application of sacrificial cathodic protection in addition to the coating. The cathodic protection will only be useful for those portions of the structure that are below water.

A coating is only as good as the quality of the surface preparation of the base material prior to application of the coating. The coating must adhere to a clean surface having the appropriate surface profile (tooth) and be within a specific moisture level. Failure to provide the proper surface preparation will greatly shorten the coating life and will result in excessively premature coating failure. Standards for surface preparation have been developed by the Steel Structures Painting Council (SSPC) and the National Association of Corrosion Engineers (NACE). These specifications should be followed for the specific type of material and environment for which the coating is to provide protection. A common basic error in surface preparation is to blast clean a metal surface to the proper specification but then wait too long before applying the coating. If the environment is at all humid, a rust bloom may form on the material surface, reducing the ability of the coating to adhere to the base material.

If a coating is to be used on steel piles or bulkhead, it is very important to have the coating process monitored by knowledgeable coating technicians. Hiring a testing laboratory near where the coating will occur is often a wise investment. Once a pile or bulkhead is installed, it is extremely difficult and costly to remove it if the coating is found to be improperly applied. It is virtually impossible to replicate a quality coating near or under the water surface.

The four basic categories of surface preparation by blast cleaning used in marine construction are: brush blast (SSPC-SP 7, NACE 4); commercial blast (SSPC-SP 6, NACE 3); near white (SSPC-SP 10, NACE 2); and white metal (SSPC-SP 5, NACE 1). Immersed metal should be cleaned to white metal, metals not immersed but in a marine environment may use near white blast. Commercial and brush blast are generally used for mildly corrosive and industrial exposure. If possible use the very best surface preparation or eliminate the coating altogether. A sand or water blasting operation in or around a marina must be performed with extreme caution. A marina is full of delicate surfaces, such as boat hulls and people who may suffer from the effects, even distant effects, of a blasting operation. Sandblasting may also involve other environmental hazards and precautions must be taken to mitigate any damage to sensitive environmental areas.

Typical coatings used for immersed steel include: coal-tar, coal-tar epoxy, epoxy, epoxy phenolic, phenolic, vinyl, and zinc rich coatings. For steel in a mildly corrosive environment which does not include immersion, coatings include: alkyds, silicone alkyds, coal-tar, coal-tar epoxy, epoxy, urethane, vinyl, and zinc rich coatings. Concrete has had a limited success with coatings. Some coatings that appear suitable with a properly prepared concrete include: coal-tar, coal-tar epoxy, and epoxy. The use of any coating system must be performed with proper surface preparation and strict adherence to the manufacturer's recommendations for application.

It is often desirable to coat or recoat a below water structure. This is difficult at best, and therefore the recommendation is made to apply a good coating before the structure is placed in the water. If underwater coating must be applied, it will be necessary to clean the surface of foreign materials and achieve a competent surface to which the coating can adhere. Underwater-curing materials are designed to displace the water film on the surface of the material to be coated. The coating product is usually a mastic, applied by hand smearing, consisting of a two part epoxy with 100 percent solids. The dry film thickness will be large, in the order of 100 to 200 mils. These compounds can be successfully used with concrete, fiberglass, metal, wood and other properly prepared substrates. Sand or water blasting may be necessary to properly prepare the surface to be coated. The cost of underwater coating application is very high and its success is quite uncertain.

Coating may be applied by brush, roller, spray, and dipping. Brush application is used for touch up work and small projects, but is usually not suitable for major coating projects. A proper brushing job, using good quality brushes and care to apply the proper coating thickness, can provide an excellent finished product. The brush action tends to help cause the coating to firmly adhere to the material surface. Rolling is a common method of coating application, again not always suitable for large work. The roller and brush are good because they minimize loss of coating during application. Spray coating, however, has considerable coating material loss during application, often between 20 and 40 percent of the applied coating quantity. Spraying is fast and not as labor intensive as brushing or rolling. It also provides a consistent film thickness if applied by a competent technician. Both conventional and airless spray equipment are available for most coating systems. Heavy pigmented coatings may not be suitable for spray application. Manufacturers' recommendations should be consulted for proper application. Dipping is usually only applicable to zinc coatings. Repetitive major industrial components, such as automobiles, may be dipped, but most marine applications do not fit into this category, except for galvanized components. If a major structural component is to be hot dipped galvanized, it is wise to ascertain before design, that there is a plant that can handle a piece of that size in their dip tank. Facilities that perform hot dip galvanizing are becoming harder to find because of the concern with the environmental effects of the galvanizing process and the difficulty of disposal of dipping process waste materials. Galvanized products are generally used for above water and splash zone applications.

Coating systems selected for use in a marine environment must be selected with care and must be suitable for the intended use. Manufacturers' specifications must be followed for surface preparation and coating application. All coating systems will, in time degrade, so constant maintenance should be provided to ensure continuing coating integrity. A proper coating system can add substantial life to a structure and the cost of a proper coating system should be factored into the overall life cycle costs of the structure. This is one case where the least expensive first cost may result in considerable maintenance cost over the life of the structure.

15
Design Load Criteria

The structural analysis and engineering design of marine structures requires an understanding of the loads imposed on the structures. Loading takes many forms. The first load to be considered is the dead load of the structure. The dead load is the weight of all components of the structure that are considered permanent. Live loads are the loads imposed by effects on the structure other than its own weight. Live loads may embody a full range of effects such as people loading on a deck, wind pressure on the dock system, loads resulting from vessels striking the facility, wind generated wave effects and forces associated with water currents. In traditional marina design live loads are often construed to be the load imposed on a deck surface by the addition of people and movable equipment. Other associated live loads such as wind, wave, impact, etc., are generally specified as such and are independently analyzed.

The determination of loading criteria may not be cookbook in nature and may require significant review of environmental conditions, vessel sizes and marina configurations. Rules of thumb may be useful but must be tempered with sound engineering judgement.

15.1 DEAD LOAD

Dead load may be defined as the weight of all structural framing and other structure components fixed to and permanently integrated into the structure. For marina dock systems the dead load may include: float or pier framing, decking, railing, flotation units, hardware, utilities, power posts, transformers, dock boxes, pile guides, cleats, fire protection equipment (if permanently affixed to the structure) and any other fixed materials or equipment. It is not generally necessary to attempt to account for every bolt or nail but rather to equate some approximate number for all hardware items as a lump sum. The same is true for estimating the weight of structural components. An estimated weight for average board lengths, etc. is usually sufficient. One factor that should be considered in estimating dock weight is the unit weight of wood. Weight of wood will vary with its species, moisture content, and added preservative treatment. If possible the simplest way to determine the unit weight of wood for a specific project is to weigh a representative sample. Using a one foot long sample of each size of wood employed in the project will easily and accurately provide an accurate unit weight which can then be multiplied by the estimated total length of that kind of wood to obtain a dead load estimate for that component. If representative wood is not readily available, reference tables for wood weights may be consulted.

The dead load weight is important in two general areas of marina design. In projects using floating dock systems, the dead load weight will be used to determine the available freeboard of the dock system under its own weight. This is the normal height of the dock system above the water level as the system will float without addition of people or other moveable weights. In combination with the freeboard calculation, will be the calculation for the amount of buoyancy material (flotation) necessary to support the dead load at the prescribed freeboard. The freeboard and buoyancy calculations work together such that an alteration to either freeboard or buoyancy will effect the other parameter. The second use of the dead load weight in marina design is for the analysis of structure foundations in fixed structures. Footing, foundation wall or pile analysis must consider the imposed dead load force on the foundation as well as any live load. Pile end bearing pressure, pile embedment length and pile stability will all relate to the magnitude of the imposed total load.

Some useful unit weights of common building materials are presented in Table 15-1 to assist in estimating dead load.

15.2 LIVE LOADS

Live loads are the active and changing loads that may be imposed on a structure. In marina design the general category of live load is people weight. Associated with people weight are any special loads that people might carry. Instead of attempting to determine an actual number of people that might use any facility and then multiplying that number by some average weight of each person, the accepted methodology is to relate people weight, live load, to a weight per square foot (square meter). This technique is standard throughout the building industry, and as a result certain criteria have been established for the relationship between types of people load and load expressed in terms of weight per square unit. Live loads when presented in terms of weight per square unit are assumed to be uniformly distributed over the entire surface under consideration. Some traditional building code live loads for various activities are shown in Table 15-2. Marina dock systems may have different criteria, but these live loads may be useful for design of upland structures. Local building codes must be consulted to assure compliance.

Table 15-1. Unit design weights for common materials.

Material	Pounds per cubic foot
Aluminum	168
Asphalt	69-94
Brass	524
Brick, common	112
Canned goods, cases	58
Cast stone masonry, (cement, sand, stone)	144
Cement, Portland	94
Clay (dry)	120-140
Clay (wet)	165-195
Concrete, plain, stone or gravel aggregate	144
Concrete, reinforced, stone or gravel aggregate	150
Earth (dry)	96
Earth (damp)	108
Earth (wet)	120
Gasoline	41-43
Glass	180-196
Gravel	100-120
Ice	57
Iron, cast	450
Iron, wrought	485
Lead	710
Manganese	460
Riprap, limestone	80-85
Riprap, sandstone	90
River mud	90

(continued)

Table 15-1. *Continued*

Material	Pounds per cubic foot
Rubber	58
Sand	90-100
Snow, loosely piled	35
Stone, loosely piled	75
Water, standard at 62 degrees F (16.7 C)	62.354
Water, Sea at 62 degrees F (16.7C)	63.975
Wood, seasoned	
Ash, commercial white	41
Cypress, southern	32
Fir, Douglas, coast region	34
Oak, commercial reds and whites	45
Spruce, red, white, and Sitka	28
Southern pine, short leaf	39
Southern pine, long leaf	48
Hemlock	30

As can be noted in Table 15-2, there is a wide range of values associated with prescribed live loads, in this example from 20 to 250 pounds per square foot depending upon the activity. There currently is no universal code or guideline for the assignment of live loads in marina structures. Local codes may have some regulations but these are often misleading or erroneous. The generally accepted minimum criteria for prescribed live loads in marina design for floating and fixed dock structures is as follows:

Floating dock system used for residential or light commercial use	20 psf
Floating dock system associated with active marina in traditional use	30 psf
Floating dock system used for public assembly, boat shows, events, etc.	40-60 psf

Gangways up to 6 feet in width	50 psf
Gangways over 6 feet in width	100 psf
Fixed pier decks, pedestrian access only	50 psf
Fixed pier decks subject to vehicle traffic	*

In addition to the uniformly distributed live load, a loading analysis should also include the effects of a concentrated load of 400 pounds applied anywhere on the surface of the structure. The uniformly applied live load and the concentrated load need not be considered as acting simultaneously.

In an attempt to make live load numbers more meaningful, let us look at an average person standing erect and see what kind of load this person represents.

Table 15-2. Typical building code live load criteria for upland structures.

Occupany or use	Live load (psf)
Places of Assembly	
Fixed seats	60
Movable seats	100
Platforms	100
Offices	50
Office corridors	100
Residential dwellings	
First floor	40
Second floor	30
Uninhabitable attics	20
Stair and exitways	100
Sidewalks, driveways subject to trucking	250

---

*The loading should be a minimum of H10-44, as specified by the American Association of State Highway and Transportation Officials (AASHTO), which represents a 20,000 pounds truck, with a wheelbase of 14 feet, a front axle load of 4,000 pounds and a rear axle load of 16,000 pounds.

Figure 15-1 shows a typical average person having a weight of 150 pounds. The plan area of this average person is roughly a rectangle having sides of 2'-2" by 1'-8" or an area of approximately 3.6 square feet. Dividing the average weight of 150 pounds by the 3.6 square feet yields a rough live load of 41.7 pounds per square foot. This practical approach demonstrates that for marinas where many people might gather, live loading in the order of 40 pounds per square foot is realistic. Of course, the average load is just that and real loads may be higher or lower. A gathering of professional football players will undoubtedly create a higher live load per unit area than will a group of school children in the same area.

It is important to note that due to the type of framing and flotation units associated with floating marina structures, an increase in live load capacity may also provide for a stiffer structure by virtue of having more flotation

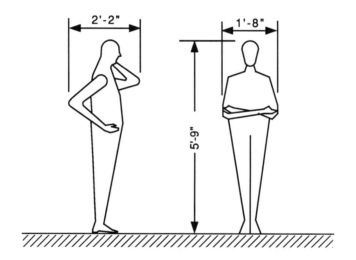

Space occuppied by an average adult standing at rest:
2'-2" = 2.17' and 1'-8" = 1.67',
therefore an average adult requires approximately
1.67' x 2.17' = 3.6 square feet. At an average adult
weight of 150 pounds, the average uniform load
per square foot = 150 lbs / 3.6 square feet = 41.7
pounds per square foot (live load).

Figure 15-1. Calculation of a rational people live load for structures, such as docks used for public assembly, people waiting areas, etc., using physical body characteristics and an average weight of person. This analysis primarily relates to float buoyancy considerations. Structural components of dock systems generally require a higher design live load and appropriate factors of safety.

units, spaced closer together. The added cost, therefore, of providing a 30 psf live load capacity over a 20 psf live load capacity will not only provide greater live load capacity but may provide a system better able to resist the effects of transiting wave regimes and offer a more stable walking surface.

There has been some discussion within the industry of reducing live load criteria for new lightweight structurally engineered glue-laminated wood beam or high strength metal alloy floating dock systems to take advantage of their higher strength to weight ratio as compared to traditional wood or metal frame systems. Most marina designers do not subscribe to this philosophy and recommend a minimum live load design capacity of 20 psf for any marina floating dock system. The argument has been that the number of flotation units needed to achieve a 20 psf or greater live load capacity reduces the desirable flexibility of the lightweight structures. Any properly designed structural system of this nature can still remain flexible with adequate flotation provided for a 20 psf live load. Also, making a system too flexible may make it too lively to be suitable for a pedestrian walking surface. To reduce the live load below 20 psf compromises the system for even the lightest marina usage and may result in below minimum suggested float freeboards.

15.3 WIND

The investigation and determination of design wind speeds has been discussed in Chapter 5. With a design wind speed and direction established, a determination of wind loading on a structure may be estimated. The actual effect of wind on marina type structures is unclear. To date no known extensive wind tunnel tests have been performed on marina and boat models to ascertain a realistic wind pressure criteria. It may be that the subject is moot in that individual vessels vary widely in configuration and wind resistance. The designer must, however, attempt to relate marina characteristics to definable load in order to assess dock system structural strength and dock mooring criteria.

The traditional methodology has been to equate wind speed to a value of pounds per square foot of surface area and develop a loading by addition of all exposed surface areas. The Corps of Engineers *Manual SR-2* (page 134) provides a curvilinear plot of wind velocity, in miles per hour, to velocity pressure, in pounds per square foot (Figure 80-B in SR-2). Two curves are presented, one for a 0-10 degree flat roof house and one for a small craft silhouette. By entering the plot with a given wind velocity and intersecting the curve, a representative wind pressure may be determined. Associated with the plot is a map of the United States showing isotachs for the predicted fastest mile of wind, 30 feet above the ground, for a 50 year return period (Figure 80-A in SR-2). For any location in the United States a generalized 50 year return period wind may be determined from the isotachs. The representative wind pressure may then be applied to areas associated with average boat profile height and length to develop a wind load per boat and subsequently an overall load for all boats in the marina. Some designers will also include contributions of exposed dock system profile in the loading calculation. *Manual SR-2* also allows the reduction of wind effect on boats shielded by other boats. *Manual SR-2* suggests a shielding effect of 80 percent for boats downwind of a wind blowing on the leading boat beam and a shielding effect of 75 percent for boats downwind of a wind blowing from bow to stern. Summation of the individual wind loads allows for the determination of the total marina wind load. The resulting load may be used for calculation of mooring forces.

Although this methodology has served the marina industry well for the past 15 years, there is considerable concern that the analysis is not representative of actual loading conditions in a marina and may result in under design of the required strength capability. A number of organizations and individuals are currently working to refine the loading analysis to reflect real life conditions. The area lacking the greatest confidence level is the effect of boat shape on wind resistance. This effect is represented by a drag coefficient. Until extensive wind tunnel tests are performed on varying types of boat profiles, a rough estimate of wind resistance attributed to recreational boats can be made. Further investigation and definition of projected wind velocity at specific marina locations is necessary to provide a reasonable wind velocity for a particular site rather than using a very course wind speed estimate from a map isotach or even from wind data gathered at well documented but distant locations. The effect of wind velocity decay as the wind passes over close proximity to water or over developed land may possibly reduce ambient wind levels at the specific site and if so should be considered in marina loading analysis.

In conjunction with H.L. Burn and Associates, Sydney, Australia and wind analysis studies performed by Prof. W. H. Melbourne of Monash University, Victoria, Australia, a wind loading analysis has been developed that attempts to present a reasonable and rational response to wind loading while keeping in mind the large number of variables associated with developing wind load criteria. The first item considered is the windage presented by the boats themselves. Figure 15-2 shows curves of boat profile height versus boat length from three different investigations. The lower curves are those presented in the American Society of Civil Engineers, *Manual 50. Report on Small Craft Harbors,* 1969. The second set of curves represents that data presented in the U.S. Army Corps of Engineers, *Manual SR-2, Small-Craft Harbors: Design, Construction, and Operation,* 1974. The third set of curves, the top set, is based on measured field data developed by the authors in the late 1980s. It can be readily seen that boat profile heights have increased over

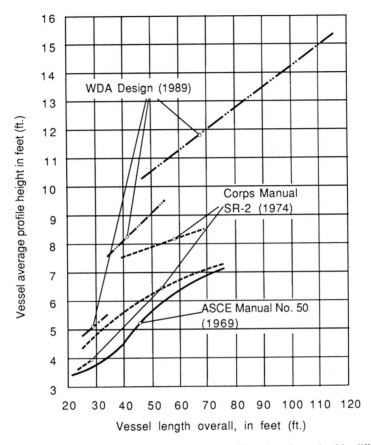

Figure 15-2. Comparison of vessel profile height vs. vessel length as determined by different studies. Note the change in vessel profile height over time.

the past two decades. In fact, the relative profile height has at least doubled in the size of boats for which marinas are currently designed. Boat profile height data may continue to change, although indications are that near maximum boat design values have been achieved and that further increases in boat profile height may result in unstable and unseaworthy boats.

The next step is to determine a terrain air velocity. This wind speed should be adjusted for local conditions, as previously discussed, to obtain a wind speed that is representative of design or maximum winds applicable for the specific site. The selected wind speed should also have some rational merit. Designing for a maximum predicted hurricane (cyclone) wind speed may grossly affect the economics of the project and may have little rational meaning. This aspect of design requires risk analysis. It may be that the potential for experiencing a maximum wind is low enough that the risk of the event occurring is assumable. That is, if the maximum wind occurs, there will be significant damage to the marina but the cost to design for this wind speed is so great that the project would not be economically feasible. If this option is selected, the marina developer must appreciate the risks and provide for an operational plan that will minimize the potential damage and protect the interests of his customers to the greatest extent possible. In general the design wind speed will be in a range between 50 knots and 64 knots, 64 knots being hurricane threshold. The prudent designer will provide for some factor of safety in the design such that the actual capacity of the system before failure will be greater than the design thresholds.

Wind speed will usually be described in knots (kts), (nautical miles per hour), miles per hour (mph), or in meters per second (m/sec). The direction from which a given wind impacts on the surface of a boat hull, or other structure, is also important. It is possible to investigate the wind effects over a 90 degree arc of wind angle, however, limited studies seem to indicate that wind striking a berthed vessel about 75 degrees off the bow or stern will present the greatest surface area over which the wind will act. Unless the marina layout is permanently fixed and the prevailing winds are extremely constant, it is prudent to design on the basis that the wind will, at some time, arrive at the maximum design velocity at an angle that will provide the greatest boat surface area. Figure 15-3 shows an orientation for wind angle versus a so-called direction factor that accounts for impacted boat surface area. The direction factor is based on a profile height of boat end area rather than side area.

The shielding effect of one boat by another is a topic of considerable

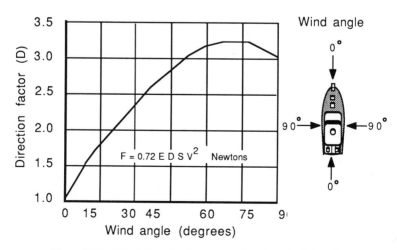

Figure 15-3. Wind direction factors as a function of wind angle.

DESIGN LOAD CRITERIA 341

debate. This analysis considers an array of berthed boats similar to the effects experienced by a saw toothed building roof structure, which has been studied for wind effects. The first panel of the saw toothed roof will feel the full effect of the wind pressure. The second saw tooth panel will experience a shielding representing a diminishment of wind pressure to 0.5 times the incident wind pressure and the saw tooth panels thereafter will experience a wind pressure of 0.3 times the incident wind pressure. Although not a pure analogy, the methodology will provide an approximate total value of wind pressure that is close to, or greater than the actual effect. The resulting structures may be a bit over designed, but then again the designer really does not know what the shape of vessels in the marina may be, or what actual speed the wind may attain. Figure 15-4 shows the relative values of shield-

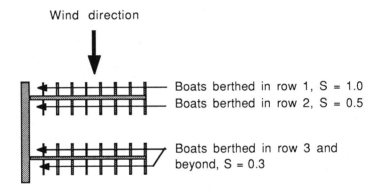

Wind blowing along fore and aft axis of berthed boats

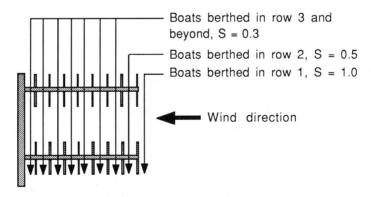

Wind blowing on side profile of berthed boats

Figure 15-4. Shielding effect of boats based on wind direction.

ing (S) used for differing angles of incident wind direction on a typical marina layout.

It is easiest to perform the following calculation in metric units and, if necessary, convert the answer to other appropriate units. The basic formula is:

$$F = 0.72 \, [E \times D \times S \times V^2], \text{ in Newtons, where 1 Newton} = 0.225 \text{ lbs.}$$

where:

E = Vessel end area, in m^2, end area equals 0.33 side area from Fig. 15-2
D = Direction factor (Fig. 15-3)
S = Shielding factor (Fig. 15-4)
V = Terrain air velocity (m/sec)
and 1 mph = 0.447 m/sec.

The application of this formula will provide the individual boat loadings for a given layout. By grouping common arrays of boats into a single unit, the combined array loading can be determined and this data input to the mooring design analysis. An array usually consists of a grouping of similar size boats.

There are a variety of other formulas and approaches to determining wind loading on berthed boats. The above formula has worked well in numerous marina designs but is subject to refinement as the state-of-the-art in this area of marina design is improved.

15.4 CURRENT

Some marina sites are significantly impacted by current. It is important to orient such a marina to mitigate the effects of the current both from the perspective of boat maneuvering into and out of a berth and the current loads against the boat, transferred to the mooring system. In general the most suitable berthed boat orientation will be with the boat centerline parallel to the current. The natural streamlining of the boat hull will present the least surface area and resistance to the current flow. The boat will also tend to remain upright without appreciable heel. If the boat is broadside to the current, and if the current is strong enough the boat may heel or list.

It is very difficult to analyze a marina for the total effect of current since every berthed boat will have a differing below water configuration and therefore load contribution. If it is important to consider a loading factor due to a strong current the following rough calculation will provide an approximate estimate of boat current loads. The current effect is related to the projected area of the below water portion of the boat hull, a drag coefficient associated with the shape of the submerged portion of the hull and the current velocity. The below water area may be roughly estimated by determining the boat waterline beam, multiplying the beam by the draft of the full portion of the hull (exclusive of deep thin keels) and applying a block coefficient to convert the rectangular area to a more representative hull shape. For power boats the block coefficient may be very close to 1.0 in both end on and side on exposures. For sailing vessels the block coefficient will be more difficult to estimate due to the wide variety of keel configurations. It is probably safe to assume a uniform load associated with representative size powerboats to get the approximate estimate of current load for a marina mooring design.

The pressure (P) in pounds per square foot against the below water portion of a hull may be approximated by using the formula:

$$P = V^2 C_D,$$

where P = pressure in pounds per square foot, V = current velocity in feet per second, and C_D = a representative drag coefficient. **For rough estimates, use a C_D for end-on profile equal to 0.6 and a C_D for a broadside profile equal to 0.8.** If more exact current forces must be considered then drag coefficients for specific vessels must be investigated. The pressure that is determined, multiplied by the appropriate below water area will give an estimate of the total force exerted by the current on the restrained vessel. This load may be considered as additive to wind and wave forces to develop a combined load analysis for mooring system design.

In general, the current contribution is often neglected in marina design, since in order to develop a significant force a very strong current is necessary, which is generally inappropriate for marina siting.

Some designers have attempted to reduce current effects by installation of current mitigating or diverting structures. This approach must be used with caution unless in-depth studies can be performed to truly assess the impacts of the current mitigation or diversion. In addition to having to resist massive current forces, the diverting structures may influence sedimentation transport, area flushing, and may result in dangerous turbulence at marina entrance areas or in other areas adjacent to the interrupted current flow.

15.5 BOAT WAKE

The effect of boat wake on marina activities is escalating as the population of boaters grows and more marinas are located in close proximity to one another, resulting in the passage of many boats past a marina. Boat wake is discussed in detail in Chapter 7. The design loads associated with boat wake effects are difficult to assess due to varying boat shapes, boat speeds, and depth of water in which the event occurs. It may be important, however, to associate some form of loading with boat wake action to assure a suitably designed dock structure. Generally the boat wake effect is an isolated event occurring independently of wind and wave effects. Although it may be additive to the imposition of other forces, it usually is a stand alone event. If the weather has so deteriorated as to produce prodigious wind and wave forces, the chances are that the recreational boater is not underway in his or her boat. The loading impact of boat wake is generally not assessed in structure design, but rather the magnitude of the boat wake wave height is analyzed and mitigation of the wake made before it becomes a force to be reckoned with in the marina. If floating wave attenuation structures are used to mitigate boat wake wave, the structures will have been designed to resist a wave of a magnitude with a force greater than the forces associated with boat wake events. If the attenuation structure is installed solely for the purpose of boat wake attenuation then it must be designed to resist the maximum boat wake waves as discussed in Chapter 7.

15.6 BOAT IMPACT

Boat impact loading, Figure 15-5, is especially important in the design of floating dock finger floats, where the potential exists for a maneuvering boat to strike the end of the finger float. Unless the boat is exceptionally large, berthing in the alongside fashion to a dock will result in lower forces than design wind/wave loads and therefore the overall system should be able to resist the imposed forces. There may be local overstressing or component failure resulting from a tough docking, but the system should stay together. To a megayacht captain, a floating dock system will appear fragile, no matter how well constructed, so most dockings will be performed with care. If dock damage does occur due to a bad docking maneuver or extreme weather conditions, the marina owner must hope that the yacht has good insurance coverage. If continual boat impacts are experienced on a particular dock, the dock probably needs to be redesigned and/or rearranged to make a more suitable berthing area.

To calculate the forces associated with vessel impacts, it is necessary to determine several pieces of information. First is the need to associate a velocity of approach to the vessel in question. The velocity should not be less than 1 foot per second (0.59 knots) with a value of 3.4 feet per second (2 knots) a fairly standard approach velocity for impact analysis of recreational boats. It is also necessary to know the weight (displacement) of the vessel causing the impact. For simplicity, vessel weight may be associated with its length (L) to roughly estimate the vessel weight. **For recreational boats the formula for boat weight is:**

$$W_{min} = 12L^2 \text{ (in pounds)},$$

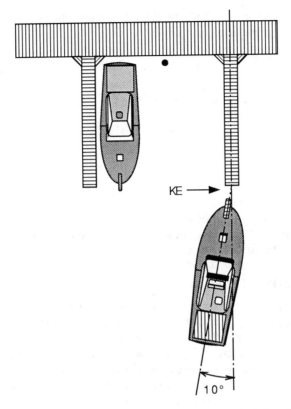

Figure 15-5. Diagram of boat impact angle on finger float for calculation of impact force.

therefore a boat 40 feet in length is estimated to weigh, 12 times $40^2 = 19,200$ pounds which is a rough approximation. For commercial vessels a representative formula for vessel weight is:

$$W_{min} = 25L^2 \text{ (in pounds).}$$

The other important piece of information for impact analysis is the angle at which the vessel strikes the object under study, perhaps the end of the finger float. The conventional angle of attack is considered to be 10 degrees off the centerline of the finger float. If the dock under consideration is a main float with alongside docking, an angle of 45 degrees might be more representative of a potential impact load. The angularity of approach will be used to develop the component load associated with the impact for load analysis.

The velocity component (V) will be equal to the approach velocity (V_{Approach}) times the sine of the approach angle (sin a).

$$V = V_{\text{Approach}} \sin a$$

The impact force, as kinetic energy (KE) can be determined by the formula:

$$KE = W V^2/2g$$

where $g = 32.2$ ft/sec^2.

By structural analysis, using unit deflections and relative structure stiffness, the structure bending moment and resulting bending stress may be developed and assessed for adequacy. A worked example of this analysis is presented in Chapter 16, Section 16.2.

15.7 HYDROSTATIC LOADS

In the assessment of the structural integrity of immersed bodies, it is often necessary to analyze the hydrostatic loading on a structure to provide adequate strength to resist the applied water or fluid pressure. The deeper a body is immersed in water the greater will be the pressure against the body. Most people are familiar with the pressure intensities associated with marine diving to great depths. The pressure exerted by the water is a linear, triangular distribution with a value of zero at the water/air interface which increases at the rate of the equivalent weight of water. Therefore in sea water at standard conditions the pressure increases by the weight of sea water equal to 64 pounds per cubic foot. Figure 15-6 shows a graphic presentation of water pressure as related to depth of immersion. Curves for both standard fresh water and sea water are presented. Figure 15-7 shows the triangular load distribution on the sides of an immersed U shaped box. Note that the pressure on the bottom of the box is uniform and of an intensity equal to the maximum pressure exerted on the sides. This information is important for the design of flotation pontoons for marina type floats or for the foundations for floating homes or other large structures. This type of analysis is critical for the design of floating dry docks which submerge deeply to accept a floating vessel for subsequent lifting out of the water for repair or service. It is also important to understand the mechanics of the pressure application to allow intelligent structural framing of the pontoon. The spacing of side stiffeners (scantlings) to effect the most economical and lightest weight structure may be based on the varying distribution of side wall pressure.

DESIGN LOAD CRITERIA 347

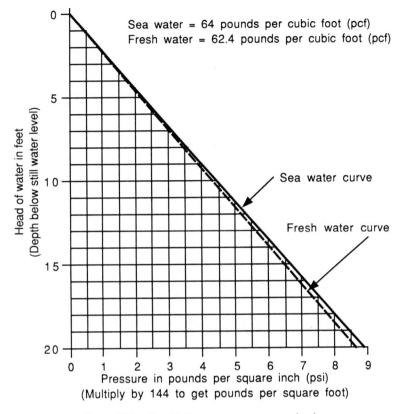

Figure 15-6. Plot of water pressure vs. water depth.

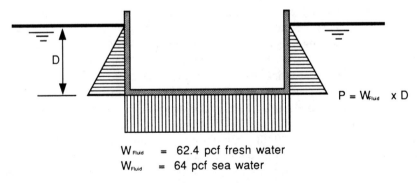

Figure 15-7. Hydrostatic pressure diagram of a rectangular box immersed in water, showing increase in water pressure with depth of submergence.

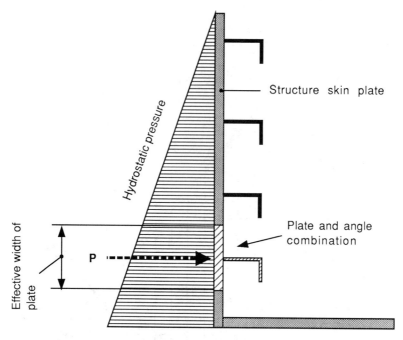

Figure 15-8. Detail of immersed box skin plate and interior angle showing analysis criteria.

Figure 15-8 shows, in detail, how the pressure loading might be calculated for the lower interior side wall stiffener (scantling) on a steel shell pontoon. In this case the properties of the combined section of skin plate and scantling angle would be used to develop a structural beam, which bends between two supports, which are usually transverse bulkheads or frames. For mild structural steel an effective width of skin plate equal to 60 × the thickness, (60t) may be used for calculation. The effective breadth of plating (60t), assumed to be acting in combination with any plate, may be determined by the formula $2\sqrt{E/F_y}\,(t)$, where E is the modulus of elasticity (psi), F_y is the material tensile yield strength (psi) and t is the plate thickness (in.). If 0.375 in. plate is considered, then the effective width of the plate equals 0.375 in. × 60 = 22.50 in. A composite beam is constructed that has an outer flange consisting of plate 22.50 in. × 0.375 in. and a web equal to the angle leg dimensions and an inner flange equal to the outstanding leg of the angle. By using standard beam analysis, section properties can be determined and the composite beam can be treated with traditional statics formulas. The applied load to the beam section will be the average of the pressure loading between the top and bottom of the effective plate width. If the effective plate width is of insufficient width to join up with the section above or below, then the piece of plate in between must be considered as a beam itself and its ability to carry the applied load must be analyzed. Since the water pressure load diminishes in the direction toward the water surface, the scantling spacing can be opened up to effect an economy of materials. An option is to change the size of the scantling angle, but this is rarely done because the continuous face of the outstanding leg of the angle forms a convenient surface upon which to build transverse frames or bulkheads. Changing skin plate thickness is sometimes considered on deep structures but if the thickness is changed, the skin plates should be aligned on the inside to again allow a continuous framing surface on the interior. A bottom or top deck may be framed similarly. Top decks usually have additional plate thickness to accommodate wear and corrosion. The extra thickness is not considered in the structural calculations and is sacrificial. In most marine structures the scantling angles are run in the longitudinal direction to provide longitudinal strength as well as local plate stiffening. The angles are also run with the outstanding leg pointed down to prevent water from standing on the angle surface. If the angle legs must be turned up, then limber holes for water drainage should be incorporated into the angle. Angles are also continuously welded, all around, to prevent corrosion between the plate and angle. The principles applied to the steel structure example will also work for many other materials including other metals, plastics, and timber.

Analysis of longitudinal strength is accomplished by summing the section properties of all continuous longitudinal members, which usually includes the structure skin and scantlings and any longitudinal frames or bulkheads or keel structures. The individual member section properties themselves must be adjusted by the transfer factor about a common baseline and then the total of the individual properties and the transfer factors must be divided by the gross area to arrive at a section property for the composite unit.

Transverse strength is usually provided through frames or bulkheads. Since the hydrostatic pressure will nominally be equal on both sides, the transverse frame or bulkhead acts like a large compression member with a uniform bending load induced from the bottom skin load and a loading of whatever form the superimposed top structure takes. For a fully immersed structure the local hydrostatic pressure on the top will be less than the local hydrostatic pressure on the bottom skin due to the depth versus load relationship.

15.8 LOAD TRANSFER

Interaction of Boats and Docks

Load transfer analysis in marina design has been essentially nonexistent, often because of the rather crude estimates that must be made in design computations to accommodate the variable forces associated with unknown vessel characteristics, wind velocities and other basic design parameters. However, the high cost associated with modern dock mooring systems may suggest further investigation of load transfer to reduce mooring loads by realistic appreciation of how the mooring forces arrive at the resisting mooring device. A marina dock system is dynamic in nature. The wind or wave action first must act on a floating vessel which then will move until the resisting mooring lines become taut and transfer the load to dock cleats, which then pass the load into the dock structure. The dock structure then will transmit the load to a pile by moving against the pile or to a bottom anchor system by moving until the anchor system becomes taut. Eventually the load is transmitted to the subsoil by pile interaction with the soil or by tension resistance of an embedded anchor. All this takes time. In the course of these events some energy loss occurs such as in moving the boat, stretching the mooring line and deflecting the mooring pile. A close analysis will show that although a wind load is considered as affecting all parts of a structure simultaneously, in real life, the applied loads will occur over time and in a staggered fashion as various mooring lines become taut and transfer the load. In response to the various forces, the mooring system may not have to accept the full load on an instantaneous basis. The consideration of these effects may allow a reduction in the total mooring capacity of the system. For most design projects, this type of in-depth analysis is probably not warranted. It is important, however, to appreciate the method by which a dock system is loaded to be able to assess its importance to the project under consideration.

Types of Pile Load Transfer

Another consideration is the load transfer between the marina pile and the bottom into which the pile is embedded (see Fig. 15-9). Piles used in marina and small craft harbor design are usually constructed of timber, steel or concrete. Whatever the material, the basic method of load transfer is similar. One form of load transfer occurs by transmitting the load axially through the pile to the subsoil. The pile then acts as a column and a reasonably direct transfer of load occurs in a vertical direction. If the pile is designed to transmit the load by resting on a solid soil layer at the tip of the pile, the pile is called an end bearing pile. If the soil makeup is such that an adequate friction component can be developed between the side of the pile and the soil, the pile is called a friction pile. Friction piles become important when solid subsoil (refusal) is encountered only at deep depths, depths which would require very long piles to reach. The length of embedment of a friction pile will be determined by the type of subsoil and the frictional resistance it can develop. It is usually necessary to perform soil boring tests at the site and possibly field or laboratory test soil samples to adequately determine the friction coefficient for the soils. Soils containing clay are often most suitable for providing good frictional resistance between the pile and the subsoil.

DESIGN LOAD CRITERIA 351

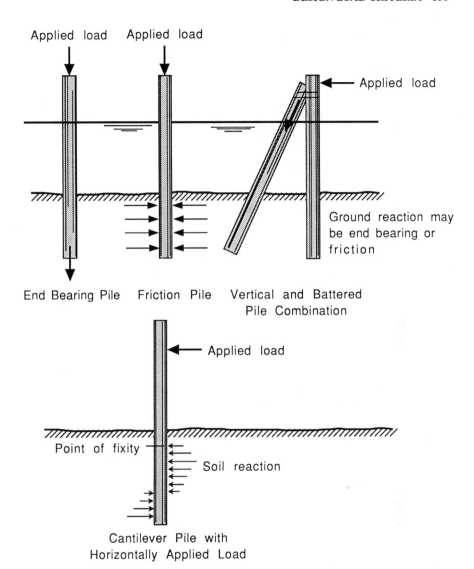

Figure 15-9. Applied loading criteria and resulting soil reactions in cantilever and battered piles.

If the load is applied in a horizontal plane, the pile must respond differently than it would to a vertically applied load. One method to accept horizontal load, such as from marina floating dock systems, is to provide a group, two or more, of piles with certain piles being arranged vertically and other piles being installed on an angle to the vertical pile (battered). The vertical pile will accept some of the applied load as a bending load in the pile and the battered pile will resolve the applied load into components and accept the load as an axial load in compression. The ground reaction of the vertical and battered pile combination may be either through end bearing or through friction, depending on soil characteristics. The battered pile solution is often appropriate for fixed structures but it may become inappropriate for floating structures. Floating structures are generally connected to a mooring pile by some form of sliding device that allows the float to ascend and descend on the pile to accommodate tidal or other water level changes. Unless the connection between the vertical and the battered pile is sufficiently beyond any possible interference with the floating dock guidance system, the potential for becoming hung up exists, with resulting damage to the dock system.

Most pile supported floating dock systems use an array of single piles to support the system. The pile is a cantilever having its embedded length fixed into the subsoil and the applied horizontal load transferred to the subsoil by the bending of the pile. The subsoil will resist the force action by pushing against the embedded pile. At some point in the embedded length there is a point at which the load action reverses and the soil on the opposite side of the pile assumes the task of pushing against the pile to keep it from rotating out of the subsoil.

The analysis of each pile restraining mechanism is a complex task requiring knowledge of the magnitude and location of the applied force, the section properties of the pile and the characteristics of the subsoil. It is important, however, to understand the various ways in which a pile transfers dock load to be able to make some basic decisions about the suitability of specific sites to accommodate a cost effective mooring solution. If such a solution is not possible, then alternative mooring systems, such as bottom anchored by cable or chain, may have to be considered.

15.9 SAFETY FACTORS

Virtually all design analysis must consider factors of safety in the design. This is especially true in marina design where a large number of variables are involved in the analysis. A structure or system must be designed for working stresses associated with normal and anticipated maximum loadings. It is always possible, however, that the system may be subject to overload and if this occurs immediate failure should not occur. The designer therefore builds in factors of safety to prevent failure if the structure is somewhat overloaded. Factors of safety will vary with different types of material and design situations. As an example, an allowable working stress in mild steel may be 85 percent of the yield stress because the material is homogeneous and the properties well defined, whereas the working stress in a subsoil may be one half or less of the predicted yield stress. Due to the nature of chain, the working stress is often one third of the yield stress and the working stress;

DESIGN LOAD CRITERIA 353

it is one fifth the yield stress for wire rope cable. Another way to provide a safety factor is to perform a calculation in the traditional manner but then increase component material size for wear, corrosion or increased loading carrying capacity. This methodology is often used in marina pile sizing when steel pipe pile is specified. Normal factors of safety associated with design in structural steel will be incorporated into the design and then an extra eighth inches of pile wall thickness added for corrosion. In addition to the extra thickness, a modern coating may be applied to provide additional corrosion protection. Although this may seem redundant and exceptionally expensive, the designer knows that the minimum extra first cost of additional steel is far less costly than having to prematurely install new piles.

15.10 SPECIAL CASE—ROWING DOCK

Floating rowing docks (see Fig. 15-10), have special conditions that warrant consideration over and above traditional floating dock design techniques. Freeboard, height of dock deck above still water level, must be such as to allow the free passage of the oars (sculls) above the deck during docking or

Figure 15-10. Northeastern University's Henderson Boathouse on the Charles River, Boston, Massachusetts, constructed in 1990. A fixed timber, sloped pier connects by a hinged ramp to a low (5 inch) freeboard floating dock used for collegiate rowing activities. A custom designed floating dock was necessary to meet low freeboard requirements while providing adequate live load capacity necessary for use by heavy weight rowing teams in launching and retrieving boats. *(Architect: Graham Gund Architects: Photo: Copyright 1990, Steve Rosenthal)*

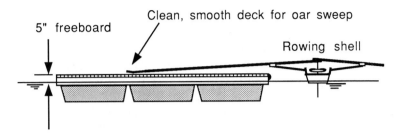

Figure 15-11. Shallow freeboard rowing float with clean, unobstructed deck surface and soft edge fendering.

undocking maneuvers. This dimension generally translates to a freeboard of between 4.5 and 5.5 inches, with 5 inches being desirable. Within this range a shell should be able to lay alongside a float without obstruction to the oars or riggers (see Fig. 15-11).

The freeboard criteria and the need to provide sufficient flotation capacity to support a number of rowers along the dock edge during boarding translates to a float that needs approximately 100 percent flotation coverage to meet these goals. The nature of the low freeboard, about 5 inches, allows a maximum live load equal to a maximum 5 inch displacement or $5/12$ of 62.4 pounds per cubic foot (pcf) (fresh water) which equals 26 psf total displacement load, with the deck awash. The structure itself will occupy some of the displacement volume, so a realistic live load for a rowing float may be as low as 10 psf. It is therefore important to properly analyze a rowing dock to provide the maximum possible live load capacity to prevent the outboard edge of the float from submerging when loaded with a crew of rowers.

Since it is also necessary to allow the oars (sculls) to pass over the float deck, no cleats or other surface obstructions should be located on the deck surface. The rowing shells are fragile craft so appropriate fendering should be installed along the berthing edge. Some crew coaches have also expressed a desire to have the float deck surface as smooth as possible to prevent damage to the oars as they pass over, and occasionally hit the float deck surface. If an abrasive surface is used, such as concrete, it may be desirable to overlay the concrete with a soft material like indoor-outdoor carpeting.

16

Design and Construction

The design of waterfront structures is different than traditional design of land structures. In waterfront design, material selection is critically important to account for the severe exposure and resulting material corrosion and degradation. Waterfront structures are often subjected to dynamic conditions associated with wind and wave loading and boat impact. In large measure, a good waterfront designer has had to learn by experience as well as by more formal book learning. A key to successful waterfront design is an appreciation of the forces at work and a translation of this knowledge into a design that attempts to work with the forces rather than just resisting the forces. Mother nature can be a hard task master and will never let us forget the latent forces existing in a marine environment. Therefore, design load criteria addressed in Chapter 15 are important considerations in design. This chapter provides planning and design guidelines and applies the design criteria previously developed.

16.1 SUGGESTED PLANNING AND DESIGN GUIDELINES

A major obstacle in marina development has been the lack of universal planning and design guidelines. Although there may never be a universally accepted common set of guidelines, certain parameters appear to be widely used and have attained de facto acceptance as being appropriate for the planning and design of marinas. The presentation of these guidelines, based on experience and rule making by a number of coastal communities, is an attempt to offer rational criteria for marina design without consideration of specific site or political ramifications. The guidelines should be used with caution because they are not site specific and certain local factors may alter the relevance of all or part of some of the material. It is also important to obtain and use the established rules and regulations of the approval agencies in the project area. If there is a lack of design guidance, then these guidelines may be of value in providing rational design criteria. Many of the considerations summarized in the guidelines are explained in further detail in Part II.

Site Investigation and Analysis

Water Level Datums. An accurate and relevant water level base datum must be established at the commencement of the project. Water level is normally related to low and high water levels in tidal areas and specified baseline elevations for rivers and lakes. The baseline water level should be established by a competent surveyor or other professional qualified to determine site specific elevations. It is not acceptable to estimate a water level based on purely visual observation or by general reference to a tide table or to other published general water level guides for a wide range of geography. If necessary, establish a water level monitoring gage (see Appendix 2, Fig. A2-1), which when measured over time will provide any water level data correction related to the specific site. It can not be stated strongly enough, that an accurate water level datum is critical to the successful planning of a marina. Common water level reference datum used in the United States includes: mean low water, mean low water springs, mean sea level or National Geodetic Vertical Datum of 1929 (NGVD), mean high water, high tide level, normal pool level, low water datum lake level, etc. In addition to normal water level changes, reference should be made to astronomical maximum predicted water levels (highest high water, lowest low water), flood levels and storm surge. Maximum flood levels generally relate to 100 year predicted still water flood elevations. The 100 year flood level, with storm surge if applicable, is particularly important for setting the base elevation for habitable structures and the upper elevation of piles used in float mooring systems.

Upland Elevation Datum. In establishing the applicable water level elevations, be careful to relate the water level elevation to the local standard upland base elevation. Many communities will have a standard base elevation which is to be used for sewerage works, bridge elevations, etc., and this elevation probably will differ from the water elevation base. Many problems are encountered when two different base elevations are used. At the least, clearly mark the plans for the appropriate elevation base and continually monitor the project for discrepancies in elevation usage.

Horizontal Control Grid. In addition to the establishment of an elevation reference base, a survey grid should be established to relate the project to a commonly used geographic plane. Many communities have established horizontal control points that relate to a community wide, or greater, area grid system. In the United States, most states have a coordinate grid system, often emanating from a zero-zero control point in or near the state capitol. By tying into the appropriate grid system, additional information such as the coordinates of channel control points can easily be added to the plans for reference and proximity assessment.

Site Topography and Bathymetry. Using the appropriate vertical and horizontal control datum, a field survey of the geography of the upland (topography) and the below water area (bathymetry) should be made and plotted on a plan to an appropriate scale. The topographic plan should locate all natural and manmade structures, above and below ground, and present the ground elevation as spot elevations and contours. Contour interval will vary with the change in elevation of the land, but are generally either; 1, 2, or 5 foot intervals. The bathymetry should be recovered by use of a grid system, usually having dimensions of $50' \times 50'$, $100' \times 100'$, or $200' \times 200'$ depending on the gradient of the below water profile. The closer the grid spacing the more accurate the survey, and the better the analysis of the effects of the existing bottom profile. If dredging is contemplated, the greater the number of depth elevations (soundings) obtained, the more accurate will be the dredge quantity estimate.

Site Property Survey. A property survey should also be made by an appropriate professional as required by local law, to establish property boundaries and any land encumbrances such as easements through the property. Water rights establishment may require further adjudication by appropriate review in a court of law. Do not assume water rights if the project may impact an area of questionable right of use.

Geotechnical Survey. Site investigation should also establish the basic types of soils into which foundation or float mooring systems will be restrained. The normal method is to perform soil boring tests that will identify the type of soil found at specific depths in the test area. From this data a competent engineer can interpolate and extrapolate data to develop an appropriate engineered solution for the soil embedment design. From an engineering perspective, there are never enough boring data provided at a given site. If possible, perform soil boring investigations on a reasonable grid system, or at the least perform a soil boring in the area of every major structure and randomly in a potential mooring field.

Environmental Studies

Wind Climate. Obtain wind data and determine the magnitude and direction of maximum winds, strength and direction of prevailing winds and percentage of occurrence (in strength and direction) of all winds. **Develop a design wind based on a 50 year return period for a wind gust having a natural period of 60 seconds.**

Wave Climate. Perform a wave climate study to assess the effects of incoming waves from all directions. The study should include fetch considerations, water depth, design wind criteria, wave refraction and wave reflection impacts.

Marina Flushing. The marina should be designed to provide maximum flushing and circulation. It may be necessary to conduct a flushing study including analysis of minimum seven day average wind speed criteria (annual basis). To assure adequate marina basin flushing, the following points should be considered.

1. The marina configuration should not include dead ends or other geometric designs that impede water circulation, create stagnant water zones, or collect debris.
2. Marina basins should have flow through circulation or be able to demonstrate adequate circulation patterns.
3. The marina basin bottom profile should be at a higher elevation than the water body to which it connects and should slope downward toward the connecting water body.
4. It should be demonstrated by analytical calculation that water quality in the marina basin will not be degraded by lack of adequate flushing times.

Shore Protection Structures

Natural vegetated shorelines and sloped rip rap revetments are the preferred method of slope stabilization and protection. Vertical walls should be avoided to prevent wave reflection within the marina and to minimize hardening of the shoreline profile. Engineered structures should be shown not to adversely affect sedimentation transport, littoral drift or cause significant changes to bottom topography in proximity to the site. Structures should not be constructed with sharp corners that may impede flushing action, collect debris, or cause shoaling.

Proximity to Navigation Channels

Navigation channels should not be restricted in any form by construction of the marina. Structures should be set back from designated channel lines by one of the following criteria, unless more stringent local, state or federal regulations apply.

1. A minimum of 25 feet between any structure or potentially berthed vessel and any point on the channel line.
2. A set back distance equal to 10 feet plus the projected top of slope required for maintenance dredging of the channel to control depth. This setback would apply to fixed structures that may impact mainte-

nance dredging operations. Vessels that may be moved for dredging could fall within this set back line.
3. Additional setback distance may be required in areas of the channel where commercial vessels must make turns or otherwise maneuver and where the effects of vessel maneuvering might compromise the safety of boats and people in the marina.
4. Marinas located in proximity to navigation channels should be designed to withstand the effects of transiting vessel wake. Vessels abiding by the applicable rules of the road and moving at allowable speed should not further be restricted as a means to protect the marina.
5. Marina access to a navigation channel should be designed to allow entering vessels reasonable visibility of oncoming boat traffic and should be as gradual as feasible to allow blending with the channel traffic flow. Obstructed, perpendicular access points should be avoided.

Dredging and Dredge Material Disposal

Dredging should be avoided. Marinas should be sited whenever possible in areas having adequate natural water depth. Adequate water depth, at lowest low water, should be considered to be the draft of the vessel to be berthed plus 3 feet. An absolute minimum water depth in a marina should be 4 feet. If dredging is required it should encompass the smallest area feasible to accomplish the project goals. The dredge profile should slope downward toward the open water area. No pockets should be created. Sedimentation infill should be estimated along with maintenance dredging requirements for the ensuing 25 year period. Dredging may only occur at such times during the year as deemed appropriate to minimize impacts on aquatic life and benthic species. The dredging technique employed should minimize short term turbidity, and if appropriate, the installation of silt curtains may be required to minimize a turbid plume in the waterway.

Dredge material analysis should be performed to determine the characteristics and toxicity of the materials. Standard tests include: grain size analysis, chemical constituents, and bulk sediment analysis.

Prior to any dredging operation, an acceptable method of dredge material disposal must be approved. Upland disposal is generally the preferred method of disposal. Disposal methodology should consider the effects of possible contamination of ground water supplies, handling and spillage effects and effluent discharge to any waterway. In saline waters it may be necessary to assess the effects of chloride concentrations. If appropriate, and approved by the cognizant regulatory body, dredge material may be used to enhance existing wetlands or create new wetlands.

Upland disposal containment having a dike structure should maintain a freeboard of a minimum of 2 feet between the top of the dike and the impounded dredge material. The containment areas should be designed to provide an maximum settling of suspended solids prior to effluent release to the receiving waterway. Effluent discharge should have a suspended solids concentration not exceeding 200 milligrams/liter. Erosion control should be provided and maintained for both the sides of the containment area and the discharge line. Existing dredge material containment areas should be utilized to the fullest extent possible. Modification of existing structures may be permitted to adjust containment area capacity.

Ocean disposal of dredge material requires deposition at a federally approved disposal site. Bulk sediment analysis and bioaccumulation analysis may be required to ascertain the appropriateness of ocean disposal. Ocean disposal may require provision of a clean cover material if the dredge material has significant toxic constituents.

Water Supply

The project should demonstrate that an adequate supply of potable water is available to service the project without compromising the existing municipal supply works. **Water usage requirements should be estimated on the basis of 25 gallons per slip per day for recreational boats and 65 gallons per slip per day for commercial charter boat operations.** These numbers represent an estimate of peak day demands and may be seasonally adjusted for overall water supply volume demands. Local studies of water usage may be substituted for the guidelines if the studies provide adequate documentation.

The water supply lines should be consistent with local code requirements and classified as suitable for potable water by the National Sanitation Foundation or by other appropriately recognized bodies. Main water supply lines should be fitted with a backflow preventer in accordance with standard commercial plumbing practice. Individual marina hose bibbs should also be fitted with screw-on type backflow preventers at each hose bibb location. Shut off valves should be located and marked to provide a rapid means of isolating the supply line in the event of a pipe or fitting failure.

A separate water supply line may be required for fire fighting. This line should be a pressure rated pipe suitable for the intended application and dedicated for fire fighting purposes. When provided, 1.5 inch diameter standpipe outlets should be placed at about 200 foot intervals along the supply line. Standpipes should be galvanized steel or bronze, well braced and located such that the centerline of the discharge valve is 18 inches above the deck surface. The area around the fire hose connection valve should be sufficiently clear to allow the free swinging of connection wrenches. The discharge valve pipe should be protected by an end cap. The standpipes should be painted red. If ancillary seawater pumps and piping are used for fire fighting, they should be prominently marked as such and labelled as nonpotable water. Fire hose and playpipes should not be provided at the standpipes. The hose and attachments will generally be provided by the responding firefighting units or by properly trained marina personnel.

Sanitary Facilities and Wastewater Systems

An adequate number of sanitary facilities should be provided on the upland for use by marina tenants, visitors and marina personnel. The facilities should be constructed and maintained in a clean and sanitary fashion. **Sanitary facilities should be provided within 500 feet from the shore end of any pier.** Facilities may be provided in several locations to meet this provision. Provision of handicapped accessible toilets should include at least one toilet location convenient to the marina. All sanitary facilities should be well marked and readily identifiable.

Sewage pumpout facilities for boats should be provided at a minimum of one pumpout per 100 recreational slips or fraction thereof. The slip count may consider only the size of boats normally fitted with sewage holding tank capability. Runabouts and day sailers not having provision for fixed sewage holding tanks need not be included in the boat count for pumpout capacity.

Table 16-1 presents the suggested number of sanitary facilities appropriate for various size marinas. **Sewage generation, for marinas having only toilets and lavatories, should be calculated at the rate of 20 gallons per slip per day.**

Table 16-1. Recommended number of sanitary facilities for marinas of varying size.

No. of Seasonal Wet Slips	Toilets F	Toilets M	Urinals M	Lavatories F	Lavatories M	Showers F	Showers M	Pumpout Stations
0 - 50	1	1	1	1	1	0	0	1
51 - 100	2	1	1	1	1	1	1	1
101 - 150	3	2	2	2	2	2	2	2
151 - 200	4	2	2	3	2	2	2	2
201 - 250	5	3	3	4	3	3	3	3
251 - 300*	6	3	3	4	4	3	3	3

*For marinas exceeding 300 slips, increase the unit requirements by one unit per 100 additional slips.

This relationship holds for marinas, up to 100 slip capacity, also having showers. Above a capacity of 100 slips, marinas with toilets, lavatories, and showers should have sewage generation calculated at 32 gallons per slip per day. Slip usage is based on two persons per slip. **Marinas having trailer boat facilities or rack storage should have sewage generation calculated at the rate of 10 gallons per rack or trailer parking space per day. Charter boat operations should be sized on the basis of 10 gallons per person per day for the licensed capacity of the vessel plus crew.** If motel accommodations, restaurants or shopping malls are incorporated into the marina, appropriate code values of sewage generation should be added to the marina contribution to obtain a total sewage flow estimate. **If water saving devices are incorporated into the sanitary facility construction, sewage flow estimates may be reduced by the ratio of water savings versus standard fixtures as stated by the manufacturer's test literature. If boat pumpout sewage is not added to the primary sewage disposal system the total quantity of sewage flow from the marina operation may be reduced by 10 percent.** Handling of boat pumpout sewage must be defined.

Boat pumpout facilities should have the capacity to remove sewage from on-board holding tanks. The pumpouts may be portable or fixed, and if fixed, located to provide easy access to users. The pumpouts should be available during normal marina operation hours. Minimum pumpout equipment should include a pump with a capacity of a minimum of 10 gallons per minute or of a size to adequately handle the volume of flow associated with the size of the facility. The pump should be capable of passing 1.5 inch diameter solids or be capable of macerating the solids prior to ingestion by the pump assembly. The pump should be of an anticlogging variety. The pump should be able to pump against the maximum head developed by elevation change and line losses. The suction connection to the boat should be a tight fit and adjustable by adapters to service boat discharge connections between 1.5 inch and 4 inch diameter. The suction end should be fitted with a one way check valve. The suction hose should be flexible, nonkinking, noncollapsible, and of a heavy duty material of the shortest length suitable for the intended service. The pump should be valved to prevent sewage discharge during any pump repair or service operation. Discharge piping should be rigid or noncollapsing flexible, with locking connections. The line should be watertight and appropriately fastened or secured to the dock or pier. The discharge line connection to upland holding tanks or municipal sewerage systems should be made in accordance with the best practice to ensure a tight and safe discharge that will not compromise public health. Washdown water should be supplied at each boat pumpout location. Potable water hose bibbs should be fitted with a back flow preventer and the hose bibb and any associated hoses clearly marked as not being for use as a drinking water source.

If boat pumpout sewage is to be kept in holding tanks for offsite disposal, the holding tanks should be sized appropriately for the volume of sewage generated and the frequency of removal of material from the holding tank. **Generally, a 1,500 gallon holding tank can serve up to 100 slips.** The tank should be of watertight construction and if buried in the ground, provision should be made to prevent buoyant uplift from ground water or tidal influence. The tank material should be suitably protected from corrosion or be constructed of corrosion resistant materials. The tank should be capable of being completely emptied and should be appropriately vented. Suitable fixtures should be incorporated into the tank to allow spill free discharge into and removal from the tank. Incorporated into the tank design, or appropriately located elsewhere, should be provision for emptying the contents of portable boat holding tanks. This device should be designed for portable tank contents disposal, with a suitable opening for discharge of contents and a tight cover. The unit should be equipped with a washdown system to allow cleaning of the portable unit. The washdown system should be clearly marked as unfit for drinking water.

Solid Waste

The proper disposal of solid waste should be encouraged by placement of trash receptacles in convenient areas in a reasonable number. Generally as a minimum, a trash receptacle should be placed at the shore end of each pier and in any area where people tend to gather such as a ship store, haul-out facility, launching ramp, etc. Container type dumpsters should also be provided to allow for the frequent emptying of smaller trash receptacles. The marina should be posted to advise users that overboard discharge of solid waste in the marina or waterways is prohibited. **Solid waste generation is estimated at 3 pounds per slip per day.** The project should demonstrate that solid waste disposal will not compromise existing solid waste disposal areas in the community or elsewhere.

An example of solid waste dumpster requirements was observed in a 210 berth marina where solid waste is collected in five, 2 cubic yard dumpsters which are emptied twice a week during the summer boating season. In winter, three dumpsters are used and emptied once a week. These numbers may be helpful in establishing initial solid waste loading, but may later be modified to suit the particular needs of the facility.

Waste Oil

The marina should provide a suitable waste oil container for disposal of used engine oil, etc. The container should be designed to allow easy placement of the waste oil into the tank and the fill pipe should be fitted with a tight cover. The container should have a tray or other structure to contain any overspillage resulting from the oil transfer operation. **In general, a 250-275 gallon waste oil tank will provide a suitable waste oil capacity for up to 150 boats.** During seasonal decommissioning and commissioning the tank may require frequent emptying. The waste oil should be disposed of in an approved manner and may be required to be handled as a hazardous material. Waste oil containers should be clearly marked for their use and location. Additional information on waste oil tanks may be found in Chapter 19, Section 19.9.

16.2 SIMPLE DOCK STRUCTURE ANALYSIS

The basic structures associated with marinas are floating docks and fixed piers. The design of these structures requires a knowledge of basic engineering design principles and physics. Floating docks are constructed of many different materials but all types must utilize similar design principles. The basic principles include calculating the dead load of the structure, calculating the residual freeboard at dead load and live load, calculating the structural strength of individual structural components and assessing the structure for exterior applied loads. To explain some of the basic principles, an example is presented of the design of a simple wood frame dock on polyethylene flotation. The principles used may be applied to other forms of construction and to differing materials.

Floating Dock Design Example

The problem under consideration is to design a basic wood frame floating dock having a dead load freeboard between 20 and 24 inches and a live load freeboard capacity of 25 pounds per square foot at a minimum freeboard of 12 inches.

Because of the desire to construct in wood and a desire to minimize splices in the wood, the float frame size is limited by the size of wood available for construction. A float size that meets the available wood criteria is 8 feet 6 inches wide by 20 feet long. These floats will be joined by heavy duty, galvanized steel hardware to provide a continuous but flexible main walkway unit. The float frame will be wood and be designed to have the deck planking run lengthwise to conserve materials and reduce fabrication labor. Wood products, in the United States, are generally readily available to lengths of 24 feet. Beyond this length the wood may be a custom order or the selection of material may be limited. Therefore it is important to know the availability of materials before beginning a float design. Labor is usually a high cost item, so it is important to design the structure to minimize cutting and fitting. Standard lengths of material are used, where possible, to avoid material waste.

The use of the float system, in the example, is specified to be in a protected harbor and to be used for traditional light recreational marina loading. This information tells us that the float needs to be strong, but it will not be necessary to have it resist high sea states or other high loading criteria. The structure will be designed as a medium duty structure.

From experience, it is known that a float will generally be subjected to its greatest loading in the lengthwise direction. Therefore the float will be framed in the longitudinal direction with interior transverse members to support both the flotation units and the deck plank. Selected wood members are: $2'' \times 12''$ members for the basic dock frame, and $2'' \times 6''$ members for deck planking. In order to provide sufficient longitudinal strength the side stringers will be doubled-up to create a member of $4'' \times 12''$. A single $4'' \times 12''$ could be used if the material is readily available at a competitive price, but large, single wood members have a tendency to warp. Galvanized steel connection hardware will be used to join the wood members. The connection hardware is very important to the success of the design. If wood members are not well fastened together, the dynamic forces on the structure will soon compromise the holding power of the fastenings and the float will begin to work itself apart. The basic float frame will be developed to appear as shown in Figure 16-1.

Once the general member layout has been established, the dead load weight of the float unit can be calculated. This is done by tabulating the

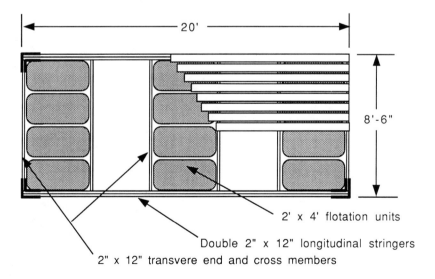

Figure 16-1. Plan view of wood floating dock used for design analysis example.

materials to be used and estimating the weight of each component. The estimated weight of the wood members per unit length can be used, or a more accurate weight can be achieved by weighing a sample piece to obtain a representative unit weight.

Dead Load Calculation:

Member tabulation: 4 pieces: $2'' \times 12'' \times 20'$ @ 4.1 plf = 328.0 lbs.
6 pieces: $2'' \times 12'' \times 8'$ @ 4.1 plf = 196.8 lbs.
17 pieces: $2'' \times 6'' \times 20'$ @ 2.0 plf = 680.0 lbs.
Total estimated weight wood = 1,204.8 lbs.

Hardware Weight: From experience, the estimated weight of hardware for this type of float is 1 pound of hardware per square foot of float or $8.5' \times 20' \times 1$ psf = 170 lbs of hardware per float unit.

The dock system will not have utilities or special deck equipment so the total of the wood weight and the hardware will be the dead load. It may be necessary to add the weight of the floatation units if the flotation unit manufacturer's load chart does not account for the flotation unit weight. The total dead load, in the example calculation is 1,204.8 lbs. plus 170 lbs. equals 1,374.8 lbs., or about 8 per square foot of float.

A foam filled, polyethylene shell, flotation unit has been selected that is approximately $2' \times 4'$ in plan dimension and has a side depth of 13.25 inches below where the flotation unit connects to the wood frame (see Fig. 16-2). From the float frame layout it is determined that 12 flotation units in 3 rows of 4 floats can be installed in each float. Now it is necessary to determine if

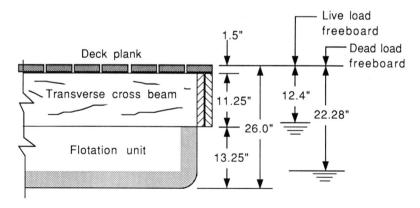

Figure 16-2. Transverse cross section of a floating dock at corner, for load capacity analysis.

the preliminary design will meet the specified freeboard criteria. The total dead load is estimated to be 1,374.8 lbs. Dividing this weight by 12 flotation units yields a unit load per flotation unit of 114.57 lbs. The flotation manufacturer has provided information that indicates that the flotation unit can provide 30.8 pounds of flotation per inch of submergence. Dividing the individual flotation unit load of 114.57 lbs. by 30.8 lbs./inch submergence yields a flotation draft of 3.72 inches. The total depth of the dock from the deck level to the bottom of the flotation unit is 26 inches. If the draft of flotation unit submergence, 3.72 inches, is subtracted from the 26 inch depth, a dead load freeboard of 22.28 inches is obtained, which is within the specified criteria for dead load freeboard of between 20 and 24 inches. If the calculated freeboard did not meet the specified criteria it would be necessary to adjust the amount of flotation provided to obtain the desired freeboard.

Now the ability of the preliminary design to meet the live load capacity and live load freeboard criteria must be determined. The live load is specified as 25 pounds per square foot of dock surface. The float is 8.5 feet by 20 feet so the dock plan area is 170 square feet times 25 pounds per square foot which equals 4,250 pounds of live load per dock unit. To the live load, the dead load is added to develop the total load, which equals 4,250 pounds plus 1,374.8 pounds equals 5,624.8 pounds. The preliminary design calls for using 12 flotation units per dock, so the load per flotation unit becomes 5,624.8 divided by 12 or 468.7 pounds per unit. The flotation unit manufacturer's literature indicates that as the unit submerges it increases its per inch buoyancy due to the tapered shape of the flotation unit. The average buoyancy at a near maximum draft is shown to be 34.5 pounds per inch of displacement. Therefore 468.7 pounds divided by 34.5 pounds per inch equals 13.6 inches of flotation unit submergence to float the dock structure with full dead and live load. The total depth of the float structure is 26 inches, so the draft of 13.6 inches subtracted from the 26 inch depth equals 12.4 inches of freeboard which satisfies the minimum criteria of 12 inches of freeboard at full load. The design works! Keep in mind that extra flotation may have to be added in areas of increased external load such as at gangway landings.

The dock structure has been analyzed for buoyancy requirements and now must be analyzed for structural frame adequacy. For the deck structure analysis a live load criteria of 50 pounds per square foot will be used even though the buoyancy live load criteria is only 25 pounds per square foot. The higher frame live load is used to assure that the structural frame will not fail before full dock submergence is achieved. The first frame component to be analyzed is the deck plank, which is a nominal $2'' \times 6''$ wood member. The wood is southern yellow pine, graded as No. 1 and planed on all four sides (S4S). Southern yellow pine, No. 1 grade, surfaced dry, at 19 percent moisture content has an allowable bending stress of 1,450 pounds per square inch (psi). The section modulus (S) of a $2'' \times 6''$ plank, planed to $1.5'' \times 5.5''$ equals 2.06 in.³. The maximum allowable bending moment (M) may be determined by multiplying the section modulus (S) by the allowable bending stress (F_b), or $M = S \times F_b$. Inserting the appropriate numbers yields, $M_{allowable} = 2.06$ in.³ $\times$ 1,450 psi = 2,987 in.-lbs. Now calculate the actual applied load and compare it to the allowable load. The transverse framing members are 4 feet apart so the span of the deck plank is about 4 feet. The bending moment formula for a beam in simple bending without end support is $M = \frac{1}{8} WL^2$, where W is the uniformly applied load (50 psf or 25 plf for a 6 inch wide plank) and L is the span. Substituting with numbers yields $M = \frac{1}{8}$ (25 plf live load) $\times 4^2 = 50$ ft-lbs. or 600 in.-lbs. which is less than 2,987 in.-lbs. allowable. The deck plank meets the allowable stress criteria for bending. The deflection of the deck plank under the design live load of 25 plf may be determined from the formula for deflection (Δ) which is: $\Delta = 5 (wl^4)/384 EI$, where $w =$ the uniformly applied load in lbs. per inch (25 plf equals = 2.08 pounds per linear inch), l = the span in inches ($4' = 48''$), $E =$ the modulus of elasticity of wood (1,700,000 psi), and $I =$ the moment of inertia of the plank section. The moment of inertia for the $2'' \times 6''$ plank equals $\frac{1}{12} bd^3$, where $b = 5.5'$ and $d = 1.5''$, $I = \frac{1}{12} (5.5 \times 1.5^3) = 1.55$ in.⁴. Therefore the deflection $\Delta = 5 (2.08 \times 48^4)/384 (1,700,000 \times 1.55) = 0.054$ inches. The allowable deflection based on the formula of $\Delta_{allowable} = L/200 = 48''/200 = 0.24$ inches which is greater than 0.054 inches. The deflection is within allowable limits.

It is also important to analyze a single, concentrated load on the deck plank. A realistic concentrated load of a person's weight of 180 pounds standing in the center of one plank will be applied to see if the plank is strong enough. The bending moment formula for a concentrated load at the center of the span is: $M = Pl/4$, where P is the concentrated load of 180 pounds. Therefore the moment $M = (180$ lbs. $\times 4')/4 = 180$ ft-lbs. The bending stress f_b, associated with this load equals 180 ft-lbs. $\times$ 12 in./ft divided by the section modulus (S) of 2.06 in.³ equals 1,049 psi, which is less than the allowable 1,450 psi. The deflection may be calculated from the formula $\Delta = Pl^3/48EI$. Substituting with numbers yields, $\Delta = (180$ lbs. $\times 48^3)/48 \times 1,700,000$ psi $\times 1.55$ in.⁴ $= 0.16$ inch, which is less than $L/200 = 0.24$ inch.

The deck plank should also be checked for its resistance to shearing from the applied loads. The shear area of the deck plank equals its cross sectional area of $5.5'' \times 1.5'' = 8.25$ in.². The allowable shear stress for the wood is 100 psi. The shear load equals 25 pounds per linear foot times the span of 4 feet or 100 pounds divided by two supports equals 50 pounds shear at each support. The allowable shear load equals 8.25 in.² $\times$ 100 psi = 825 pounds which is greater than the actual 50 pound shear load. The plank is satisfactory in shear.

DESIGN AND CONSTRUCTION 369

With the deck plank determined to be suitable, the transverse cross tie should be analyzed. The cross tie is a 2" × 12" plank on edge, running between the side stringers. The loaded area for the interior cross ties will be a 4 foot strip, 2 feet on each side of the beam. At a structure design load of 50 psf plus an additional dead load of 5 psf for the decking weight, the uniformly distributed load will be 55 psf × 4 ft or 220 plf. The bending moment formula is $M = \frac{1}{8} wl^2$. Therefore $M = (220 \times 8.5^2)/8 = 1,987$ ft-lbs. The section modulus of a 2" × 12" plank on edge is 31.64 in.^{3}. The bending stress f_b equals the bending moment divided by the section modulus or 1,987 ft-lbs × 12 in./ft/31.64 in.^{3} = 753.6 psi, which is less than 1,450 psi, satisfactory.

The shear load is 220 plf times the span of 8.5 feet divided by two supports equals 935 pounds. The shear area of the plank is 1.5" × 11.25" = 16.88 in.^{2}. The shear stress therefore equals 935 lbs./16.88 in.^{2} = 55.39 psi, which is less than the allowable 100 psi. The shear and bending analysis is conservative because in reality the flotation units will provide uplift resistance on the beam and reduce the induced stress.

The composite float assembly should be analyzed in bending. There are a number of analyses that might be used to determine the adequacy of the structure in bending. The method selected to analyze the system is to consider the 20 foot long float unit as only being supported at each end. The analogy is that of a beam, simply supported on the ends, and having a load equal to 50 psf on the deck plus a dead load of 8 psf for a total load of 58 psf (Fig. 16-3). Again the analysis is conservative in that the flotation units would, in a wave condition, produce an uplift at the ends which would reduce bending and shear stresses. In order to perform this analysis it is necessary to calculate the section modulus of the composite float frame

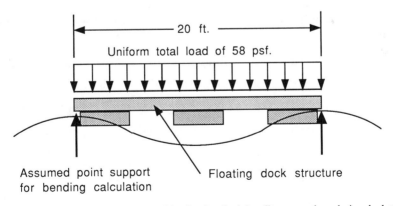

Figure 16-3. Simple beam analogy used for floating dock bending strength analysis calculation.

section. This is accomplished by using a standard engineering section modulus calculation with transfer effect. The basic frame components for this calculation will include one deck plank working in combination with two 2" × 12" longitudinal stringers. The transverse members will not add to the longitudinal strength. It is easiest to organize the calculation as shown in Table 16-2. Since the top deck will become a compressive zone in the example, the appropriate section modulus to use is $S = 109.02$ in.[3] for calculating the bending stress. First calculate the bending moment: $M = \frac{1}{8} wl^2$, where w equals the total uniform load calculated for one set of side stringers having a contributory load width of $8.5'/2 = 4.25$ ft. The dead load equals 8 psf, so 8 psf times 4.25 ft. equals 34 pounds per linear foot. Plus a live load of 50 psf times a width of 4.25 ft. equals 212.5 pounds per linear foot, for a total load of 212.5 plf plus 34 plf equals 246.5 plf. The bending moment therefore equals $M = \frac{1}{8}$ (246.5 plf) $\times 20^2 = 12,325$ ft-lbs. The compressive bending stress at the top fiber then equals, $F_b = M/S$, = 12,325 ft-lbs. × 12 in./ft./109.02 in.[3], = 1,357 psi which is less than the allowable bending stress for the wood of 1,450 psi, which may be increased by a factor of 1.33 because the induced load is caused by a temporary condition, in this case, wind/wave action. The resulting allowable stress is 1,450 psi × 1.33 = 1,929 psi. The design is satisfactory in bending.

Another case for bending consideration occurs when the water wave has moved ahead and now the float is supported only in the center (see Fig. 16-4). The uniform load will be the same as above, or 246.5 pounds per linear foot. The section modulus will now be 93.10 in.[3], reflective of the change in support mechanism, with the bottom of the float deck in compression. Each

Table 16-2. Calculation of section modulus properties for composite floating dock section. Section modulus is used to determine dock bending stress.

Piece	Area	Y dist.	Ay	Y_c	Ay_c^2	I_o
2" x 6" decking	8.5 in²	12 in	99 in³	5.12 in	216.51 in⁴	1.54 in⁴
2 - 2" x 12" stringers	33.75 in²	5.625 in	189.84 in³	1.25 in	66.26 in⁴	355.99 in⁴
	42.00 in²		288.84 in³		288.84 in⁴	357.51 in⁴

$y_c = 288.84$ in³ / 42 in² = 6.877 in. $I_t = Ay_c^2 + I_o = 640.28$ in⁴

Section modulus (S) = 640.28 in⁴ / 6.877 in. = 93.10 in³ @ bottom fiber

640.28 in⁴ / (12.75 - 6.877 in) = 109.02 @ top fiber

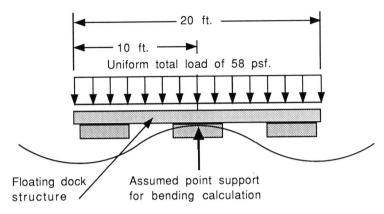

Figure 16-4. Simple cantilever beam analogy used for floating dock bending strength analysis calculation.

half of the linear length of the float will be treated as a cantilever beam, bending about a fixed point at the center. The new bending moment equals $M = \frac{1}{2} w(l)^2$, where $w = 246.5$ plf and l, the span, now equals one half the dock length or 10 feet. $M = \frac{1}{2} \times 246.5$ plf $\times 10^2 = 12{,}325$ ft-lbs. The bending stress equals, $f_b = M/S$, $f_b = 12{,}325$ ft.-lbs. $\times 12$ in./ft/93.10 in.$^3 = 1{,}589$ psi, which is less than the allowable stress of 1,929 psi, as developed above. Again the design stresses are below the allowable stresses and the design is satisfactory.

Now check the shearing stress in the double $2'' \times 12''$ side stringer for the total applied load of 246.5 plf. The shear area of the double $2'' \times 12''$ member is 33.75 in.^{2}. The load P, at each support is equal to the uniform load, 246.5 plf times 20 feet divided by 2 supports. P therefore equals, 246.5 plf $\times$ 20 ft/2 = 2,465 pounds. Simple shear is expressed as P/A, or 2,465 lbs/33.75 in.$^2 = 73$ psi, satisfactory, less than an allowable shear stress of 100 psi.

The above calculations represent a very basic analysis of structural considerations for a simple marina main float. More complex structures and structures subject to severe sea state conditions may require additional analysis to investigate torsion, fatigue, impact loading and internal load transfer. Each structural system must be analyzed for the specific prevailing conditions, structure geometry and types of materials employed.

Now continue the analysis by considering a simple finger float having a width of 4 feet 3 inches and length of 26 feet. The width of the finger was established to provide significant stability and to accommodate a 4 foot wide flotation unit. The finger will be designed to resist the imposed loads without use of knee (corner) braces.

As with the main float, the finger will be framed longitudinally, with double $2'' \times 12''$ side stringers. Transverse members will be added to keep the stringers separated and to act as an attachment location for the flotation units. Again, for simplicity, the finger is treated as a simple beam supported at each end. The loading will be the same as for the main float, or 58 pounds per square foot. The effective load width is one half the finger width for each set of stringers, so the uniform applied load (w) equals 58 psf times 4.25 ft divided by 2 (one half the width) equals 123 pounds per linear foot of stringer length. The bending moment for a simple beam supported at the ends is $M = \frac{1}{8} w (l)^2$, therefore $M = \frac{1}{8}$ (123 plf) (26 ft.)^2 = 10,394 ft-lbs. The bending stress, $f_b = M/S$, = 10,394 ft-lbs × 12 in./ft/109.02 in.^{3} = 1,144 psi, less than 1,929 psi allowable, satisfactory. Treating the finger as a cantilever beam supported at the midpoint provides a moment $M = \frac{1}{2} (w) (l)^2$, where l now equals 26 ft/2 or 13.5 ft. $M = \frac{1}{2}$ (123 plf) (13)^2 = 10,394 ft-lbs. The bending stress $f_b = M/S =$ 10,394 ft-lbs × 12 in./ft/93.10 in.^{3} = 1,340 psi, less than the allowable 1,929 psi, the design is satisfactory.

Since the design load per linear foot for the finger is 123 plf and for the main float it is 246.5 plf, and the structure is the same, the values of shear stress for the finger will be less than the main float and are satisfactory.

This particular project requires consideration of the application of a 4,500 pound load transmitted to the outboard end of the finger float (see Fig. 16-5). The finger will be analyzed as a short column to see if the finger can accept the applied load. The double side stringers act together as a column in resisting the applied compression load. The side stringers are stabilized by the finger deck planks and the attached flotation. The column length (l) is 26 feet or 312 inches, the area of the two, double stringers is 67.52 square inches and the least dimension for stability is the depth of the 2" × 12" stringer, or

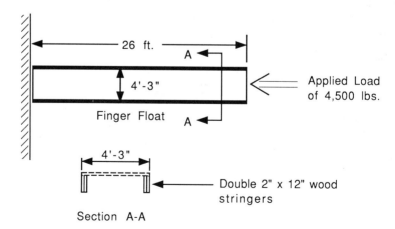

Figure 16-5. Plan view and cross section of 4'-3" × 26' finger float.

DESIGN AND CONSTRUCTION 373

11.25 inches. First, check to see if the composite section can be considered as a short column by seeing if the length (l) divided by the depth(d) is equal to, less than or greater than 80, for a main member. Therefore, $l/d = 312$ in./ 11.25 in. $= 27.8$, less than 80, satisfactory. The modulus of elasticity (E) for the wood is taken as 1,700,000 psi. The formula for the allowable stress in this column is $F_c = 0.30E/(l/d)^2$. Substituting numbers yields: $F_c = 0.30 (1,700,000)/(27.8)^2 = 660$ psi. The actual stress $f_c = P/A_t, f_c = 4,500$ lbs/67.52 in.$^2 = 67$ psi, less than 660 psi, satisfactory. The allowable stress could be increased by a factor of 2, because the applied load is an impact load and considered temporary.

The analysis must also consider a horizontal load on the finger as developed by the berthed boats (see Fig. 16-6). The example uses a 30 foot long boat, having an equivalent profile height of 3.8 feet, with a wind pressure of 20 psf. The full boat load then equals: 30 ft × 3.8 ft × 20 psf equals 2,280 pounds. Consider that the vessel on the lee side is shielded by the windward

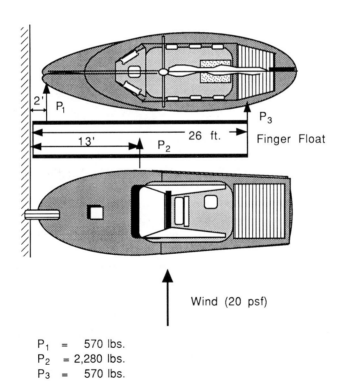

P_1 = 570 lbs.
P_2 = 2,280 lbs.
P_3 = 570 lbs.

Figure 16-6. Plan view of 4'-3" × 26' finger float with berthed boats for determination of lateral wind load stress.

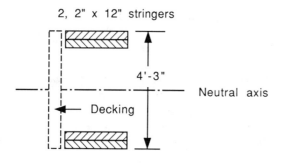

Pc	Area	Y	Ay	Ay^2	I$_o$
▭	33.75 in^2	24 in.	810 in^3	19,400 in^4	25 in^4
▭	33.75 in^2	24 in.	810 in^3	19,400 in^4	25 in^4
				38,886 in^4	50 in^4

$$I_t = 38,930 \text{ in}^4$$

Figure 16-7. Calculation for finger float section modulus treating the finger as a composite beam. Finger beam shown rotated 90 degrees to simplify the calculation.

boat and has a load of 50 percent (1,140 lbs) of the boat on the windward side. Assume that the full load of the windward side boat is applied to the finger near the vessel midship point and that the vessel on the lee side will be pulling on mooring lines attached to each end of the finger. From the loading shown in Figure 16-6 above, an equation for the bending moment may be developed by taking individual moments of the loads about a point at the line of connection of the finger with the main float. Then $\Sigma M = P_3(2') + P_1(15') + P_2(26') = 570$ lbs $(2') + 2,280$ lbs $(15') + 570$ lbs $(26') = 50,160$ ft-lbs. The section modulus for the composite beam, as shown in Figure 16-7, equals 1,527 in.^{3}, so the bending stress $f_b = M/S$, or $f_b = 50,160$ ft-lbs $\times$ 12 in./ft/ 1,527 in.^{3} = 394 psi, which is less than 1,450 psi allowable, satisfactory.

It is also necessary to perform an analysis for a vessel striking the end of the finger as an impact load. The criteria is a 30 foot boat approaching at a 10 degree angle and at a speed of 2 knots (see Fig. 16-8). The velocity (V) equals 2 knots times 1.69 to get velocity in feet per second, which equals 3.38 ft/sec. The weight (displacement) of the vessel may be determined by a general formula which states that the weight (W) equals 12(length)^2. In this case, the weight equals, 12 (30)^2 = 10,800 lbs. It is necessary to resolve the component

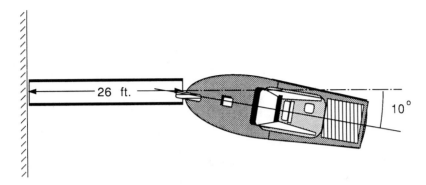

Figure 16-8. Angle of impact of 30 foot boat on finger float.

of the impact force by applying the formula that velocity, $v = V \sin a$, where a equals 10 degrees. Therefore $v = 3.38$ ft/sec. sin 10 degrees, $= 0.59$ ft/sec. The modulus of elasticity (E) of wood will be considered to be 1,700,00 psi. Determine the kinetic energy (KE). $KE = Wv^2/2g$, where g is the acceleration of gravity (32.2 ft/sec.^{2}. KE then equals, 10,800 lbs (0.59 ft/sec.)^2/2(32.2 ft/sec^2) = 58.38 ft-lbs.

The finger will be analyzed as a cantilever beam with an applied load of 1,000 lbs, or whatever unit value, and then superimpose its load on the previously calculated moment to obtain the bending stress. The deflection (Δ) associated with a 1 kip (1,000 lbs) load equals $Pl^3/3EI$, where P equals the unit load, l equals the length in inches (26 ft × 12 in./ft = 312 in.), E equals the modulus of elasticity (1,700,000 psi) and I equals the moment of inertia of the structure (38,930 in.^{4} from previous analysis). Substituting numbers, yields: $\Delta = 1^k$ (312 in.)^3/3 (1,700 ksi) (38,930 in.^{4}) = 0.153 in. The deflection, Δ, also equals the square root of the kinetic energy divided by a stiffness coefficient, k. The stiffness coefficient, k, may be determined by dividing the unit load by the deflection calculated above, or $k = P/\Delta$, = $1^k/0.153$ in. = 6.54 kips/in. For $\Delta = (KE/k)^{1/2}$, substituting numbers yields, Δ = (0.058 ft-kips × 12 in./ft/6.54 kips/in.)$^{1/2}$ = 0.33 in. Therefore the load P caused by the impact of the boat equals 0.33 in. × 6.54 kips/in. or 2.13 kips. The applied moment can be calculated by saying $M = P ×$ distance = 2.13 kips times 26 feet equals 55.5 ft.-kips. The bending stress then becomes f_b = (55.5 ft kips × 1,000 lbs/kip × 12 in./ft)/1,527 in.^{3}, = 436 psi, which is less than 1,450 psi, satisfactory.

The calculations described above provide the basic analysis needed to establish the structural integrity of the float system. Additional calculations may be necessary for complex structures and structures subjected to an aggressive sea state.

16.3 DOCK HARDWARE

The calculations presented in Section 16.2 provide an analysis for determining overall structural strength. It is very important that dock hardware also be analyzed for structural strength and ability to transmit the imposed external forces. Each dock system will have different design parameters so providing an example analysis may not be of significant value. A critical area of concern in dock hardware design is in the connection of floats, one to the other. As a general rule of thumb, the interconnections should have the same strength as the main dock units or be provided with a stress relieving mechanism, such as a hinge. If hinges are used, they must be sufficiently strong to transmit the forces and be designed for wear, fatigue and galling. The hinge should also be designed to provide a constant plane between the dock surface with no vertical play. They should be located as far apart as possible to reduce the forces in the load couple and be rigidly attached to the float, with backup material if necessary. In general, the hinge should be designed to transmit the tensile and compressive loads to longitudinal members in the dock structure, rather than to a transverse end member alone. Hinges should be protected from corrosion and material degradation. Any space created by the hinge, between docks, should be kept to a minimum, say 1 inch, or the dock should be provided with a cover plate to prevent a gap large enough to cause personal injury.

Cleats must be well designed and constructed. They must be fastened securely to the dock frame, not just to the decking. In general, cleats should be through bolted and never just lag bolted to wood members. More research is needed to define cleat loads and cleat testing procedures. One common fault in cleat design is that the diameter of the cleat horns are insufficient for the line normally fastened to the cleat. Rope manufacturers suggest that the diameter of a cleat horn be not less than twice the diameter of the line that passes around the cleat. Smaller diameter cleat horns will allow intense compressive stress buildup in the line as it passes around a small radius corner, with increased potential for premature line failure. Numerous instances of mooring line failure associated with small diameter cleat horns has been observed in marinas after major storms. An even larger problem is that the cleats on the boats themselves are often totally inadequate for the intended use, but that is a subject beyond this book.

As mentioned in reference to cleat attachment, lag bolting in wood is generally unsuitable. Timber should be fastened by through bolting, with washers placed at the bolt head and nut when they are in direct contact with wood. Bolts should be of a proper length, with a minimum protrusion of the bolt shank beyond the nut. A minimum of three bolt threads should project beyond the nut, but no more than the thickness of a nut, whenever possible. Bolt ends should be recessed flush into the wood wherever the possibility of damage or injury may result from bolt projection beyond the surface of the wood.

Smooth, round shank nails should be avoided for wood fastening, as they tend to work loose very quickly. Nails may be coated or have a roughened surface to provide good holding power. Nails with annual rings or other surface texture are satisfactory for fastening decking but should not be used for dock connection or fabrication. Hot dipped galvanized nails also work effectively for deck fastening, because the dipping process provides a rough surface which is quite resistant to pullout. Be careful of power driven nails because they are generally quite smooth to allow for an orderly flow through the nail gun mechanism. Most galvanized power driven nails are electroplated rather than hot dipped and tend to be too smooth for marine use. Stainless steel power driven nails appear satisfactory for deck fastening. It is good business to specify a quality nail and make sure it is used in the dock fabrication since loose decking and renailing can become a maintenance nightmare if poor quality materials are used.

16.4 DOCK UTILITIES

Domestic Water Distribution Systems

The primary dock utility, common to most marinas, is water. There are two types of water distribution systems: domestic water piped to the individual berths for use on the boats and a larger water main used for fire protection. The domestic water should feed from an adequate upland source, either on-site or municipal. In most cases the water supply will be municipal and regulations governing its use and installation must be followed. Prior to feeding a dock distribution system, the main water supply should be fitted with a main shut-off valve. In some marinas, in colder climates, the main shut off may also have a line drain provided to allow some drainage of the dock distribution line. At this location, there should also be a primary backflow preventer valve to ensure that no back flow of water from the dock system can enter the main water line. It also may be necessary to provide a pressure reducing valve if the municipal pressure is excessive. **Most boat plumbing systems are designed to operate at a pressure between 30 and 40 pounds per square inch (psi). The lightweight plumbing found on most boats cannot tolerate pressures of 75 to 100 psi which are sometimes found in municipal systems.** On the other hand, if the municipal water pressure is too low, it may be necessary to provide a pressure booster pump to provide adequate pressure at distant locations on the docks.

From the land, there is usually some form of flexible piping to the dock system. The pipe should be smooth bore and pressure rated. The flexible pipe often is hung on the underside of access gangways or ramps. Be sure the gangway or ramp has the capacity to carry the extra load resulting from a fully loaded pipe. The piping will be subject to ultraviolet rays and so should be rated for this type of exposure, or rapid deterioration may result. The type of piping used on dock systems will vary and may be subject to approval by the appropriate authorities. In general, it is wise to avoid hard piping (copper) if the dock is a floating system and subject to movement. The rigid piping is highly subject to cracking and joint failure from movement and thermal expansion. It is also often difficult to make good joints out over the water with the inevitable wind blowing. Plastic type pipes have been used with good success. The plastic pipe should be approved for use with potable water systems. In areas subject to freezing, the dock water piping system must have provision for positive drainage and/or fittings for blowing down the system in winter. Some marinas will charge the water line with a suitable antifreeze solution to prevent water line freeze up.

Water is provided to a berth either by a separate hose bibb (faucet) or by a hose bibb integrated into a power post assembly. If the water is incorporated into the power post, all water lines and fittings must be isolated from any electrical components. Individual hose bibbs should be fitted with screwon backflow preventers. The hose bibbs should be located so as to be convenient for connection of hoses and the filling of buckets and water bottles. The hose bibb should be located such that the stream of water is directed away from any electrical fittings and off the dock deck, if possible. Generally, a hose bibb is provided for each boat berth. Boatowner's hoses lying around on dock decks are dangerous and unsightly. It may be desirable to provide hose racks on the docks or dock boxes, or require boatowner's to store hoses when not in use.

Fire Protection Water Systems

Fire protection systems are becoming commonplace in modern marina installations. The requirements for provision of water distribution systems for fire fighting are perhaps the most varied of any utility furnished in a marina. Fire protection is most often the responsibility of the local fire department or fire marshall. There appears to be little consensus on the appropriate requirements for marinas and each fire department injects its own thinking and desires. Several important reference standards that may be helpful in designing fire protection systems are found in the National Fire Protection Association's publications NFPA 303, *Marinas and Boatyards,* NFPA 307, *Marine Terminals, Piers and Wharves,* and NFPA 312, *Fire Protection of Vessels during Construction, Repair and Lay-Up.* Some marina engineers are concerned that fighting boat fires with water may not only be generally ineffective but dangerous. Fuel fires that spread to the water surface may be expanded by a high pressure water stream, water applied to electrical fires may be dangerous, and large volumes of water sprayed on a boat may sink a boat that otherwise could have had the fire extinguished by less disastrous methods. The use of high pressure water systems by inexperienced persons is also a great danger. Volunteer fire companies will often have a summer picnic with an event in which a fully pressured fire line is set out in a field and two or more fire fighters rush out to subdue the whirling fire line, often with limited success. Imagine an unconstrained fire line whipping around a populated marina! Provision of fire protection is, however, necessary and the current recommendation is to provide suitable fire hydrants (standpipes) at locations such that a standpipe is within 500 feet of any point on the docks. The standpipe may be wet or dry, generally dry, and be fitted with appropriate shut off valves. The standpipe will be mounted through the dock deck with the hose connection between 18 inches and 3 feet off the deck. They are usually painted red and labelled as a fire hydrant. Because the fire main will be a high pressure line, it is important to provide adequate thrust blocks along the fire main at appropriate locations such as at changes in direction. Hose and nozzles are generally not provided, but rather supplied by the responding fire company. The fire lines should be tested periodically.

The use of foam to suppress class A (common combustible materials, wood, paper, plastic, fibers, rubber, etc.) and class B (flammable liquids and gases, fuels, paints, alcohol, propane, etc.) fires has become an important fire fighting technique in marine environments. Foam is made in conjunction with available fresh or salt water. It is an effective smothering agent and provides a secondary cooling aspect. Foam sets up a blanket or vapor barrier that prevents flammable vapors from rising and therefore can help isolate fires from neighboring fuel tanks, etc. It acts to prevent reignition by covering a material and absorbing heat. It can be used a a protective film on oil spills to prevent ignition. Because of the water content and therefore electrical conductivity, foam should not be used on live electrical equipment. It should also not be used on combustible-metal fires. Most foams are not compatible with dry chemical use, the exception being Aqueous Film-Forming Foam (AFFF). AFFF is an important type of foam often used on marine fires. The foam is created by mixing a water soluble surfactant with a stream of fresh or salt water. The resulting foam controls vaporization of flammable liquids by means of a water film that forms upon application of the foam. The thin water film prevents flammable vapors from reaching the flame. Since AFFF can be used with, before, or after the use of dry chemicals it is attractive for shipboard or marina fire suppression.

Another potential type of foam, that may be useful, is high-expansion foam. This foam is also generated by mixing a foam concentrate with water. An aspirating-type nozzle (air is introduced into the nozzle) is used to expand the foam in the ratio of about 100:1. This foam is usually used to suppress fires in confined spaces such as boat engine compartments, where its high expansion property fills the space and deprives the fire of necessary oxygen. By virtue of its water content, it also cools the burning materials and absorbs heat from the fire. If the fire is hot enough the foam will turn to steam which continues to cause oxygen deprivation and helps suppress the fire. Use of foam in a confined compartment requires evacuation of any personnel since, although not toxic itself, high-expansion foam deprives the space of oxygen and impairs vision and hearing.

The availability of foam fire suppression systems, mixing water, and trained personnel can provide an important safety feature in any marina facility.

In addition to a water mains system and foam application capability, and perhaps as important, is the selective placement of portable fire extinguishing equipment. Often use of the portable extinguisher can prevent the need for using the more dangerous high pressure water system. Marina personnel should be frequently trained in the use of fire fighting equipment.

Electrical Power Requirements

The design of the electrical power system for a marina is a complex task involving a number of variables and local codes. It is strongly urged that the design be performed by knowledgeable electrical consultants. Provision of dockside power is virtually a must in most cruising areas. Internationally electrical power characteristics will vary greatly, both in voltage and in cycles, therefore specific marinas will have to investigate the power characteristics common to their area.

In the United States electrical appliances generally are rated at 115 volts AC. The nominal power supply to service these appliances is 120 volts AC. Ratings of 110, 115 and 120 volts all fall within the 120 volt AC power supply category. Other supplied voltages may be 208, 220, 230 or 240 volts. A 120/208 volt, three phase power system can be used for equipment rated at 200 or 208 volts. Equipment rated at 230 or 240 volts should be supplied from a 120/240 volt single phase source. Equipment rated at 200 or 208 volts cannot be used with the 120/240 volt supply. However some 220 volt rated equipment may be operated from either a 208 or a 240 volt power supply.

In determining the marina power arrangement, the local utility company should be consulted to verify their power characteristics and the distribution system most compatible to the power company. Most power companies will have a three-phase electrical system and will balance the single phase demands on the three high voltage phases. It will generally be most cost effective to provide a power system that is compatible with the utility's distribution requirements. There are some practical limitations on loads for the different types of power service such as: 100 KVA or 400 ampere peak load on a 120/240 volt single phase, 3 wire service; or 500 KVA or 1,200 ampere peak load, three phase, 4 wire service. A knowledgeable analysis should be made of the marina's power needs and the utility's power characteristics to determine the most appropriate system at the least cost. A three phase system may result in considerable savings in electrical equipment and cable.

The basic electrical code used in the United States for determining electrical power distribution system requirements is the National Electrical Code (NEC), Article 555, Marinas and Boatyards. The code is revised from time to time so the latest edition should be consulted. The National Electrical Code presents guidance in selection of feeder cable and determination of the required capacity. Some care must be taken in determining cable size by not only considering the code's requirements for thermal loading but also the effect of voltage drop at the ends of long marina cable runs. Voltage drop is a major concern in marina electrical design due to the usual long runs of feeder cable to service distant berths. It is therefore important to analyze each marina as a discreet project and not buy into a prepackaged design that may not be suitable for the specific marina conditions.

Another area of concern in electrical systems design is the interpretation of the diversity factor, an allowance for the fact that all marina power receptacles will not be powered up at the same time. The unknowledgeable might assume that a power demand can be estimated by determining the number and rating of all the marina's power outlets and summing their total power requirements. In large marinas, the resulting number will be staggering. Instead, the code allows for a reduction in power demand as a percentage of the total rating by a sliding scale based on the number of berths. As an example, a marina having 1-4 receptacles would be required to provide power equivalent to 100 percent of the sum of the ratings of the receptacles, whereas for marinas having more than 100 receptacles, a value of 20 percent of the sum of all the receptacles may be used for calculating power requirements. A sliding scale is presented in the NEC code for marinas having between 0 and 100 receptacles. There also may be some ways to group receptacles to further modify the overall power requirements.

Modern marina design also considers using step-down transformers on the docks to reduce feeder size and minimize voltage drop. Care must be exercised in the selection of on dock transformers to assure their suitability for use in the harsh marine environment. Figure 16-9 shows a modern marina installation with transformers on the dock, on extensions to the dock system designed to carry the transformer load while not obstructing pedestrian flow on the docks. These transformers are good looking, safe, and cost effective.

The actual receptacle rating at individual berths will vary with the type and size of boats berthed. Some guidance is presented in Table 16-3 for electrical receptacle rating as general information to be customized for the specific marina.

Figure 16-9. Floating dock at a main access gangway showing modern on-dock electric power transformer, set on dock extension to avoid using space on main float walkway.

Table 16-3. General power requirements for different types and sizes of boats.

Type of Boat	Power Requirements at Power Pedestal
Sailboats and powerboats to 30 ft. in length	Single 30 amp, 125 v. receptacle
Sailboats and Powerboats 30ft. to 40 ft.	Two, 30 amp, 125 v. receptacles
Sailboats 40 ft. to 50 ft.	Two, 30 amp, 125 v. receptacles
Powerboats 40 ft. to 50 ft.	Two, 30 amp, 125 v. receptacles or one 50 amp, 125/240v. receptacle
All boats 50 ft. to 80 ft.	Two, 50 amp, 125/240 v. receptacles
All boats 80 ft. to 90 ft.	One, 100 amp, 120/240 v. single phase outlet
85 ft to 100 ft. +	Two, 100 amp, 120/240 v. single phase outlet
200 ft. +	One, 100 amp, 480 v. three phase outlet

Note: Some European built boats in the megayacht range, 90 ft. + may require one or two 100 amp, 120/208 v. three phase outlets.

DESIGN AND CONSTRUCTION 383

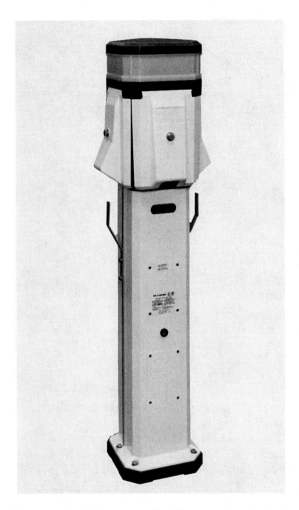

Figure 16-10. Modern marina dock power pedestal incorporating electrical outlets, low level lighting with automatic darkness sensor, cable brackets, and electrical metering capability. *(Photo courtesy: Charles Industries, Ltd., Marina Development Div.)*

The purpose of the electrical installation is to provide electrical power to the boat owner. A wide variety of power post pedestals (see Fig. 16-10 and Fig. 16-11), are available in today's market. It is important to purchase a quality power pedestal for installation ease and long term minimal maintenance. Power pedestals also can provide receptacles for cable television and telephone as well as water service outlets. They also may be purchased with

Figure 16-11. Installed marina utility power pedestal showing its location on dock in relation to two boats which it is serving. It incorporates a potable water hose faucet, CATV and telephone service outlets.

integral low level lighting to illuminate the dock walking surface without blinding maneuvering boats.

Telephone service to the individual berths has been a desired amenity in many boating areas. The rapid expansion of cellular telephone service in the United States and some areas world wide may soon eliminate the need for provision of hard wired telephone service. The cellular phone service allows the boat owner to have telephone service while underway as well as at the berth, and such service appears to be gaining in favor over telephones limited to at-berth use. If hard wired telephone service is provided on the docks, adequate space in the dock utility raceway should be provided. Some separation may be required between main electrical power and telephone feeder cables. It may also be necessary to provide telephone company distribution panels on the docks and a main panel on shore. Consult the telephone company early in the utility design to ascertain their requirements.

Cable television (CATV) is a highly desired amenity in many modern marinas. The actual CATV installation is usually performed by the cable company serving the area. The cable company should also be consulted early, and their requirements incorporated into the utility design package. Cable TV may require local panel boards on the docks, for signal boosting or interference rejection. All of these requirements can usually be easily worked into the dock design package if determined early in the design process.

16.5 DREDGING

The need for new and maintenance dredging is vital to the survival of many marinas and other harbor facilities. Dredging involves the removal of bottom sediments to provide adequate water depth to safely allow the passage or berthing of commercial and recreational vessels. The material removed must be appropriately relocated. It is the relocation or disposal of the dredged material that often creates the problem in obtaining the requisite approvals for a dredging project. A primary reason for the difficulty in obtaining dredging approvals is the inappropriate characterization of all dredge materials as hazardous wastes or toxic sludge. Although there certainly are areas where bottom sediments contain high concentrations of hazardous or toxic constituents, a study by the United States government indicates that 90 percent of dredged material in the United States is acceptable for disposal in the aquatic environment. The most active dredging organization in the United States is the U.S. Army Corps of Engineers, the same agency that must review and rule on the appropriateness of private dredging projects. The Corps of Engineers is mandated by the Congress to provide safe and accessible navigation routes for commercial maritime traffic and therefore it is empowered to dredge and maintain navigable waterways in the United States. Although sometimes an ally to private dredging interests, the Corps of Engineers does not always weight the factors in a dredge project review in the same manner for its national interests as it does for private sector projects. Powerful political lobbies can affect the outcome of a major port development dredging project while similar private interests may be ignored. In many cases, however, the Corps of Engineers will participate in assisting local interests by allowing the piggy backing of small local interest projects on to major port development or navigation improvement projects. The piggy backing may allow the deposition of dredge material at disposal sites reviewed and approved by the government for the primary project with little or no further investigation of the suitability of the disposal site for the private project. This situation may result in considerable savings in required dredge material and disposal site testing and expedite the approvals process. It is most prudent to investigate local Corps of Engineers or other government activities for possible project coordination with other proposed or scheduled dredging activities.

Dredging Systems

Dredging is performed by either onshore or floating equipment (see Fig. 16-12), designed to reach the bottom sediments and initiate a path for the transport of the captured sediment to a disposal location. Typical equipment

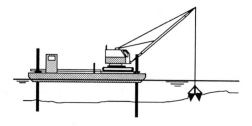
Crawler crane on barge with clam shell bucket

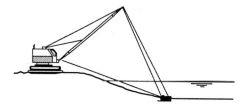
Dragline crane working from shore

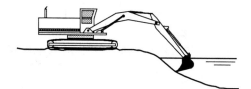
Tracked backhoe working side slope

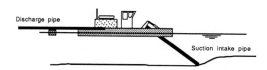
Transportable, floating, hydraulic, pipline dredge

Figure 16-12. Typical types of dredging equipment which may service marinas.

used in dredging from shore includes wheeled or tracked excavating machines which reach out to the area to be excavated (dredged) and cast the dredge material into an adjacent scow or on to the shore or trucks for further handling. Barge mounted cranes are a common dredging device, employing either clam shell or dragline buckets. The clam shell bucket operates off the boom of the crane and is a bucket that opens and can be closed watertight much like the opening and closing of a clam's shell, hence the name. The open bucket is swung into position over the area to be dredged and dropped through the water column into the bottom sediment. The bucket is then closed by the crane operator and raised with a full bucket of bottom sediment and entrapped water. The crane may swing the full bucket to a scow or barge alongside or to shore or to some other location outside the dredge envelope. If deposited in a scow, the fully loaded scow will generally be taken offshore to an approved disposal site and off loaded. The disposal site will often be an open water dumping area, where the dredge material will be allowed to disperse over the bottom. Disposal in open water approved areas requires monitoring by approval authorities, usually by having an inspector of the permitting agency ride along with the barge and assure that, in fact, the dredge material was properly disposed at the approved site. The cost for the inspector is usually borne by the party doing the dredging.

A dragline dredge operation would have a crane located onshore with an open ended box-like bucket which is cast or dragged outshore into the area to be dredged. The open end of the bucket is pulled toward the crane, filling the bucket with bottom sediment. The full bucket is tripped up by the crane to retain the sediment and maneuvered either to a scow or to the shore for dumping. The dragline operation has limitations on reach out from shore and its ability to cast the dug material to appropriate locations. The dragline method also results in high turbidity in the local waters from the agitating action of the bucket being dragged through the bottom sediments. In this era of environmental awareness and concern, dragline dredging is not an acceptable dredging technique because of turbidity and bottom biota disruption.

Another onshore method of dredging can be typical power shovel equipment with dipper buckets. These machines scoop the material into the bucket like a spoon into a bowl of sugar. The full bucket can be then swung to waiting trucks, barges or on to shore for further handling and final deposition. Dipper shovels have significant limitations when working from shore because of their restricted reach and swing. Backhoe type equipment can also be used but such equipment has similar restrictions on reach and swing.

All of the above described equipment may be barge mounted to enhance their access to the dredge area and disposal location. Variations of the above equipment, on a large scale, are used by various major dredging companies and the Corps of Engineers for major waterways dredging projects. In general the larger equipment is unsuitable for marina and small craft harbor operations and if suitable is often cost prohibitive. An additional type of equipment used in both large and small dredge projects is the hydraulic dredge. The hydraulic dredge works on the principle of sucking a slurry of bottom sediment and water through a pipe. The intake pipe is often fitted with cutter heads to assist in breaking up the bottom sediments into small enough pieces to allow passage through the discharge pipe. The desirable feature of this system is the ability to run long lengths of discharge pipe from the dredger to a distant disposal location. Distances of several miles of pipe run are possible with sufficient pump or booster pump capacity. Hydraulic dredging is carried out extensively by the Corps of Engineers in the maintenance of the nation's waterways. On a smaller scale, several manufacturers have developed hydraulic dredges suitable for use in confined marina spaces. **The marina size hydraulic dredges have suction intake pipes about 8 inches in diameter, can reportedly handle 120 cubic yards of sediment per hour, can operate in less than 24 inches of water and can reach down about 15 feet below water level.** The barge unit may be as small as 8 feet in width (for portable over-the-road travel) and 30 to 40 feet in length. This size allows the unit to reach in between marina finger floats or piers without the need to remove the marina structures. The drawback to hydraulic dredging is the fact that the slurry (bottom sediment and water) contains about 80% water. This water must be contained at the disposal site by impoundment behind dikes, with the excess liquid appropriately handled to prevent site erosion and fouling of adjacent waterbodies or watercourses. The high water content will also result in the need for long term drying of the disposed sediment before its reuse. The dredge material impoundment area may also be odoriferous depending on the chemical and physical makeup of the dredge material. In some cases where the bottom sediment is clean sandy material, approval may be obtained for hydraulic pumping of dredge material to an inwater location adjacent to the dredge area. The Corps of Engineers, itself, will occasionally use this technique in clean material areas.

The method which is used is called a sidecast dredging. In this method the material is sucked up an intake pipe and discharged through a pipe located at some distance from the side of the dredge vessel. As the dredge vessel moves forward the disposal pipe follows, creating a sort of dredge material berm alongside the dredged channel. This technique may be highly unsuitable in areas of high littoral drift and sediment transport, where the sidecast material will rapidly migrate to the depressed channel cut.

The most common dredging equipment used in marina construction is barge mounted cranes with a mechanical clam shell bucket or an hydraulic pipe dredge. Dragline buckets operated from a shore based crawler crane are occasionally used, but the high water turbidity resulting from dragging the bucket along the bottom is often not acceptable to regulatory agencies.

Bottom sediment of a course grained nature, sands and gravels, might be suitable for dragline operations, but fine grained silts and silty sands may preclude dragline consideration. If a clam shell bucket is used, the bucket should be of a watertight variety to minimize loss of material to the water column and the resulting water turbidity.

Floating dredge equipment is held on station by either pile type spuds or cable anchoring systems. Spuds are vertical piles usually driven thru a sleeve welded to the barge. The spuds are set into the bottom sediment by their own weight and generally are not driven. The spuds may then be easily picked up by the on-board crane and the rig can be moved to the next location. Barge movement is accomplished by an assisting tug boat or small workboat or by a set of mooring cables attached to deck winches on the barge. In mechanical dredging by bucket, a dump scow is generally moored alongside the crane barge into which the dredge material is deposited for future disposal at an approved site. If the digging is close to shore, the barge mounted crane may be able to swing the full bucket to shore and either deposit the material into waiting trucks or place it in temporary storage piles for later handling. If on site storage is elected, an appropriate impoundment and dewatering dike area must be created to prevent the dredge material effluent from fouling the adjacent land or waters.

Checklist for Dredging Projects

A proposed dredging project involves many physical and environmental considerations as well as appropriate design parameters and dredging equipment. In order to provide adequate consideration of the many factors involved, it is handy to develop a checklist of items to be developed during the planning stage of the dredging project. The following is a suggested list which may be helpful in securing the necessary information and providing appropriate response in design and regulatory application.

1. Develop a proposed dredge envelope, on a site plan, as necessary to provide adequate water depth in all operational portions of the marina.
2. Evaluate the site environment in terms of existing depth, current velocity and direction, access to channels or fairways, wave exposure and potential dredge material disposal locations.
3. Obtain an adequate hydrographic survey of the area to be dredged, often called a predredge survey. Area coverage will vary with the magnitude of the project and the size of the area involved. Soundings on a 50 foot or 100 foot grid are often satisfactory. The closer the sounding data, the better the estimate of dredge quantity to be removed will be. A close grid spacing may be necessary if the bottom surface is very irregular.

4. Obtain samples of the material to be dredged and have it analyzed for physical, chemical and biological characteristics. Check with local authority regarding location of sampling, number of samples and depth to which samples are to be taken. Standard tests include bulk sediment analysis, bioassy, bioaccumulation and density. In the United States, the Environmental Protection Agency has established certain protocols for performance testing which must be observed.
5. Identify local social, environmental and institutional factors that may impact dredging. Some concerns are noise from dredging operations, odor from dredged sediments, deposition of dredge materials on roadways for upland disposal, etc. Identify any shellfish or other biological activity that may be impacted by dredging. Water turbidity resulting from mechanical dredging can travel with the tidal or river flow and influence distant areas that may affect primary resource areas.
6. Assess local sedimentation transport to determine the potential for maintaining a marina basin without frequent maintenance dredging.
7. Investigate dredge material disposal locations. Some options are: on-site with direct burial or dewatering and reuse elsewhere on site; upland but off site, requiring transport to a distant acceptable location; ocean disposal at an approved location; or perhaps beach or shoreline renourishment projects. Appropriate dredge material disposal is a key factor in obtaining regulatory approval. An often used method of upland disposal for clean dredge material is for the capping of landfill sites.
8. Assess available dredge plant equipment to select the most cost effective equipment that will perform the required dredging task. Large equipment on a small project may result in quick dredging but also large amounts of overdredge if the equipment cannot be controlled adequately to maintain design depth of cut.
9. From physical dredge material characteristics determine an appropriate side slope for the dredged area that can be maintained. Normal side slopes are three horizontal to one vertical. Other slopes may be acceptable or required in nontypical material. Material characteristics can also be assessed to determine the swelling, bulking or fluffing that will occur in dredge material as it is handled in the dredge process. This aspect is particularly important in determining the size of an upland containment area for disposal. The bulking may require an initial storage volume 1.5 times the in situ dredge volume or greater. Hydraulic dredging will contain as much as 85 percent handling water by volume, so the disposition of this water must be accounted for in the design.

10. In calculating dredge material quantity, an allowance for overdepth must be considered. A normal allowance, for payment purposes, is one foot over depth. Upland disposal sites should have the capacity to handle the overdepth volume of material. On small marina projects, the overdepth quantity may be equal to the calculated in situ volume. Side slope overdredging should be considered and factored into quantity calculations and cost estimates. Again, a one foot overdredge is typical.
11. Evaluate dredge plant equipment to be used in the project. Typical methods include: backhoe machines from shore or on floating barges; cranes on shore or on floating barges using mechanical clam shell buckets; cranes with dragline buckets usually operating from shore; and hydraulic dredges vacuuming the bottom and discharging through a pipeline to shore.

16.6 HANDICAPPED ACCESS

An accommodation often overlooked in the development of waterfront facilities is the provision of barrier free access to physically challenged persons. The very nature of the ground terrain at a waterfront/land interface creates an immediate problem of transition from a stable ground base to the dynamic movements of the water body. Barrier free access in structure design only began to be implemented after World War II when the need to provide accommodation for numerous disabled veterans became evident. The process of providing barrier free access has been slow due to lack of public awareness, cost of implementation and technical obstructions. Today, however, federal mandate and many state regulations require providing barrier free access to public facilities. To ignore appropriate barrier free design may not only be morally unacceptable, but may be contrary to public law.

Accessibility in maritime structures design has most often been addressed as a minor issue, to be implemented only where an easy solution existed. This may be attributable to the feeling of many people that physically challenged people did not belong "messing around on the water." Today, many physically challenged persons regularly participate in water related activities. Often the ability to participate has been obstructed more by the lack of appropriate access to the facility than by the inability of the individual to perform the activity. An example of this is the ability of many physically challenged people to participate in recreational sailing. In fact a class of sailboat, the Freedom Independence, has been developed to give disabled persons the ability to sail and to skipper the boat. However, often the ability to get to the boat, across gravel parking lots, down steep ramps, etc., has precluded participation in waterfront recreation.

In many areas of boating activity such as lakes, rivers and coastal areas with small tidal rise and fall, providing accessibility is simply a matter of applying the appropriate handicapped access code requirements to the design effort. Other areas having large tidal fluctuation, steep ground slopes or exposed sea states may find access implementation difficult, if not impossible.

We present here information on designing for accessibility derived from a number of sources, we also define terms associated with accessibility and present a case study in which implementation of traditional accessibility guidelines has been compromised by large tidal rise and fall.

What is Accessibility?

Accessibility is the concept, implemented by codes, regulations, policies and practices, which provides an environment for the safe approach, entrance and use by persons with physical disabilities. In order to use an environment it is necessary to develop an accessible route which may be defined as a safe, continuous, unobstructed path connecting all elements of a facility that may require transiting by persons with physical disability or other disability. In many cases this access is guaranteed by federal law under the Architectural Barriers Act passed by the United States Congress in 1968. In 1984, Massachusetts Governor Michael S. Dukakis issued Executive Order No. 246 which mandates that persons with disabilities have equal opportunity to state employment, therefore creating the need to make all Massachusetts state facilities and services accessible. Other states have similar policies. Irrespective of mandated policy, it makes good business sense to provide accessibility since access is a requirement to the selling of goods and services and accessible facilities provide safer working conditions, open opportunities for greater market share and often enhance the character of the facility.

What is Barrier free?

Barrier free is the concept of providing accessibility free of impediments to safe use and passage. It generally relates to the physical obstructions in structures that inhibit or deny use or passage by persons with disabilities. Factors to be considered for developing barrier free designs include: maximum allowable slope in ramps or walkways, width of passages, height of handrails, elevation of ground obstructions, etc. Persons in wheelchairs or on crutches can often be stopped or tripped by simple changes in grade or consistency of gravel parking lots. Adherence to barrier free design principles will create a safe environment not only for disabled persons but for all users.

Designing for Barrier Free Access

No attempt is made to cover all aspects of designing for barrier free access. Several excellent publications are cited in the references and should be consulted for further information on design guidelines. This section provides information on several areas generally associated with marina and water transportation facility design and it explores problems and solutions. The primary areas of interest in marina type facilities are: toilet facilities, parking areas and gangway (ramp) access to dock systems. It should be noted that access codes, regulations and policies vary and the prudent designer should verify local criteria before embarking on a design project.

Guidelines for Barrier Free Design in Marinas

Parking Areas. Most marina facilities are accessed by automobile, so provision should be made to accommodate handicapped persons in appropriate locations, affording safe passage to necessary facilities. Regulations often state that when 15 or more parking spaces are provided, a portion of the parking spaces should be provided for handicapped use. General rules require the following:

Number of parking spaces	Spaces required for Handicapped
15-25	one space
26-40	5%, but not less than 2 spaces
41-100	4%, but not less than 3 spaces
101-200	3%, but not less than 4 spaces
201-500	2%, but not less than 6 spaces

The designated handicapped access parking spaces should be the ones closest to the activity they serve. If handicapped parking space cannot be provided within 200 feet of the activity they serve, a drop-off area should be located within 100 feet of the activity.

The width of handicapped parking spaces should be 12 feet for perpendicular or diagonal parking layouts. An alternative is to provide two, 8 foot wide spaces separated by a 4 to 5 foot wide aisle, which can service two vehicles. The center aisle must be painted or striped yellow to designate a special area, not for general vehicle parking (see Fig. 16-13).

Sidewalks associated with parking areas must have a curb cut installed at each handicapped space or pair of spaces such that handicapped persons do not have to enter the traffic flow to access the sidewalk. Alternatives may be acceptable if addressing safety and access issues. Sidewalks located adjacent

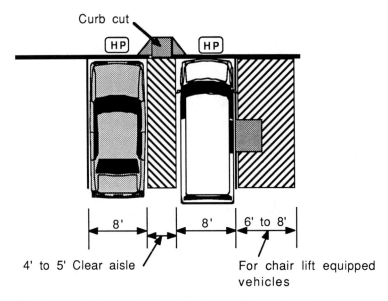

Figure 16-13. Parking layout dimensions for handicapped accessible spaces.

to a row of parked cars should have automobile wheel stops at each parking space to prevent encroachment by the car onto the sidewalk and thus denying handicapped passage.

It is important that parking areas be relatively flat, not more than 5 percent (one-in-twenty slope) and be paved or have a hard, packed, smooth surface. Cross slopes may make it difficult for a handicapped person to open car doors or transfer from wheelchair to car seat.

In laying out accessible parking it should be borne in mind that many handicapped transport vehicles have higher than normal roof heights (chair vans). A recommended minimum height clearance for chair vans is 9 feet 6 inches. It is also important to provide adequate side or end of vehicle clearance to accommodate side lift platforms associated with chair vans. Side platform chair lifts require a clear distance of 8 feet from the side or rear of the vehicle. Side rotary chair lifts require a 6 feet side clearance. Chair lift van parking spaces should be located in a level or near level area.

Curb ramps should be a minimum of 3 feet in width, exclusive of flared sides, and have a maximum slope of 1:12. Flared sides should have a maximum slope of 1:12. Curb ramps should not be located so as to be obstructed by parked vehicles.

Drainage or other gratings should not be located in handicapped pathways or sidewalks. If this location is unavoidable, gratings should have openings no greater than ½ inch to prevent entrapment of wheelchair wheels, canes, crutches or high heels. The narrow openings should be perpendicular to the direction of pedestrian traffic flow.

Appropriate signage and detectable warnings should be incorporated into the accessible parking design.

Sidewalks and Pathways. Sidewalks and pathways should provide the shortest accessible route to the activity. Minimum walkway width should be 3 feet, with 6 feet desirable. Slopes should not exceed 1:20 and cross slopes should be avoided or limited to a slope of 1:50. Drainage should be provided. Level changes should be flush or no more than ½ inch in height. Walking surfaces should be slip resistant, stable and firm. Curb ramps should be provided as necessary to appropriate design standards.

Stairs and Steps. Stairs may be used by some handicapped persons with partial or unsteady mobility, so certain precautions should be taken to provide safe access. Stairs should have a slip resistant surface and finish. The preferred stair riser height is 7 inches and tread width is 11 inches. Open risers are not permitted as they may allow feet and canes to slip between the treads and create difficulty for persons in leg braces. Risers should slope rearward no more than 20 degrees and stair nosings are prohibited as they are likely to trip persons with limited ankle flexibility and persons in leg braces. Detectable warnings of changes in elevation, such as at stair landings, should be provided for persons with limited or impaired sight. Generally a 30 inch warning strip running the full width of the stair at the top landing is required.

Handrails. Handrails are frequently in evidence in waterfront facilities and provide essential support to stair and ramp users (see Fig. 16-14). Handrails should be provided on both sides of a stair or ramp since some persons with disability favor one body side over the other for strength or support. For stairways, handrails should be set such that the top of the handrail is 34 inches above the point of intersection of the stair riser and tread. In areas where children are likely to use the handrail, a lower handrail at 19 inches above the reference point should be provided. Handrails located along walls should have the handrail extend at least 12 inches, horizontal, beyond the top and bottom riser. Handrail extensions may be preempted if its provision creates a safety hazard or unsafe condition. Handrails should be set 1.5 inches away from adjacent walls or other surfaces that will impair adequate grip of the handrail. Greater clearance may result in wedging or jamming of arms between the rail and wall. Handrails should be round or oval in cross section and not less than 1.25 inches nor more than 2 inches in diameter.

396 PART 3/ENGINEERING DESIGN

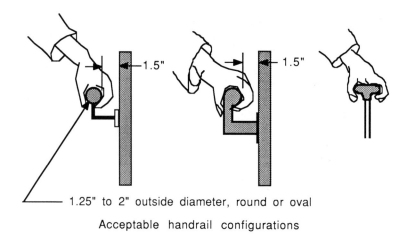

1.25" to 2" outside diameter, round or oval
Acceptable handrail configurations

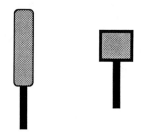

Unacceptable handrail configurations

Figure 16-14. Acceptable and unacceptable handrail configurations for handicapped access.

Handrails should be designed to allow a firm grip on the rail. Solid vertically oriented blocks of wood or metal are not acceptable.

Ramps. A ramp is defined as a pathway with a slope of greater than 1:20. Many areas of marina activity use ramps or gangways to transit a grade change. In many cases the implementation of ramp (gangway) handicapped access criteria will be the most difficult of compliance in waterfront facilities. Often the primary access to waterfront facilities requires accommodation of large tidal level changes, making maximum slope requirements difficult to implement. A case study is presented later demonstrating the problems associated with large tidal fluctuations and possible solutions.

Most regulations state that ramps should not exceed a slope of 1:12 (see Fig. 16-15). This slope, in fact, may be too steep for many wheelchair persons to negotiate. Ramps may generally be any length so long as they include a

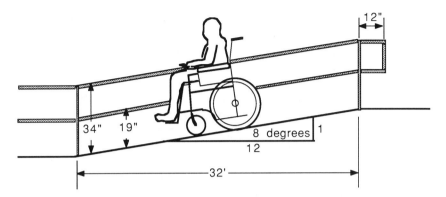

Figure 16-15. Maximum permissible handicapped ramp slope.

landing for resting every 32 feet or at any change in direction. The preferred inside to inside of railing dimension is 48 inches. This dimension allows standing people to pass while allowing the wheelchair person to grasp both rails for propulsion and restraint. Ramps having a slope greater than 1:15 should not be wider than 48 inches, clear inside dimension.

Intermediate, level platforms for turning or resting should have a minimum length of 48 inches and be the full width of the ramp. Landings at the tops of ramps should provide an area 60 inches wide by 60 inches deep, landings at the bottoms of the ramps may be 48 inches by 48 inches minimum. Both upper and lower landings should be level.

Ramp surface should be slip resistant. Special care should be taken for ramps and gangways which are subject to wet or freezing climates. Edge protection along the side of the ramp should be provided to prevent wheelchair wheels, canes or crutches from slipping over the edge. It is also a good idea to provide this protection to prevent other items such as dropped tools from falling over the edge, either into the water or onto the dock or people below.

Ramp handrails should be located on both sides of the ramp or gangway and be placed to have the upper rail at 34 inches above the ramp surface and, additionally, a handrail at 19 inches above the ramp surface. The regulations measure this handrail height vertically from the ramp surface but this is difficult on an articulated gangway. For articulated gangways it is suggested that the 34 inches be measured perpendicular to the walking surface. Where not a hazard, handrails should extend 12 inches beyond the the top and bottom edge of the ramp. With an articulated gangway, it is suggested to continue the natural line of the handrail 12 inches, since any other option will create unusual extension attitudes during gangway articulation.

Toilet Facilities. The first premise of toilet facility design is to provide for adequate maneuvering room within the toilet room (see Fig. 16-16) for a person in a wheelchair and with the room occupied by other people. A 48 inch approach path width to the toilet room is necessary, followed by a minimum 36 inch entrance door width. A minimum 36 inch clear path between toilet stalls and sinks or other obstructions should be provided. Each toilet room should provide a minimum clear space of 60 inches diameter, measured 12 inches above the floor, to allow turning of a wheelchair without contacting any fixtures or other obstructions. Each toilet room should have at least one stall that measures 60 inches wide by 72 inches deep with a 36 inch wide door which swings out and has an automatic self-closing hinge and a pull device to assist in door closure. Doors that swing into a stall may be allowed if additional space is provided within the stall to accommodate the door swing. Cognizant code may also address location and type of door lock, toilet fixture height, grab bars, coat hooks and other pertinent features.

Other fixtures include urinals and lavatories. Urinals, where provided, should include one, either wall or floor mounted, but with the rim of the basin 15 inches above the floor for wall mounted fixtures or it may have the basin rim flush with the floor. Flush controls should be mounted no more than 44 inches above the floor and not require foot operation. An area 60 inches by 60 inches should be provided for a wheelchair to turn around at a urinal.

Lavatories may be standard plumbing fixtures and in fact, special handicapped lavatories are discouraged since many designs actually impede usage by disabled persons and create an institutional atmosphere. Lavatory design features include: wall mounted without legs or pedestals, and rim located at

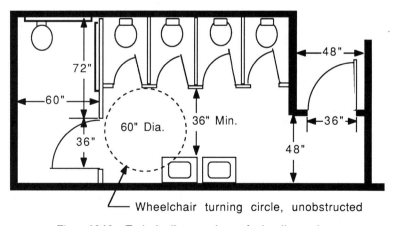

Figure 16-16. Typical toilet room layout for handicapped use.

a height of 32 inches above the floor and extending at least 22 inches from the wall. Counter type lavatories are generally acceptable if provided with clear open knee space of 30 inches in width, 22 inches clear depth and at least 27 inches of clearance between the bottom of the counter and the floor. Single lever faucets are generally the desired fixture and knob faucets are not allowed. Exposed hot water and drain piping should be recessed, insulated or otherwise guarded. Soap dispensers, towel dispensers, hand dryers, sanitary napkin dispensers and trash receptacles should have the operable mechanism mounted no higher than 42 inches above the floor. One of each device should be located within reach of the accessible lavatory. A full length mirror is best to service the needs of disabled persons, children, short and tall people.

Many marinas feature showers for their patrons, and providing an appropriate shower facility (see Fig. 16-17) for disabled persons is a desirable if not mandatory amenity. If provided, the shower should be able to accommodate both wheel-in and wheelchair transfer use. General inside dimensions of the shower stall should be 36 inches wide by 60 inches in length, with a minimum 36 inch door opening. Floors should be pitched to drain within the stall at the corner furthermost from the entrance. Floors should be nonslip finish. The shower stall should not have a curb which would deny access to roll-in persons. A noncorrosive shower seat should be attached to a side wall near the shower head for use by transfer users, or hinged out of the way for stand

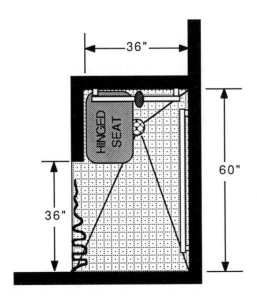

Figure 16-17. Typical shower stall dimensions and layout for handicapped use.

up showering. The shower seat should be a hinged, padded seat, at least 16 inches deep, folding upward and securely attached to the shower stall wall. The seat should be at least 24 inches long and mounted with the top of the seat 18 inches above the floor. Shower controls should be single lever type, operating a pressure balance mixing valve. All controls should be located on the center wall adjacent to the shower seat. Shower heads should be attached by a flexible metal hose on a wall mounted slide bar, adjustable from 42 inches to 72 inches above the floor. Two grab bars should be provided, one 30 inches long and one 48 inches long mounted horizontally 36 inches above the floor. Soap trays should not include hand hold provision unless they can safely support 250 pounds of load for 5 minutes.

Case Study—Boston Harbor, Massachusetts

Various regulatory agencies controlling upland and waterfront development in Massachusetts and the city of Boston wish to fully implement the regulations and codes pertaining to full access to public facilities. To date it is unclear whether marinas are or will be construed as public facilities and therefore questions arise as to whether they come under the mandate for full access. However, water transportation facilities, often included as part of marinas in urban redevelopment projects, do come under the public facility umbrella, and must be made to accommodate barrier free access. Some marinas which hold boat shows, and accommodate sailing clubs and other public assemblies may also be required to implement full access opportunity.

Designing maritime structures for full barrier free access in Boston Harbor is a difficult technical challenge. The primary obstacle is the large tidal fluctuation. Boston has a mean tidal range of 9.6 feet and a spring tide range of 11.0 feet. Actually the extremes are greater. For the calendar year 1987, the predicted highest high tide elevation was +11.7 feet (January 1, 1987) and the lowest low tide elevation was −2.1 feet (January 1, 1987). These heights represent a range of 13.8 feet. Analysis of the tide tables for 1987 reveals that there were a total of 705 high tides and 705 low tides for the year. Boston has a semidiurnal tide, that being generally two high tides and two low tides per 24 hour day. Some days have one high and two lows or two lows and one high. Tides are more closely related to the Moon than the Sun and the tidal day is about 50 minutes longer than the solar day so tides occur later each day thereby alternating periods of high and low tides on about a two week cycle. In addition to astronomical tides (predicted tides) tidal height may be augmented by meteorological influence, wind, earthquakes, etc.

For design of barrier free access to waterfront structures, it first is necessary to define the reasonable maximum tidal range in which to base the design. Because the intimate effects of tidal elevation variation is not always readily understood by agencies not familiar with marine design, the most commonly used definition seems to be mean high water and mean low water, in Boston's case a range of 9.6 feet. Further study of the tide tables reveals, however, that in 1987 of the 705 high tides, 283 or 40 percent were predicted to exceed heights of +9.6 feet and 265 of the 705 low tides were predicted to be lower than 0.0 feet. This means that using the mean tide range will result in elevations exceeding +9.6 feet or 0.0 feet during 40 percent or more of the tidal cycles. This fact is somewhat mitigated by the time of day progression of the times of high and low tide for purposes not operating 24 hours per day. Water transportation operations generally may have scheduled runs from 6:00 AM to say 8:00 PM. Each day the time of high and low tide will vary so no one time slot will always be subject to a specific tidal height or tidal elevation.

Understanding of the tidal range (see Fig. 16-18) is necessary to undertake the problem in Boston Harbor. Currently many, if not most, harbor structures in Boston have been constructed such that the land grade at the waterfront is at an approximate elevation of +16 feet, on a datum where mean low water equals 0.0 feet. During periods of major storm events these piers or other waterfront structures may flood. On the datum of mean low

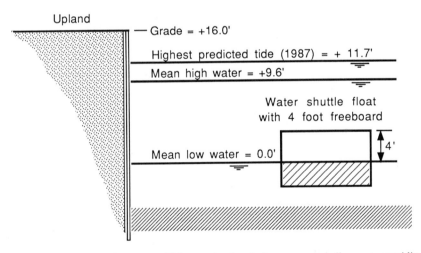

Figure 16-18. Areas with extreme tidal water level variations present challenges to providing handicapped access. This diagram presents the relative water levels in Boston Harbor, Massachusetts. A water transportation floating dock is shown with a 4 foot freeboard, the elevation of which will change during the tidal cycle. There is a need to provide a walkway transition between the upland fixed elevation and the changing float elevation that will meet handicapped access requirements.

water equals 0.0 feet, the predicted still water (not considering the effects of augmented tidal elevation or wave height) is as follows:

Return Period in Years	Storm Water Level above Mean Low Water
10	13.8 feet
50	14.6 feet
100	15.0 feet
500	15.8 feet

With the adoption of new Federal Emergency Management Agency (FEMA) flood insurance rate maps, higher waterfront structure elevations may be required than now exist. The higher elevations of these structures will serve to increase the difficulty in implementing barrier free access by increasing the difference in elevation between land at grade and floating structures at low water levels.

Retaining the land grade elevation at +16 feet yields the following elevation constraints that must be addressed to serve a high freeboard (4 feet) floating dock used as a water shuttle boat landing. The net tidal range between land grade and the water shuttle float at mean low water is 12 feet (+16.0 feet minus 4 feet of freeboard on the float). Using generally accepted access regulation guidelines of ramps not to exceed 1 to 12 and horizontal ramp length not to exceed 32 feet, the layout shown in Figure 16-19 would be necessary to meet the code.

If the water level was constantly at mean low water this system could be constructed using fixed platforms and sloped ramps. For a fluctuating tide, however, it is necessary to accommodate the rise and fall of the water level. The overall ramp length of 160 feet could be reduced by one or more

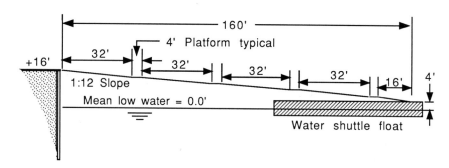

Figure 16-19. Theoretical handicapped access ramp configuration required to meet handicapped access code parameters for tidal variation in Boston Harbor, Massachusetts.

switchbacks. Use of switchbacks, however, increases plan view depth which may reduce vessel maneuvering room if the water berth is confined.

A solution developed for one water shuttle facility uses the ramp platform scenario with some modifications to accomplish a practical solution, but one that does not meet the full requirements of the regulations (see Fig. 16-20 and Fig. 16-21). The question becomes, is a practical solution that has the potential of working, yet is not in strict conformance with the regulation, better, at this time, than a solution which appears to have technical limitations that may compromise its long term use and survival?

The deviation from the code is basically to replace the first 32 foot ramp with a 50 foot gangway. The 50 foot dimension is necessary to fit the geometry of the tidal rise and fall and provide compliance as close to the 1 to 12 slope as possible for the major part of the tidal cycle. A 50 foot gangway on a slope of 1 to 12 yields a vertical drop of 4.17 feet. The total drop from yard grade of +16 feet to the deck of the shuttle float having a freeboard of 4 feet and floating at mean low water (0.0 feet) equals 12 feet. Subtracting the 4.17 feet of drop provided by the 50 foot gangway yields an additional drop of 7.83 feet to be accommodated. Each 32 foot ramp drops 2.67 feet, so the total drop remaining of 7.83 feet divided by 2.67 feet equals a requirement for 2.93 ramps of 32 feet or two 32 foot ramps and one ramp length at 0.93 × 32 feet which equals 30 feet. The initial 50 foot gangway becomes important when the system position is examined at high water levels. The 50 foot gangway reaches elevation +16 feet (yard grade) plus 4.17 feet (uphill vertical interval) or +20.17 feet before exceeding the 1 to 12 slope. The tide level when the gangway is at +20.17 feet is 20.17 feet minus the structure buildup of 7.83 feet

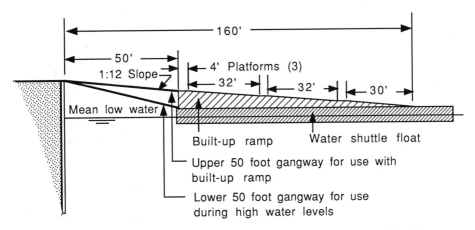

Figure 16-20. Handicapped access ramp using 50 foot approach gangways and build-up on floating shuttle dock.

Figure 16-21. Photograph of partially completed water shuttle dock with dual gangway arrangement for handicapped access.

plus the 4 foot freeboard or 11.83 feet, so the tide level is 8.34 feet. Not quite the +9.6 feet mean high tide level. In this solution a second 50 foot gangway is provided running directly from the yard grade to the float. This gangway reaches the 1 to 12 slope at elevation +16 feet (yard grade) plus 4.17 feet (1 to 12) or elevation 20.17 feet. The float has a freeboard of 4 feet so the water elevation when the second gangway begins to exceed the 1 to 12 slope is 20.17 feet minus 4 feet or 16.17 feet, greater than the highest recorded tide in Boston and exceeding the 500 year return period for tidal height.

The two gangway solution is valid at all anticipated high water levels and to mean low water, but the slope will be exceeded during times of low water less than 0.0 feet in elevation. It should be noted that the time interval when the tidal level is below 0.0 feet is small in terms of the total daily tidal cycle. Therefore the penalty of exceeding the slope at times of tide below mean low water may be acceptable in terms of providing a viable access route.

Providing accessible facilities for waterfront structures is often simply a matter of applying the appropriate design criteria. In many cases, the cost increase to provide accessible facilities will be negligible or of minor financial impact. In some areas where large water level change is prevalent, the technical solution to accessing waterborne structures may be difficult. Each access problem should be looked at closely with all major site parameters defined. It then may be possible to provide a solution that is compatible with the regulations or with only minor deviations that may be accepted within the context of an access variance. Providing accessible facilities is not only a desirable or mandated feature for handicapped accessibility but a feature that can improve the site for all users.

16.7 GEOTECHNICAL CONSIDERATIONS

The appropriate design of a structure foundation is critical to its structural integrity and long life. The first step in designing a proper foundation is to obtain as much information about the subsoil as possible. The standard methods include digging of test pits, soil borings and in situ soil testing. The interpretation of the subsoil data is a complex subject and may require the expertise of a geotechnical engineer. In addition to geotechnical investigation, it is often valuable to observe and determine the foundation conditions of similar structures in the adjacent area. Both successful and failed existing structures can tell a lot about appropriate and inappropriate foundation design. It is not recommended that the marina developer venture into the realm of foundation design unless his background is in this particular field. The consequences of an improper design can be significant and may not be immediately apparent, but may occur over a long period of time. Such consequences may include structure settlement and slope subsidence. Subsoil investigation should begin at an early stage in the design process to fully alert the marina designer to any conditions that may compromise a design or result in construction costs that may cause the project to be financially unviable.

Most marinas rely on a sufficiently deep layer of competent material in which to embed piles. It is therefore important to determine if such a layer exists, its vertical location and extent, and the capacity of the material to resist the imposed pile loads. **In general any pile will require a minimum embedment of 15 feet in competent materials. More generally piles may require 20 to 40 feet of embedment in good soils to provide sufficient capacity for modern pile marina design.** If less depth is prevalent and bedrock underlies the soil, it may be necessary to socket the piles into the rock, a costly installation. In some cases, a shallow layer of subsoil may be sufficient to allow use of a bottom anchored mooring system, if anchor holding power can be developed. If hard bottom or rock is encountered at high elevations, some form of rock socketing or pinning will probably be necessary.

Care must be taken not to place full faith in soil boring or other subsoil investigation data. A limited soil boring program may not uncover data that is representative of all the subsoil conditions, and therefore confidence placed on analysis of the soil boring recovered soil specimens may not be reflective of actual conditions. Many soil engineers will completely discount the contribution of bottom silts when in fact the silt will provide some degree of pile support. In questionable conditions, a test pile or series of test piles is highly recommended. A pile equivalent to the kind that is intended to be used can be selectively driven and loaded by a lateral pull, with the load and pile deflection or movement measured and calibrated to actual soil capacity. The cost of a test pile program can easily be offset by the savings in appropriate final pile design.

16.8 FAILURE ANALYSIS

Introduction

The lack of definitive engineering data associated with marina systems design has been exacerbated by the competitive nature of the recreational boating industry with resulting cost cutting design and manufacture to "win-the-job," often at the expense of product survivability. The marina industry began in the 1950s with build-it-yourself construction. The 1960s saw the beginning of commercial marina product manufacture, with considerable growth into the 1970s. The 1980s have seen the imposition of regulatory control, often affecting the extent and quality of marina design and construction. During this nearly three decades of growth the demand for and sale of recreational boats have steadily increased, frequently outdistancing the availability of marina accommodations. The considerable demand for marina space has, in many cases, fostered inadequate marina design and construction as developers rush to implement marina building as low cost, high profit investments.

The discussion of engineering or product failure, by necessity, is critical of certain aspects of design or manufacture. The critique in this paper is presented to educate and mitigate the recurrence of similar failures in future projects. It is not intended to be critical of any individual or any group of designers or product manufacturers. In many cases, product manufacturers and designers have corrected deficiencies and now offer acceptable products and services. In an attempt at fairness in this most sensitive subject matter, the authors have included instances of their own experience with failure, and do so as an exercise in education.

The marina industry has grown from shared experience, some good—some bad. Today's marina buyer or developer is generally sophisticated in business but often naive about the difficulty in constructing and surviving at the sea-land interface. The ultimate marina user, the boatmen and women, are both intelligent and demanding. In most areas of the world, the recreational boater enters the market looking for a leisure time activity financed with discretionary income. The recreational boater often is financially successful but short on time. The marina industry must provide suitable, safe, attractive facilities in order to sustain the demand for recreational boating interest.

As the demand for marina facilities grows, the availability of suitable sites diminishes. Factors influencing these circumstances include:

1. Most suitable sites are built on first.
2. Competition exists with land developers who can pay more for the land and use it for high return development, i.e., residential occupation.
3. Customer demand for more and better facilities at lower per vessel costs.

The results of loss of suitable sites includes development in areas exposed to significant sea stresses, effects of transiting boat traffic, use of deep water with subsequent difficulty in restraining marina systems, polluted or highly corrosive environments that significantly degrade structures over time and areas that may require dredging with its associated environmental concerns and costs.

Marina failures do not always result in catastrophic destruction. Failure can arise because of improper site selection or because inadequate protection makes a marina subject to wave, boat wake or wind action with a resulting intolerable comfort condition within the marina. Recreational marina users will rarely tolerate excessive boat movement at the dock and will move to more comfortable locations. In this case the marina may fail economically as a result of poor or deficient engineering. Failure may also lead to personal injury of marina personnel or users with resulting litigation that may impact the ability of a marina to survive. Of course there are the structural failures that are generally obvious and can result in significant financial impact to the marina owner because of repair or replacement costs, loss of revenue during repair/replacement and loss of good will to the users, if not loss of their vessels.

Marina failures may be categorized in several major groups:

1. Site Dictated Failure
2. System Structural Failure
3. Component Structural Failure
4. Layout and Planning Failure
5. Operational Failure

Site Dictated Failure

Site dictated failures occur when site selection and implementation of necessary protection works are neglected or not fully appreciated by the designer or developer. Often the lack of suitable marina sites has caused adventurous development to take place. Numerous examples exist of exposed sites without adequate protection from sea state exposure. The recent development of floating wave attenuation devices has also led many designers and manufacturers to develop products that may look impressive, but which are not adequately designed to withstand or ameliorate the imposed sea conditions. Although not always true, it seems that the lower the product's initial cost, the shorter its life cycle and the higher the annual maintenance costs.

An example of site dictated failure is Pier 39 in San Francisco Harbor, where a marina complex was initially protected from significant exposure by a floating breakwater. The floating breakwater failed and was replaced by a comprehensive, fixed vertical wall breakwater. The only beneficiary of this type of failure will by the lawyers who will litigate the action.

Exposed open water sites, both coastal and inland, must be carefully analyzed for effects of wind, wave, wake and other pertinent climatic conditions such as ice effect, littoral drift and material deposition, bottom scouring, and potential major climatic events such as hurricanes (cyclones) or tornadoes. It is incumbent on the designer to make the client aware of the risks, and in some cases to reject the design assignment if conditions can not be fully analyzed or accommodated in the design. It is also incumbent on the designer to assist the client in risk analysis to assure that a cost effective facility can be constructed.

Massive rubble-mound-type fixed breakwaters are often considered as protection for severely exposed locations. These structures are generally difficult to obtain approval for and are quite costly. The structures also require intensive engineering analysis both in construction techniques and materials as well as in orientation. Several noted failures of fixed earth breakwaters have occurred as a result of one or more of the following reasons:

1. Inappropriate armoring stone
2. Inappropriate shape
3. Inappropriate orientation
4. Insufficient height

Insufficient height appears to be a major failure mode. If the breakwater is overtopped by a wave, the wave may reform inside the protected basin with subsequent potential damage to berthed vessels and marine structures. In 1983 damage occurred at Marina Del Ray, in California, 3 miles inside the breakwater because of over topping and wave reformation.

Storm events may occur at any time and at most locations throughout the world. It is therefore unwise to develop a marina on a temporary basis, believing that the system will be adequately strengthened in the future or during the next phase of development. In 1985, hurricane Gloria struck the northeast coast of the United States in September, just as the traditional boating season was ending. Many marina and boat owners were caught unprepared with resulting major losses. One marina had installed a new marina dock system on a temporary basis destined to be relocated during another phase. The marina pile support system was installed with piles driven only 5 to 6 feet (2 meters) into the sea bottom. This embedment was insufficient to resist the loads imposed by the hurricane force winds, and the entire floating system, boats, floats and piles were driven ashore with considerable losses. In the same area, boats single point moored offshore of a fixed breakwater broke free of their mooring and were tossed over the breakwater by the overtopping waves. These vessels were constructive losses. In many cases, boats were carried so far inland from the effects of the hurricane that helicopters had to be engaged to airlift the boats to the shoreline for transfer to repair facilities or to the scrap heap.

System Structural Failure

The basic design of marina systems requires that conventional structural principles do not apply in isolation. The understanding of the dynamics of a system is critical in the design process. Unless adequate provision is made for the cyclic nature of the imposed and support forces, long term damage and excessive wear may be built into the design. One major consideration is the effect on the structure of low-height, long-period waves. These waves (swell) may result from distant far field wave generation or they may be reflected waves moving well inside a marina basin or up a river. Example characteristics might be: 5 second wave period, 9 inch (23 cm) wave height, and wave length of 150 to 200 feet (45 to 60 meters). The marina float system must be able to continually and consistently transmit or resist the imposed forces or some component of the system will fail. The system must be able to accommodate vertical and lateral forces as well as torsional forces as the system pitches, yaws and rolls. Probably the most prevalent form of marina float failure is in unit-to-unit connections. These connections may be articulated or nonarticulated, but in all cases must be designed to accommodate the imposed loadings.

It should be remembered that wave/wind action in marinas seldom, if ever, impacts the float system at right angles (90 degrees). The varying angle of attack of the wave/wind front creates an unbalanced loading on the dock system. The system, especially the dock unit interconnection, must have some limited movement to accommodate all six degrees of freedom, or the structure must be structurally rigid. Structural rigidity is difficult to achieve and generally uneconomic.

Dock mooring support systems play an important role in resisting and transferring the imposed wind/wave load. Excessive spans between mooring support locations can induce excessive dock system bending resulting in overstress and failure. Systems using continuous timber walers for longitudinal rigidity are often incompatible with pile or bottom anchored support systems. To overcome this weakness, large numbers of closely spaced mooring support points are required, often with great economic penalty. To reduce the chance of system failure, it is extremely important to analyze the mooring support system in context with the strength of the proposed dock system to assure compatibility between dock system strength and support system span. In some cases, the alleged benefits of a low cost dock system are offset by the need to provide a comprehensive mooring support system at great cost. Bottom anchored support systems (chain or cable) must be intensely analyzed to assure adequate support for the dock system. The nature of this kind of mooring system is such that large bending moments can easily be induced into the dock frame, especially in areas subject to tidal fluctuation. Two ways to mitigate the effects of bottom anchored system load support are: 1. provide definable bottom anchor points such as stake piles (pile driven into the bottom and cut off just above the bottom) and 2. to pretension the mooring support system to minimize horizontal dock excursion. Fixed bottom anchorage points, especially if they can be located such that all bottom connection points are on a constant plane, will allow use of equal lengths of mooring chain (cable), mitigating the effect of varying chain (cable) length and its response to load and vertical elevation change of the dock system. Alternative bottom mooring anchors such as mushroom anchors, concrete blocks, etc. all require time and distance to set and therefore corrupt the design analysis. Their ultimate holding power is also open to question. Pretensioning of mooring chain (cable) after installation and checking of pretension regularly results in a design preload on the dock system, minimizing the movement of the dock system due to chain (cable) straightening under load. Again with the imposed load impacting the dock system at varying angles to the system and at varying instants of time along the system, chain (cable) mooring systems are likely to come up hard on one end while unloaded at the opposite end, creating substantial bending within the dock system.

Contribution, or lack thereof, of dock finger systems to overall dock system stiffness or flexibility must also be considered. One of the greatest areas of marina system failure is in finger to main dock connection. Articulated finger docks must really be that, they must provide adequate freedom to relieve imposed stresses in all directions or they must be capable of transmitting those forces that cannot be relieved. Unfortunately, it is rare that a finger dock is only subject to relief from vertical forces, yet most articulated finger connections are designed only to relieve those forces. Lateral and torsional forces may be large enough to cause failure at the interconnection.

Another possibility for mooring support system failure occurs in pile support systems where adequate pile freeboard is not provided, so that during significant storm events, the combination of meteorological high water plus storm surge elevation, plus wave crest elevation, plus dock system freeboard total to higher than the tops of the provided pile support system. The dock system may then overtop the piles with obvious consequences, including the possibility of the entire dock and boat system coming ashore.

Pile mooring support systems are also compromised when, for various reasons, the piles are not driven deep enough to support the full system load at storm water elevations. Hurricane Gloria, which struck the northeast United States in 1985, dramatically showed the lack of wisdom in driving piles for a temporary mooring of docks (intended to later be moved in another marina expansion phase). Such piles were inadequately embedded to resist the imposed storm loading and were extracted by the wind/wave forces and the entire dock system and berthed boats driven ashore with great losses.

It is therefore prudent to design and analyze a system as a system and to assure compatibility between dock system and dock components and dock system and the mooring support system.

Component Structural Failure

Dock system component failure is probably the most routinely experienced marina failure. The comments presented above regarding system failure usually result in, or result from, component failure. A primary area of dock component failure is in dock unit-to-unit connection and in finger-unit-to-main-dock-unit connection. Rigidly fixed dock unit connections have evidenced failure generally because of either vertical dock oscillations and induced vertical bending and shear or by horizontal or lateral forces creating in-plane bending and shear. Occasionally torsion of the dock system will exceed allowable stresses and induce connection failure.

Vertical dock oscillations in nonarticulated systems result in bending stresses which must be transferred from one dock unit to another. Continuous timber walers or laminated continuous beams are often used as the main frame load transfer mechanism. Failures result when the timber properties of the section are exceeded and longitudinal fracture of the timber fibers occurs or the shear capacity is exceeded and a vertical shear failure occurs. If the timber is adequate to transmit the imposed forces, then, often, the fastening bolts will fail. Insert bolting in concrete has been shown to be inadequate to resist the pullout forces associated with timber waler strength. The localized stress around insert bolting is generally greater than the strength that can be achieved by embedment in the concrete. Through bolting can offset this weakness in many cases. Fastening bolt shearing occurs when the specified bolt size or number of bolts is inadequate to resist the imposed loading. Care must be taken to assure that a balanced load transfer system is designed, especially for composite designs such as concrete pontoon units with timber or steel walers. Too often what seems like a simple solution to minor failure may result in a multiplying effect. A case in point occurs when timber wale failure is evidenced in a concrete/timber wale system. The immediate solution appears to be to increase the timber section by using larger or more wales. The result may be that the wales resist the high imposed load but the bolts fail. Then higher strength bolts are used and the bolt and timber combination is adequate for the imposed load but the concrete holding the bolts fails, which results in a major compromise to the dock system. In this case it may be that the type of system specified was inadequate for the intended purpose, and that, rather than modifying the dock system, it would be advisable to mitigate the incoming forces (wave attenuation, etc.) or replace the system with a more suitable system. It is difficult to justify continual strengthening of a dock system, since the conditions causing the failure, if frequently occurring, must compromise the ability of the marina to safely berth the marina boats. High energy sites impacted on a regular basis must attempt to mitigate the energy forces.

Horizontal or lateral forces, in many cases, can be made tolerable by prudent design of the mooring support system. Maximizing the support span consistent with dock system structural integrity is critical. Most dock systems provide for accommodation of the vertically imposed forces but are weak in horizontal restraint, relying on the pile or chain support to intercept and transfer the horizontal forces. This is particularly true in articulated dock systems where a simple vertical hinge is provided. The hinge may work quite well vertically but it may require hinge binding and local component bending to resist horizontal forces. The solution is either to provide for the horizontal bending or to restrain the dock system adequately by mooring support points to prevent the hinge from having to transfer the imposed load in that direction. Continuous member systems must have sufficient rigidity or flexibility to transmit the load.

Dock system torsion also leads to dock component failure, particularly in continuous waler systems and on metal box truss systems. The continual racking of the system components overstresses the components, usually bolted connections, resulting in bolt pullout or fatigue at bolted connections. Bolt hole elongation in both timber and metal often fails at a connection joint. Aluminum and other soft metals are particularly susceptible to fatigue and wear. Torsion in box truss systems may induce stress reversal and turn a member designed as a tension member into a compression column, with inadequate resistance to local buckling, causing the member to fail, and distorting or fracturing the truss system. The arching that occurs as a wave train passes beneath a dock system repeatedly causes members to undergo a transition from being tension members to being compression members, and vice versa.

Layout and Planning Failure

Layout and planning failure results from the inattention to site constraints and lack of understanding of boathandling and other specific marina constraints. The result may not be catastrophic structural failure but rather failure of the marina to perform as a viable entity. An example of layout and planning failure is the architectural design of a sloped ramp designed to make use of forklift rack storage type berthing in an area requiring a large elevation difference between the yard grade and the launching water level. Since negative forklift descending travel has been limited to 10-13 feet (3-4 m) a sloped incline was felt to be acceptable to reduce the vertical difference in elevation at the boat launching point. The failure was the inability to recognize that a boat on a forklift has a finite degree of vertical rotation, in this case the forklift was incapable of positioning the boat to near horizontal for the descent down the incline, resulting in the probability that the boat would slide off the forks. A good idea gone bad because of limitations in the equipment destined to serve that facility.

Another layout failure occurs in the planning of marina fairway widths (distance between rows of finger docks). Inadequate fairway width results in difficulty in boat maneuvering in and out of slips with increased potential for boat impact upon docks or mooring piles.

Other layout and planning failures include inadequate provision of width between finger docks to safely accommodate traditional boats of the prescribed slip length. The narrowness may preclude full sale or lease of the marina berths and may also cause additional loading on the dock system by not allowing the berthed boat to utilize conventional tie up, but instead force the boat to load the dock (by fendering, etc.) in a manner inconsistent with appropriate load transfer. Narrow slips also are more prone to be impacted by the vessel entering or exiting the slip.

A major issue facing the world today is that of rising sea levels and their impact on waterfront design. Although there is much debate about the predictions of how much the sea level will rise, all knowledgeable students of the subject agree that sea level rise will continue into the foreseeable future and that its rate will increase. This reality must be considered in marina design and appropriate provision must be made for the potential rise in sea level in such features as landside access, mooring pile height, shoreside building elevation and length of chain (cable) for bottom anchored mooring systems.

Operational Failure

The most consistent cause of operational failure is inattention to marina system maintenance. Marina dock systems are dynamic in nature and therefore susceptible to enhanced wear as well as to large and continual sea state forces. Failure to address minor maintenance problems as they occur often results in major structural failures.

A marina gangway, bolted metal truss variety, was seen to be missing a bolt at a diagonal brace location beneath the handrail. The marina dockmaster was instructed to immediately replace the bolt. The pressure of other duties caused a delay in finding and installing a simple bolt and nut. A charter fishing party boarded the gangway and it failed by local buckling of the now unsupported compression handrail. No personal injuries occurred, but a totally collapsed gangway did result. Operational failure.

Mooring pile guide hardware is often unmaintained, resulting in nonrolling rollers, binding pile hoops and hung up chains. All these situations cause excessive component wear, and occasionally component failure. Eventually the combination of component failures may result in major structural failure within the system.

All marinas should have a written storm management plan, known to all marina support staff. The plan should address the active measures to be taken during various storm event thresholds. As each storm threshold is reached more active participation may be needed to secure the marina and the berthed boats, if any, from the effects of the storm and to mitigate system failure. Areas of potential system failure should be recognized and special attention given to mitigate damage in these areas.

Some marina failures are inevitable because of the exposure and the dynamic nature of the marina environment. Most observed failures, however, result from inattention to design details, poor layout or marina planning, lack of maintenance, shoddy product manufacture, or lack of understanding of what a marina is and who it serves. A true cost-benefit analysis should be made of all marina components and attractive low first cost products scrutinized for their suitability for the intended use and environment.

17

Mooring Systems

Mooring systems are an integral part of any successful marina design. Although the primary thrust of this chapter relates to floating structures, some aspects of the mooring design covered in this chapter will also relate to fixed dock, cantilever pile design and to mooring systems used in other aspects of small craft harbor design. The development of the loadings that must be resisted by mooring systems has been developed elsewhere in the text, primarily in Chapter 15, which discusses design loadings, with the essential forces being wind, wave, current, and mechanical impact.

Prior to the specific design of mooring systems it is important to develop an understanding of the structure to be supported, a single boat, a group of boats, or a dock system. It is also important to have an appreciation for the types of subsoil that will hold the mooring system, the applied loads, and the types of material best suited for mooring structures. The two primary types of mooring systems are pile supported (see Fig. 17-1), and chain or cable supported (see Fig. 17-2). Occasionally, a cantilevered or parallel arm mooring system may be suitable for individual docks or small dock systems.

The structural integrity of the dock system plays a critical role in mooring design. Before a mooring point location can be established, the dock system must be analyzed to determine the span between mooring support points that the system can tolerate. Loosely connected floating dock sections may require mooring support on each section due to their inherent lack of capacity to transfer applied boat load between sections. On the other hand, continuous, monolithic docks may be able to span 50 or 100 feet between mooring support locations. Whatever the case, the allowable span between mooring points must be established before analyzing the actual mooring design. The current trend in good marina design is to provide quality dock systems that allow the use of a high strength mooring system with minimum support points. This method usually reduces mooring installation cost and provides a more attractive marina.

416 PART 3/ENGINEERING DESIGN

Figure 17-1. A mainwalk (headpier) at the Guilford Yacht Club, Guilford, Connecticut. The floating dock system is supported by a series of cantilever, steel pipe, mooring piles. The piles are arranged on one side to facilitate future basin dredging by allowing the dock system to be temporarily moved laterally.

17.1 PILE SUPPORTED MOORING SYSTEMS

Individual, cantilever piles are probably the most commonly used form of marina mooring support system. A single pile is driven into the subsoil and used as a mooring restraint without additional support or bracing. The pile material may be steel, timber or concrete. Some firms are investigating the use of recycled plastic for fabricating a pile section, but experience with this new material is quite limited. The analysis of cantilever piles for marina mooring systems is a complex engineering task involving a variety of parameters. The methodology employed in the analysis is not universally agreed upon by active practitioners in the field of structural design. Well disciplined engineers will attempt to quantify design parameters and subject the analysis to rigorous investigation. The problem is that the design parameters are, at best, educated guesses. First there may be variable wind and wave conditions which are imposed on a variety of boat shapes and configurations. Secondly, the pile is often embedded in soil material that may have significantly different properties with depth change and location. Thirdly, timber piles

MOORING SYSTEMS   417

Figure 17-2. Except for several piles along the fuel dock, the Admiral's Hill Marina, Chelsea, Massachusetts is moored by a chain, bottom anchor system. The angled marina dock layout was designed to accommodate the alignment of a federal channel line associated with a commercial channel serving the site area. *(Photo credit: William F. Johnston, Quincy, MA)*

may have differing structural properties on a pile by pile basis. Even with the above considerations, however, it is necessary to make an attempt to engineer a suitable pile design. Because of the number of variables involved, it is often rational to use a rather simplistic mooring pile analysis and if in doubt require a driven pile load test.

Although the imposed load may not be well defined it is necessary to develop a rational pile design load. The analysis will be performed in two basic parts. One will be to develop the imposed design load. The second will be to analyze the capacity of a particular pile. Once the individual capacity is determined, the total design load can be divided by the selected pile capacity to determine the total number of piles required to accommodate the total design load. If the marina has different size boats, variable soil conditions or changing water depths, the marina layout will be divided into compatible arrays or blocks, such that each array responds to similar criteria. In some cases, it may be desirable to change pile size and/or type between arrays to effect the most economical and rational design solution. The aspect of changing pile size within a material type must be done with caution. Generally it is not recommended to change a pile wall thickness but rather to vary the pile diameter, for pipe piles or the web depth of a wide flange or H-pile section. The idea is to make any change in pile size very noticeable to prevent the wrong pile from being installed by busy contractors trying to get a job done under difficult conditions. Easily distinguishable pile types will minimize incorrect pile placement.

Figure 17-3 shows the type of information that must be considered in determining a suitable pile height above the apparent bottom. The criteria may vary in specific locations, but the point is to be sure that adequate pile height is provided and that the analysis considers the highest rational height on the pile, that may be subjected to the design load. It is also important to understand that a cantilever pile will tend to bend about some point of fixity below the apparent natural bottom. **The distance between the apparent bottom and the point of fixity will vary depending on the type of soil, but usually it is between 2 feet and 10 feet, with 3 to 5 feet often used as a presumptive trial number, if better soil data is unavailable.** If the soil is so weak that a point of fixity cannot be readily achieved, then the soil does not have the strength to support a pile and at least that layer of soil should be

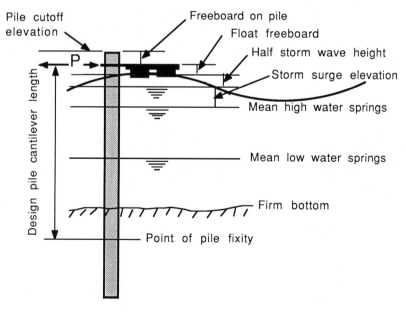

Figure 17-3. Parameters to be considered in determining mooring pile design height.

discounted in the pile restraint analysis. In loose, silty soils, it may be necessary to perform a pile load test or alternatively, completely discount any contribution of the silty soil. From actual pile load tests it has often been possible to assign a strength to soils that appear, from soil boring data, to have zero strength capacity.

Once the pile design cantilever length is established, this length will be used as the loading moment arm. Two considerations will be made. The first is to determine the bending capacity of the selected pile for the design cantilever arm length. To do this it is necessary to know the section modulus of the pile at the point of fixity and the allowable stress for the material selected. The formula used in the dock structure design (Chapter 16), $F_b = M/S$, or in words, the bending stress is equal to the moment divided by the section modulus, can be rewritten. The rewritten formula becomes $M = F_b \times S$, or the allowable bending moment is equal to the allowable bending stress of the material times the section modulus. The formula for bending moment (M) is the applied load (P) times the moment arm (L), or $M = PL$. If the allowable moment, developed by multiplying the allowable stress by the section modulus, is divided by the cantilever arm length, the allowable applied load that this pile can reasonably carry can be determined. Of course there are factors of safety that prevent the pile from failing at this applied load. This analysis has determined a maximum design load for a selected pile. Now it is necessary to investigate the capacity of the soil to resist this load.

Evaluating soil capacity is very difficult, even for experienced geotechnical engineers. Generally the best data available on soil characteristics will be obtained from soil borings and field and laboratory tests of representative soil samples. There are a number of computer oriented, pile analysis programs that can interpret the input soil data and develop the required embedment for any particular pile or configurations of piles. Without this powerful technique, it is necessary to rely on basic engineering analysis. One simplistic approach is to determine a reasonable value for the compressive strength of the soil and equate the surface area of pile bearing on the soil to determine a pile embedment length. Figure 17-4 shows example conditions for a laterally loaded cantilever pile in sand and clay bottom soils. By trial and error, or by computer, a pile embedment length can be determined by summing active and passive moments about some point below the bottom surface. Generally it is unwise to have any pile embedded less than 15 feet into the bottom, so an initial calculation at 15 feet below bottom is a starting place. In the example, a 12 inch diameter pile is selected, which will exert about one square foot of surface area per linear foot of pile against the soil. A factor of safety in the design of 2.0 is desired, so the soil capacity values, determined by soil investigation, will be divided by 2 to get an assumed, allowable soil capacity. The soil capacity used for the design will be 3 tons per square foot (TSF)

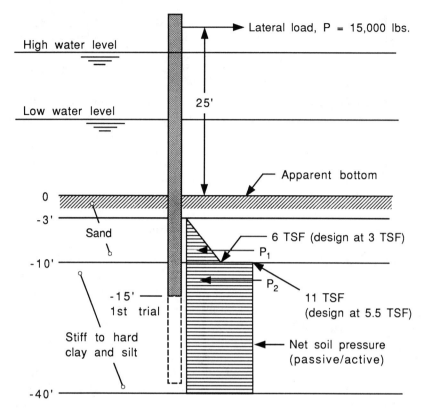

Figure 17-4. Diagram of laterally loaded pile and soil conditions as used in example pile analysis.

maximum for sand and a constant 5.5 TSF for the clay layer. Starting at the tip end of the pile at 15 feet below the bottom, an overturning moment from the applied lateral load equal to 15,000 lbs times 25 feet plus 15 feet or a moment of 600,000 foot-pounds is calculated. This is resisted by five feet of clay embedment at 5.5 TSF or 5 times 11,000 pounds equals 55,000 pounds times 2.5 feet equals 137,500 foot-pounds plus 7 feet times 6,000 pounds (3 TSF) divided by 2, for the triangular load, times an arm length of 5 feet plus 0.33 times 7 feet (2.31 ft.) or a total resisting moment of 291,010 foot-pounds. The applied moment of 600,000 foot-pounds exceeds the resisting moment so the pile embedment must be increased. For the next trial, add 5 feet to the pile length, for an embedment of 20 feet. The overturning moment becomes 45 feet times 15,000 pounds or 675,000 foot-pounds. The resisting moment equals, 10 feet times 11,000 pounds times 5 feet equals 550,000 foot-pounds plus 7 feet times 6,000 pounds divided by 2, times 12.31 feet equals 258,510 foot-pounds for a total resisting moment of 808,510 foot-pounds, which is greater than the applied moment so the pile is embedded deep enough to resist the applied lateral load. The trial and error analysis can be continued to determine an exact pile embedment length, if desired.

Another more rigorous analysis can be found in *Design Manual 7.2, Foundations and Earth Structures, NAVFAC DM-7.2,* published by the Department of the Navy, Naval Facilities Engineering Command, and available through the Superintendent of Documents, U.S. Government Printing Office, Washington, DC. Section 7, Lateral Load Capacity covers an analysis of single piles or groups of cantilever type piles and relates pile material strength to soil capacity. Soil capacity is developed by use of a coefficient of variation of subgrade reaction. Graphs and tables are presented to aid in the analysis.

Unless very good data is available on the soil characteristics, it may be wise to be conservative and provide a somewhat stronger pile, driven deeper than the analysis provides to assure that the critically important marina mooring system will not fail.

17.2 PILE GUIDE SYSTEMS

The use of a cantilever pile mooring system requires the provision of an adequate connection between the pile and the dock system to transfer the applied dock load to the pile. Most pile guide systems will be designed to allow only the application of a horizontal load, by use of a bearing or rolling guidance system rather than a rigidly attached connection that can add vertical or local bending stresses. The most common pile guide systems employ either hoops or rollers. Rollers or roller type connections are desirable to provide a smooth vertical movement and prevent the buildup of vertical forces as the dock system translates vertically. Roller systems are often made of rubber or plastic materials which also reduce noise. Noise can be a problem with metal to metal connections. If rollers are used, they must be sufficiently dense to prevent excessive compression and resulting contact of the roller frame with the pile. Roller systems should also be provided with stainless steel axle bolts to mitigate axle corrosion and subsequent seizing of the roller and axle. A nonrolling roller will soon be worn flat, never to roll again. The flat surface roller will not provide as smooth an accommodation to vertical movement and may allow the buildup of vertical forces between the dock and pile.

A loose chain hoop is also used, sometimes with the addition of short pieces of pipe on the chain to act as rollers. Metal to metal contact should be avoided so this system is usually found only in use with timber piles. Care must be taken to assure that the timber pile is smooth and without knot projections or the loose chain might get hung-up and create a dock that is unable to move with the water level change. The chain, even with pipe rollers, tends to be quite abrasive and can speedily wear the surface of soft wood or concrete piles.

Rollers have been found to wear better than direct sliding contact devices. Various dense, synthetic materials have been used for pile sliding surfaces, but the results have not always been successful. In a marine environment, biofouling materials and salt crystal buildup can cause rapid loss of material in direct sliding contact with a pile.

In theory, a pile guide should be capable of taking the maximum design load in all horizontal directions. In practice, the part of the pile guide that is attached to the dock can usually accept considerably greater applied compressive load than the hoop section can take in tension. It is therefore prudent when designing a mooring system to locate piles so that most of the piles will transmit the prevailing load direction as a compression force to the dock system. It is also generally a good idea to locate a number of piles along the main dock structure rather than having all the piles at the ends of finger floats. Piles on the ends of finger floats may tend to place the finger floats in tension or compression for which the fingers must be designed or the finger connections and structure may fail.

In addition to designing the pile guide assembly for the design load, the same as developed in the pile design analysis, the dock structure must be analyzed to assure an orderly transfer of this load from the dock structure to the pile guide. Pile guides should always be welded or through bolted to the dock system and the attachment system should be capable of resisting all tensile, compressive, and shear loads.

17.3 FIXED CANTILEVERED SYSTEMS

In areas having small water level change, it is sometimes possible to support a floating dock system by attachment to shore or some other rigid structure. In these cases, it may be desirable to utilize a cantilevered arm type of mooring system. The cantilever arms would be attached to a rigid structure such that the arm can articulate with one or more degrees of freedom. The usual practice is to provide a hinge that allows the arm to move vertically but is braced to resist horizontal movement (see Fig. 17-5). One caution in using such a system is that the floating dock will have a horizontal excursion travel as the slope of the arm changes with water level fluctuation. The system cannot generally be used in conjunction with other restraints, such as outboard piles, because of the horizontal movement of the dock. Therefore the cantilever system must be designed to accept the full design load of the structure. The actual design of the cantilever arm will vary with its configuration but will usually include considerations of tension and compression, and possibly bending and torsion. The connection of the arm to the fixed

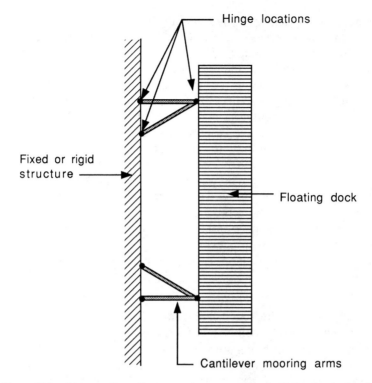

Figure 17-5. Diagram of cantilever mooring arms supporting a floating dock unit.

and floating structures must also be properly engineered. A significant bending moment may be generated by the applied load, way out at the end of the arm, and this must be accounted for in the design. In addition, the design of mooring arms should consider any tendency of the dock to have differential oscillation due to imposed irregular wave response.

Fixed cantilever mooring arm systems are generally applicable to small docks and are not suitable for mooring of large dock systems having main docks and submains with finger docks. The applicable docks are usually located parallel to a bulkhead or other extensive shore structure. In the proper environment and for a minimum length of dock, this system can be quite cost effective.

17.4 BOTTOM ANCHORED

The increased interest in the use of chain, cable and synthetic line mooring restraints within the marina industry, due to ice, deep water, poor pile conditions and aesthetic considerations prompts a re-examination of past use concepts and methods. The often neglected catenary calculations are based upon well known principles, which lend themselves well to computer design. These calculations are subject to specific interpretation as well as to misuse. Precise positioning and horizontal movement control, as they relate to pretensioning of the moored structures, is critical to the tightly controlled position requirements of today's modern marinas. Electrical cables, water lines, fire mains, pumpout force mains, and even TV cable lines will tolerate only small movements between each floating dock or fixed and floating dock combination. All floating systems must eventually be fixed to the shore and the movement slack in all these connecting lines and pipes will be limited. Rules of thumb for chain anchoring of docks and even for recreational craft swing moorings should not be used in marina design, especially in view of the ability to examine variations possible with chain sizes, lengths and anchor types along with marina space problems encountered in this highly technical, computer oriented design process.

There are several time honored rules of thumb that relate to mooring and anchoring systems:

1. Swing boat mooring rodes should have a scope of three times the depth of water, generally consisting of a heavy and a lighter chain segment along with a mooring pennant.
2. Anchoring systems with a chain rode should use a scope of three to five times the depth of water, with the smaller scope ratios related to the heavier chains.
3. Anchoring systems with a nylon rode should use a scope of at least seven times the depth of water.

The question is: does any of this make sense? Should you trust your boat on its swing mooring, your ship at anchor or your marina's position as designed by a "seaman's saying" ? How do these rules apply to the serious business of the mooring of floating docks and wave attenuators? Moored systems vary in their purposes. All systems strive to be protective and, as a most basic goal, to hold everything in place either under maximum annual conditions or under the extreme wind and wave conditions. This is best exemplified by a moored floating wave attenuator. Here the purpose is to simply hold it in place without dragging under extreme wave conditions, and perhaps to some extent minimize its horizontal motions. Floating docks may require slightly different goals where minimum horizontal movement is required, as the water level and wind or current loads vary. The goal in this case is for a comparable fixed position as experienced by pile restrained systems, but without such problems as tall piles extending skyward in large ranging tidal areas, or the pile difficulties encountered because of water depth.

MOORING SYSTEMS 425

Mooring systems are no better than their anchoring devices. The common mooring or anchoring system involves one of the many anchor designs available today. For the anchoring of ships and boats, sophisticated designs have evolved which usually incorporate lightness for ease of handling, coupled with extensive engineering of the sometimes complicated levers and surfaces to allow these anchors to hold impressive horizontal loads many times their own weight. Semipermanent mooring installations of floating structures as well as small boat swing moorings often use the ubiquitous mushroom anchor, aptly named for its shape. Often these are the anchors of choice because of price, quickness of resetting or burial as reorientation occurs, and the added conservative belief that even if the mushroom anchor does not dig in, at least it has substantial weight under water to provide a modicum of safety. To a degree this is true, but in salt water an iron or steel anchor's weight will only be 87 percent of its weight in air and it will be only slightly heavier in fresh water.

The holding power of the numerous traditional, and many of the so-called designer, anchors has been the subject of much research by the U.S. and other world navies as well as by anchor manufacturers. There are considerable amounts of data and load holding tables available that will provide this information. Regardless of its design, most self digging, self placement or gravity style of anchors hold best if the pull on the anchor is close to the horizontal and parallel to the bottom. The method of attaining a horizontal pull along the bottom, caused by a horizontal pull at the water's surface many feet above the bottom, is left to the physics of the catenary of the connecting rode. If an anchoring or mooring rode is connected to the underside of a floating structure, the weight of the rode causes it to hang vertically downward. With sufficient rode length, several times the depth of water, the other end of the rode may be laid along the bottom and connected to an anchor. The floating structure connected to this anchoring system has a downward or vertical force applied to it due to the weight of the rode. The anchor at the other end can have a horizontal force applied to it as long as the rode remains on the bottom. So the stage is set for a horizontal force to be applied to the floating structure, this force is then translated into a tension force in the anchor rode, and ultimately becomes a horizontal force translocated to the anchor. It all works well as long as the vertical section of the rode hanging down from the floating structure tries to remain vertical due to gravity and the horizontal part of the rode tries to remain lying on the bottom due to gravity. Since gravity is the major consideration here, the heavier the rode, the greater the tendency for the maintenance of this fundamental force distribution where the vertical section remains near vertical, and the horizontal portion remains on the bottom.

Ultimately the translocation of a horizontal force on the surface, down into the water along the curvilinear shape of the anchor rode, arrives at the anchor in a near horizontal orientation. This condition is attained by the chain configuration holding the shank down, even under loaded conditions, by the catenary shape formed by the chain.

Webster defines the catenary as the curve assumed by a perfectly flexible inextensible cord of uniform density and cross section hanging freely from two fixed points. Most people are familiar with the catenary shape of the cables from which a suspension bridge is hung. A light line stretched tight in the horizontal will have little noticeable sag in the middle due to the downward pull of gravity. However, in a similar situation a heavy metal chain pulled tightly, even very tightly, will have considerable sag in the center. That sag represents two resulting benefits of the catenary. One benefit for an anchoring system is derived from the characteristic catenary shape between its near vertical surface end and its near horizontal bottom end. A second benefit for an anchoring system is that the sag represents a reservoir of additional rode length, if needed. Simply stated, the heavier the cable, line or chain, the greater will be gravity's effect, the greater will be the sag and the greater will be this reservoir of excess rode or "stretchability."

If there were no such thing as a catenary or if the rode were neutrally buoyant, under a horizontal surface force the rode would simply become taut, straight, and be connected directly with the anchor. Its angle of connection with the bottom would be theoretically the same as the angle of its departure from the surface and the angle would be geometrically determined by the water's depth and the length of the rode. Shallow water or a long rode means small angles at the top and bottom. Deep water or a short rode means larger angles and therefore greater downward pull of the surface structure and an upward pull on the anchor, tending to lift it. With the principle of the catenary working, it is possible to manipulate the shape of the anchoring or mooring system to create the most beneficial condition for the holding power of the anchor. That beneficial condition occurs at the near horizontal approach of the rode to its connection with the anchor.

So the catenary does two things. It can provide the proper force lead to a set anchor by translocating the horizontal pull down the rode and at the same time allow some effective flex or apparent stretch of an otherwise inelastic mooring line. As mentioned above, if there were no such thing as a catenary, the use of an inelastic anchor or mooring rode such as a steel cable or a metal chain would be of little benefit other than for providing strength. On the other hand, a stretchable rode such as a nylon rope furnishes at least the flex and stretch properties of the catenary, but unfortunately it helps little to translocate the horizontal pull applied by a floating object at the surface to a horizontal pull at the anchor on the bottom.

The mathematical equations that describe the catenary shape are quite explicit and easily solved by hand or, especially, by computer. An array of very interesting results can be obtained which describe in detail all of the angles, forces and lengths relating to specific catenary problems. Let us look at the reaction of mooring rodes when it is desirable to have a zero degree angle intercept with the bottom for the best possible holding power, and determine the length of line needed to form the catenary so that this configuration can be attained. Figure 17-6 shows both an approximately equal vertical and horizontal scale presentation, and a more definitive but vertically distorted drawing showing various sized chain rodes and a one inch nylon rode, all placed under a 5000 pound horizontal load. Five thousand pounds of horizontal load is an average upper load for dock mooring rodes. As can be seen in perspective, the lighter the rode's unit weight, the longer the rode that is necessary, with the one inch nylon rode requiring hundreds of feet to attain the proper catenary shape for the zero degree bottom approach angle.

Looking at this in another way, consider a situation in which a one inch chain just intersects the bottom at zero degrees under a 500 pound horizontal pull, the conditions require that the chain be 51 feet long. What happens if that 1 inch chain rode is replaced with a lighter chain and finally with a one inch nylon rode of the same length, all under a 500 pound horizontal surface force? Figure 17-7 displays the comparative reactions with only the one inch

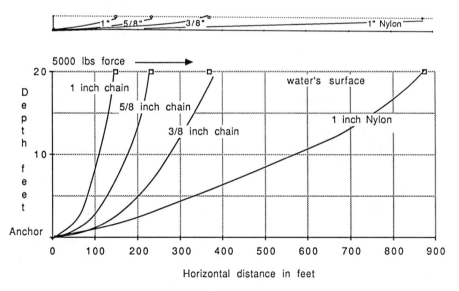

Figure 17-6. Anchor rode catenary for various chain or nylon rode sizes with a surface loading in the horizontal of 5,000 pounds. Top picture shows true scale, bottom picture uses distortion in the vertical for clarity.

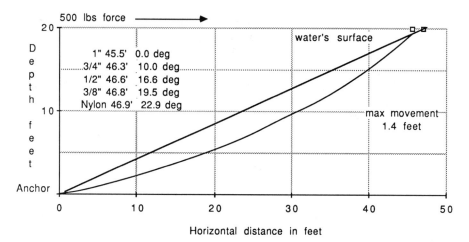

Figure 17-7. Anchor rode catenary for a 51 foot long mooring using various chain or nylon rode sizes with a surface loading in the horizontal of 500 pounds. The data shown gives the angle that the rode makes with the bottom and the horizontal distance from the anchor position to the float.

chain and the nylon rode presented as the extremes. If this situation involved the use of an anchor that absolutely required that its shank remain parallel with the bottom, lying flat against it horizontally, only the one inch chain would prove successful in realizing the full holding potential of that anchor. Even the three-quarter inch chain produces a ten degree uplift tilt for the shank of the anchor. It is also interesting to note that under this particular analysis a horizontal distance difference of 1.4 feet exists for the loaded one inch nylon rode as compared to the loaded one inch steel chain. This distance is an approximate measure of the chain length reservoir contained in the steel chain's catenary for this length of chain.

Figure 17-8 shows another interesting display of the reaction of a 1 inch steel chain as it loads up from its rest position with its float at 31 feet from the anchor's horizontal location, to a distance of approximately 47 feet for its fully loaded position. Notice how only a 100 pound horizontal pull moves the chain's surface position to within a few feet of a more fully loaded location. This demonstrates quite well how the use of pretensioning can significantly reduce the amount of dock movement between a so-called no load and a fully loaded condition. With only a comparatively small amount of initial or installed load on the system, 85 to 95 percent of any horizontal movement can be eliminated at the time of installation. This pretensioning is attained through the use of opposing rodes in the design of a mooring system for floating docks. To take full advantage of this concept, a computer analysis is required, especially for large ranging tide or lake level fluctuation areas, in

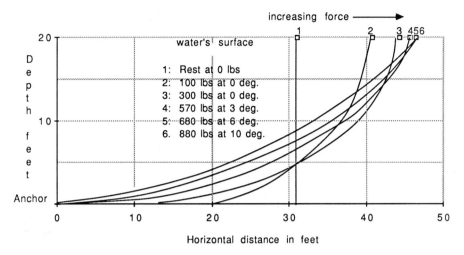

Figure 17-8. Anchor rode catenary changes as it loads up with increasing horizontal force. The angle that the rode makes with the bottom at each load level is indicated.

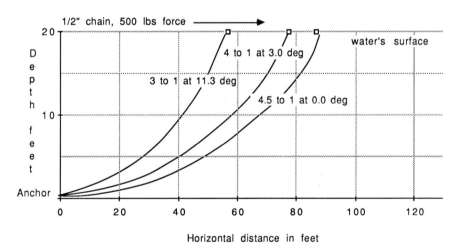

Figure 17-9. Anchor rode catenary changes for increasing anchor rode lengths with a horizontal force of 500 pounds. Rode lengths are indicated as the ratio of length to water depth.

order to optimize the rode's length and weight and the resulting degree of horizontal movement that will result.

Figure 17-9 addresses the concept of chain length rules of thumb. Using a half inch chain and 500 pounds of horizontal force the 3 to 1 rule produces a severe uplift of 11.3 degrees on the anchor. It takes a minimum of 4 to 1 in this case to maintain a 3 degree angle with the bottom and a 4.5 to 1 for a zero degree intercept. Doubling the horizontal force to 1000 pounds as shown in Figure 17-10 creates even more severe uplift problems in the same water depth. In this case a 6.3 to 1 ratio is needed to maintain the zero degree intercept. All this leads to the conclusion that the greater the pull, the longer the rode necessary to reduce this uplift problem. So it appears that before any rules of thumb on chain length versus water depth can be used, three other questions must first be answered. One question involves the holding power of the anchor for various uplift angles of the shank. All anchors are not created equal and some will break out of the bottom with only a slight, 3 to 6 degree, uplift on their shanks while others can tolerate 10 to 15 degrees or more. The second question relates to the weight of the rode being used and its ability to form the required catenary. A few feet of chain attached to an anchor and then connected to a nylon rode is often offered as a good technique for keeping the anchor's shank close to the bottom. Referring back to Figure 17-6, the ratio of water depth to rode length for a one inch nylon line with a zero degree intercept with the bottom is 44 to 1. Obviously, any added weight provided by the addition of chain will help to shorten the required rode length, but a few token feet of chain is virtually a waste of time when the load builds up on the rode. Even using all five-eighths inch chain as shown in Figure 17-6, this case requires a ratio of nearly 12 to 1, so a few feet of chain added to the nylon rode will not do much for the system's performance under load. It is obvious that the rode's weight and the anchor's holding abilities at various uplift forces must be carefully considered or your

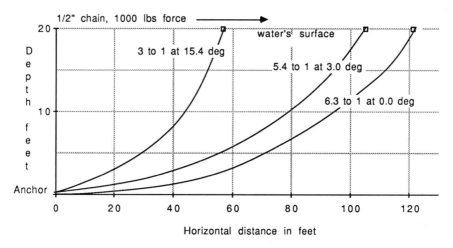

Figure 17-10. Anchor rode catenary changes for increasing anchor rode lengths with a horizontal force of 1,000 pounds. Rode lengths are indicated as the ratio of length to water depth.

bottom anchored mooring system becomes a great mystery regarding its actual holding characteristics. This may be all right for the transient yachtsman anchoring for the night, but it is not all right for a marina installation which is expected to be eventually stressed to its design loads. The last question raised is the size of the horizontal surface pull. As demonstrated, the greater this pull, the longer the rode must be, which all but eliminates any serious use of a rule of thumb.

So a major problem in anchoring floating docks and wave attenuators is the fact that in order to obtain proper holding of the anchor, a small angle of approach to the bottom is required and therefore a significant length of chain may have to be used. Many times this creates a situation which is untenable due to existing marina space or the proximity of state or federal channels or other bottom areas where anchor installation is prohibited. As pointed out above, even a 5 to 1 ratio for a floating dock rode may be of insufficient length.

In many installations the commonly used anchor is being replaced by bottom fixtures which nail the rode to the bottom at a precise location. Coupled with a good definition of the shape of the bottom, it allows the mooring to be precisely designed and installed and to perform as expected. Embedment anchors, buried concrete blocks, steel plates, stud piles, chain or cable webs and similar systems can be installed in precise locations and produce horizontal mooring distances far shorter than would be required for conventional anchoring systems. In these cases, the up angle on the anchor is less important, providing that the potential extraction force has been considered in the design of the anchoring device. Angles of 25 degrees and greater may, under some circumstances, be allowed, depending on the conditions. Thus, shorter anchor rodes, and in some cases fewer anchors, can produce space saving, compact designs.

As it turns out, shorter anchor rodes and larger bottom angles are mixed blessings because they can also create additional problems not encountered with normal anchoring techniques. Figure 17-11 compares a one inch chain mooring system with 1000 pounds of horizontal loading for changing water levels. Two chain lengths are used for demonstration purposes. A shorter 40 foot rode in 15 feet of water goes from an angle of 15 degrees with the bottom to an angle of 25 degrees as the water level rises to 20 feet. At the same time the float structure is pulled 2.8 feet in the horizontal direction closer to the anchor location. A similar situation with a longer rode of 60 feet only moves a distance of 1.5 feet closer to its anchor location. The smaller the ratio of the horizontal distance between the anchor's location on the bottom and the floating structure's location on the surface, to the depth of the water, the greater will be the amount of horizontal excursion as water levels change. This can create some interesting and challenging mooring design problems

432 PART 3/ENGINEERING DESIGN

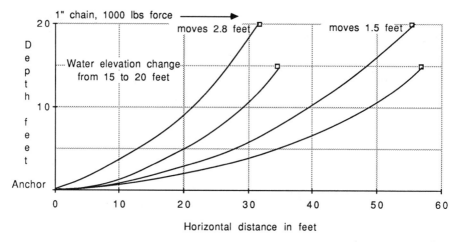

Figure 17-11. Anchor rode catenary changes for differing water surface distances to the bottom. Two different rode lengths are also shown, with the effect on the horizontal float to anchor distance indicated.

for marinas with bathymetric depth variations from one part of the marina to another (see Fig. 17-12).

The amount of the horizontal movement to be tolerated in a moored floating dock installation can often be an important design and economic consideration for the project. The movement or location change between a condition which includes full design force loaded docks at low tide, for example, and their unloaded rest position at high tide represents the maximum horizontal excursion distance that will be allowed by the mooring design. This distance becomes an important mooring design criterion that must be established before the complete mooring design can be carried out. It becomes the horizontal movement distance limitation criterion that must be scrupulously followed and applied to all parts of the marina docks and associated complex during the mooring design process or considerable warping, racking and twisting may occur as water levels and loads change.

As demonstrated in Figure 17-11, the smaller the ratio of the horizontal distance from the float to anchor, to depth of water, the larger the horizontal movement will be with changing water levels. Generally in the design process, the maximum water depths encountered, coupled with the shortest horizontal distances available between the dock and the anchor, set the upper limit of the design criterion. A reduction in horizontal movement for all other parts of the dock installation can then be completed and properly matched for the various sections. The initial object is to design the mooring system so that the floating docks will remain in exactly the same position horizontally

MOORING SYSTEMS   433

Figure 17-12. Constitution Marina, Charlestown, Massachusetts. This 260 slip marina is held by a chain, bottom anchor system in water depths that vary from 8 feet to 40 feet along a single length of dock. The mean tidal range adds 9.6 feet to the water depth. A computer generated mooring solution was necessary to design the system. *(Photo credit: William F. Johnston, Quincy, MA)*

throughout the rise and fall of the water level. It is important to understand that with opposing mooring rodes which may, and often do, lead to anchors in different water depths, the amount of horizontal movement allowed by each rode must be carefully matched so as to be nearly equal. If this is not done the docks will shift horizontally to a greater or lesser extent during water level oscillations, depending on the degree of mismatch of the opposing mooring rodes. With uneven bottom depths a compromise allowable movement must be established and all moorings fitted and designed for that allowable movement. This can be done by variations in the mooring rode lengths and mixes of chain sections and weights in order to attain what appears to the dock to be a matched or tuned system. It is an interesting computer exercise and is best left to the professionals. As an alternative method, a false uniform bottom depth horizon within the marina may be established using a combination of submerged piles and cable webs. Points of rode connection can be then established for convenience of position or the necessary rode lengths determined for water level variations. The mooring design then becomes quite simple since all depths in all parts of the marina can be made exactly the same and all the connecting rodes are therefore the same.

Regardless of bottom irregularities, consideration must also be given to pretension of the mooring system so that even in an unloaded situation, the docks remain close to their fully loaded locations. This further complicates the design and one must be careful to make sure that sufficient excess rode length is stored in the so-called catenary reservoir so that under storm tidal conditions the floating structures will not be pulled under the surface. One approach to this problem involves the use of clump weights, heavy block weights positioned at some midpoint along the rode. The idea here is that the clump weight forms an ancillary anchor and is only lifted off the bottom or repositioned during storm conditions. It aids greatly in providing a more horizontal pull on the main anchor because of its downward pull on the rode. Its use has several serious drawbacks. One involves the chafe on the rode at its point of attachment and the second is that after bottom lift off occurs during storm loading conditions, its rest location following the storm may be considerably different from where it started. This changes the geometry, and therefore the response of the whole system. In other words, the movements of the clump weights are unpredictable and they find little use in a tightly controlled or limited float movement mooring design.

More and more marina installations want to moor their floating structures. This provides for attractive, unencumbered, and clean looking installations. It also provides survivability in ice hostile and storm prone areas, high tidal fluctuation regions, and it eases the process for annual removal of docks or wave attenuators for storage elsewhere when necessary. The fixed position attributes of a pile restrained floating marina are usually desired by the designer of mooring arrays. This means that very careful mooring designs must be carried out in conjunction with the use of modern embedment type anchors in order to assure the unchanging geometry and integrity of the system. The relatively capricious, rule of thumb treatment for moored floating systems is rapidly being replaced by scientific studies and computer aided designs. Proper mooring design requires a relatively complicated computer analysis. Once done properly however, an installation should not require future tending of the rodes as conditions vary from season to season.

17.5 SWING BOAT MOORINGS

In the case of swing boat moorings, the idea is to hold the boat in place at least during the maximum annual wind and wave conditions for a specific area, with 360 degrees of freedom to swing, and at the same time, to efficiently utilize competed for water space to maximize the numbers of swing moorings by using the shortest mooring rode possible. Since the mushroom anchor is used extensively for many purposes in the marine world and especially for swing moorings, understanding its performance characteristics provides some insight into effective mooring configurations.

An accepted criterion states that if a mushroom has its dish half buried, shank horizontal allowing for a horizontal pull, it holds approximately three times its air weight in sand and two times its weight in mud. With a shank angle of between three and six degrees it will lose approximately 30 percent of its holding capacity. Beyond six degrees it will lose 50 percent or more, resulting in the holding force approximating the dead weight of the submerged mushroom anchor. The upward pull results in an additional loss of anchor weight once it breaks out. Effective use of the mushroom anchor therefore requires the shank to be held flat or near flat on the bottom which should allow automatic half burial of its dish under loaded conditions.

Many other types and configurations of anchors have been traditionally used for boat moorings. Old engine blocks, balled up heavy chain, automobile or railroad axles with a wheel attached, and junk steel or concrete blocks of many different shapes often form the bottom weight or anchor for swing boat moorings. Their use, in general, is not encouraged mainly because of the inability to assign any holding strength for these objects. If you don't know what horizontal force they may be capable of resisting, then you really cannot use them in a mooring design. It is difficult enough to assign holding values to various commonly used anchors, especially considering the mud or sand bottom type variations that can occur. However, if the oddball clump of material is to be used as an anchor, the problem of estimating the holding power becomes impossible unless in situ pull tests are carried out.

For years the yachtsman's sources of information have suggested that swing boat moorings be determined by water depth and should consist of a heavy and a lighter chain segment along with a mooring pennant. For example, it is often suggested that the length of the light chain be equal to the water's depth and the length of the heavy chain be equal to one and one half times the water's depth along with a short pennant governed by a multiple of the bow chock height. At the very least, this translates to no more than an overall mooring of three times the depth of the water. Many available sources of printed information cite mooring system specifications which have been quoted and requoted, and which may be traced back to before World War II. These mooring specifications have never changed and continue to be quoted and used by harbor management commissions, municipalities, yacht clubs, marinas and harbor masters, even though boat configurations and wind loadings have changed significantly for the given boat lengths. A narrow beamed, low freeboard 35 foot sailboat of 1940 is a completely different vessel from a contemporary vessel of similar length. The 35 foot sailboats of today are high sided, half again as wide as they used to be, have windshields, dodgers, biminis, and an array of wind steering vanes, surfboards and inflatables stored on deck. By conservative estimate, the wind loading on sailboats of today may be two to three times what it was when many of these mooring specifications were developed. A similar situation is true for power boats and may be even worse for them. There are few power boats over 25 feet without a flying bridge, tuna tower, outriggers, or top enclosures.

All of these configurations greatly increase the wind loading and wave loading on these vessels and create a surface craft that wants to swing quite widely from side to side on its mooring tether. The wind loading not only creates an increased pull on the mooring pennant as the wind drives the vessel backwards, but it also creates an even greater pull on the mooring as the vessel sails to the right and left of the wind, presenting a much larger drag profile to the wind than just its frontal outline.

A computer based set of mooring calculations using modern boat hull configurations and chain catenary determinations is shown in Table 17-1 for comparison with some commonly suggested moorings from yachting guides. An effective water depth of 18 feet, which included a storm tide and the vertical distance from the water to the cleat on deck, was used. The computer

Table 17-1. Comparison of swing mooring arrangements as suggested by traditional yachting sources and as determined by computer analysis.

	Yachting information sources			Computer determined		
	Generic	35' Cruising	35' Racing	Cruise 30'	Cruise 37'	South N.E. 32-38'
Mushroom wt. in lbs.	--	250	200	350	500	500
Heavy chain, in.	27'	30'@ 1"	30'@3/4"	24'@ 1"	31'@ 1"	15'@ 1"
Light chain, in.	18'	20'@3/8"	20'@5/16"	30'@1/2"	31'@1/2"	65'@1/2"
Pennant length	9'	20'	20'	19'	26'	15'
Scope chain/total	45/54'	50/70'	50/70'	54/73'	62/88'	80/95'

based calculations shown represent site specific, minimum mooring specifications for given wind and wave conditions for a southern New England harbor. A wind loading of 45 knots of steady wind, a 3 foot wave and 60 knot gusts were used to determine the force on each boat. They are minimum configurations based on a maximum allowable up angle of the mushroom shank of 6 degrees, and a holding power of twice the weight of the mushroom. Any smaller scope of rode or chain weight will cause the up angle to exceed 6 degrees and breakout of the mushroom will occur when the holding power drops to the weight of the mushroom anchor alone.

As shown in Table 17-1, the commonly accepted Generic column has a ratio of rode length to depth suggestion of exactly 3.0 to 1, whereas other source specifications are 3.9 to 1, and the computer based designs range between 4.1 and 5.3 to 1 depending on the boat size, using a heavier mushroom anchor in all cases. **In the absence of a site and boat specific computer mooring design, it would appear that with modern boat designs and the increased wind and wave loading which is expected, a more realistic 4 to 1 chain rode length to depth ratio should become the common minimum rule of thumb for swing boat moorings which should also add at least another 100 to 150 pounds of anchor weight in order to obtain more favorable holding power.** One additional caution should be made here. When dealing with rode lengths and depth of water ratios, make sure that a storm tide is always included in the calculations. Several feet of additional water depth can easily change a 3 degree up angle on a mushroom's shank to beyond 6 degrees with the potential for pull out.

Mooring Grid Spacing

The total length of the mooring rode plays an important role in determining the grid spacing or distance within the mooring area. The grid spacing is defined as the distance between the positions of the adjacent anchor locations. The fundamental concepts involve first computing the required mooring swing radius which consists of the length of the total mooring rode up to the deck cleat plus the length of the boat. Next, the determination of the grid spacing that will reduce the possibility for adjacent boats to collide. Nothing guarantees that two adjacent boats will not touch within a mooring grid unless each boat is separated from its neighbor by a distance equal to the sum of both of their mooring swing radii as defined above. Then, at least in theory, the boats may stream out in opposite directions from each other without touching. This is a highly unlikely situation, especially under full wind and sea loading conditions, and it is unlikely that valuable water space would be devoted to such a configuration. If this situation represents the largest grid spacing that is probable, then what is the smallest grid spacing that may be accepted?

Any grid spacing that is made smaller than the largest one just described will also present the increasing probability of adjacent boats colliding. Furthermore, any smaller grid spacing must unfortunately rely on the idea that all boats will swing identically with the wind, current and sea conditions. The smallest grid spacing to be accepted would be one that is determined by subtracting the low tide depth or lowest expected surface level of water, the pennant length, and the boat length from the previously established mooring swing radius. What you are left with is the potential length of chain that could be lying in a straight line on the bottom with the boat in a resting or unloaded condition and with an adjacent mushroom anchor placed several feet just beyond. This configuration just barely reduces the tendency for one mooring's chain from becoming entangled with the adjacent mooring's mushroom anchor or bottom chain when both boats and moorings stream in the same direction. Unfortunately, if each boat were to somehow stream their bottom chains in opposite directions, they will still intertwine on occasion. In addition, since moored boats tend to sail, swing or oscillate from side to side on their moorings, it is conceivable that two boats will sail together with their sides meeting. Usually, the higher the wind conditions, the more violent will be this sailing about on the mooring. As the wind increases, the arc radius of swing increases along with the velocity of movement, thereby increasing the collision probability and the severity of impact. So as the grid size is increased from this defined shortest distance, the defined maximum distance is approached, while reducing the probability for the boats to collide or their moorings to twist together.

There are many different approaches and philosophies which are used in determining an acceptable grid spacing. Each site differs in prevailing and storm wind directions, the presence and direction of tidal or river currents, and the wave arrival direction exposure, which may not be simply down wind. Careful site analysis and consideration of special environmental conditions will allow an optimized spacing to be established, within a series of specific agendas and compromises. What that means is that grid spacing is often legislated rather than engineered. It is usually necessary to determine different mooring sizes for different boat sizes and water depths within a mooring grid. It is often unwise to mix grid spacings. Less likelihood for collisions or chain twisting will occur if the mooring area is divided into several different but homogeneous mooring swing radius areas, or as an alternative, let the largest mooring swing radius determine the overall grid spacing. This allows the mooring owner, occupant or patron to increase their mooring and boat size when desired, without the problems of relocation, reassignment or loss of a mooring opportunity. This often works well and gives rise to fewer disgruntled people, when mooring assignments are administered by yacht clubs or municipalities, the mooring resources are limited, and the waiting lists long. However, it also means that the mooring grid must be tailored in advance.

With the requirement for a surface anchor and sufficient scope to the swing mooring rode, the design of a mooring grid becomes a space problem. Some areas may not have sufficient space to provide for all of the moorings desired. That problem may be solved by using some form of embedment anchor, metal plate, or stud pile cut off at the mud line, to affix moorings to the bottom. If this can be done, the moored boats may be tethered a good deal closer to each other and many more moorings can be placed. The only considerations now will be the upward extraction pull caused by a much shortened mooring rode and sufficient scope must be made to allow for abnormal tides. Additional advantages of embedment anchoring systems arise because heavy chains can now be replaced with much lighter chains, cables or even nylon lines. Furthermore, the periodic repositioning of errant mushrooms is no longer required. However, periodic inspection, measurement, or visual observation of the embedded system must be made, and that is generally done with a diver.

The mooring grid design can be established in many geometric patterns, and often the shape and boundary arrangement of smaller mooring areas may dictate the mooring location pattern that works best for the site. One logical pattern quite commonly used, especially for large mooring areas, involves the equilateral triangle. The equilateral triangle has three equal legs and three equal interior 60 degree angles. The length of each leg is set at the design grid spacing for the pattern. The grid is most easily laid out using parallel lines with each mooring position spaced along each line at the designed grid distance. Adjacent parallel layout lines have equally spaced but staggered locations for each mooring position. The parallel lines will be separated from each other by 86.6 percent of the designed grid spacing. The geometry of this pattern is shown on Figure 17-13.

The spacing of the mooring locations, specifically the anchor location is certainly not an exact science. The common mushroom anchor, even with its dish half buried and lying on its side will move around on the bottom. As near full load conditions occur from differing directions, the mushroom will rotate accordingly and align itself with the direction of the pull. The larger or heavy chain will straighten out and under high pull storm conditions the entire mooring configuration will form a straight line. Changing storm wind conditions will eventually rotate the entire system, straightening it out again in a different orientation. As the mushroom rotates towards a new realignment, partial or full extraction from the bottom will occur and some horizontal movement of the entire mooring may take place. Since this is a random occurrence, the direction of movement is not predictable, but the wandering of the mushroom's location on the bottom will occur. If the mooring configu-

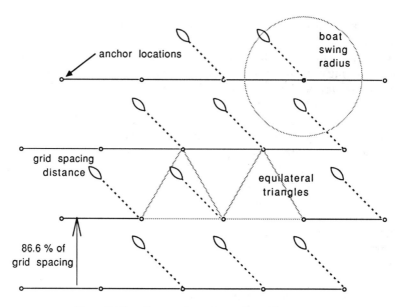

Figure 17-13. Commonly used mooring grid geometry.

ration is insufficient even for normal winds or nonstorm conditions, the daily movement of the moored boat back and forth and from side to side can cause rapid displacement of the mushroom. Any frequent movement and disarray within the mooring area's pattern is a good indicator that the specified moorings are insufficient, ether individually or collectively. Aerial photographs of the mooring buoy locations during the off seasons, or with all boats in-place during the season, can sometimes provide useful annual information about the extent of this movement problem and when it might be necessary to rectify it.

One final point. Any mooring system is only as good as the pennant's chafe protection provided by each boat owner. If chafing occurs from a poorly designed bow cleat, even a one inch nylon pennant can be cut through in a matter of minutes. That boat is threatened with disaster, obviously, but so is the whole fleet of moored vessels down wind and down current from it. The proper selection and enforcement of the use of chaffing gear is a mooring area management problem and is just as important as the proper mooring specifications or mooring grid determinations. There have been many surveys made following storms which showed that the majority of boats that were lost or damaged resulted from the parting of the mooring pennant, usually due to chaffing. This is an easily preventable happening. Simply put, proper chaffing gear is the same as the so-called nail that was missing from the horseshoe that resulted in the loss of king and kingdom.

PART 3 – INFORMATION SOURCES

1. American National Standards Institute, Inc. 1986. *American National Standard for Buildings and Facilities: Providing Accessibility and Usability for Physically Handicapped People.* New York: American National Standards Institute, Inc.
2. American Society of Civil Engineers. 1969. *Report on Small-Craft Harbors, ASCE—Manuals and Reports on Engineering Practice—No. 50.* New York: American Society of Civil Engineers.
3. Beebe, R. P. 1975. *Voyaging Under Power.* New York: Seven Seas Press.
4. Blain, W. R. and Webber, N. B., eds. 1989. *Marinas: Operation and Design, Proceedings of the International Conference on Marinas,* Southampton, UK, September 1989. Southampton, UK: Computational Mechanics Publications.
5. Brown, B. C. 1986. *Stroke!,* Camden, ME: International Marine Publishing Company.
6. Bruun, P. 1981. *Port Engineering,* 3rd edition. Houston, TX: Gulf Publications.
7. Chamberlain, C. J. 1983. *Marinas, Recommendations for Design, Construction and Management,* Vol. 1, 3rd edition. Chicago, IL: National Marine Manufacturers Association.
8. Commonwealth of Massachusetts, Division of Capital Planning and Operations, 1988. *Design for Access: A Guidebook for Designing Barrier Free State and County Buildings.* Boston, MA: Commonwealth of Massachusetts.
9. Davies, Thomas D., Jr. and Kim A. Beasley. 1988. *Design for Hospitality: Planning for Accessible Hotels and Motels.* New York: Nichols Publishing.
10. Department of the Navy. 1965. *Marine Biology Operational Handbook: Inspection, Repair and Preservation of Waterfront Structures.* Washington, DC: Department of the Navy, Bureau of Yards and Docks.
11. Dunham, J. W. and Finn, A. A. 1974. *Small-Craft Harbors: Design, Construction, and Operation. Special Report No. 2.* Vicksburg, MS: U.S. Army Corps of Engineers, Coastal Engineering Research Center, Waterways Experiment Station.
12. Gaythwaite, J. W. 1981. *The Marine Environment and Structural Design.* New York: Van Nostrand Reinhold.
13. Gaythwaite, J. W. 1990. *Design of Marine Facilities for the Berthing, Mooring and Repair of Vessels.* New York: Van Nostrand Reinhold.
14. Henry, R. G. and Miller, R. T. 1965. *Sailing Yacht Design.* Cambridge, MD: Cornell Maritime Press, Inc.
15. International Marina Institute. 1987. *Proceedings of the Marina Design and Engineering Conference,* Boston. Wickford, RI: International Marina Institute.
16. Kinney, F. S. 1962. *Skene's Elements of Yacht Design.* New York: Dodd, Mead & Company.
17. Lord, L. 1963. *Naval Architecture of Planing Hulls.* Cambridge, MD: Cornell Maritime Press.
18. Marinacon. 1988. *Proceedings of the 3rd International Recreational Boating Conference, Singapore.* Rozelle, NSW, Australia: Marinacon.
19. Marine Training Advisory Board. 1988. *Marine Fire Prevention, Firefighting and Fire Safety.* Washington, DC: Superintendent of Documents, U.S. Government Printing Office.
20. Moss, F. T. 1981. *Modern Sportfishing Boats.* Camden. ME: International Marine Publishing Company.
21. Murphy, G. 1961. *Properties of Engineering Materials.* Scranton, PA: International Textbook Company.
22. Myers, J. J., Holm, C. H., McAllister, R. F. 1969. *Handbook of Ocean and Underwater Engineering.* New York: McGraw-Hill Book Company.
23. National Fire Protection Association. 1990. *NFPA 303: Marinas and Boatyards.* Quincy, MA: National Fire Protection Association.
24. National Fire Protection Association. 1990. *NFPA 307: Marine Terminals, Piers and Wharves.* Quincy, MA: National Fire Protection Association.

25. National Fire Protection Association. 1990. *NFPA 312: Fire Protection of Vessels During Construction, Repair and Lay-Up.* Quincy, MA: National Fire Protection Association.
26. Quinn, A. D. F. 1972. *Design and Construction of Ports and Marine Structures.* New York: McGraw-Hill Book Company, Inc.
27. Robinette, Gary O. 1985. *Barrier-Free Exterior Design: Anyone Can Go Anywhere.* New York: Van Nostrand Reinhold.
28. State of California, The Resources Agency Department of Boating & Waterways. 1984a. *Layout and Design Guidelines for Small Craft Boat Launching Facilities.* Sacramento, CA: The Resources Agency, Department of Boating & Waterways.
29. State of California, The Resources Agency Department of Boating & Waterways. 1984b. *Layout and Design Guidelines for Small Boat Berthing Facilities.* Sacramento, CA: The Resources Agency Department of Boating & Waterways.
30. States Organization for Boating Access. 1989. *Handbook for the Location, Design, Construction, Operation, and Maintenance of Boat Launching Facilities.* Washington, DC: States Organization for Boating Access.
31. Tsinker, G. P. 1986. *Floating Ports, Design and Construction Practices.* Houston, TX: Gulf Publishing Company.
32. U.S. Forest Products Laboratory, U.S. Dept. of Agriculture. 1974. *Wood Handbook: Wood as an Engineering Material,* Agriculture Handbook No. 72. Washington, DC: Superintendent of Documents, U. S. Printing Office.
33. Veterans Administration. 1978. *Handbook for Design: Specially Adapted Housing.* Washington, D.C.: VA Pamphlet 26-13. Department of Veterans Benefits, Veterans Administration.
34. Whiteneck, L. L., Hockney, L. A. 1989. *Structural Materials for Harbor and Coastal Construction.* New York: McGraw-Hill Book Company.
35. Zadig, E. A. 1974. *The Boatman's Guide to Modern Marine Materials.* New York: Motor Boating & Sailing Books.

PART 4
OPERATIONS AND MANAGEMENT

18

Haul-out and Boat Handling Systems

Since earliest times, boating facilities have provided some form of boat haul-out system to allow storage or repair of boats. Primitive systems used a gradually sloping beach with log rollers and some form of pulling motion, often furnished by men or horses. A more modern version of this concept is the slipway or marine railway dry dock for large vessels and the conventional automobile boat trailer or larger hydraulic trailer for recreational boats. In areas of tidal fluctuation, a vessel was, and still is, often grounded out at high water and worked on during low water levels. In areas having little or no tidal influence, floating dry docks were developed to allow a vessel to be dry docked on the hull of another vessel. The original floating dry dock is thought to have been constructed during the time of Peter the Great when a British sea captain found himself with a leaky vessel stranded in the tideless Baltic Sea. He was innovative, and he found a larger derelict vessel, the *Camel,* whose shell he used in combination with new watertight bulkheads to form a vessel that he could place his leaky ship into and then dewater. This same principle is used today in the floating dry dock. In other areas, basins were dug in the upland and fitted with closure gates at the seaward end. A vessel was floated into the basin, the gates sealed and the basin dewatered while the vessel was braced up and kept dry. Basin or graving docks are still used for large ships but their cost is generally prohibitive for small boat operations. Cost and land or water occupation also generally rule out the use of marine railways and floating dry docks for small boat handling. Smaller more cost efficient means of small boat haul-out are required for recreational boats.

Straddle hoists became the prominent form of small boat haul-out facility shortly after they were devised in the 1950s. The system uses a structural steel frame to which fabric slings are attached. The slings are placed around and beneath the boat and raised to bring the boat out of the water. Early straddle hoists ran on railroad type rails and wheels and were limited to short distances of travel. Later versions of straddle hoists were fitted with pneumatic wheels and steering capability, allowing the hauled vessel to be moved about the boat yard or marina for storage or repair. Straddle hoists do require structural piers to run on and sufficient water depth to accept the full draft of the boat. In some cases these requirements precluded the effective use of straddle hoists so another system evolved, the hydraulic trailer.

Hydraulic trailers are a relatively new innovation, although traditional automobile trailers have been in use for decades. The hydraulic trailer has the ability to access a boat from a properly designed sloping ramp. The units have adjustable hydraulic arms with bearing pads that can be placed on the vessel bottom to provide support. The adjustable supports allow accommodation for virtually any configuration of boat hull, without the need for special blocking or trailer adjustment. Hydraulic trailers have been configured for boats up to 60 feet in length and possibly for greater lengths. The trailers can also be constructed for over-the-road use which allows a boater to have his boat delivered to almost any off site location for storage or repair. In fact, the original development of the hydraulic trailer by Brownell in Massachusetts occurred because of the need of a boat yard located some distance from the waterfront to be able to haul and store boats.

Recently there has been a rapid growth in on-land boat storage of small boats, and this has resulted in improvements to rack storage boat handling systems. Larger, greater capacity, and greater reach forklifts have been developed to enhance this form of boat handling. Forklifts having vertical extension beyond 30 feet are available as are large fixed crane systems which allow boat stacking five or more tiers high.

The state-of-the-art in boat handling is rapidly changing and improving. As waterfront land becomes more precious, new innovative methods to increase the density of boat storage and provide more rapid movement of boats from land to the water will be developed.

18.1 STRADDLE HOISTS

Dry docking of small boats with straddle hoists is a routine operation in the thousands of facilities which use this type of facility worldwide (see Fig. 18-1). In most cases, if the boat to be hauled can fit between the haul out piers it can be lifted by the straddle hoist. The factors of safety built into the hoist by the manufacturer allow for local overloads without compromising the structural integrity of the hoist or damaging the vessel. The wheel base dimensions of the hoist versus the length of vessel hoisted is generally in a range that precludes significant vessel overstress caused by large overhanging of the fore and aft ends of the vessel beyond the sling position points of the hoist.

The recent increase in straddle hoist capacity, in the range of 100 tons to 500 tons (see Fig. 18-2), creates a situation in which large vessel loads are

HAUL-OUT AND BOAT HANDLING SYSTEMS 447

Figure 18-1. Modern 70 ton capacity yacht straddle hoist executing a 90 degree turn in close quarters. *(Photo courtesy: Marine Travelift Inc., Sturgeon Bay, WI 54235)*

Figure 18-2. A major straddle hoist haul-out facility. A 500 ton capacity straddle hoist with a 200 foot long landing craft. *(Photo courtesy: Marine Travelift Inc., Sturgeon Bay, WI 54235)*

supported by a few, closely spaced hoist slings. It therefore becomes necessary to properly analyze the loading and positioning of large vessels to preclude overstressing of the hoist or the vessel and to assure adequate load balance in the hoist slings. Sling overload caused by improper positioning may create a situation in which the system is unable to safely lift a vessel that, if properly positioned, could be lifted.

It is assumed that a hoist's capacity is equal to the manufacturer's rated and certified capacity and that the hoist is well maintained and, in all respects, mechanically and structurally capable of performing the required hoisting maneuver. Heavy use and wear on a straddle hoist may limit safe lifting capacity. This is especially true of hoisting slings. Hoist slings should be regularly checked for damage and wear and replaced at manufacturer's suggested intervals. Near capacity dry dockings may require new slings to assure adequate strength.

In order to improve the safety of straddle hoist haul-out of large boats, a methodology is presented for determining vessel loading. The analysis is similar to the normal analysis used for any dry docking on marine railway dry docks, floating dry docks, vertical lifts or basin (graving) type dry docks. The procedures use simple physics and some naval architectural considerations.

Vessel Characteristics

The initial investigation for any successful vessel dry docking maneuver is to obtain as much information about the vessel as possible. Appropriate information includes:

Principal Dimensions: Length over All (LOA); Length along Waterline (LWL);

Beam (Bm); Drafts Forward and Aft, (D_F and D_A);

Vessel Displacement (W), i.e., Weight of Vessel, as floating;

Location of Centers of Weight, i.e., Longitudinal Center of Gravity (LCG) and Vertical Center of Gravity (VG);

Docking Plans, Curves of Form, Bonjean Curves, Displacement Curves, etc. for the vessel to be dry docked should be obtained for the analysis.

Vessel information may be obtained from the owner/agency controlling the vessel, from published characteristics in trade journals and magazines, from operators of boatyard who previously dry docked the vessel, and if necessary from physical measurement of the vessel in docking condition. Estimated weights and dimensions are better than nothing, but all possible attempts should be made to obtain good, accurate vessel data.

HAUL-OUT AND BOAT HANDLING SYSTEMS 449

The vessel hull should be checked for the locations of below water projections such as: transducers, keel cooling grids, rudders, shaft penetrations, bilge keels and any other projections that might interfere with sling or block positioning.

Straddle Hoist Characteristics

The straddle hoist operator (dockmaster) should verify the manufacturer's published hoist specifications and dimensions before undertaking any major dry docking effort. Hoist sling locations must be checked along with rated sling capacity, actual measured clearance between hoist piers and available water depth at hoist well for the time of the dry docking maneuver. Hoist safety device limits should never be overridden without specific approval of the hoist manufacturer.

Determination of Vessel Displacement

Verify reported vessel displacement by calculation. Estimating vessel weight, even by experienced personnel, is dangerous when preparing for a critical dry docking maneuver.

A rough estimate of vessel displacement for preliminary calculations may be made by using the Block Coefficient Method. The Block Coefficient Method assumes that the vessel below the water hull occupies a portion of a rectangular volume formed by multiplying the vessel waterline length (LWL) times its beam (Bm) times the average draft. This volume is then modified (reduced) by a coefficient (the Block Coefficient) which accounts for the reduction in volume due to the shape of the vessel's hull (see Fig. 18-3).

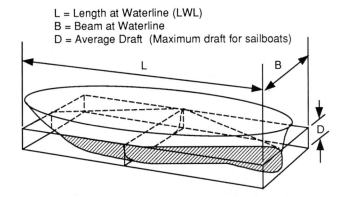

Figure 18-3. Block coefficient diagram, showing the rectangular volume to be modified by the block coefficient to determine the submerged volume of the vessel under consideration.

Table 18-1. Representative block coefficients for several types of vessels.

	Representative Block Coefficients are:
Powerboat	0.36 to 0.40
Motoryacht	0.44 to 0.55
Large Sailboat	0.19 to 0.26
Harbor Tow Boats	0.50 to 0.52
Ocean Tow Boats	0.54 to 0.59
Fishing Vessel	0.54 to 0.59
Patrol Craft	0.49 to 0.55
Ferry Boats	0.42 to 0.44

Representative block coefficients are shown in Table 18-1. The formula for displacement by the Block Coefficient is:

$$W \text{ (displacement, in pcf)} = (LWL) \times Bm \times D \text{ ave.} \times \text{Coeff.}$$

Divide displacement in pounds per cubic foot (pcf) by 35 cubic feet per long ton (CF/LT) for salt water to get the displacement in long tons.

A more accurate method of determining vessel displacement is by use of the Curves of Form for the particular vessel (see Fig. 18-4). Government vessels will usually have Curves of Form and other vessels built to a class may have the curves aboard. Determine the as-floating mean draft by adding the fore and aft drafts and dividing by two:

$$\text{Mean Draft} = \frac{D_F + D_A}{2}$$

By entering the Curves of Form on the axis marked average (mean) draft and extending a horizontal line (in this example) to the curve marked Displacement, or other such designation, and dropping a line down to the other chart axis, a direct reading of displacement for that particular mean draft can be read.

Location of Vessel Center of Gravity

The location of the vessel center of gravity is important in order to determine the distribution of the vessel load on the hoist slings and, after lifting, for determining the vessel load on the keel blocks if the vessel is set down and blocked.

The best way of determining the location of the longitudinal center of gravity (LCG) is from the Curves of Form for the particular vessel, if avail-

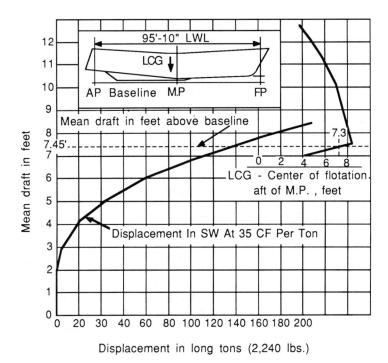

Figure 18-4. Example of displacement and other curves of form.

able. In this case enter the Curves of Form with a mean draft and intersect a horizontal line with the curve marked LCG and project a line down to the remaining chart axis. The axis will be labelled in feet or some other linear measure. Note on the LCG curve or on the axis label if the LCG is forward or aft of midships. For most vessels the LCG will be aft of midships, exceptions may be icebreakers or other vessels having engines or other large weights forward.

A very rough estimation of LCG, for recreational powerboats, can be made using the following formula applicable to average size powerboats:

LCG = 2/3 waterline length from forward.

Calculated Example

With the vessel data determined as outlined above and knowledge of the hoist characteristics, calculations for straddle hoist dry docking can be performed. The following example assumes a 150 ton capacity straddle hoist and a naval type vessel.

452 PART 4/OPERATIONS AND MANAGEMENT

Vessel characteristics:
LOA = 102 feet
LWL = 95 feet-10 inches
Beam = 21 feet
Draft fwd = 7.4 feet (navigation draft plus 2 feet-8 inches to baseline per docking plan)
Draft aft = 7.5 feet

Determine vessel displacement for dry docking:

$$\text{Mean Draft} = \frac{7.4 \text{ feet} + 7.5 \text{ feet}}{2}$$

From Curves of Form (Fig. 18-4) displacement = 140 long tons (LT = 2,250 lbs)

Location of LCG:
From Figure 18-4, LCG = 7.3 feet aft of midship point, for Displacement = 140 LT.

Calculate moments about a convenient point, in this case, Frame 72 (see Fig. 18-5), write the equation for the summation of moments and the equation for the summation of vertical loads. Solve the equations to determine the resultant loads at sling position R_1 and R_2.

$$\Sigma M(\text{FR. }72) = 16.8 \text{ feet } (140 \text{ LT}) - 8\, R_1 - 40\, R_2$$
$$= 2{,}352 \text{ LT-FT} - 8\, R_1 - 40\, R_2$$
$$\Sigma V = 0,\ 140 \text{ LT} = R_1 + R_2$$
$$R_1 = 140 \text{ LT} - R_2$$

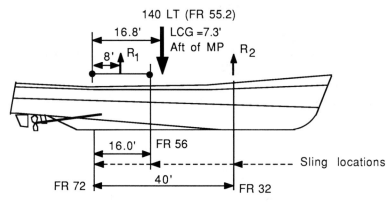

Figure 18-5. Diagram of vessel and straddle hoist sling load positions. Straddle hoist has three sets of slings, but only two resultant points on the hoist frame.

2,352 LT-FT = 8 (140 − R_2) + 40 R_2
2,352 LT-FT = 1,120 LT-FT − 8 R_2 + 40 R_2
R_2 = 38.5 LT, R_1 = 101.5 LT
R_1 = 8 slings, therefore 101.5 LT/8 = 12.69 LT/sling
R_2 = 4 slings, therefore 38.5 LT/4 = 9.63 LT/sling
Satisfactory, rated sling capacity = 60,000 pounds = 26.8 long tons.

Sling loads should be within 20 percent of each other, in this example the loads are approximately 24 percent apart and it may be necessary to shift the slings a bit to further balance sling load, if internal vessel framing allows.

Check Vessel Overhang for Overstress

140 LT/100 feet = 1.4 LT/FT approximately, therefore w
= 1.4 LT/FT × 2.25 Kips/LT = 3.14 Kips/FT

Calculate the approximate bending moment, as shown in Figure 18-6.

$$M = \frac{wL^2}{2} = \frac{3.14(32)^2}{2} = 1,607 \text{ FT-KIPS}$$

Calculate the estimated hull section modulus S. Treat the hull as a box girder of ⅛ inch thick plate for purposes of estimating the section modulus, Figure 18-7.

$S = 3,480 \text{ in}^3$
$$Fb = \frac{1,607.68 \text{ FT-K} \times 1,000 \text{ lbs/K} \times 12 \text{ in/FT}}{3,480 \text{ in}^3} = 5,444 \text{ psi.}$$

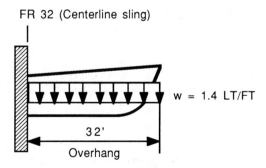

Figure 18-6. Overhang of vessel hull beyond hoist forward sling position. Treat hull overhang as a cantilever beam for hull strength analysis.

454 PART 4/OPERATIONS AND MANAGEMENT

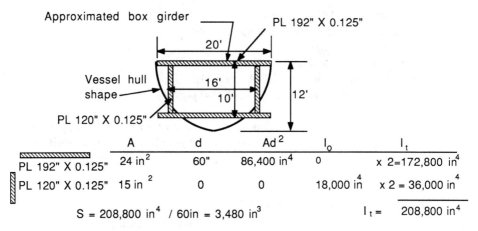

Figure 18-7. Approximation of vessel hull section modulus by treating vessel hull as a simple box girder.

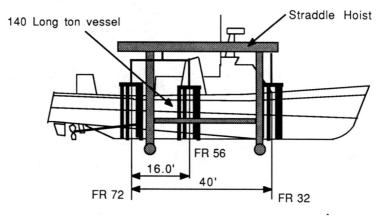

Figure 18-8. Diagram of a vessel positioned in a straddle hoist, showing sling positions as determined by analysis.

This is less than 18,000 psi allowable, hull not overstressed. Figure 18-8 shows the positioning of the vessel in the straddle hoist.

Summary

Analysis of vessel loads including the displacement and the longitudinal center of gravity related to the straddle hoist sling position will allow calculation for a safe dry docking maneuver without overload to slings, straddle hoist lifting mechanisms or overstressing of the vessel hull.

18.2 MARINE RAILWAYS

Marine railways are often referred to as railway dry docks or inclined slipways. The marine railway has been around for hundreds of years in its many forms. The simplest form is that of an inclined, smooth beach over which boats are hauled ashore by use of round logs or other rollers which reduce the friction between the boat hull and the ground surface. The hauling mechanism may be man or animal power or some form of machinery, usually a cable or chain winch. Improvements to the primitive systems originated in the 1800s when the combination of timber ways, captive rolling systems, a cradle or carriage to support the boat, and chain hauling systems were introduced. Each improvement provided greater safety in the haul-out operation and increased the size of vessel that could be hauled. Marine railways have capacities from the smallest backyard boat to boats exceeding 6,000 tons.

Modern marine railways generally consist of a track system with two or more ways which are supported by a foundation of pilings or solid substrate soil (see Fig. 18-9). Most often the track is supported on timber, steel, or

Figure 18-9. A simple modern marine railway dry dock used for yacht hauling. *(Photo courtesy: Childs Engineering Corporation, Medfield, MA)*

concrete pilings, driven securely into the bottom. The type of pile used is dependent on the imposed vessel load, the bottom soil load carrying capacity and the quality of the ambient water and soil into which the piles are driven. The track system may also be of timber, steel or concrete construction depending upon the imposed load, water quality and local cost of materials and labor. A large percentage of track and foundation systems around the world are constructed of timber. The timber is usually available, easily worked, of adequate strength and is cost effective. Care must be taken with any material to protect it from corrosion and degradation.

Attached to the track is a rolling surface and guidance system either of flat plate or structural shape rails. The rails provide for minimal friction between the moving cradle and the track. Lower capacities, under 150 tons, often use railroad type rails in combination with flanged wheels which are attached to the cradle. Higher capacity marine railways require closely spaced, solid iron rollers, arranged in captive roller frames to provide adequate load transfer between the loaded cradle and the track. In using either flanged wheels or solid rollers it is critical that they be properly designed to be round and smooth to minimize load shifting, bumping and jerking of the cradle as it moves along the incline. Captive rollers consist of groupings of rollers in frames, which if designed properly, provide for a full bearing of rollers under the cradle in the full up position as well as full roller support in the full down position. The rollers will move at a different speed than will the cradle and because of the inevitable slippage of the roller system down the track, will require adjustment from time to time. If the roller frame system is made too long, it may either fall off the outshore end of the track or be jammed into the bottom, resulting in a nonrolling system which in time will wear the nonrolling rollers flat on the upper surface or cause the roller frames to buckle and fall off the track, making vessel and cradle derailment highly likely. It is no easy task to reset a derailed cradle on the track, and a docked vessel may be damaged if the derailment is severe.

A cradle or carriage rests and rolls on the roller or wheel system. The cradle may be of wood or steel construction. Older cradles were sloped to meet the grade of the inclined way, but modern cradles are built-up to allow a level working surface, at grade, in the full up position. In order to attain a cradle declivity that closely matches the drag of a vessel's keel, modern marine railway tracks are built to a vertical curve, so that the level cradle in the full up position rotates on the vertical curve to allow a sloped cradle in the down position to accept the drag of the vessel to be dry docked. The cradle usually has two or more runners that are coincident with the track below. Across the runners are transverse structural beams that transfer the vessel load to the runners and thence to the track and foundation. The transverse beams are fitted with keel blocks and bilge blocks. The keel blocks are generally wood blocks built up on the track centerline and are set to support the vessel's keel. The usual keel block spacing is 4 to 5 feet. Outboard of the keel blocks may be a number of fixed or sliding bilge blocks which will support the vessel near the turn of the bilge and prevent the vessel from rolling over. **The usual distribution of vessel load is about 75 percent on the keel blocks and 25 percent on the bilge blocks.** Care must be taken to assure that most of the vessel's weight is on the keel blocks before the bilge blocks bear on the vessel or damage may result to the vessel hull from bilge block overload.

The cradle may be fitted with uprights along the outboard edges to facilitate vessel positioning and bilge block operation. Most often the uprights are fitted with a catwalk that allows docking personnel to assist in positioning the vessel when the cradle is at the outshore end of the track and the cradle is submerged. The cradle is moved up and down the track by a cable or chain system. The safest and most reliable system is an endless chain system that uses a special heavy hauling chain connected at both ends to a lighter outhaul or backing chain. The backing chain runs through a sheave or pulley mounted underwater near the outshore end of the track. The hauling chain is connected to the inshore end of the cradle and then passes over a sprocket or pocket chain wheel which is connected to a power unit at the inshore end of the track. After leaving the chain wheel the chain runs down the track to its connection with the backing chain. This endless system allows for smooth movement up and down the track without the need to spool cable. Chain is the preferred hauling mechanism since it is strong, wear resistant, capable of transmitting high loads through a sprocket wheel and its residual strength can be measured by simply measuring the smallest wire size and calculating the available strength. Cable, on the other hand, is often less expensive but is more difficult to transfer load through and can only be tested for its residual strength by destructive testing.

Marine railways were for many years the primary means of hauling vessels in marinas and boat yards. Today, however, marine railways seem best suited to the task only when the appropriate sloping shoreline and other geometric constraints can be easily met, and there is a need for hauling boats of capacities between 250 and 6,000 tons. Below 250 tons, straddle hoists seem to be the most popular and cost effective means of safe haul-out.

In one case, a boat yard had a sizeable yacht storage and service business but needed its marine railway for dry docking tug boats and other commercial vessels. Construction of a straddle hoist facility was not economically acceptable, so a design was produced to utilize the strengthened uprights of the marine railway as supports for a sling operated yacht haul-out (see Fig. 18-10). The marine railway can be backed down the ways to a depth over the deck suitable for the draft of the vessel to be hauled. The vessel can be

Figure 18-10. Modification of a small marine railway to accommodate a straddle hoist capability for small boats. *(Photo courtesy: Childs Engineering Corporation, Medfield, MA)*

secured in the slings and the vessel load transferred to the slings. The railway can then be moved inshore and the vessel worked on or transferred to a trailer or shuttle car for movement in the yard. This arrangement allows for rapid hauling of small vessels without the need for special blocking which is normally associated with marine railway use.

A similar use of an existing marine railway has been to remove the deck blocking and install chocks to hold a conventional hydraulic trailer. The railway cradle backs down the ways with the trailer on-board and with the boat fitted to the trailer in the conventional manner. The unit is hauled inshore, is placed high and dry and a vehicle is connected to the trailer for movement in the yard. This system requires a suitable means of access over the forward end of the railway cradle for accessing the trailer.

18.3 VERTICAL ELEVATORS

There are a number of vertical lifting elevators available worldwide and they generally consist of a platform that is raised and lowered into the water to effect boat haul-out. For large ship dockings, and to a lesser extent docking of smaller vessels, the patented Syncrolift® is a well known vertical lift. The Syncrolift uses a large structural platform raised and lowered by a series of synchronized cable winches. The system allows for distribution of large vessel imposed loads while providing a smooth and continuous lifting mechanism. Other vertical lifts utilize chains or other forms of cable to perform the lifting operation. Several vertical lifts combine displacement principles of floating dry docks with chain or cable stabilization. One small boat lift system uses a form of inclined rack and pinion to allow either vertical or inclined lifting.

A drawback of the vertical lift is the requirement for structural foundation piers or other support devices and the requirement of a uniform water depth equal to the draft of the largest vessel plus a safe margin, plus any blocking height and the depth of the lifting platform. In most cases, the vertical lift is not generally suitable for small recreational boat applications.

For personal use there are single boat vertical lifts designed to allow a boat owner the opportunity to raise his boat above water level to prevent fouling or to provide access to the below water equipment. For marina operations, however, these small lifts rarely serve effectively.

Another specialized version of the vertical lift is the patented Hydrohoist® that can hoist a single boat from the water, generally for personal use (see Fig. 18-11). These hoists use a catamaran type of arrangement of floodable and flotation pontoons which may be submerged to allow docking of a boat on the connecting structural frame. With the boat on-board, the flooded

Figure 18-11. Individual boat vertical hoist with high performance boat in dry docked position.

460 PART 4/OPERATIONS AND MANAGEMENT

pontoons are dewatered by introduction of air which displaces the water and provides buoyancy. These lifting units are used by many high performance boaters whose boats, because of their use at high speeds, cannot retain an antifouling bottom coating, and therefore will foul rapidly in coastal waters if they are not removed from the water.

18.4 CRANES

In many older marinas, cranes are still used for boat hauling operations. Certainly early marinas found available construction cranes to be a suitable means of boat hauling and launching. The cranes were readily available, could often be rented rather than bought, and were mobile. In some areas fixed cranes in the form of stiff leg derricks or shear legs have been successfully used.

The use of cranes to haul boats has several significant problems. Major problems occur because of the lifting and breaking mechanisms. Many boats have had rapid and unexpected launches because of a crane brake slipping, especially in wet weather. Cranes also lose lifting capacity as they boom out to pick up a boat. If the boat load is too large for the crane, the crane may

Figure 18-12. Mobile, telescoping, hydraulic crane with boat slings and sling support frame, in use at Grebbestad, Sweden.

topple or the boom may fail, in either case an unfortunate situation. Crane boom load restrictions also generally require a crane to be located as close as possible to the edge of the water. The combined load of crane weight and boat load being lifted may impose a large ground load along a bulkhead or seawall. This combined load may be of sufficient magnitude to cause the wall structure to fail. Although a frame may be used with slings to support the boat during lifting, the primary lifting mechanism is still the cable (see Figs. 18-12 and 18-13). This situation provides the opportunity for excessive boat movement during lifting and moving operations. Unless extreme care is exercised, and tag lines are used to restrain the boat, the boat may develop wild gyrations. This makes the boat likely to hit something, and in addition material and equipment in the boat will be moving around. In at least one case, a boat fire started during a crane lifting operation when the swinging of the boat, hanging from the crane cable, caused a boat battery to become loose and to short circuit. The resulting spark from the electrical short, ignited combustible materials in the boat. This condition may be aggravated by sloshing gasoline, gasoline fumes, and oily bilge water.

Using a conventional construction crane as a boat hauling device, is generally not desirable under most conditions. Properly designed small boat hauling equipment is readily available at reasonable cost, so there is little reason to use a system that is less than adequate.

Figure 18-13. Rotating, pedestal crane with boat slings and sling frame on Oslo, Norway waterfront.

18.5 HYDRAULIC TRAILERS

The hydraulic trailer is basically a boat cradle on wheels (see Fig. 18-14). The wheels are pneumatic providing the capability of carrying large loads at low ground pressures. The low ground pressure permits the trailer to operate on soft soils, sandy beaches and on traditional pavement. The wheels may be connected to the trailer frame by hydraulic cylinders which allow the frame to be raised or lowered about the wheel axles. This feature provides for the easy transfer of boat load from the trailer frame to the land based blocking during storage operations. With the boat and trailer combination in position on the upland, blocking may be placed close to the boat hull. Then the trailer frame may be lowered, the cross beams removed and the trailer removed from around the boat.

The trailer frame is usually of steel construction which has been galvanized or otherwise coated to prevent corrosion. The frame is fitted with removable cross members which will support the boat keel during hauling and moving, but which will be removed, as described above, for boat storage operations. The trailer frame also has moveable arms or poppets which will be placed to support the boat, generally at the turn of the bilge area. The arms are generally hydraulically operated providing a uniform, controllable support for the boat while making adjustments for each individual boat a simple matter. The flexibility and extendibility of the support arms allows

Figure 18-14. Basic hydraulic boat trailer. Note adjustable arms and pads that can be positioned against boat hull for bilge support.

universal fitting of both power and sailboats without the need for special fittings or adjustments. This feature allows rapid hauling and storing of a variety of boat configurations without the need for special equipment.

An advantage to using hydraulic trailers is that they allow boats to be hauled out of the water and moved directly to a storage or repair area in a single operation. They also allow boats to be positioned on shore very close to one another, thus increasing boat density and therefore revenue. Many trailer units are highway certified so the hauled boat may be taken off site, away from the waterfront. Because of the high cost of waterfront property and the resulting high land storage fees, many boaters are having their boats delivered directly to their homes or to other areas suitable for boat storage at reasonable cost. However many communities do not allow boat storage in residential zones. Hydraulic trailers also have the ability to move a boat, often a large boat, from coast to coast without having the boat touching the water. Because many Americans frequently relocate throughout the country, many large boats are moved to new locations in this fashion. This type of major move is often less expensive, safer and more insurable than are traditional water crossings. Hydraulic trailers may be usable at public launching ramps (usually during off peak times), thus allowing an otherwise land locked or inaccessible marina to offer hauling and launching services.

Although the appearance of hydraulic trailer rigs may make the casual observer wonder about their safety, safety is well supported in the design. The use of several independent load supporting arms and independent keel supporting cross beams provides a good safety factor against a major system failure, the kind of failure which may occur in single point support systems. Boats can be supported directly in areas of load concentration without side thrust loads which may occur in sling and cable lifting systems. Controllable, pilot operated lock valves are used in the hydraulic system to prevent function loss in event of hose failure. Remote control operation allows the operator to observe first hand the operation which is being performed. However, successful docking and hauling using an hydraulic trailer is only as good as the operator, so only experienced and knowledgeable operators should use these types of facilities.

18.6 LAUNCHING RAMPS FOR HYDRAULIC TRAILERS

The development and application of hydraulic trailers to haul small boats has raised questions about the appropriate ramp slope which would be necessary to facilitate trailer haul-out for a wide variety of boat configurations. Traditional automobile trailer launching ramps may be suitable for hydraulic trailer use, if the ramp slope is between 12 and 15 percent, and the length of ramp under water provides sufficient water depth to accommodate the draft of the vessel to be hauled. **Automobile launching ramps are optimally designed for a slope of 12-15 percent. Some states will opt for a broken slope, with the below water portion having a slope of 20 percent.** Grades approaching 20 percent may be difficult for automobiles and hydraulic trailer tractors to climb with a loaded trailer.

Hydraulic trailers are available in high capacity ranges for boats up to 60 feet in length and drafts to 9 feet. Trailers range in size from 20 feet to 50 feet in length with capacities to 85,000 pounds and more. The required steepness, 12 to 15 percent, for hydraulic trailer ramps relates to several issues. One issue is the draft and declivity of the keel of the vessel to be hauled. A brief study of representative power boats indicates keel declivities (slope of the keel from a horizontal line) from horizontal, or a slight drag to the stern, of about 13:1 (horizontal to vertical, or about 4.5 degrees). These types of boats can easily be hauled on low slope grades of 10-12 percent if adequate draft is available. Sailboats, however, are compromised by deeper keel configurations. In looking at sailboats a line may be constructed from a point on the curve of the bow to the forward most and deepest point on the keel. This line represents the maximum slope of the trailer on the launching ramp to physically allow the vessel's keel to ground out on the trailer bed. As the trailer moves forward, the vessel will be rotated such that the position of the boat waterline is close to parallel to the trailer bed. This rotation of the vessel is of concern since a high load concentration may develop at the point of keel contact with the trailer bed. Also if bilge support pads (hydraulic pads) are placed too soon in the operation, high bilge loads may result during the rotation process.

If the ramp slope is constructed steep enough to allow dockage of deep draft sailboats, Figure 18-15, the slope may be in excess of that needed to accommodate the slope of powerboats, Figure 18-16. The result is that high loads may develop at the forefoot (location where the boat stem meets the keel), between the boat and the trailer keel support. For some boats a high concentration of load at this point could damage the hull.

The apparent solution would be to construct ramps which are sufficiently long with reasonable slopes to accommodate shallow draft powerboats, while further runout of the ramp would provide deeper water for the sailboats. This condition is attainable when using an hydraulic trailer that can be separated from the tractor (see Fig. 18-17). In a typical operation, the trailer operator will back the tractor-trailer unit down the ramp until the tractor wheels are at the water's edge. The trailer will then be separated from the tractor and allowed by gravity to run further down the ramp until adequate water depth is achieved for the draft of the vessel to be docked. The tractor and the trailer are connected by a steel cable which is controlled from a winch on the tractor. The trailer unit may be separated from the tractor by as much as 60 feet. The front end of the trailer has rectractable "landing gear"

HAUL-OUT AND BOAT HANDLING SYSTEMS   465

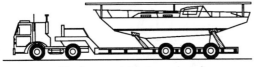
Trailer and boat on level

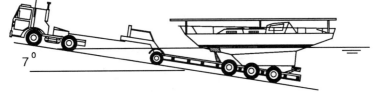
Trailer on 8:1, (12.5%, 7 degree) slope, boat floating
(preferred slope, with detached trailer)

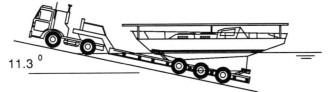
Trailer on 5:1, (20%, 11.3 degree) slope, boat floating
(maximum slope for hydraulic trailer use)

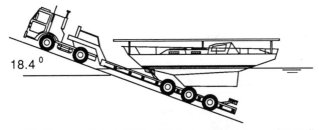
Trailer on 3:1, (33%, 18.4 degree) slope, boat floating
(excessive slope, high loads will develop on boat and boat
may become unstable during haul-out)

Figure 18-15.   Hydraulic launching ramp slope vs. floating sailboat alignment with trailer.

wheels to support the inshore end of the trailer. The boat is positioned and the trailer winched forward, as necessary, to ground the vessel. The trailer mounted hydraulic bilge supports are placed against the hull and the trailer is winched inshore and reattached to the tractor. The connected unit can then proceed up the ramp and to its final destination.

If a detachable trailer is not used, then the trailer ramp may have to be

Trailer and boat on level

Trailer on 8:1, (12.5%, 7 degree) slope, boat floating
(preferred slope, with detached trailer)

Trailer on 5:1, (20%, 11.3 degree) slope, boat floating
(maximum slope for sailboats, may be excessive for powerboats)

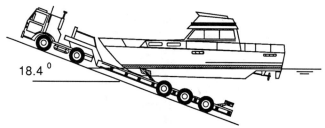
Trailer on 3:1, (33%, 18.4 degree) slope, boat floating
(excessive slope, high loads will develop on boat and boat
may become unstable during haul-out)

Figure 18-16. Hydraulic launching ramp slope vs. floating powerboat alignment with trailer.

HAUL-OUT AND BOAT HANDLING SYSTEMS 467

Figure 18-17. Hydraulic boat trailer unit, with trailer detached, hauling a small power boat on a ramp with an approximate 8 to 1 slope. *(Photo courtesy: Brownell Systems, Inc., Mattapoisett, MA)*

steepened to achieve the required draft over the immersed trailer. Extreme care must be exercised in the use of ramps steeper than 15 percent to prevent concentrated loads on the vessel, as discussed above, and also to prevent a partially grounded vessel from becoming unstable before being properly supported on the trailer. **A graphic analysis of representative boat keel profiles indicates that a desirable slope for hydraulic trailer ramps may be in the range of between 8: 1 (12.5 percent) and 7: 1 (15 percent).** It is also preferable to design the ramp with a short piece of vertical curve in the area of the transition between the ramp slope and level grade. The recommended hydraulic trailer ramp design is shown in Figure 18-18.

Launching ramps may be constructed of different materials, if appropriately designed. Local conditions and material availability will govern material selection. Frequently used materials include: portland cement concrete, bituminous concrete (asphalt pavement), gravel, stone, steel, and timber. Besides structural strength, the selected material should have an anticipated useful life appropriate to its cost and future requirements, and the material should have a surface treatment that will allow adequate traction for the tractor vehicle. The ramp should terminate in a level area that provides good

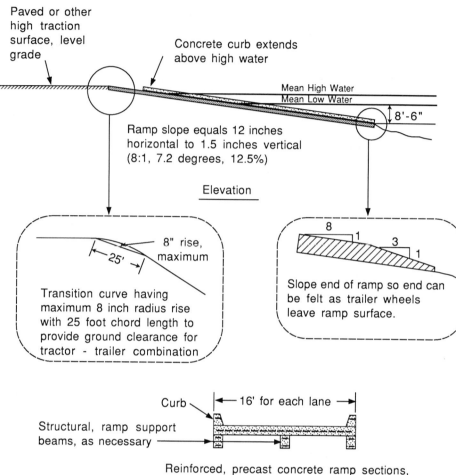

Figure 18-18. Recommended launching ramp criteria for hydraulic trailer use, using a detachable trailer unit. *(Adapted with permission from a design by Brownell Systems, Inc., Mattapoisett, MA)*

traction, preferably a paved surface. This area should be large enough to allow easy maneuvering of the tractor-trailer combination.

It is especially important that the launching ramp be constructed in an area free of overhead obstructions, especially electrical power lines. Hydraulic trailers can be expected to haul large boats which may have masts, fishing towers and bridge structures giving them elevated height. Prior to over the road hauling, all equipment that exceeds the legal limit for combined trailer and boat height or equipment that might project beyond the trailer in a dangerous fashion must be removed.

Ramps should be sited after thorough consideration of the prevailing winds and or current. Cross winds or currents may make routine docking very difficult. Protection and access docks may be necessary to position boats and secure the boats before and after trailer operations.

The upland area servicing the launching ramp should be free of obstructions and have an adequate area available for turning and maneuvering the trailer rig. Sharp turns should be avoided and good visibility provided in the launching area. It may be necessary to cordon off the launching area during launch and retrieval operations. It is desirable to provide a parking area adjacent to the launching ramp if several trailers are to be used during peak hauling or launching times. Any parking or boat storage area should be reasonably level but with enough pitch to allow surface water drainage.

18.7 AUTOMOBILE TRAILERS

Automobile boat trailers have been around a long time and provide a valuable resource for the public to gain access to the nation's waterways. The automobile trailer consists of two or more pneumatic wheels connected to a steel frame which has pads or rollers to support a boat. The trailer rig is attached to the automobile through a hitch. The trailer hitch is usually a ball and socket device that allows considerable freedom of rotation so that the trailer can comfortably track with the primary vehicle. The trailer hitch is securely fastened to the automobile, usually to some part of the automobile structural frame. Bumper hitches have been used in the past but they are now considered unsafe when they are used in conjunction with today's low impact, shock absorbing bumpers. There are several classes of trailers and hitches based on gross trailer weight. Trailer load should be adjusted to provide the recommended tongue weight at the trailer hitch. Improper loading may cause the driving vehicle to sway or fish tail and may elevate the front of the vehicle causing blinding headlight glare to oncoming drivers at night.

Most marinas have limited access for automobile trailer launching. If properly controlled, automobile trailered boats can bring substantial business to the marina. They also can cause confusion in parking areas and use valuable parking space. Storage of the autos and trailers can be land consuming. It seems that everybody always wants to go boating at the same time, so launching ramp operations can be very busy, resulting in long lines and unhappy customers.

In some boating areas where in-water slips are at a premium, land areas have been set aside to allow trailer boaters the opportunity to store their boats on their trailers in a protected yard. When they desire to use the boat they can drive in, connect up the trailer and launch the boat, storing the car and trailer in their storage location. This system provides the trailer boater with the opportunity to keep the boat near the water, to drive to the marina in a conventional fashion and often to participate in normal marina activities without having to travel with the trailer rig and store the unit in the front yard.

18.8 LAUNCHING RAMPS FOR AUTOMOBILE TRAILERS

Small boats of trailerable size constitute a major percentage of the boats used in recreational boating. An estimated 5.5 million boat trailers were in use by United States boaters in 1988. To service these trailerable boats there were an estimated 6,600 formal boat launching ramps. Many small boat launching ramps were probably not covered in the survey, including private ramps, informal road endings at the waterfront, and stable shingle beaches. In many areas a shortfall of in-water berthing facilities combined with escalating slip fees has increased the numbers of trailerable boats resulting in the overcrowding of the available launching ramps. New launching ramps are needed and many are currently under construction or consideration. A large number of small boat launching ramps are constructed by local and state governments.

In order to be effective, safe, and efficient, small boat launching ramps must be well planned, appropriately designed and adequately constructed. A number of planning and design considerations must be addressed for the development of a useful and effective facility. Among the issues requiring consideration are: site location, site exposure and protection, existing site grades, ramp orientation and length, ramp design criteria, materials of construction, ancillary facilities and upland parking.

Site Location

In general, launching ramps should be located in areas that will best serve the public need. This sounds trite but it has meaning. Many observed launching ramps, especially public ramps created by governmental agencies, are located by expediency rather than by good planning. Often waterfront land associated with other public works projects is developed into a launching ramp facility. Some examples are: land under or adjacent to a bridge abutment, land alongside a public beach or tagged on to a port development. These sites might be politically desirable but they may not serve as safe and efficient locations for boat launchings.

Some features that influence good location include: good access from feeder highways; adequate approach roads that can handle the volume, size and weight of trailer rigs; adequate maneuvering areas for the auto-trailer combination; reasonable site topography to accommodate an appropriate ramp design and related facilities; adequate parking areas for cars and trailers; areas that have or can be provided with adequate security; a gradually sloping shoreline that will avoid large cuts or fills to create the ramp; and a site that offers reasonable protection from wind, wave and passing boat wake effects.

Site selection should avoid obvious conflicts such as proximity to sewer or drainage outfalls, areas that may be historically significant, residential areas where noise and traffic may be a problem and public or private swimming beaches. If a launching ramp has to be located near a swimming beach, buoyage or other protection must be provided to assure an adequate separation between swimmers and boat maneuvering activities.

Often a conceptual marina layout will show a launching ramp adjacent to a straddle hoist haul-out facility. The thinking is that all haul-out and launching activities will be close to each other. This is generally not good planning since the launching ramp requires a gradual sloping beach while the hoist well requires a reasonably deep hole to facilitate haul out at all water levels. The construction of a launching ramp adjacent to a hoist well may result in enhanced siltation of the hoist well basin or accelerated erosion of the launching ramp slope. It is therefore prudent to provide adequate separation between the facilities to prevent proximity effects.

Site Exposure and Protection

Site exposure relates to the severity of ambient wind, wave and passing boat wake effects. Ramp siting and orientation should consider these issues and an attempt should be made to mitigate their effects. In general, wind wave or boat wake should not exceed 6 inches in height, wave trough to wave crest, at the launching area. Even a 6 inch wave can cause considerable difficulty and danger during boat launching and retrieval. It may be necessary to provide wave/wake attenuation to achieve a safe and usable launching ramp.

Currents in the area of the launching ramp can also cause difficulty and danger during launching and retrieval operations. In most cases currents will be oriented parallel to the shoreline and therefore will be perpendicular or angled to the axis of the ramp. This orientation is generally acceptable if the current is not too strong. Excessive currents will cause a boat to yaw off the centerline of the trailer making launching and especially retrieval difficult and possibly dangerous. In areas of strong currents the boat operator may try to power on and off the submerged trailer. The ramification of this action is obvious and should be avoided. If strong currents are unavoidable at the selected launching ramp site, it may be necessary to construct some form of current diverting structure. These structures can be very disruptive to the local environment, causing silt deposition, erosion or other bathymetric changes. Currents which are diverted may also change direction and velocity near the ends of the diverting structure thus inhibiting boat maneuvering when the boats are approaching or leaving the launching area. Care must be exercised in the placement of these structures, and a full understanding of the ramifications of their placement must be developed by competent professionals. If excessive current is apparent and mitigation is unattainable, serious consideration should be given to abandoning the site in favor of one with more acceptable site conditions.

Launching ramps should be close to navigation channels but not so close as to interfere with navigation or, for vessels using the navigation channel, to create hazards in the boat launching and retrieval operation. Signage and buoyage may be necessary to assist in maritime traffic control and safe boat operation.

Existing Site Grades

A site may have many of the necessary attributes for launching ramp development but lack appropriate transition between the upland and the water. In many cases it may not be prudent to create a launching ramp under such conditions. In general it is easier to cut, fill or otherwise modify the upland area above high water than it is to make significant modifications to the below water topography. The dynamic effects of water movement can rapidly alter man-made intrusions in a water environment. However, stabilized cuts or fills in the upland can be more easily created and maintained if properly engineered. Existing beaches or shorelines having slopes of between 10 and 20 percent are most suited for launching ramp construction. Upland site elevation is also important to assure ramp and parking area operation during periods of high or super high water levels. Parking areas and approach roads should not be excessively steep. Excessive steepness may result in launching, parking, and maneuvering difficulty and may reduce the efficiency of ramp operations.

Ramp Orientation and Length

Ramp orientation will depend on site geometry, as previously discussed. Additional factors include the ability to provide as direct a connection and alignment between the ramp and the set-up area as possible to minimize trailer maneuvering and to expedite launching and retrieval time. Another environmental consideration is that of orientation with the prevailing winds. A combination of contrary wind and current can turn a simple launching or retrieval operation into a dangerous and time consuming venture.

Ramp length will depend on the selected ramp slope, upland elevation and lowest design water level from which boats can reasonably be expected to be launched. The launching ramp should be straight from the top of the ramp to the furthest offshore end.

Ramp Design Criteria

Launching ramps may be composed of a single or of multiple lanes. The number of lanes will be determined by the anticipated use of the facility. **A general consensus among marina designers is that one lane can adequately service 50 launchings and 50 retrievals per day.** Another consideration is that a properly designed and operated ramp can average one boat launching or retrieval every 10 minutes. At 6 operations per hour for a 12 hour day, 72 launching or retrieval operations may be achieved, something less than the 100 operations generally assumed. Unfortunately, launching ramps often experience significant peak periods so that extra lanes may be needed to handle the peak loads. Critical to any launch ramp production rate is the ability to have a setup/tiedown area removed from the launching lane, so that the ramp lane is not occupied for an excessive time when it is not associated with the actual launching or retrieval operation. Congested ramp areas are becoming more commonplace and aggressive boaters can and do become unpleasant or even violent if long waits for lane space occur. If in doubt as to the correct capacity, add lanes to provide for a smooth launch ramp operation.

Ramp design calls for a slope of between 12 and 15 percent, (approximately 8 horizontal to 1 vertical, and 7 horizontal to 1 vertical, respectively). Flatter slopes may result in insufficient water depth at the end of the trailer to float the boat off the trailer, or the necessity for backing the towing vehicle excessively far into the water with resulting damage to the towing vehicle and lack of traction to remove the loaded trailer. Steeper slopes may result in difficulty with the towing vehicle climbing the grade with a fully loaded trailer. Steep slopes may also create difficulty and hazards in placing the boat back on the trailer since the boat must rotate a considerable distance to come in full contact with the trailer.

An expeditious manner in which to construct a launch ramp is by preparing a site and installing a series of interconnected precast concrete ramp units. Figure 18-19 shows a stack of precast units, upside down, at a manufacturing facility in Wallhamn, Sweden, ready for shipment worldwide. The control of concrete and surface texture assures a strong and usable facility. The side curbs keep the vehicle and trailer from falling off the ramp and provide a conduit for unit connection hardware.

Figure 18-19. Precast concrete launching ramp modules (upside down) stacked for shipment in Wallhamn, Sweden. Note holes in curb to allow interconnection.

Figure 18-20. Basic yard dolly for moving small boats around the marina.

18.9 YARD DOLLIES

Yard dollies are valuable devices for moving small boats around the boat yard (see Fig. 18-20). The dollies may be as simple as a timber or metal beam with casters, or more complex cradles and frames with pneumatic tires. In each case the idea is to provide a moving device without using a hydraulic trailer, auto trailer or a straddle hoist. The yard dollies are often used to move boats from one shop to another or to hold boats for washdown after they have been picked up by a forklift and before they are replaced in their upland storage racks. The dollies may also be used in a new boat showroom to allow easy changes to the displayed inventory. Care must be taken in the use of dollies to appropriately block or disable the rolling mechanism, especially if stopped on an incline, to prevent the boat and dolly from unintended movement. The same caution is offered if the dollies are used in a showroom, where a person leaning against a boat might cause it to move, resulting in possible personal injury.

18.10 BOAT STANDS

Boat stands (see Fig. 18-21), have become an important piece of equipment for the rapid and efficient storage of boats on land. In the past, boats would be hauled, transferred to a storage area and either placed on a prefabricated boat cradle or blocked in place by the construction of a wooden or metal

Figure 18-21. Sailboat supported by wood keel blocks and fabricated metal boat stands.

support system. Today, boat stands provide a quick means of blocking a sail or powerboat. Boat stands have been used, in conjunction with wood keel blocking, to support 85 foot naval patrol crafts. Using boat stands with large crafts or megayachts may require analysis of the overturning moment of the boat if it may be subjected to high winds. If boat stands are used on soft material or bituminous concrete (asphalt) pavement, it may be wise to place plywood or steel plate under the boat stand legs to prevent settling of the boat stand.

18.11 FORKLIFTS

Forklifts have been used for years to handle small boats. Initially these devices were forklifts developed by modifying industrial forklifts for marine use. Today, special forklifts (see Fig. 18-22) have been designed for marine use and these are the ones that should be used in marinas and boat yards. A major difference between industrial fork trucks and marine fork lifts is in the position of the load center. Marine forklifts have a rated load center at 96 inches versus an industrial fork truck which has a rated load center at 24 inches. This difference allows for the distant center of load associated with the boat on the fork. The rated capacity of an industrial fork truck will diminish rapidly when compared to the required load center for marine use.

Figure 18-22. Marine forklift preparing to withdraw a boat from third tier rack, inside an enclosed rack storage building.

Load balance is also important. The marine forklift is designed to distribute the load in a manner that provides balanced loading on the drive and steering wheels. Marine forklifts have extendable masts to allow greater vertical lift and negative drop, to get the boat off the storage rack, over the seawall and into the water. Negative lift (drop) is generally limited to about 12 feet, so forklift use is limited to a site than can accommodate this maximum negative lift. In areas with large tidal range, the forklift may not be able to deposit a boat into the water at all tidal stages. One creative solution to negative lift, is the use of a forklift with slings supported by the fork (see Fig. 18-23). Some inventive people have tried to create inclined ramps for use with forklifts to reduce the elevation change between the upland and the launch area. Extreme caution is suggested when using an incline, since a number of boats have slid off the forks when moving downhill. Maneuvering in a backward direction, with the boat pointing uphill, is a method used in some yards, but this technique must be used with extreme caution. A relatively level surface should be designed for use by forklift boathandling equipment. **The general gradability of forklifts is from level to about a 4 percent grade. Between 4 and 6 percent grades, the forklift must be used with caution. Grades up to 9 percent can be accommodated, if the forklift is fitted with a tiltback mast.** The weight component of the boat must be directed toward the forklift.

Figure 18-23. Forklift truck with boat in slings slung from the forks to achieve a negative drop without having a negative extension on the fork truck mast. In use on the Gothenburg, Sweden waterfront.

Ground condition is important for the safe and rapid use of a forklift. The ground should be relatively even and well compacted. In any circumstances where precise maneuvering is necessary, such as when placing a boat in a dry storage rack, the ground must be level and firm, or the operator will experience difficulty in positioning the boat in the rack. This is especially true when placing boats in upper racks with extended mast forklifts, where ground variations will be magnified by the height of the mast.

Marine forklifts generally are limited to a vertical lift of 30 feet, unless the forklift is specially modified for greater heights. All marine forklifts will be appropriately counterweighted to provide safe stability for the unit in the maximum extended position with a boat on-board. Small sailboats may be lifted by a marine forklift but such an application is limited, especially if the mast remains stepped. In one observed instance, it was necessary to have a yard hand stand on the bow of the sailboat, next to the forklift, to prevent the sailboat from falling off the forks because of a center of gravity located beyond the center of balance of the forks. This type of operation is an accident waiting to happen and should never be allowed.

Marine forklifts are designed with special features such as: telescoping masts, stacked masts for negative lifting, extended forks, padded forks, positive control hoisting and braking, overhead viewing from the cab, and anticorrosion materials and coatings.

A rule-of-thumb for forklift operation is that one forklift can service up to 200 boats in a dry storage facility. This number is highly variable, depending upon the type of operation and the frequency of boat use. An efficient operation might use two forklifts, one to remove or launch boats, and require only negative lift and a low height positive lift. The removed boat might then be placed in a washdown area and a second forklift used for fetching and returning boats to the dry storage building. Positive-lift-only-forklifts are less costly than forklifts having both positive and negative capability, so the cost of two lifts may not be substantially greater than a single multipurpose lift. An alternative to two forklifts is to have a permanent boat hoist installed at the water's edge for transition between the yard grade and the water and a positive-only-lift to move in the yard and service the dry storage facility.

The launching area at the waterfront should be level and of stable ground material. There should be enough clear room to allow the operator to backup away from the bulkhead and make a 180 degree turn and maneuver the rig in a forward direction to its destination. The launch area should be clear of overhead wires and above grade obstructions should not be present in the maneuvering area. Automobile parking adjacent to the launch area should be discouraged. The launch well should have a structural guard rail to prevent the forklift wheels from passing over the bulkhead. The launch well bulkhead must be designed to accommodate the maximum wheel load at the bulkhead. Bulkheads are directly influenced by loads in this position so a proper design is critical to the safe operation of the unit. Do not assume that an existing bulkhead is satisfactory for forklift use even if large vehicles have parked in close proximity without noticeable effect. The forklift and boat combination surcharge load can greatly exceed the ground pressure exerted by a truck or other commercial vehicle.

The general range of boat lengths associated with forklift use is between 16 and 26 feet, although capability of up to 40 foot boats is not uncommon. It may be unwise and costly, however, to oversize the forklift equipment for a small number of specialty boats. Remember that rated capacity decreases with load height increase. If nighttime operation is contemplated, and it would seem that it would be, the fork carriage should be fitted with adequate lighting to provide the operator with maximum visibility when placing boats in high racks.

18.12 STACKER CRANES

An attractive alternative to the use of forklifts for dry stack storage is the stacker crane. The stacker crane is generally associated with an upland or floating dry stack storage building and is an integral part of the building frame. A typical stacker crane (see Fig. 18-24) will run along structural beams

Figure 18-24. Niantic Bay Boat Valet, Niantic, Connecticut. A new, modern, dry stack storage facility using a pre-engineered, metal storage building, stacker crane running on overhead rails: and separate, fixed frame, negative lift forklift at the bulkhead for placing the boat in the water.

suspended from the building ceiling frame and running lengthwise between rows of racks. The crane itself is attached to a cross beam that moves as a unit along the runway beams. Suspended from the cross beam is a vertical shaft to which a moveable fork mechanism is attached. The forks can then move in and out of the building, can move up and down and can rotate 360 degrees. One major advantage of this system is that the operator rides with the forks so he has excellent visibility in placing the boats in the racks. The crane assembly, with boat aboard, can move out over the water to lower the boat to the water. There is virtually no height restriction except that of an allowable and safe building height. The aisle space between racks may be reduced from forklift requirements since the crane does not require the forklift chassis. In general this system can provide more boat storage per square foot than can conventional forklift systems. The cost for a stacker crane is greater than the cost for a forklift, but if the storage volume of the facility is large, the system may be cost effective.

A variation of this system is the use of a hammer head pedestal crane, like those used in building construction projects. The crane pedestal becomes the core of a circular storage building, allowing the crane to pick a boat out of a rack slot, rotate through a clear circular fairway, and lower and launch the boat through an opening in the building shell. It has been claimed that the system offers the potential for storing over 300 boats on an acre of land, assuming the required building height can be attained.

18.13 DRY STACK (RACK) STORAGE

Dry stack storage is a prominent type of marina facility in many parts of the United States and throughout the world. It offers the potential for storing large numbers of boats on a minimum footprint of land area. With increased competition for valuable waterfrontage, this method of marina operation may be highly attractive in some areas. The ability to sell (lease) cubic space versus square footage can provide an attractive rate of return on investment, especially if land values are high. Many dry stack storage facilities are incorporated into fully enclosed buildings, providing protection from the effects of weather exposure, protection from theft, and requiring less boat maintenance (see Fig. 18-24). The boat owner may be saved the trouble and cost of providing antifouling coatings on the racked boat and the boat is subject to less corrosion. The boating season may also be extended if the boat is racked on a year round basis. The marina operator has the ability to offer year round rental, providing a more stable income which is less affected by seasonal trends. **The cost per rack, around $3,000 (1989) compares favorably with the cost per wet slip of between $6,000 and $12,000.** Construction costs are more definable because of dry land construction. Major wet berth marinas may require dredging and pile mooring operations, both of which have large potentials for unanticipated construction costs.

Dry stack storage has the advantage of storing many boats in a minimal plan area. As an example most dry stack facilities have racks which can stack boats 4 high. For a 22 foot boat length, a rack width of 9 feet-3 inches, an aisle width of 52 feet (half the width attributed to each row of racks), and a system which is 4 racks high, a boat density, per acre, can be calculated. The plan area per boat equals: [(22 ft + 52 ft/2) × 9.25 ft] divided by 4 boats = 111 square feet per boat or about 392 boats per acre. **A conventional small boat, wet berth marina would have a density of about 84 boats per acre or a ratio of about 4.6 to 1 in favor of the dry stack storage.** The ratio will be even higher for comparison with swing moorings.

There are a number of problems that often surface when dry stack storage is proposed in a community. A major problem is the necessary building height and mass, which often is in conflict with local building and zoning codes. In addition the construction of such a mass on the waterfront is a major objection of abutters, residences or businesses that are in the sight corridor to the water. Most dry storage buildings are some form of pre-engineered metal building which is perceived as being unattractive and of industrial appearance. Clever use of building orientation, building facades and landscaping can often overcome these objections. Auto parking requirements may also be a major problem. Many unknowledgeable authorities will require one or more parking spaces per racked boat capacity plus parking facilities for employees. This number is unrealistic because never, to our knowledge, has a fully occupied dry storage facility had all its boats in use at any given time. **A peak day use of 33 percent of capacity is closer to reality.** Provision of excessive parking is a gross corruption of waterfront space utilization. It may be prudent, however, to provide for overflow parking in other areas of the project or in adjacent available areas. In some marinas, the lower rack boat spaces have been designed to allow auto parking in the vacant boat space. There is some question of the desirability of this concept, and serious consideration of the consequences must be given before providing this feature. The owner of a new luxury automobile may not be too happy to have his vehicle parked beneath three or more boats.

Specific site and market needs will determine the necessary dry stack storage layout. A variety of sizes are available or may be custom tailored to meet the project requirements (Table 18-2). Rack heights greater than 4 high will generally require a special extended forklift mast height to reach the upper racks. Even the top rack will be difficult to use because of the significant height above the forklift operator position and the resulting lack of visibility for boat positioning. In practice it is found that the second and third racks are the most efficient and high use boats should be located in these racks. The bottom and top racks are for low use boats. In general about 33 percent of the rack storage capacity will be high use boats with the remainder only infrequently used. Considering this use factor, it is recom-

Table 18-2. Minimum guideline dimensions for typical enclosed rack storage building.

Length of Longest Boat to be Racked		Building Dimensions		Aisle Width	Boat Load Capacity		Boat Rack Levels	Rack Depth	Standard Bay Width[a]
		Width	Eave		Levels 1 & 2	Above			
22' Boats	3 High	96'	23'0"	52'	4,000 lb.	4,000 lb.	1',9',16'	11'	18'6"
	4 High	96'	29'0"	52'	4,000 lb.	4,000 lb.	1',9',16',23'	11'	18'6"
	5 High	96'	35'0"	52'	4,000 lb.	4,000 lb.	1',9',16',23',29'	11'	18'6"
26' Boats	3 High	110'	29'0"	58'	8,000 lb.	4,000 lb.	1'2",10',19'	12'	20'
	4 High	110'	35'0"	58'	8,000 lb.	4,000 lb.	1'2",10',19',28'	12'	20'
	5 High	110'	42'0"	58'	8,000 lb.	4,000 lb.	1'2",10',19',28',34'	12'	20'
30' Boats	3 High	120'	31'0"	60'	10,000 lb.	5,000 lb.	1'4",11',20'	13'	20'
	4 High	120'	37'0"	60'	10,000 lb.	5,000 lb.	1'4",11',20',29'	13'	20'
	5 High	120'	43'0"	60'	10,000 lb.	5,000 lb.	1'4",11',20',29',36'	13'	20'
34' Boats	3 High	135'	33'0"	67'	13,000 lb.	6,000 lb.	1'6",13',24'	14'	20'
	4 High	135'	39'0"	67'	13,000 lb.	6,000 lb.	1'6",13',24',32'	14'	20'
	5 High	135'	49'0"	67'	13,000 lb.	6,000 lb.	1'6',13',24',32',39'	14'	20'

a. Bay width is for 2 boats.

Adapted from: Steelex Marine Storage/Division Star Building Systems mended that a waterfront area be developed to accommodate about 33 percent of the system boat capacity as holding slips for boat launching and retrieval operations.

Building lighting should be focused to illuminate the aisle area with only minimal lighting over the racks. The building floor surface should be hard and have good traction characteristics, but it should not be overly rough. Rough surfaces tend to wear out forklift tires. The floor should be level and any drainage pitch developed so as not to interfere with the attitude of the forklift while placing or removing boats from the racks. Aisle width between rows of racks will be determined by the size of the boats to be racked and the size of the forklift to be used (see Fig. 18-25). It is important to investigate aisle widths during the building design and be assured that the appropriate width is selected, as it is very difficult to add aisle width once the rack system has been installed. The storage area should be located close to the launching area to minimize travel time and reduce wear on the forklift.

Fire protection in dry stack storage buildings is a question of great debate. Most codes, when they exist, refer to the National Fire Protection Association's standard NFPA 303, Fire Protection Standard for Marinas and Boatyards and to NFPA 13, Standard for the Installation of Sprinkler Systems. Many knowledgeable people have a problem with the installation of automatic sprinkler systems in dry stack storage buildings because the initiation of the sprinkler will often only deluge the upper most boats, and this may not be the location of the fire. More importantly, the sprinkler may cause the upper boats to fill with water, the additional weight of which may cause collapse of the rack structure. Sprinklers directed so that filling of the racked boats is not possible may be appropriate when coupled with provision of water standpipes, hoses and nozzles. An alarm system would be most appropriate. Probably rate of temperature rise alarms would be more useful than would

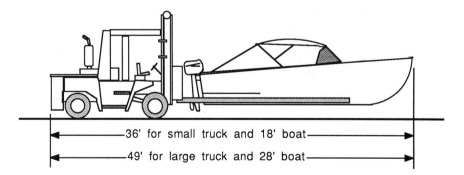

Figure 18-25. Approximate overall length of forklift truck with small boat on forks.

be smoke detectors which might be triggered by boat handling equipment or than high temperature alarms which might be triggered by normal high internal building temperatures. Alarm systems should have an independent power source. The building should also have provision for venting at knee and ridge locations. A typical fire fighting station might include a water standpipe for 1.5 inch diameter hose, a hose cabinet with 75 feet of 1.5 inch diameter hose, play pipes, an ordinary nozzle and a fog nozzle. The appropriate hydrant wrench and spanner should also be provided. It is also desirable to have Aqueous Film Forming Foam (AFFF) available (see Section 16.4 Dock Utilities, Fire Protection Water Systems for additional information on AFFF foam). Sufficient standpipes and hose cabinets should be provided to reach all areas of the storage facility. It is also prudent to provide several readily accessible portable ladders which may be used to access minor fires or other problems in boats in the racks. These ladders should be dedicated for this purpose and secured to prevent general yard use. The forklift should be fitted with a minimum of two 20 pound, classes ABC rated, portable fire extinguishers.

A dry stack storage facility must have the capability of washing down the hulls of boats just retrieved from the water and also be able to dry the hulls prior to racking. If this capability is not available, boats below other boats may become soiled by the boat above. Generally a washdown station is provided near the waterfront and dollies are provided to set the boat upon. Dollies should be provided to handle about 3 to 4 percent of the boat capacity of the storage system. The washdown area may be designed to prevent biofouling residue or wash detergent from entering the adjacent waters or wetlands. Generally an area that provides vertical percolation of the washdown water, such as occurs in a graveled area, may be adequate. Some jurisdictions may require area drainage to a large holding tank, with settling chambers to allow clean water to be runoff to the ground water system, while heavy materials will settle out for future removal and disposal in an appropriate waste disposal area. Such facilities are generally only of concern if large amounts of toxic antifouling bottom paint may be accumulated in the washdown process. Most dry storage boats do not require an antifouling paint application because of the small amount of time they spend in the water.

Security must be provided for the storage area. There are many ways to provide security and the method selected is best determined by the operator. The security force should also function as a fire watch and should be trained in first responder fire fighting techniques.

Accommodating boat launching requests is a labor intensive process and such requests often occur at sporadic intervals such as on Friday afternoons

HAUL-OUT AND BOAT HANDLING SYSTEMS   485

and on weekends. The operator must be capable of providing the customer service requirements and launching schedule demands if the facility is to be successful. During peak periods, customers are generally advised to phone ahead to give the staff enough time to properly launch and prepare the boat. Some dry storage operations will provide more than just storage, launch, and retrieval facilities. Services such as boat polishing, fueling, provisioning, and complete maintenance programs are good profit makers for the marina, and may serve to keep the operations crew busy during slack periods.

An interesting adjunct to a dry stack storage operation is an innovation that allows the placement of boats in spaces vacated by boats which are in use. The system uses a metal and wood pallet to hold either the boat or the boat owner's automobile. The pallet with boat or car can be racked in the conventional manner (see Fig. 18-26). This system is initiated when the boat owner calls for the boat to be launched. The boat is already on the special pallet in the rack. The operator picks the pallet and boat from the rack and travels to the launch area for water launch. The pallet is corrosion protected so it can readily be submerged. The empty pallet is then raised and brought on shore, where the boat owner's auto can be driven on to the pallet and

Figure 18-26. Palletized system for storing boats and cars in conventional dry stack storage facility. *(Photo courtesy: Car Pal, Inc., Lake George, NY 12845)*

chocked in place. The lift truck then returns to the rack area and places the pallet, with auto on-board, in the assigned rack. The process is reversed when the boat returns. The pallet is fitted with a drip pan to prevent dirtying of boats or cars beneath.

The auto racking system may have considerable merit in areas where insufficient parking exists to justify a rack storage complex. The system would be coupled with a valet, short term parking area to hold autos until the racking operation is ready for the vehicle. Large size vehicles may not be able to be accommodated without special size racks, but most regular size autos appear to be suitable. One available auto racking system has a rating of 10,000 pounds capacity and has been tested to 15,000 pounds. A service fee may be charged for the operation, either as part of the rack fee or as an extra fee. In areas having a high parking space to boat ratio and little parking space, the concept of rack storing automobiles may be attractive. The system also provides for patron auto security in areas of high theft and vandalism. A secondary use might be for cruising boaters who leave their autos for long periods of time in the marina parking area. The undercover, secure location in a rack may be very attractive. An enterprising marina operator might also combine the car racking with an auto wash and polish service as an additional profit service.

Figure 18-27. Straddle hoist fitted with a small, hydraulic jib crane, attached to upper, hoist beam and controlled by the hoist operator. Crane can raise and lower cable hook and swing into area of hoist well. It can be used for lifting masts and other heavy equipment from small boats.

18.14 BOAT MAST HANDLING EQUIPMENT

There are a variety of devices used to step and unstep sailboat masts. The old-time standard was a stiff leg derrick which was permanently fastened to a fixed dock or other substantial structure. The sailboat was brought in under the derrick and the mast was lifted by a set of block falls and with a winch. This method is still in use at numerous boat yards. A more modern and efficient system uses a small hydraulic, telescoping boom crane attached to one of the main upper beams of a straddle hoist frame (see Fig. 18-27). Mast work can be done in the hoist well or with the boat in the slings. The crane can also be used for lifting material from trucks which are ashore, or for lifting or placing a replacement boat engine, with the boat being afloat or ashore. Other special types of lifting devices have been developed to lift masts (see Fig. 18-28 and Fig. 18-29). A common and useful piece of marina

Figure 18-28. Mobile, hammerhead crane used for lifting masts in sailboat harbor of Lake Vattern, Sweden.

Figure 18-29. Small fixed derrick for work on small boat masts. Located on Gota Canal, Berg, Sweden.

equipment is a self-propelled, wheeled, crane with a hydraulic telescoping boom (often called a cherry picker, see Fig. 18-12 for illustration). These mobile telescoping boom cranes can be used for hauling boats, lifting boat masts, removing and installing engines in boats, moving timber boat blocking, and a variety of other lifting and moving jobs around the yard. Whatever the device configuration, it should be structurally adequate and efficient.

19

Upland Facilities and Amenities

A quality marina will provide appropriate upland facilities and amenities in addition to well designed and properly constructed in-water structures. The upland facilities may even have a greater impact on marina marketing success than the in-water facilities. Many boaters perceive a marina as a recreation facility rather than as a boat parking lot and desire to have adequate and convenient parking, clean and numerous sanitary facilities, availability of fuel, ship store, convenient stores, swimming pools and other amenities that make life at the marina more convenient and enjoyable for the entire family. If constant trips must be made offsite to fulfill domestic needs, the rational of boat ownership decreases and boat ownership becomes a chore rather than a recreational pursuit.

Not all marinas need to provide all of these services or facilities, and like every other aspect of marina design the provision of upland facilities should be tailored to the specific site and economics of the project. The two upland facilities that generally must be provided are sanitary facilities and parking.

19.1 PARKING

The automobile parking issue in marina design has not been resolved between designers/developer and regulators. On the one hand the local community desires that all patrons that may ever visit the marina be provided with adequate parking so that local roads, residences and businesses are not adversely affected. On the other hand, the provision of parking costs money and parking areas occupy valuable space. The problem rests with the use of marina facilities. Most marinas enjoy only a sporadic use, usually on good weather weekends. To provide a parking capacity equal to or greater than a one or two day annual peak demand seems unrealistic for the value of the land involved, both from a cost perspective and a land use perspective. A number of communities will require provision of parking at the rate of two parking spaces per boat berth, whether a wet berth or a dry storage berth.

This allocation of facilities is based on the idea that the boat owner will arrive in one auto and either the remainder of the family or a guest will arrive in another auto. To assure sufficient capacity, the community also says that this ratio will be applied to the total of all boats in the marina plus an allowance for staff parking. In reality, it would be an extremely rare event to have all boats in a marina in use at the same time. A number of studies have been performed in an attempt to determine the number of boats in use or occupied during peak and off-peak times to help establish rational parking estimates. In 1988 a significant survey was conducted by the International Marina Institute (Neil W. Ross, International Marina Institute, Wickford, Rhode Island, 1989). The survey was based on a study of its member facilities and other interested respondents. The need for this study was felt to be so important that a number of regulatory agencies assisted in soliciting responses from marinas in their areas of jurisdiction. Responses were received from 169 private and public owned marinas located in 23 states and 1 territory. Of the total response, data supplied by 142 reporting facilities was deemed sufficient in content to be further analyzed. The survey was based on a physical counting of actual autos in marinas and boats in use or underway. The survey was conducted over the July 4th weekend, the major boating holiday in the United States; on a midsummer, no event weekend; and during a midsummer weekday. Although not universal in scope, the data which was received and analyzed appears to show a consistent trend of use that is valid for establishing a rational guideline for automobile parking requirements in marinas. **In his conclusion of the survey results, Mr. Ross states, "The significance of this study was that it quantified, for the first time on a broad national base, that auto parking standards exceeding 0.5 cars per boat may be excessive in most sites." Taking into account some variables regarding boat size, boat mix, seasonality of use, transient use, etc., Ross further states as a recommendation that, "One car parking space for every two boat slips seems quite adequate as a national guideline for most high use weekends."** The report goes on to recommend that marinas investigate and secure the use of unused or underused marina land areas for overflow parking during peak use times and even look into the availability of unoccupied adjacent office or business parking space to accommodate any overflow needs. Office and business parking space is often available for use during times of greatest marina need such as on weekends and holidays.

The recommendations presented must be adjusted for high use amenities such as restaurants or shopping facilities open to the general public. There is also no guarantee that the recommended guidelines will be accepted by all reviewing regulatory agencies, so consideration of higher ratios may be appropriate in initial design development.

Generally, parking should be provided at no charge to marina tenants and their guests. In some cases, such as high value urban waterfronts, it may be necessary or desirable to charge a fee for parking or at least prime parking. If a fee for parking is charged there may be certain legal responsibilities that are incurred, so consult competent legal counsel. It also may be necessary to control parking areas with signage or physical devices. Card key systems are popular where control is required, with the user being charged a minimum fee to cover the cost of the parking control devices. Controlled parking has the advantage of reserving prime areas for the marina tenants to the exclusion of guests or general visitors. Boat owners are often upset when they have to lug their boating gear an excessive distance, when parking near the dock is occupied by sightseers who could easily walk the distance to see the boats. This attitude is often frowned upon by some regulators who feel that the waterfront should be open and available to all the public. Many boat owners believe that those who pay the fees that create the public amenities and who pay the property taxes should have some rights over the general public as long as public access is not denied.

Parking areas may be paved or unpaved. A certain amount of paved parking is desirable to prevent boat owners and guests from tracking dirt and stones onto the docks and boats. Paved parking can also be striped with lines to provide the most efficient parking arrangement. Overflow parking can be

Figure 19-1. Permeable parking area using soil filled, concrete pavers at Changi Yacht Club, Singapore. Porosity allows vertical percolation of rain water for recharging the ground water supply and minimizes point discharges into the harbor waters.

gravel or other permeable materials. Hard paved parking areas using bituminous concrete or portland cement concrete will have an impermeable surface causing sheet flow of surface runoff water. The runoff water may be collected in a drainage system and diverted to a permeable ground area or to the adjacent water body. There is considerable concern about point discharges into water bodies and wetland areas, especially runoff water containing oily residuals associated with parking areas. Permeable parking areas may help mitigate the parking concerns by allowing the soil to filter the runoff before it enters the ground water system. One way to achieve a little bit of the benefits of both hard and permeable parking surfaces is to use a paving block that provides a reasonable traction surface but allows the vertical percolation of runoff water (see Fig.19-1).

19.2 RESTROOMS AND SHOWERS

Appropriate restrooms and shower facilities are a must in virtually every marina facility. A great deal of concern has been expressed about boats in marinas fouling the area waters by discharge of waste materials. One proven method to mitigate boater discharge of sanitary waste is to provide clean and accessible toilet facilities in numbers suitable for the volume of use by patrons in the marina.

Toilet rooms should be light, airy, and reasonably spacious (see Fig. 19-2). Provision for accessibility by the handicapped is recommended. Additional information on handicapped accessible facilities may be found in Chapter 16. Sanitary facilities should be located within 500 feet of the shore end of every pier. Additional details on design considerations for sanitary facilities may be found in Chapter 16.

Restrooms are often incorporated into structures that also house the marina office or ship's store. Separate facilities may be necessary in marinas that have a large upland area, to accommodate the proximity requirements. Although marina design guidelines require a specified number of toilet facilities per number of boats, experience has shown that owners of powerboats 36 feet in length and longer will rarely use the shoreside heads while owners of boats 30 feet in length and smaller will most often use the shoreside heads. Sailboaters with boats 35 feet in length and smaller will generally use the shoreside heads while boats of any type 50 feet and over will rarely used the shoreside facilities. This means that small boat marinas must take special care to provide adequate facilities that will be used frequently. Heads may be restricted to marina tenants but should never be coin operated. The resentment resulting from pay toilets is not worth the small return of income. In fact, the restrooms should be the best that the project can afford and should be used as a marketing tool for prospective customers.

A well run marina, during peak periods, may have to clean the heads three or four times a day to maintain a high standard of cleanliness. Standard

UPLAND FACILITIES AND AMENITIES   493

Figure 19-2. Spacious and airy restrooms and showers in an upland facility at the Wentworth by the Sea Marina, New Castle, New Hampshire.

commercial cleaning habits apply to marina heads. Use of urinal deodorizers, disinfectants, and constant cleaning are recommended. The building design should provide for good ventilation. Natural ventilation as well as power ventilation is recommended. Waste baskets should not have tops but should be lined with appropriate plastic (biodegradable) or paper liners, to facilitate cleaning. Sinks should be shallow to prevent the filling of water bottles and also to prevent them from being used for washing dishes or other food preparation utensils. One deep sink, or slop sink, should be provided for the filling of bottles, buckets, and for general housekeeping needs. The marina tool inventory should include an electric snake for use in unclogging toilets and sinks.

Lighting over sinks should be plentiful but the remainder of the head need not have excessive lighting. Heads should be designed for easy washdown and cleaning. Epoxy paint is a good choice for interior finishes, as it is durable and easily cleaned. Toilet room floors should be washable and have floor drains to allow major washdown. Toilet tissue should be of the two ply variety rather than single ply. The two ply tissue is not suitable for most boat head systems (it jams the works) and the use of two ply will cut down on the pilferage of toilet tissue for use on boats. Larger marinas have experienced considerable savings by switching to two ply toilet tissue. Continuous roll paper toweling is favored over cloth towels for hand drying, although environmental concerns may suggest recyclable materials. Rather than providing electric hair or hand dryers, many marina operators prefer to furnish ground fault equipped electric outlets near the sinks. The women's restroom should have sanitary napkin dispensers. A full length wall mirror is a desirable feature in the women's room. Makeup lighting and mirrors over the sinks or in a separate sitting area is also desirable. Interestingly, women's restrooms will always require less cleaning than the men's restrooms!

The number of showers provided will again be related to the number and types of boats serviced by the marina. Peak hours for shower use are generally from 6:00 AM to 10:00 AM and from 6:00 PM to 8:00 PM. Showers also need to be cleaned constantly. Little or no tile should be used for shower stall areas. Tile tends to have moisture buildup with resulting biofouling. Fiberglass is a preferred construction material. Showers should be capable of accepting a power washing at repeated intervals. Floors should slope to a floor drain for ease of cleaning and for relief from emergency overflows. Shower curtains are preferred over shower doors for safety, ease of use and replacement cost. A dressing area in the vicinity of the showers is recommended. The dressing area should have cedar (or other appropriate wood) benches for sitting and dressing. Clothes hanger hooks should also be provided. For sanitary reasons, the use of shower shoes is recommended over the use of floor mats.

Even with the best efforts of the marina staff, heads and showers should be professionally cleaned twice a year. Heads and showers require good security. A card key system seems to work quite well. Access is limited to tenants holding a valid card key. Sophisticated card key systems can monitor time of use and card control number to track usage and to help prevent theft or unauthorized use. This may seem a bit excessive but such precautions may be necessary in some situations. Perhaps just a sign indicating that the premises are controlled will help keep the facilities safe, clean and well maintained by the users.

19.3 LAUNDRY ROOMS

Providing laundry equipment in a marina can be a money maker and a positive marketing tool for attracting customers. Outright purchase of laundry equipment is often favored over leasing. The quality of commercial laundry machines is such that down time and repair costs are minimal. The machines are generally coin operated. **Although the number of laundry units may vary by climate and type of boat mix, a rule of thumb is to provide 8 washers and 10 dryers for 300 boats and transient trade.** Money changing machines may also be desirable to keep users from bothering the marina office staff for change and also for operation beyond the hours of manning by the marina staff. The laundry facility should be located in a safe, high traffic area of the marina. Security is a major problem and must be addressed in the marina master plan and overall marina security plan. The laundry room should be well lighted and ventilated. Although the laundry machines are relatively maintenance free, the laundry room is not. The walls and floors should be designed for easy cleaning. Epoxy paint is a good choice, and again refrain from tile surfaces. The floor should pitch to a floor drain to accommodate cleaning and the occasional machine overflow or broken hose. The laundry room should be heated in winter and be well ventilated in summer. The laundry room often is forgotten in the daily cleaning and maintenance schedule. It should have a preeminent position in the marina housekeeping schedule and it should be kept clean. A clean, appropriate and well operated laundry can be a positive marketing tool.

Many commercial laundromats have seating and other amenities. A new trend is the development of singles laundromats, complete with bar services to attract young singles to meet and possibly establish friendships, or more. This is not the intent at a marina and combining laundry facilities with other entertainment might cause the laundry room to become a hangout with all the negative connotations. It may be desirable to not include seating in the laundry room, to prevent people from getting too comfortable. The laundry room can, however, become a communication area (bulletin board) to inform patrons of upcoming events, social activities, and general information. Peak times for laundry use are generally 7:00 AM to 10:00 AM during summer months and between 6:00 PM and 11:00 PM in winter.

Along with providing the laundry machines comes the burden of a lost and found department. Provision must be made for how to handle this problem with the least input and disruption of normal marina operations. Since some machine failures and other operational problems are bound to occur, a sign should be prominently placed to inform users of how to obtain repair service.

19.4 MARINA OFFICE

The marina office is the heart of the marina operation. From here the marina manager must meet and greet the public, solve customer problems, provide operational guidance to marina staff and perform the administrative aspects of marina operation. Since, from time to time, all marina users will visit the marina office, its location should be central to the marina (see Fig. 19-3). It should be easily accessible and well marked. This is especially true if a transient trade is part of the marina operation. Preferably the marina office, or parts of it, should have a commanding view of the entire marina facility. It should have public and private spaces. The public should be welcomed into an area that is comfortable but arranged to expedite people on their business rather than providing a hangout. To accomplish this goal, a counter or dutch

Figure 19-3. Centrally located marina manager and dockmaster office at Shipyard Quarters Marina, Charlestown, Massachusetts. Note the excellent visibility of the marina operations from the control tower.

door is appropriate. Both convey a brief encounter and rapid completion of business.

The marina manager should have a private office to provide the proper climate for airing patron or employee grievances. The manager will also need quiet time to complete paperwork and scheduling. There is no given size that a marina office must be, nor any specific guidelines for its form, shape or extent, other than to fulfill the above mission. In or near the marina office there should be one or more pay phones for public use to prevent tying up the marina office phones. The marina office lobby is a good place for a bulletin board and message center. Some marinas will also have a weather station readout in the lobby and other weather related information available. Vending machines may also be appropriate in this area. It may be desirable to have a portion of the marina lobby open around the clock, by marina patron gate key, to allow access to pay phones in inclement weather, perhaps one set of marina heads and showers should be included in this area, and also, perhaps, vending machines. These additional services may require enhanced security coverage.

The marina office should have a good array of communications equipment suitable for the area of marina location. This generally includes landline telephones, marine VHF radio telephones (operational and backup), citizens band equipment, if used in the area, and local use walkie talkie type radio units. The office should also have a status board with the marina layout painted or inscribed on the board together with slip assignments and other operational information. In lieu of the traditional status board some marinas have computer generated slip layouts presentations, that can provide a full range of information on slips, tenant data, utility availability, water depth, etc. Whatever system is selected, it should be useable by the marina staff and provide useful and rapid information. This aspect is especially important in busy transient boat operations. Other computer software applications will also be necessary to perform bookkeeping, inventory, invoicing, and word processing functions.

Marina office space should include a "mud room" for marina staff to get out of the weather and to remove wet and dirty foul weather gear. Depending on marina size, it also may be desirable to have a staff lounge or lunch room to prevent staff from "hanging around" the public lobby or distracting the marina manager. Somewhere in the marina office or in an adjacent area there should be a stock and supply room for storing the many pieces of equipment necessary to operate a marina.

The marina office may include a section for tenants' mail. Although marina business tends to be seasonal, many customers will use the marina address as a mailing address. If this practice is accepted, some form of mail box should be installed. The mail box is usually a small array of slots (pigeon holes), one per customer. Mail handling can become a horror show if left uncontrolled. To minimize the use, a small charge should be levied. The charge may be enough to amortize the cost of the mail box construction over three years. Mail boxes can also be used as a customer message center and for holding the keys to the customers' boats, similar to a hotel registration desk.

Boat sales offices, repair operations or other ancillary services should not be conducted directly out of the marina office public space. Separate offices within the marina office complex may be acceptable, if their use does not conflict with the basic mission of the marina operations center.

19.5 SHIP'S STORE OR CHANDLERY

Most marinas will provide some form of ship's store. While at the marina, boaters are a captive audience and many boaters will buy from a ship's store if the supply of goods is representative and the prices fair. The big question is how much stock to inventory? At some point the store will require one or more employees dedicated to its operation. At this point, the store must be able to generate enough income to support this staff. This is often difficult in a seasonal boating area. If the area has other retail boating equipment supply stores, it probably will not be profitable for the marina to compete with these stores, especially if one or more of them offer sizable discounts. In this case, the marina may offer only those products that are routinely needed by boaters such as bottom paint and painting supplies, line, fastenings, and similar products. This range of products may be able to be handled by the marina office staff as part of their other duties. If the marina has a sizeable boat repair facility, the ship's store may be operated as a division of the parts department. Care must be taken, however, not to let the ship's store operation conflict or delay the parts department operations which is usually a high profit center. Judgement and local conditions must govern the suitability of providing a ship's store and if provided, the range of its operation.

19.6 BOAT BROKERAGE AND NEW BOAT SALES

Offering boat brokerage services at a marina has both pros and cons. On the plus side, boat brokerage provides traffic in the marina by prospective marina users. By seeing what the marina has to offer, a prospective boat buyer may decide to rent or buy a slip or to use transient services. The boat brokers also need to have a place to keep and show boats so the broker will probably rent or buy slips at the marina. They may also use marina repair and service facilities to commission boats and haul out facilities to perform boat surveys. All of these activities can provide additional income to the marina.

On the negative side, boat brokerage brings many people to the marina who are not necessarily paying customers. These people will take up parking space, use the marina facilities and possibly roam around the docks. The brokerage boats may not be in the best of condition and may detract from an otherwise attractive marina. The boats may be left unattended for long periods of time with resulting neglect and create the need for forced care by marina staff to prevent damage to the boats and at worst, sinking. People offering their boat for sale are often getting out of boating and may not be receptive to paying marina storage or maintenance bills.

New boat sales can provide a good revenue stream for marinas, in good times, and can seriously impact operating capital in bad times. The advantage of new boat sales, beside direct income from the boat sale, is the after sale berthing and service work. In some markets, new boat sales often depend upon the ability of the prospective boat owner to locate a slip to berth the boat. If the marina/new boat sales dealer can provide both the boat and the slip, a sale may be completed. New boat sales will generally require large showrooms which will cost a lot to construct and maintain. The structures may also take away from other upland operations such as parking, dry stack storage, etc. Additional parking and other services may be required to support the new boat sales operation.

If boat brokerage and/or new boat sales are provide, the activities and administration should be kept separate from the marina operation so that marina services are not compromised by the demands of the other operations.

19.7 FOOD SERVICES

Provision of food services at the marina may be another significant profit center. Food services may be in the form of convenience foods (fast foods) or in the form of a full service restaurant. The site location, ambience of the area and market for food services will determine the need. Convenience foods and snack bars may be staffed by part time seasonal workers. They can be a low overhead, high profit operation in the right context. They also can be a nuisance generating wastes and litter, becoming a hangout for the younger crowd and spoiling the overall appearance of the marina. If operated in conjunction with a swimming pool or other recreational amenity, and limited to the use of marina tenants, they can be a welcome amenity for the family cook and bottle washer. Since preparing food on all but the larger boats is a real task, the availability of prepared food may be a significant feature of the marina and enhance its marketability.

Full service restaurants are generally separated from the marina operation. They may be located within the marina complex but should be divorced from the marina operation except for provision of dockage for restaurant customers and the thread of ownership. The marina operational manager should not be responsible for food preparation, food service or employee management. The restaurant may compete with the marina for parking space and may create a nuisance to marina users because of the abundance of strolling restaurant patrons and the volume of noise associated with any large public gathering. If the restaurant has a bar, patrons may get boisterous and unruly, especially late at night, which may have an impact on marina tenants. Live entertainment may also be disruptive. Noise tends to carry large distances over water. Restaurants also are large waste generators. Provisions must be made for frequent waste disposal and enclosed waste storage areas. Although many marina tenants may patronize an on site restaurant, the success of a restaurant will depend on attracting off site patrons. The increased automobile and pedestrian activity at the site may require enhanced marina security control measures with attendant costs.

Restaurants and other nonwater dependent use along the waterfront may require the marina developer to provide significant public amenities and benefits in order to obtain regulatory approval. Typical public amenities such as public boardwalks and marina overlooks may detract from the privacy and security of the marina tenants. A restaurant or other nonwater dependent activity may be a desirable feature to the marina development but its ramifications must be fully explored and the benefits and costs honestly evaluated.

19.8 SWIMMING POOLS AND OTHER AMENITIES

Marinas have become family recreation centers in many areas of the world. In order to attract upscale tenants, many facilities have included amenities such as swimming pools, tennis courts, lawn bowling greens, and even golf courses. All of these amenities can be a useful marketing tool to attract those people who enjoy a variety of outdoor recreation in addition to boating. All of these amenities require a significant land area occupation and continual maintenance. Perhaps the most controversial of these amenities is the provision of swimming pools. Swimming pools are a most desired feature in a marina, since they are a family activity and can keep the family occupied while the boat owner prepares or services the boat down at the dock. There is, however, a major liability involved in providing swimming facilities. The facility must generally be monitored by a competent marina staff individual to maintain proper conduct and, hopefully, prevent personal injury or drowning at the pool. It may be expensive to obtain insurance to operate a swimming pool and expensive to staff and maintain the facility. The provision of a swimming pool may, however, be a significant marketing tool, especially in a highly competitive marina area.

Tennis courts can also be a desired amenity to occupy the time and energy of the boater's crew when the boat is not underway. Tennis courts have less liability than a swimming facility but also require a significant amount of maintenance. A run down, neglected tennis court area is a highly visible facility which may reflect on the perception of the quality of marina operation.

One amenity that is highly desirable is a well planted, shady area that may be used for general sitting, perhaps picnicking and as a protected play area for youngsters. Many marinas only provide for boat parking and neglect to provide suitable facilities for those times when weather or mechanical failures or other reasons keep boats at the dock. A boat in its slip, baking in the sun is a mighty inhospitable place. A few shade trees, picnic tables, benches and swing sets can make the marina a pleasant place for all the family, not just the ardent boater.

The acceptability of pets in a marina is an issue that should be squarely addressed before the marina goes into operation. In general, pets do not belong on boats. To many people, however, a pet is part of the extended family and the owner will want the pet aboard the boat and in the marina. To maintain control and to keep the marina clean, the marina should have a rule that all pets must be leashed or under the direct control of the owner. An area should be provided and maintained for pets to relieve themselves, a pet toilet area. The area should be removed from the heavy marina traffic area and be well marked. Its use should be enforced. Pets must also be kept quiet and not create a disturbance for other marina tenants. Refusal to conform to pet regulations should be a cause for berthing contract termination.

19.9 WASTE OIL TANKS

Marinas may have differing policies on the amount of work allowed by the boat owner while in the marina, but most marinas do allow the owner to change engine oil and perform other routine engine and boat maintenance. The disposal of waste oil from engines has become a major issue in the desire to clean up the environment and prevent ground and water pollution. Every marina should have provision for accepting waste oil products. The volume of waste oil generated at a marina will vary with the type and size of the berthed boats and the frequency of use of these boats. Obviously a marina catering to large multiengine, offshore powerboats may have a significant quantity of waste oil generated during the boating season as compared to a marina with only outboard powered craft or with a predominance of sailboats.

As a general rule, one 275 gallon waste oil tank is suitable for up to 150 boats. The waste oil tank should be located conveniently near the docks, but it should also be convenient for access by the waste oil transporter. The waste oil tanks are often unsightly, so they may be hidden by screens or plantings. The waste oil tank should not be near a trash receptacle as the customer may be tempted to deposit the waste oil in the trash receptacle instead of taking the time to properly dispose of the oil in the tank. The tank should be filled with a funnel or other fill device that allows easy pouring of the oil into the tank. The fill pipe should have a cover to prevent addition of rain water or other foreign matter. The tank should be clearly marked as "WASTE OIL" on all visible and accessible sides. The tank may be free standing, if local regulations allow, but it should be within a concrete base that is filled with sand to catch any overflows or spillage (see Fig. 19-4). Absorbent pillows should be available in the event of spillage and the proper authority should be notified of any spill incidents.

Handling of waste oil from the storage tank may require a licensed transporter. In the United States the Environmental Protection Agency (EPA) has jurisdiction over hazardous waste handling and transport with certain aspects handled by the individual states. Waste engine oil alone may be readily disposed off, but if other materials such as engine degreasers, paints, solvents and thinners are included in the waste oil, the material may be classified as a hazardous waste and involve additional reporting and handling requirements. Large fines and penalties may be associated with violation of waste oil and hazardous waste disposal regulations.

Figure 19-4. Waste oil tank, mounted in concrete vault for overspill or tank leakage protection. The unit can be enclosed by fencing or plantings to provide a more attractive atmosphere. Signage should also be provided.

19.10 FUEL STORAGE

Provision of fuel storage and dispensing facilities has always been an issue in marina planning and development. In many economic cycles, fuel sales have been a desirable profit center. Fuel sales provide a necessary service to the vessels residing in the marina. Fuel sales also bring in transient clients and draw people from other marinas which do not offer fuel. This is a good source for attracting potential new customers for seasonal or long term slip rentals or sales. In economic cycles in which fuel supply has been limited, marinas selling fuel were often in an advantageous position to capture new slip tenants by offering a valuable commodity to its customers. This technique has to be tempered with caution and common sense since retailer regulations may not allow selective offering of fuel supplies. Limited availability can backfire, however, and alienate a potential customer. Generally the marina that offers fuel is in a better position to successfully market its slips and boat repair services than a marina that has no fuel service. In 1990, mark ups of $0.30–$0.50 per gallon were common and appeared to justify the expense of fuel storage and dispensing facilities, if the local market could produce sales in the order of 100,000 gallons per year or greater.

In the United States and elsewhere, environmental awareness has prompted new considerations regarding the sale of fuel. These considerations must be well understood before embarking on a fuel storage and dispensing program. Investigation of many traditional roadside automobile gas stations found that many had below ground tanks that leaked. The leaked gasoline and diesel fuel often was found to have travelled large distances from the leak source and in many cases the leaking fuel contaminated ground water supplies. The distance that leaked fuel can travel is affected by the type of soil around the tank and the flow pattern of fluids within the soil. If flow patterns lead to a close ground water aquifer, contamination is certain. Other soils may entrap the fuel and result in a saturated ground. This can also be dangerous as vapors may enter structures in the vicinity and result in explosions.

A result of many leaking below-ground-fuel-tanks has been the imposition of strict regulations governing the storage of fuel. Some of the new regulations, in the United States, include testing of all below-ground-tanks for leaks. Such regulations cover private oil storage tanks if older than 10 years and buried below ground. All below-ground-tanks must have corrosion protection added by the year 1998. All tanks installed after December 1988 must meet requirements for leak detection, corrosion protection, overfill prevention and installation design. For budget purposes, the cost to install a pair of 10,000 gallon buried fuel storage tanks in 1990 was between $100,000 and $150,000 dollars.

In addition, the federal government will require vendors having from 1 to 12 storage tanks to acquire insurance or provide other financial guarantees up to $1,000,000. Such guarantees are to be activated in the event of leakage or fuel spill. By 1993 all marinas having below-ground-storage-tanks must have an approved leak detection system. A leak detection system usually includes the monitoring of wells around the tanks from which samples may be taken and analysis for contamination by leaking fuels.

Vapor recovery is also a requirement. However, at this time, no standards have been promulgated on how to effect vapor recovery while fueling the boats themselves. The fuel fill configuration on boats varies widely with the on-board tank vents often far away from the inlet fill pipe which makes vapor recovery difficult. It is conceivable that some form of standard may be developed for boat manufacturers that will allow future vapor recovery during boat fueling.

Fuel Tank Installation

The actual design of fuel tank installation should be only performed by competent professionals well versed in the current design requirements imposed by the governing body. It may be useful, however, to look at a typical installation (see Fig. 19-5), acceptable for an east coast United States marina in 1990. The basic fuel capacity is 8,000 gallons each of gasoline and diesel

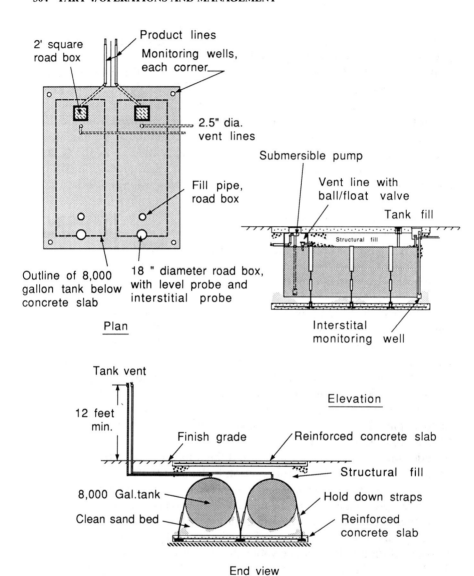

Figure 19-5. Typical installation details for twin 8,000 gallon buried fuel tanks. *(Adapted with permission from a design by Paul E. Donahue Associates, Weymouth, MA 02190)*

fuel. The typical marina will vary between capacities of 4,000 and 10,000 gallons, depending on the market for fuel sales and the proximity of fuel replenishment services to maintain marina fuel volumes. In-ground capacity generally reflects the ability to accept a full over-the-road tanker truck, which often reduces fuel costs as opposed to a split load. Check for local availability and delivery truck sizes.

Current approved fuel storage designs include double wall steel (see Fig. 19-6) and fiberglass wall tanks. There is a need to provide a corrosion resistant tank with a reasonable life expectancy. Cost will vary, so check current pricing. It may be useful to check with your fuel supplier especially if it is a large company and learn of their current policy for installation of fuel storage tanks. They will obviously have researched the economics and life factors for your area and may be able to recommend a suitable design. In the example, double wall, steel tanks are the selected tank style. The tanks will be protected by a coal tar epoxy or urethane coating to prevent corrosion. The tanks are specified for a 30 year corrosion warranty life. In addition, the tanks will be fitted with anodes which will further prevent corrosion if the protective coating is violated. The anodes will have a monitoring kit to allow

Figure 19-6. A new double wall, coated and cathodically protected, steel, 8,000 gallon fuel storage tank being prepared for below ground installation.

506 PART 4/OPERATIONS AND MANAGEMENT

checking of their condition. The anodes are welded to the tank structure and in intimate contact with the base tank steel.

The fuel tanks are to be buried in the ground near the shore of the marina with fuel product lines run below grade to the bulkhead and thence to the floating docks. Because of the proximity of the tanks to the tidal influenced ground water, or if they were to be placed in any area having a ground water level that can reach buried tank level, the tanks must be anchored. The anchoring selected is to be mass concrete. A large, reinforced concrete slab will be placed near the bottom of the tank excavation and the tanks will be strapped to the slab. When the ground water attempts to float the tanks (even

Figure 19-7. Fuel storage tank leak detection monitor mounted in the marina office. A printout of the volume pumped and the status of all detection systems is automatically prepared each morning. A violation of preset leak thresholds sounds an alarm.

full tanks can float since gasoline and diesel fuel are less dense than water) the mass of the concrete will restrain the tanks in place. In calculating the necessary weight of concrete to hold down the tanks, remember to calculate the concrete weight as a submerged weight, that is the weight of a cubic foot of concrete in air will be reduce by the equivalent weight of water, 62.5 pounds per cubic foot for fresh water and about 64 pounds per cubic foot for sea water. Since reinforced concrete weights about 150 pounds per cubic foot (pcf), and if it is assumed that the tanks are subject to sea water intrusion, the submerged weight of concrete will be about 86 pcf. The anchor slab should be bedded on an undisturbed layer of competent soil or adequate structural bedding should be provided. Between the concrete slab and the tanks there should be a bedding layer of clean sand or pea stone. The top of the tanks will be about three feet below finish grade. The area above the tanks should be filled with a structural fill material, such as clean gravel. A reinforced concrete slab should be placed on the structural fill to protect the fuel fill and vent structures.

Leak Detection and Monitoring Systems

The tank is fitted with an overfill protection device in the form of a ball float located in the tank top. Before overfill can occur, the ball float will be floated by the rising fuel in the tank into the shut-off position causing the pump in the tanker truck to shut down. Tank leak detection is provided by three monitoring systems. An interstital system consists of a probe which monitors the void space between the double tank walls for fuel intrusion. At the four corners of the tank slab are monitoring wells that sample the ambient ground water for indications of fuel contamination. A third monitoring system is a tank volume monitor which is compared with a volume monitor at the dispensing unit. The combination of these three systems provides a relatively reliable means of monitoring tank integrity, surrounding soil quality, and volume of fuel control. A remote readout of tank control data should be provided in the marina office (see Fig. 19-7). The readout can be programmed to provide regular, daily reports of system status and fuel pumpage. The system can also be connected to various alarm devices to alert marina staff or security of problems or impending problems.

20

Operation

No matter how well sited, designed and constructed a marina is, if it is not well operated it probably will not be successful. Marinas are part of the service industry and the safety and needs of the marina patrons must be provided for in a knowledgeable and efficient manner. Most marinas are operated by a team of individuals that must function effectively and with a plan of action. As a team, there must be leaders and followers with each team member having specific duties and responsibilities. In order to manage the team effectively, a well thought out plan of action must be developed. The plan of action will contain administrative aspects, operations guidance, rules and regulations for marina customers and protocols to handle emergency situations. If a suitable administrative plan is developed, many other marina functions and personnel expectations will easily follow.

20.1 ADMINISTRATIVE PLAN

A marina is a business and as such should follow solid business practices. The basis for administering a successful marina operation requires an administrative plan that formalizes the policies and responsibilities of management. The plan should be written and begin by defining the operational entity by specific name and follow with an organizational chart and chain of command. It is very important for employees to know and observe the chain of command and organizational structure of the management to allow proper delegation of responsibility and prevent misdirection of work effort.

Job descriptions are helpful to define responsibility and provide the basis for an appropriate chain of command. Some representative job descriptions for marina operations include: General Manager, Service Manager, Operations Manager and Controller. In large facilities, the General Manager may have an assistant and secretarial and clerical staff. The Service Manager may direct a staff of technicians, mechanics, bookkeepers and office staff. The Operations Manager may have a staff including a Dockmaster, dock hands, maintenance staff, security details, retail store staff, restaurant staff, and data entry specialists. The Controller will have a staff of data entry personnel, cashiers and bookkeepers. The number and degree of speciality of the staff will, of course, vary with the type and size of marina operation.

The administrative plan should also enumerate management policy with regard to employee hiring and firing, employee performance criteria and review, wage rates, pay periods, overtime, vacation, sick leave, and holiday benefits. Additionally, the plan should address policy for, excused and unexcused absence and tardiness, time keeping and work effort recording, lunch and break scheduling, use of company equipment and facilities, and any discounts or charge account policies available to employees. Sexual harassment, drug and alcohol abuse policy and testing should be clearly defined and the right's of the employee stated. Special rules for general housekeeping, personal appearance, abusive language and client relations should also be discussed.

20.2 STAFFING AND PERSONNEL

As a service business, it is imperative that an employee be competent, knowledgeable and helpful to marina patrons. Boating is a recreational pursuit and the marina staff should strive to make the marina environment a comfortable place for the tenants and to create an atmosphere appropriate for the level of service provided. Abusive or uncaring attitudes do not belong in a service oriented business such as a marina. Consideration of the customer and a sincere interest in the customer's well being will go a long way to making a smoothly operating marina and a pleasant work place for the marina staff.

Although many people dislike the requirement for uniforms, uniformed staff are often very important in maintaining order in the marina. Distinguishable marina staff are very important in marinas offering transient services where the transient boater has no other way of identifying marina staff. Uniforms may merely be shirts or blouses of consistent color with a marina logo or other identifying mark. The marina personnel who deal with the public should be neat, well groomed and knowledgeable. They should be prepared to help but not dictate to the customer. They should be assertive enough that, when necessary, their orders carry the weight of the marina manager. They should be accommodating to the needs of the customers but should not fraternize with the customers. It is very difficult for a dock hand to issue a difficult order to a customer if the dock hand has spent the previous night socializing with the customer. This type of policy may not apply to upper marina management where the marina manager may desire to participate in customer activities for the good will of the marina. This also, however, must be tempered with judgement so as not to violate the management—client relationship. Collecting unpaid bills or denying special favors from social friends may be difficult.

A marina is a dynamic operation with significant opportunity for personal injury, either to employees or guests. A desirable attribute of marina personnel is the ability to function effectively under the pressure of emergency conditions and especially during medical emergencies. It is highly desirable to have at least some marina personnel trained in basic life support skills, such as cardiopulmonary resuscitation (CPR) and first responder medical care. Many excellent courses are presented by the American Red Cross, the American Heart Association, and community hospitals. Formal or informal association with a local hospital is also a good policy. It should be a high priority in all marinas to train staff and execute drills for a variety of emergencies until such time as the staff is proficient at the required skills.

Many marinas will hire summer help to provide assistance during the peak boating season. A major problem has been retaining these people to the bitter end of the active season, Labor Day in September in the United States. Many student hires will leave early to have a few last flings before school starts and be lost to the marina during one of its busiest periods. To offset this problem, some marinas will offer a bonus only to those people who will continue to work through Labor Day or at least on weekends through Labor Day. The bonus will often be sufficient enticement for the students to work a little extra and thus prevent the marina from being shorthanded at the end of the busy season.

20.3 MARINA MANAGEMENT SYSTEMS

Berth and Mooring Management

Marinas are complex facilities often having hundreds or thousands of berths and moorings that have individual specifications and are occupied by distinct individuals. It is impossible for a marina manager or the marina staff to know the details of each berth and specifics about each boat and owner in the marina. The traditional method of maintaining some sort of order is with the use of static graphic presentations on "write on" boards or magnetized wall displays. The boards usually show the marina layout with berth identification and boat name. The system is useful but lacks the presentation of the full depth of information that may be required by the marina staff to assign a berth, locate a boat or provide the name and information on the boat owner. One or more card index files might be required to develop the additional information. This can be a time consuming operation and is further complicated when information requires changes.

An improvement on the static wall board display is a computer managed system that allows for the display of the marina general arrangement on the computer monitor. A large marina may desire to have a large monitor display or to use the computer scroll function. By use of a computer mouse, the marina arrangement can be easily scrolled to present the desired location. In addition, the software can be configured to provide various information displays by clicking on the berth or mooring for which information is desired. One window (see Fig. 20-1), may display complete information on the berth such as, berth dimensions, water depth, utilities available, maximum size boat allowed, etc. Another window may contain complete information about the berth tenant such as, boat name, boat characteristics, boat owner's name and address, and other pertinent data. The system can be very useful for transient dockage, displaying specific berth information, maximum boat size, available draft and utilities offered. Transient reservations can easily be input to the display and changed at will.

In addition to the visual display of marina arrangement and windowed data, the system can provide status reports on a variety of topics such as transient reservations, boats by size, boats by berth, reports on who is in port, vacant berths, etc. The database can be custom tailored to include invoice and accounting records, customer follow-up letters, confirmation of tran-

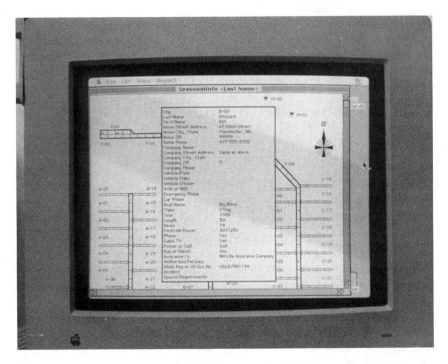

Figure 20-1. Computer monitor display of a marina schematic layout and a window showing specific data on an individual slip. MarinaOpr computer software by Gary J. Hardy Incorporated, Sunnyvale, CA 94087.

sient reservations, marina store accounts, accounts receivable, accounts payable, mailing lists, maintenance reporting and virtually any other computer definable function. The system can also be integrated with other computer base systems to provide compatibility with other in-house systems.

These types of database systems can provide the marina manager with readily available information which is needed to efficiently operate a large or small marina complex. The time saved and the information available will allow the manager and his staff to concentrate on providing the services and attention to business required of a service orientated enterprise.

Billing and Collection

Even with all the appropriate design and operation issues discused in previous chapters properly addressed and implemented, a marina is subject to failure if it does not efficiently bill and collect for the services provided. This is onc arca that marinas, in the past, have been lax in implementing. Owner/operators have been notorious in neglecting to properly and timely bill customers. They were also lax in letting payments slip and having virtually no collection arrangements. As a result, many marinas were essentially nonperforming in the financial sense. Accounts receivable soared and the owners suffered the plight of poor cash flow.

The maturing of the marina industry has resulted in a more professional approach to billing and collection. This is a must in today's climate of high interest rates with resulting cost of money and the high capital cost to construct and operate a marina. Numerous specialized computer software programs are tailored to marina and boatyard operations. The use of computer based inventory control, labor effort control, and billing procedure has taken the pain out of the necessary bookkeeping activities. Timely printouts of accounts receivable, follow-up collection reminders and cost control can greatly enhance cash flow in both large and small marinas. Marinas have specific requirements and no one of-the-shelf software may be applicable to the needs of all marinas. It is easy to custom tailor computer software to the specific needs of the individual facility to achieve a useful business tool.

The secret to financial success is first to offer the products and services desired by the customers and then to be paid in an equitable and timely manner. The marina operation must be looked upon as a business with implemented payment and collection policies that foster the financial viability of the enterprise. With all this said, however, it must still be remembered that recreational boating is a leisure pursuit and that overly aggressive or abusive billing and collection techniques may result in a nonproductive response. A positive technique is to occasionally include a marina newsletter or other pieces of information with billing statements to show an interest and concern for the customer. This may be especially important for bills submitted in the off-season when boating may not be foremost in the customer's mind. A newsy note or discussion of improvements for the upcoming season may create a positive atmosphere in the mind of the customer and result in a more rapid payment. Or a personal note, during the off-season, that the marina has checked the customer's boat recently and it looks just fine, will demonstrate a sincere concern for the customer and hopefully prod the customer to early settlement of the bill because the customer wants to maintain the marina's interest in the customer's welfare. There are, of course, many other more sophisticated methods for successful billing and collection presented in numerous texts on the subject. The ideas presented here are but a few of the ones we have experienced in our personal association with marinas and as boatowners.

20.4 INSURANCE

This is a marina development text and will only superficially address insurance issues. The information presented here should help create an atmosphere that will mitigate liability and therefore enhance insurability and reduce insurance costs. Each marina must negotiate its own insurance coverage based on specific needs and physical structure. There are, however, some common issues that relate to all marinas. Insurance is a protection against liability from all sorts of damage and injury. It is therefore prudent to minimize the potential for damage and liability related incidents.

Every marina owner or manager should be on constant patrol for areas that create high insurance risks. The physical plant should be routinely scrutinized for defects. Missing handrails, tripping hazards, loose planks, potholes, exposed electrical wiring, slippery floors, etc., should be repaired immediately. An old, weak dock system that might not survive a strong force may require strengthening or replacement. General housekeeping should be monitored. Loose tools, improper storage of flammable or hazardous materials, exposed sharp objects, piles of rubbish, etc., may compromise an otherwise favorable insurance rating. Look at your insurance claim history or ask your agent for claim histories of other marinas to see where areas of high claims exist and then attempt to reduce your exposure to those high claim areas.

Another way to reduce insurance costs is to assume certain risks. The first place to assume risk is in the policy deductible. It may be that assuming the risk associated with a higher deductible makes good business sense and can greatly reduce the overall insurance costs. Self insuring for definable risks may also be to your advantage. Educate your insurance carrier about your operation and the positive steps you have taken to minimize exposure to damage and personal injury. Most insurance is based on claims history and classification of risk. Assure your insurance carrier that you are in the most favorable category possible and that you should not just be lumped into a high exposure group that is not representative of your facility. Provide your insurance carrier with engineering documents, permits issued to your facility, documentation on the reliability of your equipment and staff, and any special certifications, commendations, or professional reviews of your facility that may warrant special consideration and lead to a determination of your facility as a lower risk operation than the average marina facility. If you minimize your exposure there is a good chance of minimizing your insurance cost. As marinas become more and more visible and form a larger industry segment, insurance underwriters are taking a greater interest in such facilities and becoming more knowledgeable in marina operation.

20.5 SERVICE ORIENTATION

In the desire to make a fair return on investment and survive in a difficult business environment, marina operators often forget that they are in a service business. The success of the marina relies on the ability to attract and retain customers. Many of these customers will consider the price they have to pay for marina service is exorbitant for the value received. The customer often will not appreciate the high capital cost of constructing a marina, the cost of prime real estate, taxes, or the cost to properly maintain a facility at the dynamic sea—land interface. It is therefore very important to provide quality services at a fair price and in a friendly manner. Customer neglect will ultimately lead to a dwindling customer base.

A good marina operation will anticipate the needs of its particular customers and attempt to provide those services to fulfill this need. Marina personnel who deal directly with the customers should be knowledgeable, friendly and helpful. The customer may not always be right, but the customer is ultimately the person who provides the revenue to operate the marina. The best advertisement is by word of mouth from satisfied customers, and this aspect of business promotion should be actively pursued.

Concern and assistance to the customer can also lead to additional income. An example of this is service selling, where the marina service manager might recommend work on and equipment for the customer's boat to enhance boating enjoyment or prevent vessel breakdown. Some marinas have a continuing program of providing a free inspection of the customer's vessel and recommending work or modifications. This type of selling, of course, must be done diplomatically and high pressure selling or scare tactics must be avoided. A sincere and conscientious appraisal of the customer's boat and appropriate suggestions will most often be taken in the spirit offered and may lead to substantial service work. Checklists and suggested programmed off-season work can also become a significant income generator.

Many marina customers, especially new boaters, may not be experienced boat handlers, so providing competent dockhands to assist in docking and undocking is a service well appreciated by many boaters. The dockhands assisting customers should be knowledgeable and helpful but not dictatorial nor attempt to embarrass the customer because of their lack of boathandling skill. Docking assistance is very important in marinas that service transient customers. The transient boater will generally not be familiar with the marina layout or with the effects of wind and current at the marina location. The dockhands may represent the first impression of the quality of the marina and they should act and respond to the customer's requests accordingly. The dockhands can, in addition to helping dock the vessel, provide information on registration and marina policy and direct the transient boater to other marina or concierge services.

Marina personnel, like the rest of us, desire to have free time on weekends and holidays, but it must be remembered that these are also peak times of marina use and the marina must be appropriately staffed to service the needs of the customers. The inability of the customer to obtain routine or emergency service during weekends or holidays will result in an unhappy customer who may take his or her work elsewhere. Just as the buzz word for real estate marketing is location, location, location; it might be said that the buzz word for a successful marina operation is service, service, service.

20.6 SECURITY AND SURVEILLANCE

A marina obviously contains a sizable financial investment in property, both property owned by the marina and the tenant boats. The marina is also a congregation of many people of different age groups and financial backgrounds. The protection of people and property is a major concern, and adequate security must be provided to protect people and property. Boats and their equipment are susceptible to theft. Access to marinas is often easy and virtually unrestricted, compounding the problem of providing adequate security. Only recently have law enforcement agencies pushed for boat titling laws to help reduce boat theft by recording the ownership history of vessels. Stolen boats and equipment can easily be resold on the open or black market. The high cost of boating often leads people to purchase equipment of questionable origin in order to save a few dollars.

Providing adequate security is difficult because people legitimately using the marina often do not want to be subjected to intense security clearance to obtain access to their boat or other marina facilities. Security issues have been exacerbated by many permitting agency requirements to provide limited or unlimited access by the general public to waterfront areas. Hopefully, permitting agencies will appreciate security problems and restrict general public access to safe areas and limit access to daylight hours. General public access to dock areas should be limited with access only allowed for boat owners and their guests. Boat yard areas and haul-out facilities are dangerous areas and should be off limits to anyone not having business in the areas. Fueling docks and upland storage tank and dispenser areas are also hazardous areas and should be properly protected. Rack storage facilities and areas where racked boats are transported within the marina can be hazardous and should have restricted access.

Limiting access is difficult in a marina because of the service and recreational nature of the facility. Security provisions must be visible and instructive but not too intimidating. Visibility is a primary positive component of a security system. Distinctive and instructional signage coupled with pleasant but firm uniformed personnel will generally reduce security problems. As in other law enforcement activities, presence or assumed presence is a key to a successful security system. Surveillance cameras and other sophisticated devices may be excellent for security purposes but may alienate the average customer. People use marinas for recreational pursuits and the idea of being watched by cameras or other intense surveillance may detract from the reason people want to be on the water.

Adequate lighting is another deterrent to illicit activity. Lighting should be carried throughout the facility including toilet facilities, laundry areas and all places where the marina customers may have occasion to pass or use during darkness. Other areas should also be sufficiently well lighted to allow observation of unusual activity. Care must be taken however, not to create a lighting pattern that interferes with boat navigation or distracts from activities aboard the berthed boats. Low level lighting on the docks should be designed to illuminate the dock surface and immediate areas for safe passage but should not be directed to the water berth or in any way blind the vision of a vessel operator maneuvering in the darkness. Haul-out facilities should be individually lighted to assure easy observation on any activity in, on, or around these dangerous areas.

Many marinas will require a secure access to the roadways and parking areas of the marina. A card key or other keyed gate may be sufficient to control access to these areas. Some marinas will be able to use only signage to direct vehicles to the appropriate destinations. Other facilities may require manned control booths to provide the required security. Security fencing around the marina upland perimeter may be required in some areas but should be discouraged unless absolutely necessary to provide the required level of security. Access to the marina docks will often require some form of control gate. If a locked entrance is provided, it must have the ability for direct exit from the docks without need for a key or other lock opening device. In an emergency people must have free passage through the gate from the docks. This may require protecting the lock system so that unauthorized people can not just reach through or around a gate to open the lock. It also may be necessary to provide security wings around a dock gate to prevent people from swinging around the gate and thereby gaining entrance. It should be kept in mind that any determined individual can compromise virtually any security system if the desire is great enough. The purpose of most security systems is to deter the average unauthorized individual from easy unimpeded access. Special effort should be made to create a welcome and attractive gated entrance. The boater should not be made to feel he is transiting through a prison gate in order to get to his boat.

Beyond the security provided by the marina in its overall facility protection system, is the security associated with the boat itself. Several companies now offer sophisticated security systems that can be used by the individual boat or monitored by the marina. Some boat owners have adapted home security systems for boat use with external alarms or automatic telephone dialers to alert them or other parties to security problems. This solution can get out of hand if some marina control is not exercised. False alarms on boats due to boat motion, power surges or power failure can be very distracting to neighboring boats if external alarms are incorporated into the system. A well run marina may want to investigate one of several security monitoring systems that would offer some degree of control by the marina staff. A boat may be outfitted with a variety of security detection devices such as, door or hatch contacts, signal mats, mooring line contacts, canvas cover removal switches, motor locks, thermal fire sensors, and high water bilge alarms. These devices may be connected to a wireless transmitter which can send a violation signal to either a dock mounted signal light or noise emitter or to a silent signal at a marina security panel in the control or security office. With each boat having an identifiable code, the specific boat that has been violated can be readily determined and the nature of the alarm identified. The data can be presented as a printed document including information about boat name and location, date, time of day, and nature of alarm. This system can also be integrated into other marina facilities to provide a comprehensive security surveillance system. The use of wireless transmitters and receivers and solar powered relay units minimizes installation costs and effort. The marina may desire to offer the boat protection services on a lease basis and generate additional revenue as well as provide additional security throughout the marina. Use of a comprehensive security system may qualify the marina and the covered boats for insurance premium discounts.

20.7 CONCIERGE SERVICES

A current trend in the marina industry is to enhance the number and quality of services provided to the resident and transient customers. Following the lead of other hospitality industry providers, marinas have begun to offer concierge services. Concierge services are most often associated with major hotels where a designated individual(s) is assigned the function of assisting the guest to accommodate their every need. The services provided are often beyond those that can be directly offered by the usual scope of the facility, and the procurement of the services may be difficult for the recent arrival or the guest unfamiliar with the local area. A fee for concierge services may or may not be levied by the provider. Arrangements are often made with the actual provider of the service to pay the concierge a commission for recommending the service.

Concierge services are varied, but a representative sample related to marinas is presented. Services may include: assisting in obtaining provisioning, making reservations for restaurant dining, obtaining tickets to major cultural or sporting events, arranging for medical assistance, providing for catering or florist services, arranging for local tours or excursions, assisting in making travel arrangements, and servicing other personal needs of the guests.

The provision of concierge services implies a first class establishment, thus positioning the marina as an attractive destination for upscale transient boaters and seasonal residents. Of course this image may work against a marina which is attempting to attract blue collar residents who may feel pressured by the availability, and unaffordability, of the services.

20.8 OPERATIONS MANUAL

An operations manual should be prepared to define marina operations policy and define actions that operations personnel are expected to execute in the day-to-day operation of the marina. All operations personnel should be given a copy of the operations manual and be required to familiarize themselves with its content and goals. It is a good idea to have periodic staff meetings and workshops to critique marina operations and update or modify the operations manual. A start-of-season workshop is constructive, especially if short term hires are brought on to service the peak season trade.

A good operations manual will be comprehensive and tailored to the specific needs and requirements of each marina. The operations manual may begin with a copy or content of the marina's administrative plan followed by sections relating to operations. Some typical operations manual categories include:

- Hours of operation and manning levels
- Personnel orientation
- Daily and weekly routine and work assignments
- Dress code and housekeeping
- Use and operation of marina equipment
- Security policy
- Fire prevention and control procedures

- Marine radio procedure
- Cash receipts
- Credit card and charge account policy
- Procurement of supplies and equipment
- Handling of outside contractors and vendors
- Allowable work which may be performed by boat owners
- Transient vessel reservations and vessel handling
- Fuel dock procedures and fuel dispensing
- Fueling of vessels other than at fuel dock
- Vending machine operations
- Marina parking areas and regulations; customers, guests and employees
- Procedure for arrival of foreign vessels
- Liaison with Coast Guard and other enforcement agencies
- Storm management plan (see Section 20.9)
- Emergency procedures (see Section 20.10)

20.9 STORM MANAGEMENT PLAN

Most areas where marinas are located are subject to storm events of one form or another. It is desirable to have a written plan prepared to cover the operations and responsibilities of the boat owners and marina staff during a storm. The actual storm management plan will vary from marina to marina as conditions warrant, but all should have a common goal of initiating early and positive actions to mitigate the effects on any storm that impacts the marina. A sample outline for a responsible storm management plan is provided for guidance and should be tailored to the specific needs of the area and marina operation under consideration.

Marina operations staff should be continually and constantly aware of weather changes that may occur and of how the changes may effect the marina. Weather awareness should be a principal component of every marina operations personnel job description. Any detectable weather changes should be brought to the attention of the marina manager or the designated person in charge. Upon determination that a significant weather condition exists or is pending, a storm management plan should be placed in effect and the appropriate actions taken. In addition to physical preparations, a log should be kept documenting the initiating of the actions, the date, time, personnel involved and the weather condition. If possible a timely recording of significant weather changes should also be logged. Of primary interest would be wind direction, wind speed, barometric pressure and wave heights. At the onset of the weather conditions, the local weather forecasting agency should be monitored or called to obtain the latest forecasts.

Before any actions are taken to secure the marina or berthed boats, all personnel should be briefed as to their duties and responsibilities. All personnel working outside should be required to wear a personal flotation device (life jacket) and any other lifesaving devices or equipment appropriate for the working conditions. No exceptions should be allowed.

A storm management plan may be delineated by several classes of events with accelerating conditions requiring greater preparation activity. Since most marinas may be subject to wind related storm conditions, the example focuses on a wind storm event. Other storm management plans may be appropriate for major events such as earthquake, fire, and flooding. The following actions will be instituted, as appropriate, for the anticipated conditions.

Storm Condition I: Winds to 40 miles per hour.

All boat owners and guests at the marina shall be advised of expected storm conditions.

Storm warning notices will be posted at all dock entrances, at the marina toilet facilities and in the marina office.

All berthed vessels will be checked for proper tie up and refastened if necessary.

The marina workboat will be checked and made ready for use.

Marina personnel will be issued personal flotation devices and they will wear them as appropriate.

Storm Condition II: Winds in excess of 40 mph but less than 74 mph.

Complete all items in storm condition I.

Advise the marina manager of pending conditions and preparation status.

Advise the marina or attendant security staff of conditions and possible marina evacuation.

If evacuation of the marina becomes necessary, only the marina manager or his designated representative will give that order.

Storm Condition III: Winds in excess of 74 mph (Hurricane conditions)

Complete all items in conditions I and II.

Request all unauthorized people in the marina to vacate the premises.

Call marina customers (see separate calling instructions) and alert them to pending conditions and responsibilities.

Assure that all items in storm condition I and II have been completed.

Secure all boats and docks—add lines, anchors, etc.

Test emergency generator and fuel supply.

Move vehicles away from waterfront.

Secure all loose equipment.

Issue portable radios to marina staff and test the radios to make certain that they are operable.

Tape exposed glass and otherwise secure buildings.

Prepare to evacuate the marina.

Take inventory of berthed boats and marina equipment.

Advise boat owners to leave the marina. Log in any owners electing to stay on board.

Instructions for call to customers
Mr., Mrs., Ms. (Boatowner). This is XYZ Marina calling to advise you of impending weather conditions. The marina has instituted its storm management plan and requests you perform the following actions prior to the arrival of the storm.

- As soon as possible you are required to reduce the wind loading from your vessel by removing sails, canvas covers, bimini tops and all removable topside equipment.
- All inflatable dinghies are to be deflated and safely stored. Rigid dinghies will be removed from the boat and securely stored away from wind exposure.
- All dockside connections (electric, telephone, water, television, etc.) shall be disconnected and stored on-board the vessel.
- Your vessel should be tied with heavy duty line that is in good condition. The line should be doubled up where possible, with a minimum of two lines bow and stern and two spring lines fore and aft.
- Your vessel should be facing bow to the sea, where possible.
- The marina staff will be available to inspect your vessel and assist you PRIOR to storm arrival.
- The marina will be monitoring VHF channel XX at all times during the storm condition. If you are planning to remain on your vessel it is mandatory that you advise the marina office and monitor channel XX at all times.
- All preparations should be completed before the arrival of the storm. No marina personnel will be allowed on the docks during the storm.
- We appreciate your participation in our storm management preparations and stand ready to offer assistance as may be appropriate to protect your property. Thank you.

The storm management plan should also include a section containing all the pertinent emergency telephone numbers for the area such as: police department, fire department, hazardous waste coordinator, spill notification agency, Coast Guard or local water law enforcement agency, weather reporting agency, security details, and key marina management personnel.

After the Storm
When the storm has abated, an immediate evaluation of the marina property and berthed vessels will be conducted. Any damage or injuries will be documented by written report with photographs and sketches, as appropriate. If necessary an underwater inspection will be made of below water marina components.

If the situation warrants, marina customers will be notified about the condition of the marina and the customer's vessel. A post storm debriefing session will be held to assess marina response to the conditions, evaluate procedures and recommend future actions to mitigate damage or enhance response.

20.10 EMERGENCY PROCEDURES

All marina personnel should be aware of the special procedures associated with any emergency occurring at the marina. To prevent confusion and elicit a prompt response, a written and routinely discussed emergency procedures plan is required. Each marina will require a procedure tailored to its special situation and circumstances. A guideline emergency procedure is presented to assist in preparation of a specific set of procedures for any individual marina facility.

Accident Procedures

Accident prevention shall be the responsibility of all marina personnel. All observed safety hazards shall be reported immediately to the dockmaster or marina manager, and if appropriate, steps should be taken to mitigate the hazard. When accidents do occur, they shall be reported to the dockmaster or marina manager as soon as possible. This includes accidents to marina personnel and others, and to any property within the marina's jurisdiction. All circumstances surrounding the accident will be presented to the marina manager in writing, stating all facts about the incident, including times, location, names, injuries, property damage and response.

All appropriate steps will be taken for the protection of life and property without undue hazard to the marina personnel involved.

Medical Emergency

Medical emergencies may be life threatening, therefore immediate and appropriate response is essential. Upon assessment that a medical emergency exists, the first action is to initiate, or have initiated, a call to 911 (or other emergency response telephone number) to request fire, police or ambulance response. Alert the senior available marina staff person and if necessary get a boat underway to the scene of the emergency. Direct appropriate marina staff or bystanders to the marina entrance or other areas to facilitate the giving of directions to emergency response personnel. Provide comfort and assistance to any injured people within the limits of the responders knowledge and qualifications. DO NOT UNDERTAKE EMERGENCY FIRST AID UNLESS QUALIFIED TO DO SO. Control of the situation should be directed by the senior marina operations person available. In addition to providing assistance to the injured party(ies), marina personnel should control the situation to prevent secondary injuries and congestion at the scene. They should clear the areas of people and equipment that might impede the free access of emergency personnel and they should limit access to bystanders.

Immediately upon return of normal conditions subsequent to the incident, a full report of the incident, in writing, should be made to the marina manager. The report should include the information required in any accident reporting. The marina manager will provide incident follow-up and advise the appropriate authorities and the marina's insurance company.

Fire

Fire in the marina or aboard vessels constitutes a major threat to the marina. All marina personnel must be aware of potential fire hazards and report apparent hazards to the marina manager at once. If possible take steps to reduce the fire hazard if such action does not place the personnel in jeopardy. All marina personnel must be aware of the locations of fire hydrants, extinguishers and other fire prevention and fire fighting equipment.

If fire is detected, immediately call the fire department or initiate a fire alarm prior to any attempt to fight the fire. Assist in the evacuation of any people immediately affected by the fire and then initiate action to prevent the spread of the fire, if such actions are reasonable and prudent. If possible remove any adjacent boats from the fire source. Deploy a marina work vessel to prohibit any boats from entering the fire area and direct any waterborne fire fighting equipment to the scene. Limited fire fighting may be appropriate with available fire extinguishers, but care must be taken not to jeopardize the personnel involved. Although water may be available for fire fighting, its use must be assessed by competent personnel to avoid spreading the fire, as might occur if the fire is a fuel fire on the water surface. The best policy is to evacuate people and isolate the fire until professional assistance arrives. If the fire is at a fuel dock, immediately close all fuel shut off valves to prevent a continuing fuel supply to the area of the fire.

Bomb Threats

In the event of a bomb threat call, the following action should be initiated. Write down everything the caller says and any detectable information such as age, sex, accent, background noise, etc. Notify the marina manager immediately of the call. Notify local security, the police department and the fire department.

If necessary, the marina manager may conduct a search of the marina, under the guidance of local authorities. If a search is authorized, only individuals familiar with an assigned area should conduct the search. Strange or foreign objects discovered during a search should not be handled, but reported to the marina manager or an on-scene bomb disposal technician. Any evacuation or re-entry to the marina will be as directed by the marina manager after assessment of the situation and evaluation of the threat. If evacuation is initiated, the pre-arranged evacuation plan will be instituted and conducted in an orderly fashion.

20.11 SUGGESTED RULES AND REGULATIONS FOR MARINA USERS

Marinas are active places with large groups of unrelated people and a significant investment in boats and equipment. Order must be maintained and regulations instituted to assure the safety and comfort of the marina tenants and the marina operator. Each marina will need to develop its own set of rules and regulations, tailored to its specific needs. For guidance, a generic set of rules and regulations is presented as compiled from several first class marinas. Any rules and regulations should be provided to each marina tenant, seasonal or transient, at the time of contract closure, and the rules should be readily available at all times from the marina office. The rules should be reviewed on a yearly basis to assure continued validity and they should be modified for any changes in marina facilities or in the policy of the marina.

Typical Rules and Regulations for Marina Users

1. *Vessel Identification.* All boats berthed or stored in the marina must be properly registered or documented as required by law. Registration numbers or name and hailing port, as applicable, shall be prominently displayed according to the regulations of the issuing agency. A copy of the registration document shall be submitted with the application for berthing or storage contract and maintained on file at the marina office.

2. *Compliance with Applicable Laws.* The boat owner and authorized user shall comply with all applicable laws, ordinances, rules, and regulations of cognizant authority in the use and operation of the boat in the vicinity of the marina. The boat shall be equipped with all required safety and life saving devices required by law.

3. *Insurance.* All boats berthed or stored in the marina will have and maintain in effect, for the duration of the contract, an "all risk" insurance policy, including hull insurance, in an amount at least equal to the actual value of the vessel and its contents. A certificate of insurance shall be filed with the contract application and the certificate shall be maintained at the marina office.

4. *Dock Use.* Only assigned docks or berths shall be occupied. Unless otherwise accepted by the marina manager, all boats shall be used for

pleasure purposes. No commercial or business use shall be made of the docks or berths without the express consent of the marina manager.

5. *Living Aboard.* The marina may, at its discretion, accept a certain number of live aboard tenants. Living aboard shall be construed as long term residence rather than casual usage. The marina manager may require provision of additional equipment or facilities to assure compliance with applicable laws, rules, and regulations. Living aboard, without the consent of the marina manager, may be cause for termination of the berthing or storage contract.

6. *Operation of the Boat.* Boat owners or their authorized representatives, are responsible for the operation of the boat in the marina area. The boat shall be operated with due care and diligence to prevent injury to any person, damage to other boats and/or the marina facilities. The boat shall be operated in a safe and seaman like manner and in compliance with all applicable Rules of the Road and local ordinances. The boat owner shall be held liable for any damage or personal injury resulting from the operation or use of the boat.

7. *Exchange or Subdivision of Berths.* No swapping, exchange or subdivision of berths shall be made without the written consent of the marina manager.

8. *Boat Tie Up.* All boats shall be properly secured to the docks with bow, stern and spring lines as necessary to provide a safe and secure connection to the dock. Lines shall be of a size suitable for the boat and the lines shall be maintained in good condition. Chafing gear or other line protection equipment shall be provided as necessary. If in the opinion of the marina manager, a boat is improperly secured, the marina may, at its discretion and without liability, refasten the boat and charge for this service. Marina personnel will, without charge, be available to advise on the proper method of boat tie up.

9. *Safety of Children and Guests.* Young children and nonswimmers shall be encouraged to wear personal flotation devices on boats and in and around the docks. Young children must be accompanied by a responsible adult at all times. No running on the docks, horseplay, swimming or fishing is permitted in the marina.

10. *Fire Prevention.* Marinas are exposed to the potential of fire and therefore it is the obligation of all marina users to use their best efforts to prevent the occurrence of fire. No open fires are allowed on boats or docks within the marina. The only exception is for approved boat stoves or heaters, permanently attached to the boat. Charcoal burners, grills, hibachis or other outdoor cooking appliances may be used in designated upland areas, with the permission of the marina manager. Refueling of boats from the docks or other boats is prohibited, except as authorized at the fuel dock. The fuel dock is a no smoking area. Posted fuel dock rules will be followed when at or

near the fuel dock. No blow torches or other open flame devices shall be used for paint removal or other boat repair use.

11. *Electrical Safety.* All electrical connections to boats will be made at the adjacent berth power posts. Cables connecting between the power post and the boat shall be of an approved type and suitable for marine applications. All connections shall be made through approved twist lock connectors. All cables and connectors shall be maintained in good condition. All electrical cables shall be kept out of the water. No special equipment shall be connected between the power post and the boat without the permission of the marina manager. The boat electrical system shall be maintained in good and safe condition. Special care will be taken to assure the boat's shore power connection is well secured and free of corrosion and loose wires. Many boat fires have been shown to originate at the shore power connection. The marina manager may, from time to time, check the electrical safety of the boat to assure that correct polarity is maintained and that the boat is not introducing stray current into the marina waters.

12. *Sanitary Facilities.* The marina prohibits the overboard discharge of boat sewage wastes. The marina has provided a pumpout facility for removal of sanitary wastes from boat holding tanks. The marina will endeavor to provide convenient hours of operation to facilitate pumpout. Shoreside sanitary facilities are also provided for the convenience of marina users. Although the marina will provide cleaning services for the upland toilet facilities, every user is responsible for maintaining the cleanliness of the facilities.

13. *Garbage and Trash.* No garbage, trash or other debris shall be thrown, placed or discharged into the marina waters. All such waste materials shall be placed in appropriate, marked containers on shore.

14. *Waste Oil.* Waste oil, inflammable liquids, or oily bilge water shall not be discharged overboard. Waste oils and other related products shall be placed in appropriate, marked containers provided by the marina on shore.

15. *Boat Appearance.* All boats shall be maintained in good condition and shall not be allowed to become unsightly or be reduced to a dilapidated condition. Trash shall not be stored on deck nor shall laundry be hung from the boat at any time.

16. *Dock Lockers and Steps.* No dock lockers or boarding steps or stairs shall be placed or attached to the docks without the consent of the marina manager.

17. *Dinghies.* Dinghies or tenders shall be stored on their associated boat whenever possible. Small boats may be stored in the water berth alongside the parent vessel, with the permission of the marina manager. No dinghies or tenders shall be stored on the docks. Dinghies or tenders of 10 feet in length shall be required to apply for a separate berth. Upland rack

storage of dinghies may be provided, at a fee, at the discretion of the marina manager.

18. *Pets.* Pets must be leashed and under the direct control of their masters at all times. Pets shall not be allowed to relieve themselves on the docks or other marina property except in designated pet toilet areas. If pets cause a disturbance to other marina tenants they shall be required to be removed from the marina.

19. *Noise.* The marina is a recreational facility with many people in close proximity. Noise shall be kept to a minimum at all times. Engines shall be effectively muffled or have other noise abating devices and comply with local noise abatement regulations. Excessive running of engines while at the dock will not be permitted. Radios, television sets, and other playback devices shall be operated at reasonable levels so as not to disturb other marina tenants. Sailboat halyards and other lines shall be tied off to prevent slapping against the mast or other structures.

20. *Soliciting.* Soliciting or advertising is not permitted in the marina.

21. *Disorderly Conduct.* Any willful violation of these rules and regulations, obnoxious or disorderly conduct by a marina tenant, boat crew or guests, that constitutes a breach of the peace, or might cause bodily injury, damage to property, or demean the reputation of the marina shall constitute grounds for termination of the offenders contract and removal of the offending owner's boat from the marina property.

22. *Dock Housekeeping.* The dock walkways shall be kept clear of all boat owner's supplies, materials, accessories and debris. No mooring lines or cables shall be laid across the main or finger walkways. The bitter ends of all lines shall be kept short and close to any cleats or tie off fixtures. No boat pulpits, anchor brackets, boomkins, or other boat structures shall overhang or obstruct the walkways.

23. *Outside Contractors and Vendors.* All outside contractors and vendors must obtain permission to work on any boat in the marina. The marina may grant permission to such contractors or vendors upon their compliance with certain conditions, regulations, and insurance requirements. Any contractor or vendor granted permission to work in the marina must check in and out with the marina office for each day at the marina.

24. *Owner Work.* The owner or charterer of a boat properly registered with the marina may perform normal maintenance work on their own boats. Extensive work, beyond normal maintenance procedures, will require permission of the marina manager and the boat may be requested to relocate to an area suitable for the intended work.

25. *Unoccupied Berths.* The marina reserves the right to make unoccupied berths available to transient boaters or for other uses. The boat owner shall notify the marina manager of any planned departure from the marina

which is in excess of three days. The marina will endeavor to have the berth open for the boat owner's return if timely notification of return is made.

26. *Removal of Personal Property at Contract Expiration.* Upon the expiration of the rental contract, all personal property must be removed from the docks. If, after thirty days of contract expiration, personal property remains at the marina, the marina will consider the personal property to be abandoned and will dispose of it as the marina deems appropriate.

27. *Storm Conditions.* The marina has a storm management plan that will be instituted in event of predicted or actual storm conditions. The marina will attempt to provide general preparation and damage mitigation measures, if possible and practical, but accepts no responsibility for the safety and care of the berthed boats. The sole responsibility for boat preparation and protection rests with the boat owner. The marina will assist boat owners when feasible and if it does not conflict with the marinas storm preparations. The marina recommends that all boats be removed from the marina in the event of storm conditions. The marina may, at its discretion, offer to haul boats, at standard rates, if time and manpower allows. Boat owners may refer to the marina's Storm Management Plan for further information.

28. *Automobile Parking.* The marina provides parking for tenants and guests in designated areas. Only those areas marked for parking shall be used. No parking is permitted in the area of the straddle hoist well, at the heads of piers, in the area of fuel tank fill manholes, or in front of vending machines. Handicapped parking is provided in designated spaces and is reserved for legitimate handicapped use. Parking in front of the marina office is limited to times when doing business at the office. Automobiles parked in violation of the rules may be moved at the owner's expense.

20.12 SIGNAGE

The use of appropriate signage can be a valuable asset in directing and controlling marina operations. Beginning with the marina entrance, signs should be well placed, easily read, and descriptive. The entrance sign should be distinctive to allow visitors positive assurance that they have arrived at the marina and are entering its premises (see Fig. 20-2). Directional signs for traffic flow patterns and appropriate parking locations are also important. Areas to be used for short term parking, to off load autos or for visits to the marina office, should be well marked and any time limits enforced.

Other signs of importance may be restrictive in nature, such as: prohibition from service or haul-out areas, dock access limitations (boat owners and guests only), rules for use of the fuel dock (see Fig. 20-3), use of work dock areas, use of solid trash receptacles, recycling areas for solid waste, prohibition of disposal of wastes in the water, pet toilet areas, handicapped parking areas, etc. Information signage should be provided for the marina office,

Figure 20-2. An attractive and easily read marina entrance sign. Prickly plantings around the sign were selected to discourage vandalism. This sign was actually carved by a marina boatowner, with some professional assistance, to help beautify the marina.

Figure 20-3. Clear and concise signage for rules and regulations at a marina fuel dock.

530 PART 4/OPERATIONS AND MANAGEMENT

Figure 20-4. A gazebo shelter at the head of a marina dock, with trash receptacle and pay telephone. Good signage indicates location and use.

Figure 20-5. A distinctive marina logo used throughout the marina to identify marina facilities and equipment.

stores or shops, toilet facilities, location of solid waste areas, waste oil receptacles, marina operation hours, emergency telephone numbers, pay telephone locations (see Fig. 20-4), laundry and other marina service facilities.

It is helpful if the marina adopts a corporate logo and uses the logo with primary signage to create a visual reference to the marina activities (see Fig. 20-5). Where appropriate, signs may be lighted to assure visibility after dark. Lighting should be discreet and not overpower the message the sign offers. Lighting should also be considerate of the privacy requirements of the tenants.

Signs should be well maintained both to serve the information function they have been installed to provide and also to present the appearance of a well maintained marina facility. Signs which are no longer applicable should be removed to avoid confusion. Care should be given to any plantings around signs so that plant growth will not obscure the sign, nor should the plantings require unreasonable maintenance to prevent obscuring of the signs.

An often neglected signage location is one that provides identification of the marina from the water. Approaching the marina from the water, especially in an area having numerous marinas in close proximity, can be difficult if proper signage is not apparent. Passing vessel control signs can also be helpful such as, no wake or steerage way only signs. Any signage that is to be read from the water must have a short message that is highly visible from a distance. You do not want boats coming close aboard the marina just to read a sign.

21
Maintenance

Maintenance is the cornerstone of facility safety and long term survival. It should be a continuing process with defined areas of scrutiny and scheduled follow-up repair and monitoring. Maintenance monitoring should be part of the daily routine, at least in so far as observations of facility components and recording of deficiencies that are observed. Maintenance may be generally grouped into two categories: unscheduled and scheduled.

Unscheduled maintenance is of a nature that requires immediate attention to protect life or property. Examples might include uneven or missing deck planks that cause an immediate tripping hazard or a broken power post with exposed wiring that might electrocute the unsuspecting user. Scheduled maintenance might include complete deck refastening or the installation of new mooring chains on a periodic basis.

Marina and small craft harbor facilities are composed of numerous components, many of which are subject to a harsh environment. The exposure may result in accelerated deterioration of certain components enhancing the potential for system failure. The best way to address maintenance is to develop a program that works for the particular facility and personnel. Some guidelines are presented which are appropriate for developing a facility maintenance plan.

21.1 MAINTENANCE PLAN

The maintenance plan is usually developed by the marina manager with input from the owner and discussions with the marina staff. The drafting of the plan should be a team effort to assure that all items requiring scrutiny are evaluated for inclusion in the plan. The plan should also include time frames for inspection surveys, staffing of inspection teams, forms for reporting data, task initiation forms, and task completion forms, as well as cost itemization for maintenance tasks.

Suggested Procedures for Developing a Facility Maintenance Plan

Item 1. Obtain an up-to-date set of plans of the facility. If the facility is new there should be designer's or dock manufacturer's drawings available. If definitive plans are unavailable, permit documents may be used for reference and a plan created to a suitable scale. It may be necessary to create a set of drawings either using in-house talent or to have the work performed by a local surveyor or marina engineer. The intent is to create a site plan which is a bird's eye view of the property showing significant features such as buildings, piers, floats, etc. If possible this plan should also show upland topography (ground elevations) and bathymetry (soundings) in the water areas. An often used scale for such plans is $1'' = 40'$ or similar scale. A scale of $1'' = 20'$ is better but if the site is large a $1'' = 20'$ scale may result in an unwieldy plan for working or field purposes. If scaled drawings are just not available, then freehand sketches, made as accurate as possible, may suffice. The purpose of the drawings is to be able to define and label areas or components in the marina to allow long term monitoring of conditions at specific locations.

Most marina facilities already have in place a slip designation system, such as Dock A, slip 23, etc. If such a defined system is in place it should be used for maintenance operations. If no specific designation system is used, one should be created. The designation system should be simple but definitive. Many marinas use compass points, i.e., north dock, slip 12, west pier, berth 11, etc.

It also may be helpful to obtain or create simple sketches of major components such as typical float constructions or typical anchorings or mooring details. These sketches can then be reproduced in quantity and labelled for specific items that require maintenance or repair.

During the research for available plans, etc., it is suggested that all data relating to the marina structures be orderly filed for subsequent recovery. Permits should be filed together; dredging analysis, quantities previously removed, material types, disposal locations, etc., should be filed as separate maintenance files, while upland soil boring reports, construction photographs, etc., should be readily available for future reference in maintenance reporting.

Item 2. After appropriate drawings and other data have been collected, then it is necessary to define which components should be inspected as part of a maintenance monitoring program. This may require several levels with some items requiring more frequent and/or intense monitoring than others. Time intervals for inspection may vary. Initially inspect all items to establish a baseline condition with which future observations may be compared. Photographic records can be of great assistance as conditions and personnel change.

Item 3. Develop a checklist to be followed during maintenance inspections. The checklist should be as comprehensive as practical to provide adequate data upon which to base decisions of required action and to determine the extent of necessary maintenance work. A sample checklist is presented in Section 21.3 of this chapter.

Item 4. Perform actual maintenance inspections. Most marinas have seasonal operation, often with a spring season startup and fall season slow down. The beginning and end of the predominant boating season are good times to perform maintenance inspections. These inspections may only be of a general overview nature with detailed inspection being performed during periods of less activity. Often the best time for detailed inspection is during the peak boating time when marina activity may be less active in terms of hauling and launching and the largest staff may be on hand. Whatever the inspection timing, the inspection should be in accordance with a written plan, with all items looked at on a scheduled basis to allow the determination of long term actions which will be required to properly maintain the facility.

Item 5. Upon completion of each maintenance inspection, a prioritized list of actions should be prepared and an implementation plan and schedule established. Cost estimates of maintenance actions must be prepared and these costs figured into the marina operations budget or special allocations must be requested.

21.2 STAFFING

Generally the marina manager will be the person in charge of preparing and implementing a maintenance plan. The manager, however, may not be familiar with the day to day conditions of every component of the facility. In larger marinas, the various department heads should participate in the development and execution of maintenance operations. All personnel should be aware of maintenance and reporting procedures and they should be encouraged to report deficiencies for consideration in the next inspection or to raise items that might require immediate action. The overall maintenance plan should be monitored by a person experienced enough in the design and construction of the facility components to be able to assess if a specific condition requires maintenance. It may be necessary to obtain the services of a marine diver to inspect below water portions of the facility. This area is often left uninspected because the marina manager is not a trained diver and what is out of sight is out of mind. This is a bad oversight in marinas located in sea water, as the most significant deterioration of metal and wood components may occur below water. A regularly scheduled underwater inspection is a must in sea water environments. The marine diver should not just be the local fellow who will do the diving for some perk or low marina use fee. The diver must be able to understand the construction of the structure, the materials of construction and be able to determine if problems exist or may develop. A trained diver may cost some money but can easily save much more money by timely reporting of problem areas before the problem becomes difficult or impossible to correct.

21.3 ROUTINE MAINTENANCE CHECKLIST ITEMS

It is useful to create a checklist to be used as a reference for maintenance reporting. The following items may be adapted as topics to create a site specific checklist for each facility or structure.

Upland Structures
- Roofing
- Siding
- Doors and entrances
- Insulation
- Structural condition
- Electrical gear
- Mechanical (HVAC) equipment
- Utilities—electric, water, gas
- Telephones
- Laundry
- Toilet facilities
- Security
- Fire protection
- Landscaping
- Drainage around structures
- Lighting

Upland Land Areas
- Obstructions or hazards
- Drainage—catch basins, grease traps
- Lighting
- Security
- Fencing
- Utilities—open trenches, overhead wires
- Erosion
- Public telephone areas
- Trash receptacles
- Fire protection and hydrants
- Landscaping
- Gates and entrances
- Portable sewage holding tank emptying station

Fixed or Access Piers
- Structural integrity
- Hazards
- Foundation condition— piles, bulkheads, etc.
- Fastenings
- Safety railings
- Security

Gangways (brows, ramps)
- Decking—fastenings, nonskid, cleats
- Railings—sound, smooth
- Chord members if truss type, weld or bolt condition
- Transition plate at rolling end—smooth movement, hangups
- Cables and pipe hanging from gangway

- Fixed end fastenings—hinges
- Rolling end—freedom to roll, condition of roller or wheels
- Lateral movement—sideways
- Signage—capacity plate

Docks and Floats
- Decking—condition and fastening
- Connection hardware—unit to unit
- Flotation
- Freeboard—consistent, level
- List and trim
- Cleats, bollards, bullrails, tie rings—condition and fastening
- Fendering condition
- Utilities—water, electrical, CATV, telephone
- Utility hatch covers and pipe chase
- Finger float connection
- Mooring hardware—pile guides, chain stoppers, winches
- Signage—security and load limit

Dock or Float Utilities
- Power post condition
- Metering operable and accurate
- Lockouts operable
- Dock lighting—burned out bulbs
- Electrical receptacles—operable
- Circuit breakers operable
- Correct polarity
- Condition of tenants electric cable
- Water main shutoff valves
- Water system drain
- Water pressure and flow
- Water system backflow valves
- Water system leakage
- Water system hose bibbs
- Cable and pipe conduits

Mooring Piles
- Corrosion in steel
- Marine borer activity in wood
- Treatment in wood
- Coating on metal
- Below water condition
- Obstructions to pile guide travel
- Vertical alignment—plumbness
- Pile butt protection
- Dents, chafing or splintering

Fueling Facility
- Properly posted with safety regulations
- Dispensing pumps and hose condition

- Emergency fire and spill plan posted
- Fire protection equipment operable, proximate and correct number
- Fuel line shutoffs operable
- Tie up line cutter—axe, knife, wire cutters
- Alarm systems operable
- Fuel spill booms and absorbent material available
- Lighting
- Electrical connections approved and safe
- Main fuel line condition
- Storage tanks properly located and installed
- Storage tank venting and monitoring

Haul-out Facilities
Straddle hoists
- Unit maintenance plan followed
- Sling condition
- Runway and pier condition
- Operator qualifications
- Moving or backup alarms functional

Marine railways
- Hauling machine condition
- Chain or cable condition
- Wheels or rollers condition
- Cradle condition
- Debris or obstruction on rails
- Track and foundation condition, especially below water

Vessel Grids
- Foundation and structure condition
- Fendering
- Signage

Hydraulic and auto trailer launch ramps
- Slope
- Condition of ramp structure
- Condition of ramp surface
- Nonskid surface
- Access ladders

- Obstructions or hazards, on deck or overhead
- Boat washdown areas
- General and security lighting
- Signage
- Fendering
- Service ladders and floats

- General and security lighting
- Transfer area condition
- Outhaul sheave condition
- Keel and bilge block condition
- Silt over outshore track end

- Lighting and utilities
- Access ladders or walkways
- Silt buildup
- Tie off hardware

- Signage
- Boarding floats
- Deadman pulling points, upland
- General and security lighting
- Transition of slope at grade

- Barrier against unauthorized or unintentional use

Hydraulic trailers
- Follow unit maintenance plan
- Movement areas free of obstructions

- Obstructions or hazards—surface or overhead
- Signage
- Safe storage when not in use

Wave Attenuation Structures

Rubble mound breakwaters
- Condition of armor stone
- Core stability and slope maintenance
- Erosion

Floating attenuators
- Float integrity, above and below water
- Unit to unit connections
- Freeboard and flotation
- List and trim
- Mooring hardware condition and tightness

Fixed pile or wood barriers
- Structure condition and integrity, above and below water
- Bottom scour at mudline
- Connection hardware, condition and tightness

- Signage
- Navigation aids
- Security
- Lighting

- Deck condition
- Maintaining position
- Lighting
- Signage
- Navigational aids
- Utilities condition
- Fendering

- Loose planking or piles
- Maintenance of vertical alignment
- Signage
- Navigational aids
- Lighting

Dredging
- Bathymetry plan—soundings, before and after dredging and periodically
- On site dredge material disposal sites—ground settling, runoff, compaction

- Siltation rate—buildup of material over time to estimate future dredging
- Flushing—changes in water circulation, adjacent new construction, etc.

Parking Areas
- Surface treatment condition
- Striping
- Lighting
- Signage

- Drop off areas—signage, painting, handicapped, enforcement
- Curb cuts and access slopes

- Security
- Gate control
- Obstructions or hazards
- Grease traps
- Traffic flow patterns and congestion
- Height restrictions, overhead obstructions
- Utility protection—hydrants, utility poles, transformers

Slopes and Embankments
- Slope stability
- Erosion
- Vegetation
- Drainage and runoff control
- Stone size and stability

Bulkheads
- Material of construction condition—decay, corrosion or mechanical damage above and below water
- Tie back condition
- Waler condition
- Visible holes, above or below water
- Settlement holes at grade—indicate holes in bulkhead and fill loss
- Coating condition
- Bottom scour at toe of bulkhead
- Siltation buildup at toe of bulkhead
- Upland surface runoff at or through bulkhead
- Upper cap condition
- Condition of bulkhead penetrations—drain pipes, conduits, etc.

Facility Safety and Housekeeping
- Stairs and walkways clear of obstructions
- Trash in designated areas
- Unauthorized areas secure
- Railings and handholds secure
- Decks clear of debris and trash
- Chemicals and flammables secured and appropriate signage
- Safety lighting operable
- Dangerous areas barricaded and posted
- Electrical lines and fixtures safe
- Slippery areas surfaced with nonskid material

21.4 RESERVE FUND

In order to appropriately apply the results of a good maintenance plan, it is necessary to have available funding to implement the maintenance operation. A budget line item is suggested for maintenance and reserve funding (see

Chapter 2). A maintenance account should be developed and updated from the conditions discovered in the maintenance surveys. Major maintenance items should be scheduled to be implemented in a time frame acceptable from the financial as well as from the need perspective. Assessing maintenance needs will allow for increment funding creation in a reserve account and will not require major capital expenditure on short notice. Of course, major emergency items will require immediate attention, but these can be minimized by conscious application of preventative maintenance and periodic maintenance surveys and inspections.

21.5 CONCLUDING REMARKS

The subject of this last chapter has been the maintenance of a facility and it may appear to be the end of the discussion on marinas and small craft harbors. In fact, it is just the beginning. The book has provided a great deal of information on regulations, planning, environmental issues, design, materials selection, ancillary facilities, operation, and maintenance. The challenge that now confronts the reader is to translate this information into a working, practical implementation of the ideas and concepts. The single most important point to keep in mind is that marinas are site specific. The marina developer/designer/regulator must tailor the marina concept to suit the parameters of the particular application and not arbitrarily apply concepts or regulations that corrupt or conflict with good design.

To a large extent good marina design will result from a combination of the application of engineering and scientific expertise mixed with a great deal of practical experience. Each problem or design decision may have a variety of possible solutions, the key is to develop the appropriate solution that best suits the needs of the particular site or operational goal.

In Chapter 1, a statement was made about the intent of the book to be a bridge between the unknown and the unexplored island of marina development. If we have done our job, you have now crossed that bridge and can successfully begin the most interesting part of the journey, the implementation of ideas and concepts that will open the gate to the island of marina development, whether it is the construction of a marina facility, evaluation of a marina project, or simply the use and enjoyment of marina facilities.

PART 4 – INFORMATION SOURCES

1. Adie, D. W. 1984. *Marinas a Working Guide to Their Development and Design*, 3rd edition. New York: Nichols Publishing Company.
2. American Society of Civil Engineers. 1969. *Report on Small Craft Harbors, ASCE – Manuals and Reports on Engineering Practice – No. 50*. New York: American Society of Civil Engineers.
3. Barnes, H. 1982. *The Backyard Boatyard*. Camden, ME: International Marine Publishing Company.

4. Blain, W. R. and Webber, N. B., eds. 1989. *Marinas: Planning and Feasibility. Proceedings of the International Conference on Marinas, Southampton, UK 1989.* Southampton, UK: Computational Mechanics Publications.
5. Blain, W. R. and Webber, N. B., eds. 1989. *Marinas: Design and Operation. Proceedings of the International Conference on Marinas, Southampton, UK 1989.* Southampton, UK: Computational Mechanics Publications.
6. Bruun, P. 1981. *Port Engineering,* 3rd edition. Houston, TX: Gulf Publications.
7. Chamberlain, C. J. 1983. *Marinas, Recommendations for Design, Construction and Management,* Vol.1, 3rd edition. Chicago, IL: National Marine Manufacturers Association.
8. Dunham, J. W. and Finn, A. A. 1974. *Small-craft Harbors: Design, Construction, and Operation. Special Report No. 2.* Vicksburg, MS: U.S. Army Corps of Engineers, Coastal Engineering Research Center, Waterways Experiment Station.
9. Gaythwaite, J. W. 1981. *The Marine Environment and Structural Design.* New York: Van Nostrand Reinhold.
10. Gaythwaite, J. W. 1990. *Design of Marine Facilities for the Berthing, Mooring and Repair of Vessels.* New York: Van Nostrand Reinhold.
11. International Marina Institute. 1987. *Proceedings of the Marina Design and Engineering Conference, Boston.* Wickford, RI: International Marina Institute.
12. International Marina Institute. 1988. *National Dry Stack Marina Handbook.* Wickford, RI: International Marina Institute.
13. International Marina Institute. 1989. *1989 National Marina Reserach Conference Proceedings.* Wickford, RI: International Marina Institute.
14. International Marina Institute. 1990. *1990 National Marina Reserach Conference Proceedings.* Wickford, RI: International Marina Institute.
15. Naranjo, R. 1988. *Boatyards and Marinas.* Camden, ME. International Marine Publishing Company.
16. Marinacon. 1988. *Proceedings of the Third International Recreational Boating Conference, Singapore.* Rozelle, NSW, Australia: Marinacon.
17. Quinn, A. D. F. 1972. *Design and Construction of Ports and Marine Structures.* New York: McGraw-Hill Book Company, Inc.
18. Phillips, P. L., ed. 1986. *Developing with Recreational Amenities: Golf, Tennis, Skiing, Marinas.* Washington, DC.: The Urban Land Institute.
19. Ross, N. W. ed. 1983. *A Special Literature Search for the Marine Boating Industry.* Narragansett, RI: University of Rhode Island Marine Advisory Service, University of Rhode Island.
20. Tsinker, G. P. 1986. *Floating Ports, Design and Construction Practices,* Houston, TX: Gulf Publishing Company.
21. Webber, N. B., ed. 1973. *Marinas and Small-craft Harbors. Proceedings of the University of Southampton Conference, April 1972.* Southampton, UK: Southampton University Press.
22. Wortley, C. A. 1989. *Docks and Marinas Bibliography.* Madison, WI: University of Wisconsin Sea Grant Advisory Services.

Appendix 1

Conversions

APPENDIX 1

Conversion Factors

Dimension

1 inch (in.)	= 25.400 mm	1 cm	= 0.394 in.
1 foot (ft.) = 12 in.	= 0.305 m	1 m	= 3.281 ft.
1 yard (yd.) = 3ft. = 36in.	= 0.914 m	1 m	= 1.094 yds.
1 in./ft.	= 8.333 cm/m	1 cm/m	= 0.120 in./ft.
1 square inch (sq.in.)	= 6.452 cm^2	1 cm^2	= 0.155 sq.in.
1 square foot (sq.ft.)	= 0.093 m^2	1 m^2	= 10.764 sq.ft.
1 square yard (sq.yd.)	= 0.836 m^2	1 m^2	= 1.196 sq.yd.
1 sq.in./ft.	= 21.167 cm^2/m	1 cm^2/m	= 0.0472 sq.in./ft.
1 cubic inch (cu.in.)	= 16.387 cm^3	1 cm^3	= 0.061 cu.in.
1 cubic foot (cu.ft.)	= 0.028 m^3	1 m^3	= 35.315 cu.ft.

Weight - Force - Pressure - Specific Weight

1 pound (libre. lb.)	= 4.450 N	1 N	= 0.225 lbs.
1 lb./in.	= 0.175 N/mm	1 N/mm	= 5.708 lb./in.
1 lb./ft.	= 14.599 N/m	1 N/m	= 0.068 lb./ft.
1 lb./sq.in.	= 0.690 N/cm^2	1 N/cm^2	= 1.450 lb./sq.in.
1 lb./sq.ft.	= 47.897 N/m^2	1 N/m^2	= 0.021 lb./sq.ft.
1 long ton/sq.ft.	= 107.289 kN/m^2	1 kN/m^2	= 0.009 long ton/sq.ft.
1 lb./cu. in.	= 0.272 N/cm^3	1 N/cm^3	= 3.683 lb./cu.in.
1 lb./cu.ft.	= 157.142 N/m^3	1 N/m^3	= 0.006 lb./cu.ft.
1 pound (libre. lb.)	= 0.454 kg	1 kg	= 2.205 lbs.
1 lb./ft.	= 1.488 kg/m	1 kg/m	= 0.672 lb./ft.
1 lb./sq.ft.	= 4.882 kg/m^2	1 kg/m^2	= 0.205 lb./sq.ft.

Moment of inertia

1 $in.^4$	= 41.622 cm^4	1 cm^4	= 0.024 $in.^4$
1 $in.^4$/ft.	= 136.558 cm^4/m	1 cm^4/m	= 0.007 $in.^4$/ft.

Section modulus

1 $in.^3$	= 16.387 cm^3	1 cm^3	= 0.061 $in.^3$
1 $in.^3$/ft.	= 53.763 cm^3/m	1 cm^3/m	= 0.019 $in.^3$/ft.

Moment

1 in.lb.	= 11.302 Ncm	1 Ncm	= 0.088 in.lb.
1 ft.lb.	= 1.356 Nm	1 Nm	= 0.737 ft.lb.
1 in.lb./ft.	= 0.371 Nm/m	1 Nm/m	= 2.697 in.lb./ft.
1 ft.lb./ft.	= 4.450 Nm/m	1 Nm/m	= 0.225 ft.lb./ft.

Nautical Units

1 nautical mile = 1.152 statute miles = 6076.1 feet = 1.853 kilometers
1 knot = 1.152 miles per hour = 1.688 feet per second = 0.515 meter per second
1 fathom = 6.0 feet = 1.829 meters
1 long ton = 1.12 short tons = 2240 pounds = 1.016 metric tons

Appendix 2
Useful Information

APPENDIX 2

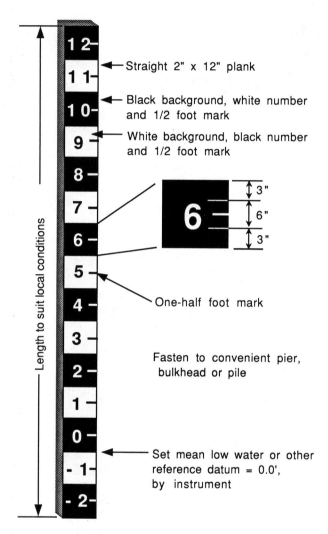

Figure A 2-1. A standard tide board for recording tidal data.

APPENDIX 2 *(continued)*

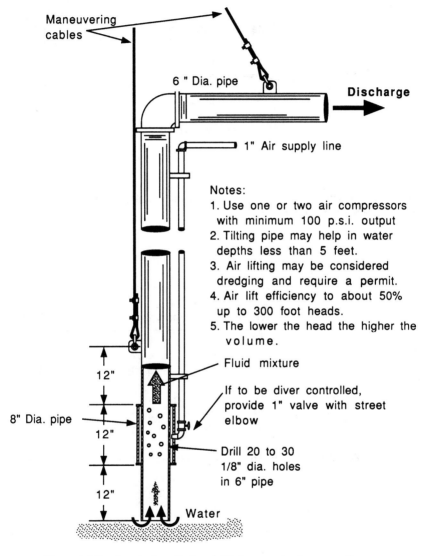

Figure A 2-2. A typical small pipe air lift for removing bottom sediments.

APPENDIX 2 *(continued)*

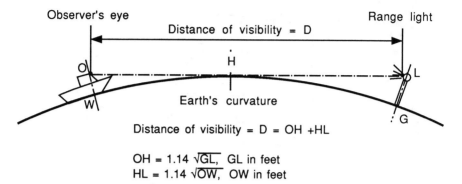

Distance of visibility = D = OH + HL

OH = 1.14 $\sqrt{GL}$, GL in feet
HL = 1.14 $\sqrt{OW}$, OW in feet

Figure A 2-3. A method of calculating the distance from an offshore observer to an object onshore, considering the curvature of the earth.

	Percent grade	Horiz. to Vertical	Degrees
Autos with trailers	12	8.33 : 1	6.85
	15	6.67 : 1	8.53
	20	5 : 1	11.31
Hydraulic trailers	25	4 : 1	14.04
	33	3 : 1	18.44
	50	2 : 1	26.57

Figure A 2-4. Comparison of various launching ramp slopes.

APPENDIX 2 *(continued)*

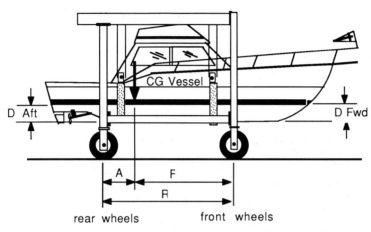

Calculation for rough estimate of vessel weight:

$$\frac{L \times B \times D \times 0.7}{31 \text{ (or 32 for fresh water)}} = \text{Weight in tons of 2,000 lbs.}$$

W = Weight of vessel
L = Waterline length
B = Waterline beam
D = Average draft = $\dfrac{\text{Draft Fwd} + \text{Draft Aft}}{2}$

Calculation for load estimate on straddle hoist wheels:

$$\frac{W}{2} \times \frac{F}{R} = \text{load on rear wheels}$$

$$\frac{W}{2} \times \frac{A}{R} = \text{load on front wheels}$$

CG = Center of gravity of vessel
R = Distance between straddle hoist wheels
F = Distance from CG to front wheels
A = Distance CG to rear wheels

If vessel center of gravity unknown, a rule of thumb for powerboats is: CG = 2/3 waterline length from forward.

Figure A 2-5. Rules of thumb for estimating boat weight and straddle hoist wheel loads.

APPENDIX 2 *(continued)*

Rules-of-Thumb and other Useful Information

This compilation of rules-of-thumb and other useful information is presented for guidance in planning and designing marina facilities. No guarantee is offered as to the correctness or applicability of the data for any particular application and caution is urged in use of the material.

1. Conditions and services found in a full service marina.

Adequate water depth for draft of boats
Secure boat tie up system
High capacity electrical system
Fresh water
Fuel, oil, propane, alcohol, kerosene
Sewage pumpout
Fire protection and fire fighting equip.
Telephone on docks
Cable TV access on docks
Ample and close auto parking
Security
Dock locker boxes
Dock carts
Dinghy rack or other storage area
Clean and ample heads and showers
Laundry or laundry service
Ice - block and cubes
Fish cleaning station (away from docks)
Bicycle racks
Freezer lockers
Information on outside services
Vending machines, soda, candy
Trash receptacles
Waste oil disposal containers
Ship store/parts

Marine VHF monitoring
Dockhands to assist in docking
Pay telephones
Mail and package acceptance
Message board
Weather condition board
Marina landscaping
Picnic area
Swimming pool
Deicing system (when appropriate)
Daily newspaper availability
Library or book exchange
VCR tape library
Recreation or lounge area
Tennis courts
FAX and office services
Posted marina rules
Concierge services
Boat haul-out facilities
Engine/mechanical shop
Hull/carpentry shop
Paint shop
Fiberglass repair
Rigging shop
Electronic sales and service

Reference: Chapter 1.8.

APPENDIX 2 *(continued)*

2. **Primary elements of a Business Plan.**

   Statement of Purpose
       Purpose of the Plan
       Nature of the Business
       Financial Information
   About the Business
       Description of the Business
       Location of the Business
       Market
       Competition
       Management
       Personnel
   Financial Information
       Operating Statement
       Balance Sheet
       Breakeven Analysis
       Loan or Investment Request Documentation

   Reference: Chapter 2.2.

3. **Typical capital cost items that may be anticipated in marina development projects.**

Land acquisition	Perimeter protection (wave attenuation)
Engineering and scientific studies	Utility installation
Legal fees	Haul-out facilities and equipment
Dredging or filling	Boat repair shops
Bulkheads or slope stabilization	Boat storage buildings
Docks and piers	Dry stack storage facilities and equipment
Dock mooring systems (piles, anchors)	Auto parking areas
Sewage disposal systems	Workboats and equipment
Marina office and dockmaster buildings	Landscaping

   Reference: Chapter 2.3.

4. **Typical financing sources for marina projects.**

Commercial Banks	Industrial or Recreational Development Bonds
Savings and Loan Associations	Venture Capital Funds, Public and Private
Mutual Savings Banks	Limited Partnerships
Life Insurance Companies	Real Estate Investment Trusts
Investment Banking Institutions	Family or Friends
Credit Unions	Money Market Funds (short term debt)
Pension Funds	Capital Markets (long term debts)

   Reference: Chapter 2.5.

APPENDIX 2 *(continued)*

5. Basic desired characteristics for marina site selection.

Appropriate zoning and Master Plan conformance
Sufficient water depth (no dredging)
Adequate upland area (no filling)
Adequate waterfrontage
Exposure protected
Not in a wetland area
Access to utilities

Not in a designated port area
Not in an area of restricted historic preservation
Not adjacent to a public beach
Near a metropolitan area or other market
Appropriate land elevation (above flood hazard areas)
Adequate transportation infrastructure

Reference Chapter 4 Introduction.

6. Currents of more than one knot are usually too swift to sustain a safe and viable marina.

Reference Chapter 4.1.

7. Checklist of marina development goals.

Year around in-water berthing
Seasonal in-water berthing
Dry stack storage
Seasonal land storage
Transient dockage
Boat yard services
   Boat haul-out facilities
   Engine and mechanical services
   Carpentry
   Painting
   Electronics
   Fiberglass repair
Boat fueling facility
Ship store

Food services
   Restaurant
   Snack bar/fast food
   Dining room
   Take-out service
Clubhouse
   Swimming pool
   Tennis courts
   Health club
Lease or sale of slips
   Seasonal lease
   Dockominium
   Cooperative
   Associated with upland residences or offices

Reference Chapter 4.2.

8. Minimum number of slips for financial viability.

It is generally perceived, in the marina industry, that a minimum number of around 250 slips is needed for a marina to be financially viable, the validity of this argument has yet to be established.

Reference: Chapter 4.2.

APPENDIX 2 *(continued)*

9. **Rule of thumb for water area requirements versus land area requirements for development.**

    Water area equals one to one and one-quarter times the land area.
    Approximate percentages for various uses:

Boat slip and dock area	= 32 percent
Channel and fairway access to slips	= 24 percent
Auto parking area	= 8 percent
Buildings and support facilities	= 6 percent
Boat storage, haul-out, overflow parking	= 30 percent

    Reference: Chapter 4.2, Figure 4-2.

10. **Plan criteria for site surveys.**

    Plan scale: 1" = 20', 1" = 40', 1" = 100', 1" = 200', as applicable.
    Upland contour interval: 1', 2', or 5' depending on site grades.
    Soundings grid interval: 25' x 25', 50' x 50'. Or 100' x 100' depending on bottom slope.
    Plan datum: National Geodetic Vertical Datum (NGVD) or other suitable water level related base.

    Reference: Chapter 4.6.

11. **Corps of Engineers (New England Division) recommended guideline for set back of structures from federal channel lines.**

    Setback from federal channel lines equal to a horizontal distance of three times the control depth of the channel.

    Reference: Chapter 4.7.

APPENDIX 2 *(continued)*

12. Typical information included in a traffic assessment study.

Project definition
    Project description
    Area Maps
    Site Plan
    Zoning Map
Existing Conditions
    Roadway Network
    Traffic Volumes
    Other Transportation Modes
    Air Quality
    Accident Reporting
    Level of Service and Capacity Estimates
Trip Generation
    Institute of Traffic Engineers (ITE) Rates
    Alternative Trip Generation Rates
Trip Distribution
    Site-generated Trips
Future Conditions
    Traffic Volumes
    Other Modes
    Proximate Area Development
    Capacity Analysis
    Signal Warrant Analysis
    Summary
Mitigation
    Primary Analysis
    Additional Analysis
    Mitigation Implementation
    Site Analysis/Geometric Design
    Development Options
    Construction Schedule and Impacts

Reference: Chapter 4.8.

13. Wind gusts

Wind gusts may cause a variation of 30 to 40 percent above the so-called mean or average or steadier wind.

Reference: Chapter 5, Introduction.

APPENDIX 2 *(continued)*

14. NOAA/Saffir/Simpson scale for hurricane categories.

Category	Winds (mph)	Storm-surge (ft)
1	74 -95	4 - 5
2	96 - 110	6 - 8
3	111 - 130	9 - 12
4	131 - 155	13 - 18
5	greater than 155	greater than 18

Reference: Chapter 5.3.

15. Converting "fastest observed one minute wind speed" to "fastest mile" wind speed.

It is generally accepted that the fastest observed one minute wind speed can be converted to the fastest mile, or peak wind, by adding 10 miles per hour, where $v_f = v_m + 10$.

Reference: Chapter 5.4.

16. Typical wind produced wave heights for some general conditions.

Weather Conditions	Wave Heights
Fair	1 to 2 feet
Moderate	3 to 5 feet
Severe or exposed sites	6 to 8 feet
Open ocean	30 to 50 feet
Highest reported	112 feet

Reference: Chapter 6.1.

17. Typical wave periods for some general conditions.

Location	Conditions	Period
Bays and estuaries	moderate weather	2 to 3 seconds
	storm conditions	4 to 5 seconds
Open coast	deep water	6 to 8 seconds
Deep ocean		9 to 15 seconds
Deep ocean	extreme conditions	20 seconds
Boat wakes		to 4 seconds
Most marinas		usually not greater than 5 seconds

Reference: Chapter 6.1.

APPENDIX 2 *(continued)*

18. Effective depth of wave action.

Waves effectively don't exist at a depth which is greater than half the distance between wave crests.

Reference: Chapter 6.1.

19. Height at which a wave will "break".

When the ratio of a wave's height (H) to the wave's length (L) becomes greater than 1 to 7, the wave will tumble or break.

Reference: Chapter 6.1.

20. Estimate of wave period if wave height is known.

If you can measure or estimate the height of a wave at a particular site, the height of the wave in feet is approximately equal to the period of the wave in seconds.

Reference: Chapter 6.2.

21. Estimating maximum wave height from FEMA still water elevations and base flood elevation wave heights.

The height difference between the still water and base flood elevations represents the wave crest height which by rule of thumb is 70 percent of the actual maximum wave height expected to arrive at the shoreline.

Reference: Chapter 6.2.

APPENDIX 2 *(continued)*

22. Relationship of wave speed, wave length and wave period.

The speed of the wave's travel across the surface of the water (C) equals the wave length (L) divided by the wave period (T). $C = L/T$

An estimate of the wave length (L) can be made if the period (T) is known by the formula $L = 5.12T \times T$.

The depth of influence of a wave in the water column may be estimated by dividing the wave length (L) by 3. $L/3$ = depth of wave influence in water column.

An estimate of the wave speed (C), in knots, may be found by multiplying the period (T) by 3. $C = 3 \times T$

A wave will break when the wave's height is 78 percent of the water's depth.

The wave crest length of a wind produced wave is approximately equal to 3 times its wave length.

A wave cannot sense the presence of an object in its path which is smaller than one-quarter of its wave length.

Reference: Chapter 6.2.

23. Wave height definitions.

The significant wave height represents the average of the highest one-third of all waves for a given set of prediction assumptions. The expected height of the highest one out of ten waves, H_{10}, is statistically determined to be 1.27 times the significant wave height, with the highest one out of one hundred waves, H_1, determined to be 1.69 times the significant wave height.

Reference: Chapter 6.4.

24. Water depth consideration for wave height determination.

A common practice is to use an average depth for wave height determination which is representative of the last 25 percent of the wave's travel and growth distance as it arrives at the site.

Reference: Chapter 6.4, Water Depths.

25. Determining theoretical hull speed of a vessel.

An approximate value of the hull speed of a vessel acting as a displacement craft, not planing through the water, can be determined by taking the square root of its waterline length and multiplying it by 1.3. The speed determined will be in knots.

Reference: Chapter 7.1, Vessel Wake Production.

APPENDIX 2 *(continued)*

26. Vessel wake wave decay.

The wakes start out next to the boat's hull in their highest form, reducing to half this height at a distance of approximately one wave length from the hull.

Reference: Chapter 7.1, Wake Travel and Behavior.

27. Tidal time cycles.

Although tides usually exhibit what is called a semidiurnal variation of 12 hours and 25 minutes, some areas have only a diurnal variation of twice that or 24 hours and 50 minutes.

Reference: Chapter 7.2.

28. Tidal data recording over a 19 year interval.

There can be many combinations of the moon, sun and earth positions so that basically the tides will take as long as 18.6 years to repeat themselves. That is why when tides are recorded for purposes of making statistical predictions and for the preparation of the tide tables for publication, the recording must be done over a 19 year period, which is referred to as a tidal epoch.

Reference: Chapter 7.2, How the Tides are Produced.

29. Wave produced current.

As a rough estimate, one-tenth of a knot of wave drift can be assumed for each foot of wave height. So if a swell system is estimated at 2 to 3 feet high, it may produce a current of 0.25 knots at the water's surface.

Reference: Chapter 7.3, Wave Currents.

30. Estimating local ice thickness.

Ice thickness can be estimated to a reasonably accurate degree by utilizing the relationship of expected ice thickness in inches, which is equal to the product of a locality factor, times the square root of the sum of the freezing degree days (FDD) during the winter. The locality factor ranges from 0.33 to 1.0 where the lowest value relates to heavy snow cover on the ice and 1.00 relates to bare ice.

Reference: Chapter 7.6, Local Ice Conditions.

APPENDIX 2 *(continued)*

31. Ice forces from thermal expansion.

Thick clear ice tends to crush and raft when subjected to pressures greater than 400 pounds per inch, with snow ice, or recently formed salt water ice crushing and rafting at 100 to 200 psi. Ice forces from thermal expansion are on the order of 5,000 to 10,000 lbs/ft, for an ice sheet several feet thick. Ice thickness of 3 inches or less may not create a thermal expansion problem because the ice tends to buckle, break and raft easily, thus relieving the pressures.

Reference: Chapter 7.6, Ice Forces.

32. Boat beam versus slip width.

It is recommended that in determining the amount of room allocated for a boat in a slip that if possible a minimum of 4 feet plus the boat beam be used.

Reference: Chapter 10.3.

33. Rule for selection of fixed or floating docks based on tidal level change.

As a general rule areas having tidal change in the range of 0 to 4 feet may select either fixed or floating docks. Between 4 feet and 7 feet of tidal range, both fixed and floating docks are used but floating docks are generally a better choice. Areas with tidal ranges greater than 7 feet are pretty much forced to use floating docks.

Reference: Chapter 11.1.

34. Estimate of berth costs and fixed and floating dock costs in U.S. dollars, 1990.

In 1990 dollars, U.S., the cost range for a berth is between $8,000 and $15,000, both fixed and floating. In 1990 dollars U.S., the square foot cost for floating dock systems ranges between $15 and $35 per square foot. Depending on the type of construction, fixed docks may cost in the range of $25 to $100 per square foot.

Reference: Chapter 11.6.

35. Channel entrance width.

In general, the entrance channel should have a minimum width of 75 feet with full control depth over this width. A 100 foot wide channel is a more preferable design criteria and should be used as the minimum where possible.

Reference: Chapter 12.1.

APPENDIX 2 *(continued)*

36. **Channel control depth.**

    The channel control depth will be based on the deepest draft of a potential user vessel plus a minimum of 3 feet additional depth plus an estimated trough depth of any significant wave action that may occur in the channel.

    Reference: Chapter 12.1.

37. **Clear width of fairways.**

    The general rule of thumb for fairway sizing has been to make the clear distance between boat extremities no less than 1.5 times the longest boat length, and often 1.75 times boat length if maneuvering conditions warrant.

    Reference: Chapter 12.3.

38. **Width of main docks.**

    The minimum width of fixed or floating main docks is considered to be 6 feet. Generally a 8 foot wide main floating dock is the favored width for most marina applications. Fuel docks should be 10 feet to 12 feet in width. Floating wave attenuators will be in the range of 16 to 24 feet in width, or greater.

    Reference: Chapter 12.5.

39. **Width of finger floats.**

    The width of a finger float is generally equal to 10 percent of the length of the finger, with a minimum width of 3 feet.

    Reference: Chapter 12.5.

40. **Float freeboard.**

    Floating docks will generally have a freeboard of between 12 and 26 inches.

    Reference: Chapter 12.5.

41. **Sewage pumout motor horsepower and discharge pipe sizes.**

    Sewage pumout device motors will generally require $1/2$ horsepower motors for heads up to 20 feet and $3/4$ horsepower motors for heads to 30 feet. Discharge piping will be from 2 inch to 3 inch diameter.

    Reference: Chapter 12.9.

APPENDIX 2 *(continued)*

42. Concrete aggregate mix.

For most work a 3/4 inch or 1 inch size aggregate is adequate unless pumping is involved, then a 3/8 inch aggregate size may be specified.

Reference: Chapter 13.2, Aggregate.

43. Water-cement ratio for concrete.

A water-cement ratio of between 0.5 and 0.55 is common for general purpose concrete.

Reference: Chapter 13.2, Water-Cement Ratio.

44. Slump test for concrete mix.

A normal slump for general purpose concrete is 4 inches. A 1 inch slump represents a very stiff mix whereas a 10 inch slump represents a very wet mix consistency.

Reference: Chapter 13.2, Testing.

45. Riprap Stone.

Generally in riprap stone structures, the greatest dimension of a stone should be no more than three times the least dimension. Stone used for riprap should have a density around 2.65. Water absorption of riprap stone should be limited to about 2 percent to provide resistance to weathering.

Reference: Chapter 13.7.

46. Unit load developed by an average person standing at rest.

An "average" person weighing 150 pounds occupies an area of approximately 3.6 square feet, yielding a live load of 41.7 pounds per square foot.

Reference: Chapter 15.2.

47. Estimate of current drag coefficient for the below water portion of recreational boats.

The current drag coefficient C_D for end-on boat profiles may be assumed to be 0.6 and 0.8 for broadside boat profiles.

Reference: Chapter 15.4.

APPENDIX 2 *(continued)*

48. Estimate for determining recreational and small commercial boat weight.

Recreational boat weight (W) may be estimated by multiplying 12 times the boat length squared. $W_{min} = 12L^2$ (in pounds). For a small commercial vessel the formula is $W_{min} = 25L^2$ (in pounds).

Reference: Chapter 15.6.

49. Design wind criteria.

An often used "design wind" would be a wind gust based on a 50 year return period and having a natural period of 60 seconds.

Reference: Chapter 16.1, Wind Climate.

50. Estimates of water supply usage in marinas.

Water usage requirements should be estimated on the basis of 25 gallons per slip per day for recreational boats and 65 gallons per slip per day for commercial charter boat operations.

Reference: Chapter 16.1, Water Supply.

51. Location of upland sanitary facilities (restrooms).

Sanitary facilities should be provided within 500 feet from the shore end of any pier.

Reference: Chapter 16.1, Sanitary Facilities and Wastewater Systems.

52. Number of sewage pumpout facilities to be provided.

Sewage pumpout facilities for boats should be provided at a minimum of one pumpout per 100 recreational slips or fraction thereof.

Reference: Chapter 16.1, Sanitary Facilities and Wastewater Systems.

APPENDIX 2 *(continued)*

53. Recommended restroom facilities for marinas.

No. of Seasonal Wet Slips	Toilets F	Toilets M	Urinals M	Lavatories F	Lavatories M	Showers F	Showers M	Pumpout Stations*
0 - 50	1	1	1	1	1	0	0	1
51 - 100	2	1	1	1	1	1	1	1
101 - 150	3	2	2	2	2	2	2	2
151 - 200	4	2	2	3	2	2	2	2
201 - 250	5	3	3	4	3	3	3	3
251 - 300**	6	3	3	4	4	3	3	3

* For determining the number of pumpout stations provided, only those slips or moorings having boats with permanently installed sewage holding tanks should be used to calculate the number of boats to pumpout station ratio. Boater use characteristics and other factors, such as quality and number of upland restrooms, may reduce the number of pumpout stations to be provided.

** For marinas exceeding 300 slips, increase the unit requirements by one unit per 100 additional slips.

Reference: Chapter 16.1, Sanitary Facilities and Wastewater Systems.

54. Estimates of sewage generation at a marina.

Up to 100 slips at 20 gallons per slip per day. Greater than 100 slips at the rate of 32 gallons per slip per day. Figure 10 gallons per dry stack rack per day and also 10 gallons per boat trailer parking space per day. For charter boats figure 10 gallons per person for the licensed capacity of the vessel plus crew. If sewage pumpout is provided but the sewage not disposed of into the upland marina sewage disposal system, the sewage flow from the marina operation may be reduced by 10 percent.

Reference: Chapter 16.1, Sanitary Facilities and Wastewater Systems.

55. Sewage pumpout holding tank capacity.

Generally a 1,500 gallon holding tank can serve up to 100 slips.

Reference: Chapter 16.1, Sanitary Facilities and Wastewater Systems.

APPENDIX 2 *(continued)*

56. Estimate of solid waste generation.

Solid waste generation is estimated at 3 pounds per slip per day.

Reference: Chapter 16.1, Solid Waste.

57. Waste oil tank capacity.

In general, a 250 -275 gallon waste oil tank will provide a suitable waste oil capacity for up to 150 boats.

Reference: Chapter 16.1, Waste Oil and Chapter 19.9 Waste Oil Tanks.

58. Water distribution system pressure at the boat slip.

Most boat plumbing systems are designed to operate at a pressure between 30 and 40 pounds per square inch (psi) and cannot tolerate pressures of 75 to 100 psi sometimes found in municipal water systems.

Reference: Chapter 16.4, Water Distribution Systems.

59. Characteristics of small hydraulic dredges used in marina dredging projects.

The marina size hydraulic dredges have suction intake pipes about 8 inches in diameter, can reportedly handle about 120 cubic yards of sediment per hour, can operate in less than 24 inches of water and can reach down about 15 feet below water level.

Reference: Chapter 16.5, Dredging Systems.

60. Acceptable slope for handicapped access facilities.

Most regulations state that ramps should not exceed a slope of 1:12 (1 vertical to 12 horizontal).

Reference: Chapter 16.6, Ramps.

61. General minimum pile embedment length.

In general any pile will require a minimum embedment of 15 feet in competent materials. More generally piles may require 20 to 40 feet of embedment in good soils to provide sufficient capacity for modern marina pile design.

Reference: Chapter 16.7.

62. Estimate of the point of pile fixity in soil.

The distance between the apparent bottom and the point of fixity will vary depending on the type of soil, but usually it is between 2 feet and 10 feet, with 3 to 5 feet often used as a presumptive trial number, if better soil data is unavailable.

Reference: Chapter 17.1.

APPENDIX 2 *(continued)*

63. Bottom holding capacity of a mushroom anchor.

An accepted criterion states that if a mushroom has its dish half buried, shank horizontal allowing for a horizontal pull, it holds approximately three times its air weight in sand and two times it weight in mud. With a shank angle of between three and six degrees it will lose approximately 30 percent of its holding capacity. Beyond six degrees it will lose 50 percent or more, resulting in the holding force approximating the dead weight of the submerged mushroom anchor.

Reference: Chapter 17.5.

64. Chain length ratios for single swing boat moorings.

In the absence of a site and boat specific computer mooring design, it would appear that with modern boat designs and the increased wind and wave loading which is expected, a more realistic 4 to 1 chain rode length to depth ratio should become the common minimum rule of thumb for swing boat moorings which should also add at least another 100 to 150 pounds of anchor weight in order to obtain more favorable holding power.

Reference: Chapter 17.5.

65. Location of recreational boat longitudinal center of gravity.

A very rough estimation of longitudinal center of gravity (LCG), for recreational powerboats, can be made by considering the LCG as $2/3$ the waterline length from forward.

Reference: Chapter 18.1, Location of Vessel Center of Gravity.

66. Distribution of vessel load on a marine railway dry dock.

The usual distribution of vessel load is about 75 percent on the keel blocks and 25 percent on the bilge blocks.

Reference: Chapter 18.2.

67. Launching ramp slope for automobile trailers.

Automobile launching ramps are optimally designed for a slope of 12 - 15 percent. Some states will opt for a broken slope, with the below water portion having a slope of 20 percent.

Reference: Chapter 18.6.

68. Launching ramp slope for hydraulic trailers.

A graphic analysis of representative boat keel profiles indicates that a desirable slope for hydraulic trailer ramps may be in the range of between 8:1 (12.5 percent) and 7:1 (15 percent).

Reference: Chapter 18.6.

APPENDIX 2 *(continued)*

69. Capacity of launching ramps in numbers of boats launched and retrieved.

A general consensus among marina designers is that one lane can adequately service 50 launchings and retrievals per day.

Reference: Chapter 18.8, Ramp Design Criteria.

70. Gradability for forklift trucks.

The general gradability of forklifts is from level to about a 4 percent grade. Between 4 and 6 percent grades, the forklift must be used with caution. Grades to 9 percent can be accommodated if, the forklift is fitted with a tiltback mast.

Reference: Chapter 18.11.

71. Number of boats that can be served by one forklift.

A rule-of-thumb for forklift operation is that one forklift can service up to 200 boats in a dry storage facility.

Reference: Chapter 18.11

72. Estimated cost per rack (boat) for dry stack storage.

The cost per rack in 1989 dollars was around $3,000 U.S.

Reference: Chapter 18.13.

73. Ratio of number of boats in dry stack versus conventional wet berth marina.

About 4 or 5 boats may be stored in a dry stack storage system for the same area required for one boat in a conventional wet berth marina.

Reference: Chapter 18.13.

74. Estimated number of boats in use on a peak day for a dry stack storage system.

A peak day use of 33 percent of capacity appears to be a realistic number for dry stack storage systems. Local conditions or special events may alter this percentage.

Reference: Chapter 18.13.

APPENDIX 2 *(continued)*

75. Car parking spaces per boat slip.

One car parking space per two boat slips is justified by recent studies of marina parking space utilization. Put another way, 0.5 to 0.6 cars per boat slip.

Reference: Chapter 19.1.

76. Number of washers and dryers in a laundry room.

Although the number of laundry units may vary by climate and type of boat mix, a rule of thumb is to provide 8 washers and 10 dryers for 300 boats and transient trade.

Reference: Chapter 19.3.

APPENDIX 2 *(continued)*

Marina Design Information Checklist

Project name: _____
Facility name: _____
Street Address: _____ City _____ State _____
Property owner of record: _____
_____

Waterway name: _____
Latitude and longitude of site: _____
Chart or map reference: _____

Boating season months: J F M A M J J A S O N D

Type of Water Body:

[ ]	Salt/brackish	[ ]	Ocean/coastal	[ ]	Bay
[ ]	Freshwater	[ ]	Great Lakes	[ ]	Harbor
[ ]	Clean/clear	[ ]	River/stream	[ ]	Open Coast
[ ]	Turbid	[ ]	Pond/Great Pond	[ ]	Cove
[ ]	Tidal	[ ]	Non-tidal	[ ]	Other _____

Water level data:

[ ] Tidal, mean tide range _____ft/m [ ] Diurnal [ ] Semi-diurnal
Highest high water _____ft/m   Lowest low water _____ft/m
[ ] Fluctuating river or lake level Normal high _____ft/m, Normal Low_____ft/m
Water level datum:
[ ] Mean Low Water, [ ] Mean Lower Low Water, [ ] Other _____

Navigation Access:

Distance to channel _____ft/m.   Control depth _____ft/m
Min. vertical clearance _____ft/m   Current velocity _____knots

Exposure:

Fetch distance _____miles/km Direction ____ Deg. True to ____Deg. True
Water depths over fetch (at what water level datum?) _____
_____
Wind data: Prevailing direction _____ Storm direction _____
Normal Maximum wind speed _____ kts. Recorded Max. _____kts.
Wind wave activity: [ ] None [ ] Mild [ ] Moderate [ ] Significant [ ] Severe
Wave length _____ft/m   Wave period _____sec. Wave height _____ft/m

Boat wake activity: [ ] Mild [ ] Moderate [ ] Significant [ ] Severe
Type of vessels generating wake: [ ] Displacement powerboats to _____ ft in length
[ ] Tugs   [ ] Large commercial vessels [ ] Other _____

Flood hazard elevation _____ ft/m (datum?_____)   [ ] Velocity Zone

APPENDIX 2 *(continued)*

Ice Conditions: [ ] None [ ] Mild [ ] Moderate [ ] Severe
[ ] Moving ice [ ] Non-moving   Normal thickness _____ in/ft/m
Ice prone months: J F M A M J J A S O N D

Siltation: [ ] None [ ] Slight [ ] Moderate [ ] Considerable
Estimated silt accumulation per year _____ in/ft/m

Storm damage history: Year _____ Wind speed _____kts/mph Direction _____
Description of storm damage _____
_____
_____

Exposure protection: [ ] None   [ ] Rubble mound   [ ] Vertical barrier
[ ] Cellular/caisson   [ ] Floating
Incident wave design height _____ ft/m Transmitted wave height _____ ft/m

Site Characteristics:

Upland area _____ acres/hectares   Number of buildings _____
Ground coverage of buildings _____ sf/sm
Car parking area _____ sf/acres, [ ] Paved [ ] Permeable, No of cars _____
Roadways _____ sf/acres   Undeveloped area _____ sf/acres
Restrictions or easements _____
_____
[ ] Overhead wires or obstructions, Height _____ Location _____
   [ ] Bridge   [ ] Power lines   [ ] Other _____

Terrain: [ ] Level   [ ] Slight grade   [ ] Moderate grade   [ ] Steep grade
[ ] Vegetated _____ %   [ ] Non-vegetated _____ %
[ ] Paved _____ %   [ ] Permeable soil _____ %

Soil data:
[ ] Gravel [ ] Sandy [ ] Clay [ ] Silt [ ] Hardpan [ ] Rock
[ ] Other _____
[ ] Soil borings available, Borings logs attached [ ]
[ ] Test pile data available, Report attached [ ]
   Depth below mudline to competent soil _____ ft/m
[ ] Ground contamination, report attached [ ]

Shoreline characteristics: Existing
[ ] Sloping beach   [ ] Reveted slope, Type _____, Slope angle _____
[ ] Rip Rap, Size of armor stone _____ core stone _____ Top elev. _____
   Outboard slope _____, Inboard slope _____
[ ] Bulkhead, Type of material _____, [ ] Tied back, [ ] Cantilever
[ ] Other, describe _____
_____
_____

[ ] Topographic plan available, Plan attached [ ],
   Scale _____ Contour interval _____
[ ] Bathymetric (soundings) plan available, Plan attached [ ]
   Scale _____ Grid spacing or contour interval _____

APPENDIX 2 *(continued)*

Environmental:

[ ] Contaminated upland soil, History and characteristics _____
[ ] Contaminated dredge material, History and characteristics _____
[ ] Endangered species, Terrestrial _____
    Benthic _____
[ ] Wetland impact
[ ] Coastal Zone Impact, [ ] Dunes [ ] Barrier Beach
[ ] Other _____

Existing In-water facilities:

[ ] Boat slips, Number _____, Size range _____ Age _____
[ ] Fixed docks, _____ sf/sm [ ] Floating docks, _____ sf/sm

Dock construction materials:
[ ] Wood, treated   [ ] Wood, untreated   [ ] Steel   [ ] Aluminum   [ ] Concrete
[ ] Fiberglass   [ ] Other _____

Pontoons: [ ] Exposed foam [ ] Encapsulated foam [ ] Polyethylene [ ] Metal [ ] Fiberglass
[ ] Filled   [ ] Unfilled   [ ] Other _____

Utilities:
[ ] Electricity, Voltage _____, Total amps supplied _____ [ ] Metered
    [ ] Step down transformers, on land [ ], on docks [ ]
    Transformer characteristics _____
[ ] Water, Pressure _____ psi, Flow _____ gal/min. [ ] Backflow preventers
[ ] Telephone   [ ] Cable TV   [ ] Other _____
[ ] Fuel dock   [ ] Sewage pumpout   [ ] Fire protection, Type _____

Dock mooring:
[ ] Piles [ ] Timber (treated, untreated)   [ ] Steel   [ ] Concrete   [ ] Other _____
    Size: _____ Embedment length _____ ft/m
[ ] Bottom anchored [ ] Chain [ ] Cable [ ] Line [ ] Other _____
    Type of anchor _____

Hauling and Launching Facilities:
[ ] Launching ramp [ ] Paved [ ] Unpaved, Slope _____% No. of lanes _____
    Width per lane _____ ft/m   Depth at end _____ ft/m
    [ ] Public [ ] Private

[ ] Haul-out facility [ ] Straddle hoist, Capacity _____ tons, Model _____
    Clear width at hoist well _____ ft/m Length hoist well _____ ft/m
[ ] Forklift, Capacity _____ lbs/kg   Extension heights +___ ft/m , -___ ft/m
[ ] Hydraulic trailer [ ] Yard only [ ] Highway,   Capacity _____ tons

Upland facilities:

[ ] Drystack, Size boats _____, No. boats _____, Bldg. size. _____
[ ] Outside storage, Size boats _____ No. boats _____ Size area _____
[ ] Inside storage (other than drystack), Boat size _____ No. boats _____

APPENDIX 2 *(continued)*

Buildings:
[ ] Office, size _____   [ ] Retail store, size _____   [ ] Laundry, size _____
[ ] Showers, size _____ [ ] Restrooms, size _____
[ ] Repair shops, size _____   [ ] Restaurant, size _____ no. seats _____

Toilet facilities:
Male: No. toilets _____, No. urinals _____, No. Lavatories _____, No. showers _____
Female: No. toilets _____, No. lavatories _____, No. showers _____

Sewerage System:
[ ] Municipal sewer [ ] Cesspool [ ] Leaching field [ ] Holding tank [ ] Pkg. treatment
Capacity _____

Fire protection:
[ ] Sprinkled, [ ] Dry pipe [ ] Wet pipe   [ ] Alarm system [ ] Standpipes, No. _____
Describe: _____
_____

Security:
[ ] Fenced [ ] Patrolled [ ] Night only   [ ] Keyed entry, [ ] Property [ ] Docks only
[ ] Other _____

Other Information:

_____
_____
_____
_____
_____
_____
_____
_____

Acknowledgement is made to the International Marina Institute, Wickford, Rhode Island for permission to adapt portions of the Institute's "Marina Survey Worksheet"

APPENDIX 2 *(continued)*

Typical Preliminary Parameters to be Determined for Dredge Material Analysis

1. Grain size analysis

Preliminary analysis may require a single pooled representative sample from the material to be dredged. Additional samples may be required depending upon the results of the preliminary analysis.

Size Fraction % of total by weight

coarse gravel	64 mm	_____
fine gravel	2 -64 mm	_____
sand	0.063 - 2 mm	_____
silt	0.004 - 0.063 mm	_____
clay	0.004 mm	_____

2. Chemical Analysis of Sediment

Initially a single, pooled representative sample is analyzed for preliminary investigation. Additional samples may be required depending upon the results of the preliminary analysis.

total solids	%	_____
total volatile solids	%	_____
total oil and grease	mg/kg	_____
mercury	mg/l	_____
cadmium	mg/l	_____
lead	mg/l	_____
chromium	mg/l	_____
copper	mg/l	_____
arsenic	mg/l	_____
barium	mg/l	_____
selenium	mg/l	_____
silver	mg/l	_____
total PCBs	mg/kg	_____

In the United States, testing generally follows protocols established by the Environmental Protection Agency. These guidelines are presented as useful information only and appropriate testing methods and parameter requirements should be determined by consulting the cognizant agency responsible for authorizing dredging permits.

Appendix 3
Associations and Organizations

APPENDIX 3

Aluminum Association
818 Connecticut Avenue., N.W.
Washington, DC 20006

American Association of State Highway
and Transportation Officials
444 North Capitol St., N.W., Ste 225
Washington, DC 20001

American Boat Builders & Repairers
Association
Building 4
c/o Yacht Haven Marine Center
P.O. Box 1236
Stamford, CT 06904

American Boat and Yacht Council
P.O. Box 747
405 Headquarters Drive - Suite 3
Millersville, MD 21108

American Concrete Institute
22400 W. Seven Mile Rd.
Detroit, MI 48219

The American Institute of Architects
1735 New York Ave., N.W.
Washington, DC 20006

American Institute of Timber Construction
333 W. Hampden Ave., #712
Englewood, CO 80110

American Planning Association
1776 Massachusetts Ave., N.W., 7th Floor
Washington, DC 20036

American Society for Testing and Materials
1916 Race Street
Philadelphia, PA 19103

American Society of Civil Engineers
345 East 47th Street
New York, NY 10017

American Society of Landscape Architects
1733 Connecticut Ave., N.W.
Washington, DC 20009

American Society of Mechanical Engineers
345 East 47th St.
New York, NY 10017

American Society of Metals
Metals Park, OH 44073

American Society of Sanitary Engineering
2030 17th Street
Boulder, CO 80302

American Wood Preservers Association
P.O. Box 5283
Springfield, VA 22150

Asphalt Institute
Asphalt Institute Building
College Park, MD 20740

Associacao Brasileira de Marinas (ABRAMAR)
Av. Antonio Carlos Magalhaes, 846, S. 259
Edf. Maxcenter - Salvador - Bahia - Brazil

Associated General Contractors of America
1957 E St., NW
Washington, DC 20006

Barrier Free Environments
P.O. Box 30634
Raleigh, NC 27622

Concrete Reinforcing Steel Institute
933 N. Plum Grove Rd.
Schaumburg, IL 60195

Construction Specifications Institute
601 Madison St.
Alexandria, VA 22314

Environmental Protection Agency (USA)
401 M St., S.W.
Washington, DC 20460

Institute for Urban Design
Main P.O. Box 105
Purchase, NY 10577

International Association of Plumbing
& Mechanical Officials
5032 Alhambra Ave.
Los Angeles, CA 90032

International Boat Industry
1 St. Johns Court
St. Johns Street
Farncombe, Surrey GU7 38A, England

APPENDIX 3 *(continued)*

International Marina Institute
35 Steamboat Avenue
Wickford, RI 02852

International Road Federation
525 School St., S.W.
Washington, DC 20024

Land Development Institute
1401 16th St., N.W.
Washington, DC 20036

Library of Congress
Architectural Reference Librarian
Prints and Photographs Division
Washington, DC 20540

Lloyd's Register of Shipping
Yacht & Small Craft Department
69 Oxford Street
Southampton, Hants SO1 1DL, England

Marinacon (Australia)
P.O. Box 194
Rozelle, NSW, 2039
Australia

Marina Association of Texas
Rt. 3, Box 279
Kemp, TX 75143

Marine Retailers Association of America
155 N. Michigan Avenue
Chicago, IL 60601

Mechanical Contractors Association
5530 Wisconsin Ave., Ste. 750
Chevy Chase, MD 20815

Meteorological Office (UK)
London Road
Bracknell, Berkshire RG12 2SZ, England

Mortgage Bankers Association of America
1125 15th St., N.W.
Washington, DC 20005

National Asphalt Pavement Association
6811 Kenilworth Ave.
Riverdale, MD 20737

National Association of Dredging Contractors
1625 I St., N.W., Ste 321
Washington, DC 20006

National Association of Marine Surveyors Inc.
305 Springhouse Ln.
Moorestown, NJ 08057

National Association of Real Estate Brokers, Inc.
1101 14th St., N.W., 10th Floor
Washington, DC 20005

National Association of Women in Construction
327 S. Adams St.
Fort Worth, TX 76104

National Bureau of Standards
Gaithersburg, MD 20899

National Electrical Contractors Association
7315 Wisconsin Ave., 13th Floor, West Bldg.
Bethesda, MD 20814

National Fire Protection Association
Batterymarch Park
Quincy, MA 02269

National Forest Products Association
1619 Massachusetts Ave., N.W.
Washington, DC 20036

National Institute of Building Sciences
1015 15th St., N.W., Ste 700
Washington, DC 20005

National Marine Electronics Association
P.O. Box 130
Accord, MA 02018

National Marine Bankers Association
401 N. Michigan Avenue
Chicago, IL 60601

National Marine Manufacturers Association
401 N. Michigan Avenue
Chicago, IL 60611

National Society of Professional Engineers
1420 King St.
Alexandria, VA 22314

APPENDIX 3 *(continued)*

National Stone Association
1415 Elliot Pl., N.W.
Washington, DC 20036

National Trust for Historic Preservation
1785 Massachusetts Ave., N.W.
Washington, DC 20036

National Yacht Harbours Association (NYHA)
Hardy House
Somerset Road
Ashford, Kent TN24 8EW, England

Permanent International Association of
 Navigational Congresses, (PIANC)
Palace, Quartier Jordaens
rue de la Loi, No. 155-1040
Brussels, Belgium

Plastics Pipe Institute
355 Lexington Ave.
New York, NY 10017

Plumbing Manufacturers Institute
800 Roosevelt Rd., Bldg. C, Suite 20
Glen Ellyn, IL 60137

Prestressed Concrete Institute
201 N. Wells St., #1410
Chicago, IL 60606

Royal Architectural Institute of Canada
Chamberlain House
328 Somerset St., West
Ottawa, Ontario K2P 0J9

Royal Institute of British Architects
66 Portland Pl.
London WIN 4AD, England

Society of American Registered Architects
320 N. Michigan Ave., Ste 1001
Chicago, IL 60601

Society of Real Estate Appraisers
645 N. Michigan Ave., #900
Chicago, IL 60611

Society of Women Engineers
345 East 47th St.
New York, NY 10017

Southern Pine Inspection Bureau
4709 Scenic Hwy.
Pensacola, FL 32504

States Organization for Boating Access
P.O. Box 25655
Washington, D.C. 20007

Steel Structures Painting Council
4400 5th Avenue
Pittsburgh, PA 15213

The Society of Naval Architects and
 Marine Engineers
601 Pavonia Avenue
Jersey City, NJ 07306

Union of International Architects
51 Rue Raynouard
75016 Paris, France

Urban Land Institute
1090 Vermont Ave., N.W., Ste 300
Washington, DC 20005

Waterfront Center
1536 44th Street, N.W.
Washington, D.C. 20007

Western Wood Products Association
1500 Yeon Bldg.
Portland, OR 97204

Wire Reinforcement Institute
8361-A Greensboro Dr.
McLean, VA 22102

Woven Wire Products Association
2515 N. Nordica Ave.
Chicago, IL 60635

Yacht Architects and Brokers Association Inc.
24 Ships View Terrace
Bourne, MA 02532

Members of the International Council of
 Marine Industry Associations

World Body: ICOMIA (International Council of Marine Industry Associations)
Boating Industry House
Vale Road, Oatlands,
Weybridge,
Surrey, KT13 9NS England

Argentina: CACEL
Avenida del Libertador 433
(1646) San Fernando

APPENDIX 3 *(continued)*

Australia:	ABIA (Australian Boating Industry Association) P.O. Box 388 Manuka, ACT, 2603	Italy:	Unione Nazionalo Cantieri e Industrie Nautique ed Affini Via Vincenzo Renieri 23 00143 Rome
Belgium:	ASPRONAUBEL Grote Steenweg Zuid 28, 9710-Zwijnaarde (Gent)	Japan:	Japan Boating Industry Association 5-1, 2 Chome Ginza, Chuo-Ku Tokyo
Canada:	Allied Boating Association of Canada 5468 Dundas St., Wost Suite 324 Islington Ontario M9B 6E3	Netherlands:	Nederlandse Vereniging Van Ondermemingen In De Bedrijfstak Watercreatie HISWA Gebouw Metropool Weesperstraat 93 Amsterdam
Denmark:	Sosportens Brancheforening Borsen 1217 Kobenhavn K		
Finland:	The Finnish Boat and Motor Association FINNBOAT Marlankatu 26B 19 SF-00170 Helsinki 17	Norway:	Norske Batbyggeriers Landsforening Norwegian Boatbuilder's Association Huitselds gt. 16 0253-Oslo 2
France:	Federation des Industries Nautiques Port de la Bourdonnais 75007 Paris	South Africa:	BTASA P.O. Box 1047 52 Loveday Street Johannesburg 2001
Germany:	Bunderswirchaftsvereinigung Freizeit-Schiffahrt e.V. Postfach 25 03 70 Agrippina-Werft (Haus Caponiere) D-5000 Cologne	Sweden:	Swedboat Svenska Gatindustriforeningen Ljusstoparbacken 20 S-117 45 Stockholm
	Deutscher Boots und Schiffbauer-Verband 2 Hamburg 36 Jungiusstrase 13, Mossohaus	Switzerland:	Schweizerischer Bootbauer-Verband P.O. Box 74 CH 8117 Fallanden-Zurich
Greece:	SEKAPLAS Association of Greek Plastic Boat Manufacturers 115 Galatsiou Avenue Galatsi, Athens	United Kingdom:	The Ship and Boat Builders National Federation Boating Industry House Vale Road, Oatlands Weybridge, Surrey KT13 9NS
Ireland:	IFMI (Irish Federation of Marine Industries) Confederation House Kildare Street Dublin 2.	U.S.A.:	National Marine Manufacturers Association 401 North Michigan Avenue Chicago, IL 60611

Index

Abrasion, 249, 324, 325
Access (public), 4, 7, 53-55, 74-75
AFFF foam, 379, 484
Air
   entrainment (in concrete), 302, 303
   flow, 88, 97
   quality, 86
   stability factor, 124, 125, 129, 130
   terrain velocity, 339-342
Aluminum
   dock systems, 247, 252, 258-259
   properties, 310-312
Anchor
   burial (mushroom), 435
   chain, 423-440
   holding power, 423-437
   mooring, 423-440
   slab for fuel storage tanks, 506, 507
Ancillary facilities
   laundry, 494, 495
   restaurant, 71
   restrooms, 398-400, 492-494, 526
   ship's store, 71, 497, 498
   showers, 492-494
   swimming pool, 71, 500, 501
Anode, 320
Astronomical tides, 144-155
Atmosphere
   corrosion, 319, 323, 325
   pressure, 144
Attenuation (see Wave, attenuation)
Automobile
   boat launching ramp (see Launching ramps)
   parking (see Parking)
   traffic assessment, 85, 86

Bank
   financial, 38
   side slope protection (see Revetment)
   suction, 84
Basin
   marina, 199-225
   turning, 220
Bathymetry (bathymetric), 52, 62, 77, 78, 116, 357, 533
Batter piles, 351, 352
Beach, 67, 68, 224, 225
Beam
   boat, 231, 234-236
   structural, 278, 348, 368-375
Bent (pile arrangement), 249
Berth (berthing)
   arrangements, 232, 233, 235, 236, 271-277
   clear width, 233-236, 275-277
   length, 232-234
Block coefficient, 343, 449, 450
Blocking, 456, 457
Boat
   beam, 234-236, 275
   berth width, 235, 236
   brokerage and sales, 498, 499
   design, 231, 232
   freeboard, 239, 240
   length, 232-234, 271-275
   mast height (sailboat), 240-242
   multihull, 242
   profile height, 236-238, 337-342
   stands, 475, 476
   storage, 280-282
   weight, 238, 239
Boatyard services, 71
Bombardon floating breakwater, 196

INDEX 579

Borers (marine), 321-324
Borings (see Soil, borings)
Breaking wave, 175, 178-180
Breakwater
  fixed vertical, 182-186, 270
  floating, 187-196
  rubble mound, 174, 175, 180-182, 196, 210-215, 270
Bridge (ship's), 83
Brow (see Gangway)
Bulkhead
  anchored, 245, 246, 309
  cantilevered, 245, 246
  forklift loading, 478, 479
  maintenance checklist for, 539
  sheet pile, 245, 246, 309
Buoyancy (of pontoons), 367
Business plan, 31-35

Cable
  moorings, 423-440
  television, 384
  utility, 381
Catenary, 426-434, 436
Cathodic protection, 249, 328
CCA preservative, 326, 327
Chain
  anchor, 423-437
  mooring systems, 423-440
Chandlery (see Ship's store)
Channel
  depths, 81, 269, 270
  entrance, 268-270
  lines, 80-81, 82
  side slopes, 80, 81
  visibility, 80, 83, 84
  widths, 269
Circumpolar sea, 148, 149
Clapotis, 113
Cleats, 376
Containment area
  dredge material, 359, 360
  waste oil tanks, 501, 502
Coastal
  regulations, 45-48, 50, 51, 53-59
  zone management, 50-51
Coating, 321, 328-330
Concierge services, 517, 518
Concrete
  admixture, 303, 304

cement, 301-303
floating dock system, 252, 256, 257, 258
mix, 304, 305
piles, 247, 248
testing, 305
Corrosion
  crevice, 320, 321
  galvanic, 321, 322, 323
  general, 306, 319-330
  protection for fuel tanks, 505
  seawater environment, 319-325
  stress cracking, 321
Costs, 35, 36, 265-267, 480
Cradle (marine railway), 456, 457, 458
Cranes, 460, 461, 479, 480, 487, 488
Creosote, 301, 324, 326
Crest (see Wave, crest)
Current
  forces, 342, 343
  general, 70, 158-164
  tidal, 161-164
  velocity, 70, 158-161
  wave, 159-161
  wind driven, 158-159

Dead load, 331, 332, 364, 365, 366, 367, 369, 370
Deadman (anchor), 245
Degradation (material)
  biological, 321-326
  general, 319-330
  wood, 299
Design
  criteria, 249-251
  life, 250, 251
  loads, 331-354, 364-375
  vessel, 231-242
Deterioration (see Degradation)
Diffraction (see Wave, diffraction)
Displacement
  dock, 367
  floating wave attenuator, 188, 194
  vessel, 238, 239, 448, 452
Diurnal tide, 144, 145
Dock
  analysis and design, 364-375
  hardware, 376, 377
  maintenance checklist, 536
  special layout considerations, 290-292
  utilities, 377-385

Mooring
  bottom anchored, 423-434
  devices, 262-265
  fixed cantilever, 422, 423
  floating docks, 251
  floating wave attenuator, 197, 198
  general, 87, 144, 290-292
  pile maintenance checklist, 536
  systems, 415-440
Multihull boats, 242

Navigation
  aids, 48
  channels, 70, 80-82, 358, 359
  regulations, 46, 47, 49
NGVD, 77
NOAA, 78, 89, 90, 92, 94, 95, 96, 99, 100, 154

Orthogonal (*see* Wave orthogonal)

Pallet (dry stack storage), 485, 486
Parking
  area maintenance checklist, 538, 539
  automobile, 73, 74, 489-492, 528
  handicapped, 393-395
Penetration
  pile, 350-352, 416-421
  wood treatment, 326, 327
Period (*see* Waves, period)
Permits, 46-53, 75, 533
Pile
  bents, 249
  cap, 249
  design considerations, 405, 406, 416-421
  guides, 421, 422
  H-pile, 248, 308, 309
  sheet, 245, 246, 309
  steel corrosion, 321, 323, 324
  test pile, 406, 417
  timber piles, 324, 325
Pile analysis
  lateral load, 416-421
  load transfer, 349-352, 416-421
  soil characteristics, 418-421
Pontoon, 257, 313, 346
Preapplication meeting, 60, 61
Preservatives (wood), 300, 301, 324, 326, 327
Prevailing winds, 69
Profile height (vessels), 236-238, 337-342
Propeller wash, 80

Public access (*see* Access)
Public trust doctrine, 53, 54, 55
Pumpout station (*see* Sewage pumpout facilities)

Rack storage (*see* Dry stack storage)
Ramps
  boat launching, 463-474
  handicapped, 396, 397
Railway dry dock (*see* Marine railway)
Reflection (*see* Wave reflection)
Refraction (*see* Wave refraction)
Regulatory requirements
  general, 45-63
  length of term, 58, 59, 60
  meetings, 60, 61
Restrooms, 398-400, 492-494, 526
Return period (wind), 93, 96, 100, 111, 118, 122-124, 130, 132-134, 180, 181, 203, 357
Revetment, 114, 316, 318
Riparian rights, 55-58
Riprap, 182, 314, 315
River floods, 166-167
Rode (anchor), 423-440
Rollers (pile guide), 421, 422
Rowing dock, 240, 353, 354
Rubble mound breakwater, 174, 175, 180-182, 196, 210, 211, 270

Safety factors, 352, 353
Salinity, 320
Sand movement, 219-225
Scantlings, 348, 349
Scour, 225
Seasonal storage, 71
Section modulus, 368-370, 374, 375, 453, 454
Sediment
  analysis, 359, 360
  dredge, 385-391
  transport, 173, 181, 199, 219-225
Sedimentation, 62, 79, 181, 221
Seiche, 165-166, 169
Semidiurnal tide, 144, 145, 152, 153, 163
Service docks, 289, 290
Sewage pumpout facilities, 262, 285-289, 361-363
Shallow water waves, 108, 111, 119, 125, 129, 150
Sheet pile bulkheads, 245, 246, 309
Ship's store, 71, 497, 498

Shoreline, 56, 77, 111, 112
Showers, 492-494
Significant wave, 104, 109, 123, 125, 130
Siltation, 201, 219, 222
Single point mooring 423-440
Slip, 72
Slipway (*see* Marine railway)
Slopes
   bottom, 77
   dredge, 390
   maintenance checklist, 539
   ramp, 463-470, 473, 537
Softwoods, 294-296
Soil
   borings, 52, 62, 249, 350, 357, 405, 419
   considerations, 405, 406, 419-421
   survey, 79
Solid waste, 363, 526
Spalling (concrete), 248, 257, 325
Spring tides, 148, 157
Stacker cranes, 479, 480
Steel
   floating docks, 254, 255
   fuel storage tanks, 502-507
   piles, 247, 248, 322, 323, 324
   stainless, 310
   structural, 305-309
Stone, 180-182, 314-316
Storms, 88, 92, 101, 105, 110, 117, 118, 128, 129, 131, 133, 519-522, 528
Straddle hoists, 242, 280, 281, 446-454, 537
Stringers (dock), 249, 369-372
Structures
   maintenance checklist, 535
   permitting of, 46-63
   shore protection (*see* also Wave attenuation), 358
Surge, 95, 123, 132-134, 155, 279
Surveyor, 52, 76, 77, 356
Surveys, 76-79, 357
Swell (ocean), 110, 160
Swimming pool, 71, 500, 501
Swing moorings, 423-440
Synthetic materials, 316-318

Teredo, 322, 323
Tides
   datum, 156, 157
   epoch, 148, 154
   flow, 79
   measuring, 156-157
   meteorologic, 155-156
   perigean, 148, 157
   prism, 218, 219, 222
   variation vs. docks, 279
Tied back bulkhead, 245, 246, 309
Timber
   docks, 245-255
   piles, 247, 249, 416-421
   properties, 295-297, 299, 300
   treatment, 300, 301
Topography, 52, 77, 357, 533
Trailers
   automobile, 469-474
   hydraulic (*see* Hydraulic trailers)
Travelift™ (*see* Straddle hoists)
Tsunamis, 144
Tugboats, 82
Turbidity, 359
Typhoons (*see* Hurricanes)

Upland (elevation datum), 356
Urban waterfronts, 16-18
Utility systems
   cable television, 384, 385
   electric, 261, 380-385
   maintenance checklist, 536
   sewage, 262, 285-289, 361-363
   water, 261, 262, 360, 361, 377, 378

Velocity (*see* Wind, Wave, or Current)
Velocity zone (V zone), 132
Vertical elevators, 458, 459, 460
Visibility, 80, 83, 84, 270

Wake (boat), 62, 80, 82, 103, 112, 135-144, 201-207, 343, 344
Waste
   oil, 363, 364, 501, 502, 526
   solid, 363, 526
Water
   depth in marinas, 279, 280
   level datum, 356
   quality, 215-219
   supply, 261, 262, 360, 361, 377, 378
Water dependent use, 74
Waterline (boat), 138, 448, 450, 451
Wave
   action, 211-215
   attenuation, 82, 112, 114, 172-198, 200-211, 270, 271, 538

barrier, 182-186
blocking, 113, 173-175, 190
boards, 173, 183-186, 270
breaking, 175, 178-180
climate, 92, 103-134, 357, 358
crest, 103-106, 112
crest elevation (height), 105, 111, 112, 115, 117, 132-134, 143, 148, 150
current, 106
development, 107-108
diffraction, 200, 212-215
dissipation, 175, 179-180
energy, 140, 175-177, 191
height, 92, 97, 99, 103, 104, 107-114, 116
histograms, 119-121
length, 104-107, 111-113, 140, 150, 175, 176
monochromatic, 214
orthogonal, 114-117, 122, 141-143
period, 97, 99, 105-112, 115, 119, 121-124, 126, 129, 150-153, 174, 203-208, 214
protection, 172-198
reflection, 113, 114, 150, 151, 175, 177, 178, 191, 200, 213, 214
refraction, 114-117, 122, 133, 141, 142, 200, 203, 204, 207, 209, 210
significant, 104, 109, 123, 125, 130, 203, 223
spectra, 119, 123, 124
trough, 105

Wetlands
 delineation, 77
 general, 45-48, 50, 63, 67, 77, 79
Wharfing statutes, 54
Whitecap, 105, 107, 108
Wind
 areas (*see* Profile height)
 data collection, 89-102
 exposure, 69
 extreme, 94-96, 99-101
 fastest mile, 97
 fetch, 108-110, 124-131
 forces, 88, 98
 gusts, 87, 94, 98, 99, 125
 loads, 337-342
 maximum, 92-94
 peak, 97-99
 shielding, 338, 340-342
 sustained, 100
 velocity, 89-101, 108, 119, 120, 122, 126-127, 238, 337-342
 waves, 103-134
Winter boat storage, 280-282
Wood (*see* Timber)
Wood floating docks (*see* also Timber), 252, 253, 254, 255

Y-booms, 262, 263
Yachts (special considerations), 261, 262
Yard dollies, 475